From Representation Theory to Mathematical Physics and Back

CONTEMPORARY MATHEMATICS

817

From Representation Theory to Mathematical Physics and Back

Workshop in Honor of Igor Frenkel's 70th Birthday
From Representation Theory to
Mathematical Physics and Back
May 31–June 4, 2022
Simons Center for Geometry and Physics,
Stony Brook, New York

Mikhail Khovanov
Joshua Sussan
Anton Zeitlin
Editors

AMS AMERICAN MATHEMATICAL SOCIETY

2020 *Mathematics Subject Classification.* Primary 20D08, 17B37, 13F60, 57K20, 83C45, 18G70, 15B48, 20H99, 57K16, 57K18.

Library of Congress Cataloging-in-Publication Data

Contemporary Mathematics ISSN: 0271-4132 (print); ISSN: 1098-3627 (online)
DOI: https://doi.org/10.1090/conm/817

∞ The paper used in this book is acid-free and falls within the guidelines
established to ensure permanence and durability.
Visit the AMS home page at `https://www.ams.org/`

10 9 8 7 6 5 4 3 2 1 30 29 28 27 26 25

Contents

Preface

The theme of the conference is the continued influence of representation theory and physics on each other and the impact of the work of Igor B. Frenkel on these fields.

This workshop at the Simons Center for Geometry and Physics from May 31–June 4, 2022 highlighted progress in the areas of vertex operator algebras, conformal field theory, categorification, low dimensional topology and representation theory of affine Lie algebras, loop groups, and quantum groups. Some constructions in string theory were used in the realizations of representations of affine Lie algebras and the Moonshine module of the Monster group. The structure of the Moonshine module has led to the mathematical theory of vertex operator algebras and connections to 2- dimensional conformal field theory. These mathematical and physical theories have been subsequently related to 3-dimensional quantum gravity. In another direction, the discovery of the Jones polynomial led to a physical construction of 3-dimensional TQFTs, which in turn advanced many mathematical developments in quantum groups and low dimensional topology. Louis Crane and Igor Frenkel introduced the categorification program with the goal of upgrading 3-dimensional TQFTs coming from representation theory of quantum groups to 4-dimensional TQFTs. They suggested that one should construct categorical quantum group actions to achieve this aim. These ideas gave rise to the development of link homologies constructed from representation-theoretic, algebraic-geometric, combinatorial, and physical structures.

The organizing committee consisted of:

- Pavel Etingof (MIT)
- Mikhail Khovanov (John Hopkins University)
- Alexander Kirillov (Stony Brook)
- Anna Lachowska (EPFL)
- Ivan Loseu (Yale)
- Andrew Neitzke (Yale)
- Joshua Sussan (CUNY)
- Anton M. Zeitlin (Georgia Tech)

The organizers of the conference would like to thank the Clay Mathematics Institute, the Simons Center for Geometry and Physics, the National Science Foundation, and Yale University for their generous support.

Mikhail Khovanov

Joshua Sussan

Anton M. Zeitlin

Contemporary Mathematics
Volume **817**, 2025
https://doi.org/10.1090/conm/817/16339

Perspectives on the past, present, and future of representation theory and mathematical physics: An interview of Igor Frenkel by Joshua Sussan

Igor Frenkel and Joshua Sussan

In this conversation, Igor Frenkel gives his perspectives on the future of representation theory and mathematical physics. This discussion between Igor Frenkel and Joshua Sussan began at Leet Oliver Memorial Hall in Yale University in the summer of 2023 and continued at the math department's new home at Kline Tower, bridging the development of many foundational results over the last few decades to conjectures about the future of these subjects.

JS How did you become interested in mathematics and physics in St. Petersburg?

IF My first interest in mathematics arose from olympiads. When I was 12 years old, I won the first prize in the St. Petersburg olympiad. And then every year after that I received first prizes – five of them. I also won several second prizes in the Soviet Union olympiad. And I was pretty sure that mathematics was the direction I wanted to go. Besides, I thought that in mathematics, as in chess, one could stay away from politics.

JS Who were your early mathematical influences?

IF My first serious mathematical influence was Ramanujan. After I had won my first mathematical olympiad, I began to read books on mathematics. So, I read calculus, and it was relatively interesting. Then I read complex analysis, it was also interesting; but then when I read functional analysis, it was too hard and also boring. At that point I was 15 or 16 and I was kind of disappointed in theoretical mathematics as it was. But then suddenly I found a book about Ramanujan with all his crazy formulas, and I tried to prove them, and I realized that it was beyond my abilities. This led me to more conceptual mathematics. For example, some of the identities for hypergeometric functions that Ramanujan had could be understood using representation theory. I think that was what I realized at the end of high school. And since then, I was already beginning to move in the direction of representation theory.

JS Could you tell me about your mathematical experiences in college?

IF While in college, I studied representation theory mostly by myself because the leading school of representation theory was in Moscow. I also read about applications of representation theory in physics, for example, about Poincaré group and unitary symmetry. I wrote my master thesis in representation theory of compact groups, which essentially was an already closed subject in the 70s. The first problem was given by my undergraduate adviser Anatoly Vershik and the second problem I found myself. Later I gave a talk in Moscow about these two results at Kirillov's seminar and he suggested that I publish them in two articles. And it was at the end of college, and I wasn't able to go to graduate school for various reasons, so it was not clear for me whether I really needed to publish these papers. In fact, they were published by other people much later with some references to me.

JS Could you describe more about your mathematical studies after college?

IF After college, since I was not pressed for publishing as people are in graduate school, I was thinking about philosophy of mathematics quite a lot, as well as about significance of mathematics and physics. I thought I understood the answer to Wigner's question concerning "the unreasonable effectiveness of mathematics in physics" as some kind of natural selection à la Darwin of mathematical structures. This is the point of view that I still stand on when thinking about theoretical physics. Right after college, I was also reflecting on how to develop the theory of compact Lie groups. And then I found the paper by Macdonald where he proved a certain generalization of Weyl's identity, this led me to infinite-dimensional groups and their representations. Then I came across the works by Kac and Moody on construction of a class of infinite-dimensional Lie algebras and of their representations. And it became clear to me that an enormous field was opening up.

By that time, I also heard some talks about integrable models at Faddeev's seminar. Calogero came to St. Petersburg and gave a talk and also Perelomov gave some talks about integrable systems. Some of these systems could be obtained from finite-dimensional Lie algebras and Lie groups. And it was natural for me to expect that if we applied not just finite-dimensional algebras, but also loop algebras and loop groups, then we could describe a much larger class of integrable systems. In particular, I wrote my first paper with two other mathematicians in St. Petersburg about application of loop groups to closed Toda lattice, which was a generalization of Kostant's result about open Toda lattice. I had an idea about affinization program: that everything one can do in finite-dimensional case, one can kind of extend to this infinite-dimensional class of algebras, and then one will have many more connections to mathematics and physics.

JS Why did you decide to come to the U.S. and how was your first experience there?

IF For four years after my graduation from college, as I mentioned before, I could not go to graduate school for various reasons. But in 1979, I was able to immigrate to the United States. And then finally I wanted to realize my initial dream to go to graduate school. I had the choice of several places where there were professors working in infinite-dimensional Lie algebras which I was already studying. Kac was at MIT, Kazhdan was at Harvard, and at Yale there were two people, Garland and Lepowsky, though by the time when I arrived, Lepowsky had already left. But still, I went to Yale, and I had a fantastic welcome there.

I worked very hard during the summer of 1979 on the application of affinization program to Kirillov's orbital theory. And I derived the characters of representations of loop groups from the heat kernel and Brownian motion, the latter appeared naturally by itself following the analogy with finite-dimensional theory. Then I had no excuses for not writing down my results, and I had essentially finished my thesis before my term in graduate school officially started. So, I was not really a graduate student in the U.S., but for different reasons than in the S.U.

In 1979 I also wrote a paper with Victor Kac on vertex operator representations of loop algebras, which later led to vertex operator algebras. This work found another important application in string theory. At the end of the same year, I began yet another paper about reinterpretation of two-dimensional quantum field theory as representation theory. So, 1979 was a very productive year for me.

JS How were vertex operator algebras (VOAs) discovered?

IF A little history before the discussion of vertex operator algebras. Soon after our work with Kac, it was noticed by several people including Kac and Lepowsky, that if one took three copies of the E_8 representation, of our construction, then a graded dimension of the representation space would be given by the j-invariant of Klein with constant term 744. On the other hand, by that time there appeared some fantastic paper of Conway and Norton about monstrous moonshine, where it was clear that coefficients of j-function were somehow related to representations of the monster group, and that there should exist a representation of approximately the same size as our construction with Kac for three E_8 algebras, but with zero constant term.

Then I began to work with Jim Lepowsky and Arne Meurman on the construction of this infinite-dimensional space. And finally, we succeeded, and we constructed the correct graded representation for the monster. However, since this was just a graded space without any structure, it wasn't sufficient to distinguish the monstrous automorphisms. So therefore, we invented a certain multiplication, and constructed a certain strange algebra preserved by the monster action. We finally published an announcement in 1984 that automorphisms of this graded vector space with this algebra structure were given exactly by the monster group. And then, Richard Borcherds noticed that one could also add other operations and the resulting structure would still be invariant under the monster action. But this addition of other operations was quite unusual because we then had not just one algebra multiplication, but infinitely many, and with infinitely many relations between them. That was what he called a vertex algebra. So, for us it was kind of immediate to add this structure to our already existent very big manuscript, and we wrote this as a book just a little later. Borcherds only published a small paper on this subject without proofs.

It's interesting that in 1984, physicists Belavin, Polyakov and Zamolodchikov also came to a construction of some strange algebra related to 2-dimensional conformal field theory, and they had a special emphasis on the Virasoro algebra which was also important in our 1984 construction and which Borcherds did not consider. But they did not state that there were infinitely many operations; however, now by post-factum, we can see that vertex operator algebra multiplications could be one of natural interpretations of their strange algebra. So, they came to essentially the same notion from a different point of view, though they did not have all the axioms. They stated an axiom of associativity, which was actually not quite associativity, but can now be interpreted as certain relations for these infinitely many VOA operations. At the end, one can say that vertex operator algebras (the name we used in our book) were discovered both from just pure mathematical point of view studying monster group, and from physics point of view studying 2-dimensional conformal field theory (CFT).

JS What was the development after this of vertex operator algebras in mathematics and physics?

IF Within the mathematical community the progress was relatively slow because vertex operator algebra was such a new object which algebraists had never seen before. It is a structure with infinitely many multiplications and infinitely many relations. But in physics it was recognized as the correct axiomatic point of view in two-dimensional CFT, and there was a lot of related work in mathematical physics.

The most important result concerning VOAs in pure mathematics was the dissertation of my student Yongchang Zhu about modularity of graded dimensions of irreducible representations of vertex operator algebras. That was a very difficult and foundational result which played an important role in the future development of this theory. Unfortunately, he did not feel that there was much interest among mathematicians, and after he had graduated from Yale, he changed his subject to number theory.

I also wrote a paper with Huang and Lepowsky on further foundations of VOA theory and we could not publish it at any leading journal since nobody was interested: it has more than a thousand citations now. In physics literature an important paper was written by Moore and Seiberg.

JS What were the big further developments in vertex operator algebras?

IF After the work of Zhu in 1990, the next breakthrough was by Borcherds who was able to prove in 1992 the genus zero monstrous moonshine conjectures by Conway and Norton. Later he received a Fields Medal for that. He employed some construction from string theory which I had developed before in relation to the studies of Kac-Moody algebras – an application of no-ghost theorem for the multiplicities of roots of these algebras. He used it in this monster case, and he was able to prove certain identities that implied the Conway-Norton conjectures. This was a big and important result. However, the mysterious genus zero property was not completely explained, this happened later.

Another breakthrough was an amazing geometric formulation of two-dimensional conformal field theory, which is a close relative of vertex operator algebras, by Graeme Segal. This led to a similar understanding of three-dimensional topological quantum field theory (TQFT) by Atiyah and Witten.

JS What is your view on the progress on the relationship between sporadic groups and vertex operator algebras?

IF After the realization of monstrous symmetry of a vertex operator algebra, it was a natural idea to look at other sporadic groups, and other simple groups, as possible symmetries of vertex operator algebras.

For Lie-type groups such a realization was relatively clear; and for groups involved in the monster, there are about 20 of those, it could be derived from our construction. But there are 6 pariahs – the sporadic groups not related to the monster at all. The largest of them is the so-called Janko group J_4. Therefore, my next question was – could this group be realized as symmetries of some vertex operator algebra. In the early 00s, I had a student John Duncan with whom we tried to solve this problem, but it was too difficult. I still consider this as the most important problem if one wants to investigate the general relation between sporadic groups and VOAs. We should remember that in the case of compact Lie groups, complex Lie groups, classification is achieved by transforming it to the classification of Lie algebras and then classifying the Lie algebras. Here, the situation is rather strange: to a finite group we may be able to attach a flat infinite-dimensional structure, possibly a vertex operator algebra, and then to classify those vertex operator algebras. But still, this program is in its initial stages. Duncan himself was able to realize some of the pariahs as symmetries of VOAs, but it did not help to understand the general relation. I still consider the J_4 example as crucial.

JS What are the big open questions in vertex operator algebras?

IF There are still many general open questions concerning the structure of vertex operator algebras. People study logarithmic VOAs and their representations. The problem of classification of VOAs has not been solved. There was some progress in the classification of rank 24 self-dual VOAs, but even uniqueness of the monster vertex operator algebra has not been proven yet. Among more modern problems I would point out the question of geometrization and categorification of VOA's. While geometrization of representations of loop algebras was obtained by Nakajima via homology of instanton spaces, the VOA structure has never been understood in these terms. The categorification of VOA's based on the Nakajima geometry must yield 4d VOA's and 4d conformal field theory, but it is still a long way to go. Another question concerns the genus zero Conway-Norton conjecture: it should be understood from some sort of new realization of vertex operator algebras. There seems to exist a conjectural realization of the monster VOA by means of 3-dimensional quantum gravity. That's a special question and I could talk more about this.

JS Could you discuss your work on 3-dimensional quantum gravity (QG).

IF It has a long history. After our book had been published on vertex operator algebras and the monster, a couple of years later, a very interesting paper appeared in the Notices of the AMS about Rademacher's formula for the j-function. In fact, this formula was based on a previous work of Hardy and Ramanujan, which was an additional attraction for me. I immediately realized that this formula might reflect certain facts about the structure of the monster and the corresponding vertex operator algebra. Clearly, it looked like a fixed-point formula for some theory, but I had absolutely no idea what kind of theory that could be, and I asked some physicists about it. Nobody could recognize anything familiar. Then several years later, there appeared the work of Maldacena and other physicists about relation between quantum field theory in dimension D and quantum gravity in dimension $D+1$. What was especially interesting for me was the case $D = 2$. About 10 years after the Maldacena paper, Witten wrote a preprint where he asked a question – what was the conformal field theory counterpart for the simplest, pure 3-dimensional quantum gravity? And his answer was that it should be exactly our vertex operator algebra with the monster symmetry. That was an astonishing conclusion, because, if you remember, the monster was discovered in the process of classification of finite simple groups, and it was the most exotic structure at the time of its discovery. Then we realized it as the symmetry of a VOA, which turned out to be a very natural algebraic structure in general, also related to physics, but this particular one was a very exotic VOA. And finally, Witten was saying that the most natural 3-dimensional pure QG was this exotic VOA with the monster as its symmetry. That was amazing! However, Witten's work was kind of nonconstructive and it was not clear from his paper what the actual states of quantum gravity should be. So where was the monster and where was the representation of the monster coming from this point of view? And then again, I asked several physicists, and nobody could answer this question.

Physicists sometimes have doubts about the existence of such 3-dimensional gravity. However, the Rademacher formula for the j-function with the right constant term clearly can be interpreted as a sum over black holes for a certain chiral gravity – just half of the full 3-dimensional QG; and the corresponding chiral CFT has to be precisely our monster vertex operator algebra. For me, as a mathematician, this formula is sufficient evidence that there must be some theory behind it. I mean, either this is what is called 3-dimensional quantum gravity, or its slightly modified, or seriously modified version, there must be something.

Since then, I periodically work in this direction. And I wrote a paper with Duncan where we explained the origin of genus 0 of the Conway-Norton moonshine under the assumption that there was a kind of gravity as we have just discussed. If there existed such a gravity, then there would also exist twisted such gravities, and then the genus 0 property would follow from our work with Duncan. So, in this sense, we can now say that Borcherds proved the genus 0 property though did not explain it. We explained it, but we didn't have a proof, so therefore it's still an interesting open direction for research. I wrote several papers with different collaborators on various approaches to this fascinating subject.

The main lesson that I learned from my work on 3-dimensional quantum gravity is that it is deeply related to the most basic number theory, and first of all, to the structure of rational numbers. In particular, the monster group should appear at the end from the absolute Galois group of rationals.

JS When did you realize the importance of quantum groups in representation theory?

IF At first, when I heard that some people had managed to deform simple Lie algebras, I was absolutely astonished, and I thought it was completely impossible. Then I understood that it was a deformation not of the Lie algebra, but of the universal enveloping algebra. Still, it seemed a strange result to me. I was finally convinced that it was a natural and canonical structure after I had seen the paper by Mikio Jimbo about the q-deformation of Schur-Weyl duality – because deformation of the group algebra of symmetric group into Hecke algebra was already standard by that time in mathematics. Here, Hecke algebra is dual to some other algebra, which must be important, and that's essentially the quantum group of $GL(n)$ type. And then from the works of Jimbo and Drinfeld, it followed that one could define it for any type.

But for a long time, I did not feel comfortable that quantum groups were not groups, and therefore were not really as fundamental. Later I realized that representation theory of quantum groups, as well as quantum groups themselves could be viewed as representation theory of loop algebras with all the complex structure erased and only the topological structure preserved.

JS Could you talk about your earlier work in quantum groups?

IF After the first surprise and strange feeling about these objects, gradually, I became more and more interested and involved in the study of quantum groups. And at some point, I developed what they called a q-disease, that was when one wanted to q-deform everything. And really, we can do it: I mean, we can q-deform octonions, exceptional Jordan algebra, Clifford algebra and the oscillator algebra.

All we need is to have q-deformation of the category of finite-dimensional representations of a classical group and then to take certain small representations. Then we take intertwining operators between certain tensor products, and we can q-deform any of the algebraic objects that could be obtained from pure representations theory. Much later, I wrote a paper with my former student Jintai Ding about Clifford and oscillator algebras in this spirit. But then I tried also to q-deform some more sophisticated structures. One of them was the representation I had developed with Kac, and we did that with my other former student Naihuan Jing. Then with Kolya Reshetikhin, we managed to construct a q-deformation of Knizhnik–Zamolodchikov equations. We were able to derive a commuting system of difference equations, which in the simplest case had solutions in terms of q-deformed hypergeometric functions. The analogs of monodromies were not elements of the quantum group like in the Knizhnik–Zamolodchikov case, but of a certain elliptic group, which also was a very interesting development.

JS Could you talk about your work on elliptic hypergeometric functions?

IF It was related to my work with Reshetikhin – because the analogs of monodromies of our difference equations were Jacobi elliptic functions, and in particular, they looked like elliptic deformations of certain hypergeometric functions. It led me to ask a question whether there was such elliptic deformation of hypergeometric functions. But when we started to work on this subject with Volodya Turaev, it turned out that the most elementary hypergeometric functions, like Gauss' ${}_2F_1$, did not have this deformation, and the next class didn't have it either. However, when we went much further, then at some point, by some kind of miracle there was an elliptic deformation. It turned out that the most complicated hypergeometric functions related to $sl(2)$ (and their analogs for other types are difficult to write down explicitly), which were disguised forms of $6j$ symbols depending on a parameter, did possess elliptic deformation. So, they were presented by elliptic functions and then we could prove a certain summation formula and some kind of symmetry formulas for these functions. This was a big surprise for us because we were thinking about this new class of functions in relation to applications to 4-dimensional topology. But even to the hypergeometric community, these elliptic deformations were completely new, so they were immediately included in the Encyclopedia of Mathematics, and there were many conferences on this subject.

This work made me imagine a meeting 200 years back, when Jacobi visited Gauss in Gottingen. It was when Jacobi had just invented Jacobi functions that were crucial in this elliptic deformation and Gauss by that time had developed significantly the theory of hypergeometric functions, which was his very famous work and was quite popular at that time. So, I just imagined that there was a possibility that in their conversation they could find these elliptic hypergeometric functions which we discovered only two centuries later.

JS Could you tell me about your work on real quantum groups?

IF I had this philosophy from the time of my life in St. Petersburg, when I was reading papers by Gelfand and his coauthors, that for whatever construction we had for compact groups, we should also have a counterpart for split real groups. So, in the first case we had irreducible representations parametrized by a discrete set; and in the second case the principal series were parametrized by a continuous set, but in the same cone. Therefore, I always expected that there should exist something similar in the quantum group case. But I only realized that it was the reality after reading a work of Jorg Teschner who constructed explicitly a principal series representation for quantum $sl(2)$. It also was related to papers by Ludwig Faddeev who first discovered the modular double of a quantum group. And this was something completely new in comparison with the ordinary quantum group case, in particular because there was no obvious quasi-classical limit. One could not just send q to 1 because another q in the modular double would blow up. So, real quantum groups had some very interesting new features, but also many similarities with the previously studied unitary case. For example, tensor products were closed, which was unusual for principal series, compared to the classical case where it was not true. This indicated that principal series representations of real quantum groups were definitely an extremely interesting object, and with my student Ivan Ip we were able to generalize it to other types. And Ip developed this work much further, later he and other mathematicians related combinatorics behind our representations of real quantum groups to cluster algebras, which is a large area of mathematics.

JS Classical quantum groups have had great success in topology. Will there be applications to topology of real quantum groups?

IF Certainly, I expected applications to topology, and I gave a talk about this in 2011 at the conference "25 years of quantum groups" when I spoke first about principal series representations. Since then, there already have appeared some results in this direction. Extremely interesting phenomena happens here: when we consider a tensor product of two representations of principal series, we get a continuous direct integral instead of a direct sum. In this case we need to introduce some topology on categories which will make them some kind of "continuous categories". This reminds me very much of Grothendieck's dream and maybe here, in the theory of representations of real quantum groups, we have some very interesting class of examples of Grothendieck topology on categories. But I never worked in this direction, so I'll just leave it for young mathematicians. Moreover, categorification of these "continuous categories" must be even more tremendous. But we first need to recall the initial categorification program.

JS What led you to introduce the categorification program?

IF The success with 3-dimensional invariants and 3-dimensional TQFT by Witten, Reshetikhin and Turaev certainly led me to think about higher dimensions. And I wanted to consider generalizations of this theory in the simplest situation, namely Yang-Baxter equation in arbitrary dimension, both in classical and quantum cases, and we wrote a paper with Greg Moore. In this work, we showed that somehow 3 and 4 are very special dimensions, higher dimensions are not so interesting for quantization. After we had written this paper with Moore, I remember that Misha Kapranov, who was interested in our work, began interpreting it in terms of higher categories. By that time, at least at the heuristic level, it was clear to many people that TQFT in dimension $n+1$ was related to the so-called n-categories. And n-categories with 1 element, in their turn, are isomorphic to $(n-1)$-categories with monoidal structure. The case of $3d$ TQFT's and its relation to the monoidal categories with a certain extra structure was already well known. So therefore, it was a question to find the next example after $3d$, namely $4d$, and our paper with Moore had shown that it might exist. Another indication came from the canonical basis in representation theory. Everything pointed out that we should be able to formulate the theory in terms of positive integral structure coefficients. This indicated that the current 3-dimensional TQFT was only a shadow of a much more powerful structure.

We began our discussions with Louis Crane, who was a postdoc at Yale, on all these topological and categorical questions. And on the other hand, I had worked on quantum groups by that time long enough to know about canonical basis and tensor categories. We also thought about other known $3d$ and $4d$ invariants studied by Donaldson, Floer, and others and we came to a conclusion that they all had to be related, and the representation theoretic approach should unite them in one magnificent picture. But it was too huge a task for us at that time to develop all our observations into a gigantic opus, so we combined our knowledge in a very heuristic paper containing conjectures on 4-dimensional topological quantum field theory; in addition, we had some kind of assurance that nothing particularly interesting was happening in higher dimensions. Actually, some already famous results in topology told us that dimensions 3 and 4 were very special. So, our paper was dedicated to the bridge from the topology in 3 to 4 dimensions, and this became the first example in the categorification program. But at that time, I did not expect that this kind of heuristic program would find a realization soon, and at such a grand scale.

JS Could you talk about the early progress in categorification?

IF I slowly continued to work on this program. With Misha Khovanov who still was my student at that time, we wrote some papers on quantum groups, which emphasized the integrality and positivity of the canonical basis. We considered a certain class of examples where canonical basis could be understood combinatorically and geometrically, and then the next step was actual categorification of these structures. I wanted to approach categorification from the point of view of representation theory. And eventually it resulted in the paper with Joseph Bernstein and Misha Khovanov on categorification of just a tensor product of 2-dimensional representations of $sl(2)$, which was the very first explicit example. I was very pleased to see that the basic generators of $sl(2)$ could be obtained using the most elementary translation functors or Zuckerman functors, and it was the starting point of categorification using representation theory. However, we were missing some kind of grading structure because we did not know how to obtain it from representation theory. It was done only much later by Catharina Stroppel, who proved all the conjectures from our paper.

But meanwhile, Khovanov had a more direct approach to this categorification, and he wrote his now famous paper introducing the Khovanov homology. At that time, I was already absolutely sure that it was a very strong invariant because the mathematics behind it was extremely rich and beautiful. And that was my main argument when I spoke with topologists who tried to convince me that one could not construct invariants of smooth manifolds using combinatorics. Only several years later, it was shown that the Khovanov homology actually did capture smooth invariants. And since then, it became more and more popular and important.

JS Is progress towards 4-dimensional TQFTs slower than you anticipated?

IF The answer is that my own progress was slower than I anticipated, but then it was a big surprise that other people made an amazing progress in this direction in developing both Khovanov homology and representation theoretic approach following our work with Bernstein and Khovanov and the work of Stroppel. So, our program with Crane turned out to be a lot more realistic than even we originally had expected. It is an enormous field today, which is still being developed at high speed by many people.

JS Could you discuss quaternionic analysis and its ramifications in math and physics?

IF First, a little about the history of my involvement in quaternionic analysis. Recall that the relation between 2-dimensional conformal field theory and representation theory of loop algebras was quite spectacular in the 80s. And they both were based on complex analysis. But when I started thinking about generalizations to representation theory of higher dimensional algebras, I met with some serious difficulties. Note that some very strong mathematicians and physicists tried to work on 4-dimensional generalizations of vertex operator algebras, and they could not go very far either. The previous analogy suggested that both conjectural theories should be built on a 4-dimensional analog of complex analysis, which can be nothing but quaternionic analysis. I began to look in this direction, but it turned out that quaternionic analysis was very underdeveloped in comparison to complex analysis. What was known was an analog of holomorphic functions, though without the algebra structure, together with Cauchy formula for a single pole. That part existed, so it was not an empty field, but it still was a very underdeveloped field. With Matvei Libine, we tried to advance quaternionic analysis as much as possible using representation theory of the conformal Lie algebra, which came from conformal transformations of quaternions. Therefore, the logic was that in order to understand the representation theory of new large infinite-dimensional Lie algebras, I needed quaternionic analysis; and to develop quaternionic analysis, I needed representation theory of a much smaller algebra which was the conformal Lie algebra.

We wrote several papers that were interesting because of the unexpected relations with Maxwell equation, which appeared naturally when we considered Cauchy-type formulas for the pole of second order. A relation to Feynman diagrams also began to appear; but it's still an open program to relate quaternionic analysis with very extensively studied 4-dimensional conformal field theory, which is an enormous field in theoretical physics. So, I would compare the present status of quaternionic analysis with a tiny head of a dinosaur while 4-d CFT can be compared with its enormous body but without the head, and it is an urgent problem now to connect the two.

There is an advantage of quaternionic analysis over physics in that we can develop it following the analogy with the complex case and its intrinsic structure. For example, we can ask what is the quaternionic version of the Riemann mapping theorem. I believe it is related to smooth structures on $\mathbb{R}^4$.

JS What is your perspective on modern physics?

IF First of all, I should say that I'm not a theoretical physicist, so this will be a dilettante point of view. Looking historically at the last 100 years, we can see that the first 50 years were very productive in terms of the development of theoretical physics, starting from quantum mechanics and quantum field theory et cetera, until the success of the standard model. So, it seems like practically everything important was done during those 50 years. But the last 50 years were kind of disappointing because if we look from the viewpoint of basic theoretical/experimental physics, much less has been achieved, and mostly the previous theory has been fully approved. But nevertheless, I still consider the last 50 years as amazingly productive in theoretical physics, which evolved though in very strange ways. Sometimes certain parts of this theoretical development would give some miraculous results. For example, 4-dimensional quantum field theory led to very beautiful mathematical structures arising from scattering amplitudes. String theory, which was a very important part of theoretical physics during this time, gave a lot of interesting predictions and outcomes both in mathematics, and in other areas of theoretical physics; most notably it emphasized the importance of holographic principle. Nevertheless, these 50 years were completely unpredictable and often disappointing. In my opinion they have shown for the first time in history the limits of power of the human mind. I mean that the most important unsolved problem was quantum gravity, and the best theoretical physicists could not solve it in 1 year as they often had done before, not in 5 years, not even in 10 years, and not even in 50 years. This is not surprising; I am convinced that even the solution of Riemann Hypothesis is only a small part of this structure.

So, what can we expect for the next 50 years? I am an optimist and though the main problem is still the same, namely 4-dimensional quantum gravity, and is no less difficult, I believe that there will be a great progress in its solution in several different directions from several different points of view. Certainly, it's always important to have experimental approval when we are talking about theoretical physics. In this case, we might not have results of experiments for a long time, but we will be able to see if the theory will be mathematically correct, consistent and beautiful. And that will be our best kind of confirmation that the future 4-dimensional quantum gravity will be the true one. Remember that Einstein knew perfectly well when his theory of 4-dimensional classical gravity reached the point of perfection.

JS Do you see string theory as a realistic model for unified theory?

IF String theory is a very interesting development in theoretical physics. If we look at string theory from a big distance, say, as at a certain quantization of 2-dimensional conformal field theory, then yes, I believe that it's a realistic model. However, the current realization of string theory, which is based on supersymmetry and compactification of extra dimensions, is less convincing for me. Although, the way it was applied to traditional physics, it still produced very important results in 4-dimensional quantum field theory, and gave us some glimpses into quantum gravity, and I am not even talking about mathematical applications. So, it's definitely very important but it should not be made into a dogma. I would compare the role of string theory to that of the enigmatic information in the book of Stanislaw Lem, "His Master's Voice". People would receive some kind of message from the outer space, and they could not decode it completely. But nonetheless, they could use parts of this message for certain applications. This led to various philosophical and political discussions. Something similar happens with string theory.

One can try to look at string theory from a future perspective as follows. Since 2-dimensional conformal field theory is already known to be related to 3-dimensional quantum gravity, we can think of string theory as of a certain quantization of 3d QG. And one of the forms of quantization is categorification, namely replacing c-numbers by q-numbers in physics terminology. On the other hand, categorification of 3d QG must be 4d QG. So, if we consider quantization not of 2-dimensional conformal field theory but of the corresponding 3-dimensional quantum gravity, then we can expect that this quantization will give an extra dimension, which will bring us somehow to 4-dimensional quantum gravity. This is similar but much wilder than categorification of 3d TQFT as 4d TQFT. Note also that the critical dimensions 26 (and 10) will continue to play their structural role but not as the dimensions of space-time anymore. This is certainly a very heuristic idea about how a variant of string theory could be a realistic model for a unified theory, and it shows the key importance of understanding of 3-dimensional quantum gravity, which is still missing.

JS What is the most surprising mathematical result by others during your career?

IF The most surprising result is probably the work of Hiraku Nakajima in 1994 where he realized representations of loop algebras as homologies of moduli spaces of instantons on the simplest possible solutions of self-dual Einstein equations called ALE spaces. That was a big surprise for me, and it connected many things together. It was known before, that the classification of simple Lie algebras was related to the list of finite subgroups of $SU(2)$, but the direct relation was never very clear before the work of Nakajima. Also, representation theory, which as I said before, had been developed from the finite-dimensional to the affine case due to the influence by Macdonald's paper, now could be extended even further and became related to a lot of new mathematics such as algebraic geometry, homological algebra, instanton theory, etc. Here for the first time, we saw that 2-dimensional conformal field theory could be understood in terms of field theory in 4-dimensions. In this case, it was not quantum field theory in 4d yet, but classical field theory of Yang-Mills equations. This result indicated that representation theory could be advanced even further and unite a lot of mathematics into a resulting big structure. How much more mathematics can be understood as some kind of representation theory? Maybe the famous phrase of Gelfand's, "Mathematics is representation theory", is not as crazy as it sounds.

JS What are the most surprising results in theoretical physics during your career?

IF Certainly, there were many interesting and surprising results in theoretical physics during the last several decades. But probably, if I were to choose one, it would be Witten's paper from 2007 titled "3-dimensional quantum gravity revisited" that I have mentioned before. He gave a very strong argument that the simplest pure 3-dimensional quantum gravity should have a 2-dimensional conformal field theory counterpart that is our vertex operator algebra with the monster group symmetry. The reason that it's so surprising, as I said before, is that the monster was a very exotic structure found as a result of the classification theorem for finite groups. We made it better understood by realizing it as the symmetry of a particular vertex operator algebra, but this vertex operator algebra was still exotic. And what Witten was essentially saying in his paper, was that the most natural 3-dimensional quantum gravity already had the monster group as symmetry. This gives evidence that the monster group is built into the most basic structure in theoretical physics!

However, as I understand, this paper was slightly controversial in the physics community, and was never published in physics journal; nevertheless, it has quite a lot of citations and influenced many people. It must be highly nontrivial to see explicitly the symmetry of this 3-dimensional quantum gravity given by the monster group, and so far, nobody has seen it directly. Its indirect presence is hidden in the Witten's argument, and in the Rademacher formula which can be interpreted as a summation over black holes that yield the partition function of the chiral $2d$ CFT with the monster symmetry. By now, physicists are already used to the E_8-symmetry, but it may take them a while to realize that the monster has something to do with the ultimate symmetry of the universe. It is interesting to note that existence of three generations of particles points to a triplication of E_8- symmetry which, as I explained before, was the first hint to the monster symmetry.

For me, Witten's paper results in a program of finding this elusive 3-dimensional gravity and its monster symmetry or some kind of mathematical modification of such a structure. I believe that the solution of this problem will be very important for further development of string theory and quantum gravity. During his remarkable career, Edward Witten has produced a number of surprising results, but I am sure that this paper will have the greatest impact in the far future.

JS What do you consider your most influential results?

IF When I check the statistics, certainly the works in vertex operator algebras seem to be the most influential with thousands of citations. Our construction of vertex operator algebra with the monster symmetry was used and developed further in many papers, so that's probably the most influential direction as well as other related works in VOAs. Second, there are my topological categorification paper with Crane and my representation-theoretic categorification paper with Bernstein and Khovanov which initiated an enormous program. I believed in this program, but still was surprised when it began to develop so extensively and became so popular. Then another work is the quantum version of Knizhnik-Zamolodchikov equation that we constructed with Reshetikhin. It was a new class of equations, and at first it was studied by various people directly and in relation with statistical mechanics, and then it appeared again at the next level of categorification.

On the other hand, I am disappointed that my paper on orbital theory for loop groups, and other two papers on central extension of surface algebras with Etingof and with Khesin did not lead to similar big directions in mathematics. I am pretty sure that my works on $3d$ quantum gravity and on quaternionic analysis will become more influential after I, or somebody else, have passed a certain critical point of development in either subject.

JS How did your colleagues at Yale influence your research?

IF When I came to Yale, as I said before, I had a fantastic reception. I was expecting to meet Howard Garland and Jim Lepowsky, but Lepowsky had already left, and Gregg Zuckerman had just become a member of faculty. We constantly engaged in discussions on mathematics and life, and we had a perfect understanding of each other. It was very stimulating for me to have them around. Also, they knew quite a lot of mathematics and they helped me both with my dissertation and with other research. Then we wrote a joint paper, which was about interpretation of some string theory results in terms of semi-infinite cohomology. We introduced a certain infinite-dimensional Kähler structure which was developed later in string theory papers. I used many results of Garland's and Zuckerman's in my research. For example, Garland was already working in the affinization program, in particular, he studied topology of loop spaces by analogy with finite-dimensional flag varieties. I used Zuckerman's work on functors in representation theory, which turned out to be very important in categorification. I also spoke with other members of the department, with Roger Howe, Piatetski-Shapiro, Feit, and Mostow. I always felt extremely comfortable at the Yale mathematics department where I spent the first three years of my life in the US and then many more years when I returned back as a tenured professor.

JS How has the Yale mathematics department changed over the years?

IF When I first came to Yale, it was primarily pure mathematics that dominated the department. Now in the Yale mathematics department, as well as in many departments across the United States and all over the world, applied mathematics plays an equally important role. From my point of view, it's a different subject, it has some intersection with mathematics, but physics, or chemistry, or biology also have intersections. Motivations for applied mathematics are very different. This field has become a lot more powerful with the use of computers, and it's very important in the analysis of networks and data, etc. There are also some interesting structures and problems from applied mathematics used in pure mathematics, however, the two subjects still have very different motivations. Pure mathematics is related to theoretical physics in some kind of mysterious ways that I tried to explain before. I believe that the present directions in pure mathematics will always be important, at least while theoretical physics will still be developing. Then it might happen that new trends in pure mathematics will become very different, and some results from modern applied mathematics such as complexity theory will be part of the future pure mathematics. But I believe that we still have at least another 100 years to prove the Riemann Hypothesis and to understand the mathematics of $4d$ Quantum Gravity before the pure mathematics changes its main themes.

From the window of my new office, I now look at the old department building with some regret. And I find it very symbolic.

JS What are your views on the direction of representation theory?

IF Modern representation theory certainly has achieved a great progress. However, I would want to see more new algebras, groups and their representations as well as new relations with mathematics and physics. At this moment, a lot of work is dedicated to the old groups, and the old representations, and their various realizations like categorification and geometrization. So where to look for new algebras and groups? One natural example is hyperbolic Kac-Moody groups and Lie algebras. These are still mysterious groups and algebras which are very large in size; and even though some representations have been constructed, it's not clear what their geometric origin is and what is behind their structure. Physicists tell us that these hyperbolic Kac-Moody algebras can be very important in quantum gravity: in a kind of ultimate quantum gravity, 4-dimensional or even higher dimensional. But this work is still at the preliminary level. So, my suggestion here is to split the problem into small steps, as small steps as possible, in particular, consider one of the simplest hyperbolic Kac-Moody algebras having a double of modular group as its Weyl group, which we originally studied with Feingold. There is a conjecture concerning the bound for multiplicities of weights for this algebra given by partition function. This conjecture has not been proven yet and its solution may shed more light on the origin and structure of hyperbolic Kac-Moody algebras. The next question will be just understanding of the multiplicities themselves. Because they have number theoretic meaning, these multiplicities are some functions over quadratic forms – it must be something very important.

In a different direction, I wrote several papers which generalized loop algebras to surface algebras. It turns out that these surface algebras have a central extension by the Jacobian of the Riemann surface. They show full analogy with central extension of loop groups. Also, orbital theory for these surface algebras is related to some classical facts about vector bundles on Riemann surfaces. And an analog of topological field theory also makes sense in terms of some invariants of Calabi-Yau manifolds and relations between them. Ordinary 3-dimensional linking has generalization to the linking of complex curves in 3-dimensional complex space. This analogy shows that there should also exist an interesting representation theory for these groups and algebras of functional dimension 2. But it's still missing, and I would consider this a very important problem.

In the theory of loop algebras, some problems are left unsolved such as construction of representations of split real loop groups. The question is what the analog of unitary principal series for this class of groups is. This problem is closely related to the recent work on the so-called analytic Langlands program. But the main question about construction of these new classes of representation is still open as I know. From a conjectural isomorphism with the category of representations of split real quantum groups, we can already guess the structure of the tensor category of representations of split real loop groups. The problem of developing harmonic analysis on split real loop groups is another important related direction. But as a first step, one needs to prove a version of the Peter-Weyl theorem for unitary loop groups. This is closely related to the so-called Wess-Zumino-Witten model which is based on the corresponding geometry of a central extension of loop group. But the analog of Haar measure which should be some kind of 2-dimensional Wiener measure is not yet constructed rigorously. It is interesting to note that the natural measure on orbits of coadjoint representations of loop groups is precisely the classical Wiener measure on compact finite-dimensional groups, which appears in direct analogy with the finite-dimensional case. So, this part is well understood and goes back to my dissertation.

But my favorite directions of modern representation theory are based on the development of quaternionic analysis and of $3d$ quantum gravity. They may lead to substantial breakthroughs in representation theory, and these directions have high priority on my own list of problems.

JS What advice can you give to young mathematicians?

IF I'm probably not going to be very original here, but just several pieces of advice that I have in mind are the following. First, you have to choose your field – algebra, geometry, or analysis depending on your talent. In my experience, when people can be discouraged when trying to solve problems that are not from the field at which they are really great. But all the areas in mathematics are related. So, when you develop your own direction and go far enough, sooner or later you will have a great perspective and eventually you'll see a lot of other mathematics. But technically you still have to start from one of those fields. It's important to choose an advisor who can recommend to you some problems that are not technically impossible. On the other hand, these problems should be exciting to you and their solution should bring some advance to the area. In principle, you can choose a problem yourself. It was more popular maybe 50 years ago when I was a young mathematician myself, but it's becoming more and more difficult in modern mathematics. It's still a possibility, but I don't recommend it any longer. Also, it's a good strategy to work with several problems: if you cannot make progress in the first one, you can still change it – you should feel free to consider another problem. And sometimes it's even good to have a couple of problems from the very beginning and to develop whatever can be developed. That's the advantage of pure mathematics.

After you have decided on the subject, it's good to hold the rope by two ends, by which I mean abstract formulation – the most general and abstract theory, as well as very explicit examples. If you have only one or another, you're in a much weaker position than when you have both ends. You should not expect that you will write some original paper right away. It may happen that the first results that you'll have obtained will already be known, but it's perfectly okay. And it may even be considered a good indication that you're moving in the right direction. So just keep working until you reach the current development of the subject, and then you'll be unbeatable and will be able to continue as far as you like. Do not be afraid to be different from your adviser and everybody else; all the famous mathematicians have developed their own unique approach to mathematics. Always remember that miracles still happen in mathematics and do not reject them if you have the luck to be a witness.

JS What is your approach to mentoring graduate students?

IF I always try to give to a student as difficult a problem as they could possibly solve at the limit of their abilities, and very often it worked extremely well. But in some cases, my problems turned out too difficult, and then we had to change the problem to a simpler one. Sometimes, my students would solve the problems that I originally gave them several years later when they would become postdocs. This happened even to some of my best students. My idea was always not just to give them some technical problem but rather to open some kind of a new direction for each student. This certainly sounds very ambitious, but it gave results in most cases.

For the most of my mathematical life, I worked on representation theory and mathematical physics, but I changed the specific direction several times during my career. Each time, I had to learn almost everything in the new topic from the very beginning, and I usually gave graduate lectures, graduate courses on this new subject. It often happened that my students would learn the field and become more fluent than myself in it. So that after the graduation, they would really become experts in the subject that I was thinking about, and then they would explain to me the further developments in these areas, and that was extremely beneficial for me.

I always encouraged interaction between students and tried to make sure that there was a creative atmosphere in my group, which sometimes reached six students and several postdocs. I remember that when I would sometimes come to the department late at night, there would always be a group of my students in the lounge discussing mathematics. I only had to watch that nobody would be stuck, and everybody knew what to do next. Then everything was advancing well; and it continues to advance well with almost two hundred of my former graduate students and grandstudents.

Contemporary Mathematics
Volume **817**, 2025
https://doi.org/10.1090/conm/817/16332

Homological link invariants from Floer theory

Mina Aganagic, Elise LePage, and Miroslav Rapcak

Abstract. There is a generalization of Heegaard-Floer theory from $\mathfrak{gl}_{1|1}$ to other Lie (super)algebras ${}^L\mathfrak{g}$. The corresponding category of A-branes is solvable explicitly and categorifies quantum $U_q({}^L\mathfrak{g})$ link invariants. The theory was discovered by the first author, using homological mirror symmetry. It has novel features, including equivariance and, if ${}^L\mathfrak{g} \neq \mathfrak{gl}_{1|1}$, coefficients in categories. In this paper, we give a detailed description of the theory and how it is solved in the two simplest cases: the $\mathfrak{gl}_{1|1}$ theory itself, categorifying the Alexander polynomial, and the $\mathfrak{su}_2$ theory, categorifying the Jones polynomial. Our approach to solving the theory is new, even in the familiar $\mathfrak{gl}_{1|1}$ case.

Contents

1. Introduction

In 1928, Alexander constructed a polynomial invariant of links in $\mathbb{R}^3$ [**11**]. In one of its formulations, it is defined by picking a projection of the link to a plane and the skein relation

$$\mathbf{q}^{n/2}\, P_{\times} - \mathbf{q}^{-n/2}\, P_{\times} = (\mathbf{q}^{1/2} - \mathbf{q}^{-1/2})\, P_{)(} \tag{1.1}$$

with $n = 0$. The same skein relation with $n = 2$ instead defines the polynomial link invariant discovered by Jones in 1988 [**48**].

Witten showed that these invariants originate from Chern-Simons theory with gauge group based on a Lie algebra ${}^L\mathfrak{g}$ [**99**]. In particular, the Jones polynomial

This research was supported by the NSF foundation grant PHY2112880 and by the Simons Investigator Award.

comes from choosing ${}^L\mathfrak{g} = \mathfrak{su}_2$, with links colored by the defining, two-dimensional representation. The Alexander polynomial, comes from the same setting by taking ${}^L\mathfrak{g}$ to be the Lie superalgebra $\mathfrak{gl}_{1|1}$. The resulting link invariants are the $U_{\mathfrak{q}}({}^L\mathfrak{g})$ quantum group invariants of Reshetikhin and Turaev [**85**].

1.1. The categorification problem. The quantum invariants of links are Laurant polynomials in $\mathfrak{q}^{1/2}$, with coefficients which are integers (up to an overall normalization). Crane and Frenkel [**28**] used this to suggest that link invariants may be shadows of a deeper, more fundamental theory, in which one can associate to a projection of the link to a plane a bi-graded complex of abelian groups

$$C^{*,j}(K) = \dots C^{i-1,j}(K) \xrightarrow{\partial^{i-1}} C^{i,j}(K) \xrightarrow{\partial^{i}} \dots,$$

such that each homology group $\mathcal{H}^{i,j}(K) = \ker \partial^i / \operatorname{im} \partial^{i-1}$ is an invariant of the link projection and whose equivariant Euler characteristic

$$\chi(K) = \sum_{i,j} (-1)^i \mathfrak{q}^j \dim \mathcal{H}^{i,j}(K),$$

is the quantum group link invariant. Independently, existence of such a theory was predicted by string theory. This was explained by Gukov, Schwarz and Vafa in [**42**], following Ooguri and Vafa in [**72**].

1.1.1. A toy model for categorification comes from a Riemannian manifold M, whose Euler characteristic

$$\chi(M) = \sum_{k\in\mathbb{Z}} (-1)^k \dim \mathcal{H}^k(M)$$

is categorified by the cohomology $\mathcal{H}^k(M) = \ker d_k / \operatorname{im} d_{k-1}$ of the de Rham complex

$$C^* = \cdots C^{k-1} \xrightarrow{d_{k-1}} C^k \xrightarrow{d_k} \cdots .$$

The Euler characteristic is, from the physics perspective, the partition function of supersymmetric quantum mechanics with M as a target space: $\chi(M) = \mathrm{Tr}(-1)^F e^{-\beta H}$, with the Hamiltonian being the Laplacian $H = dd^* + d^*d$ and $d = \sum_k d_k$ the supersymmetry operator. If h is a Morse function on M, the complex can be replaced by a Morse–Smale–Witten complex C^*_h with the differential $d_h = e^h d e^{-h}$. The complex C^*_h, as Witten showed [**98**], is the space of perturbative ground states of a σ-model on M with potential h. The action of the differential d_h is generated by instantons.

1.1.2. Both the Alexander and the Jones polynomial have known categorifications; however, they have very different flavor.

A categorification of the Jones polynomial was discovered in 1999, by Khovanov [**50, 51**]. Khovanov's construction is remarkable for being simple and completely explicit. However, unlike in our toy example of categorification of the Euler characteristic of a Riemannian manifold, the construction is purely algebraic – it does not originate from either geometry or physics. If we replace the Lie algebra $\mathfrak{su}_2$ by an arbitrary (ordinary) Lie algebra ${}^L\mathfrak{g}$, there is an abstract algebraic framework for categorification of quantum link invariants due to [**95**], based on a derived category of modules of an algebra known as the KLRW algebra, constructed in [**52, 87, 95**]. Unlike Khovanov's construction, Webster's categorification exists only formally.

The Alexander polynomial is categorified by Heegaard-Floer theory, discovered in 2002 by Ozsvath and Szabo [**79–83**]. It is based on Floer theory, or the A-model, with target $Y_{\mathfrak{gl}_{1|1}}$ which is a symmetric product of Riemann surfaces $\mathcal{A}$. Floer theory is modeled after the Morse theory approach to supersymmetric quantum mechanics, applied to a theory in one higher dimension. The vector spaces underlying the chain complexes are spanned by intersection points of a pair of A-branes constructed from the link. The differential is defined by counting disk instantons. Unlike a generic A-model, Heegard-Floer theory turned out to be solvable explicitly [**64, 66**].

1.1.3. The link categorification problem is to find a general framework for link homology theories that works uniformly for all Lie algebras and ideally originates from physics or geometry. The theory should explain what link homology groups are and why they exist. In addition, it is not unreasonable to hope that it may be solvable.

Chern-Simons theory is a rare example of an explicitly solvable yet non-trivial quantum field theory. If one is fortunate enough, the theory whose Euler characteristic Chern-Simons theory computes should be solvable as well. Previous attempts [**24, 25, 54, 55, 92, 101**] gave theories that were either not general, not solvable, or both.

1.2. The solution. There is a generalization of Heegaard-Floer theory from $\mathfrak{gl}_{1|1}$ to other Lie super-algebras ${}^L\mathfrak{g}$, discovered in [**1, 2**]. The theory manifestly categorifies quantum group link invariants. It provides a rare example of a category of branes which is solvable explicitly, yet highly non-trivial.

In this paper, we give an explicit description of this new Floer theory and how it is solved. We give an algorithm for computing link homologies that applies for any ${}^L\mathfrak{g}$. We describe the theory in detail in the two simplest cases, corresponding to $\mathfrak{gl}_{1|1}$ and to $\mathfrak{su}_2$. We prove explicitly Thm. 1, that the theory gives homological link invariants and Thm. 2, that it its Euler characteristics satisfy the skein relations (1.1). The approach to solving the theory is new, even in the familiar context of the $\mathfrak{gl}_{1|1}$ theory.

In the course of solving the theory, one also discovers that KLRW algebras arise naturally as algebras of opens strings in the Floer theory associated to simply laced Lie algebras ${}^L\mathfrak{g}$, as proven in [**5**]. We will describe the extension to other ${}^L\mathfrak{g}$, including to non-simply laced Lie algebras, and Lie superalgebras $\mathfrak{gl}(m|n)$ and $\mathfrak{sp}(m|2n)$ elsewhere.

2. Overview of the theory

The section gives a brief overview of the theory for any ${}^L\mathfrak{g}$. In later sections, we will describe in detail the theories in the two special cases, corresponding to ${}^L\mathfrak{g} = \mathfrak{gl}_{1|1}$ and $\mathfrak{su}_2$.

We will study framed links in $\mathcal{A} \times \mathbb{R}$ where $\mathcal{A} = \mathbb{C}^\times$ is an infinite complex cylinder. For this choice of $\mathcal{A}$, the theory will have two descriptions, related by mirror symmetry. Mirror symmetry will make complementary aspects of the theory easy to understand.

2.1. Homological link invariants from A-branes. Pick ${}^L\mathfrak{g}$, which is either a simply laced Lie algebra, or the Lie superalgebra $\mathfrak{gl}(m|n)$. Pick also a collection of marked points on $\mathcal{A}$, or rather pairs of them, labeled by a minuscule representation V_i of ${}^L\mathfrak{g}$, and its conjugate V_i^*. The geometric structures below are associated to the zero weight subspace of $\bigotimes_i V_i \otimes V_i^*$, where i runs over all pairs of punctures.

The theory extends to non-simply laced Lie algebras ${}^L\mathfrak{g}$, and to $\mathfrak{sp}(m|2n)$, with some minor modifications. It is possible it exists more generally yet.

2.1.1. A "matching" is a collection of curves on $\mathcal{A}$ which have endpoints on a set of marked points, colored by pairs of conjugate representations. For every two such matchings, colored red and blue, we get a link in $\mathcal{A} \times \mathbb{R}$ by taking the red matching to overpass the blue. Arbitrary links arise in this way.

To assign a pair of matchings to a link, recall that any link has a representation as a "plat" closure of a braid, with a sequence of d caps at the top and d cups at the bottom. The result is due to Birman [**22**].

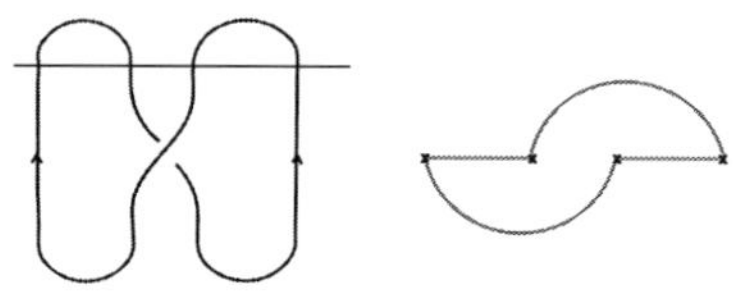

FIGURE 1. A pair of matchings derived from a plat closure representation of the unknot, for $d = 2$.

To the top closure, we will assign a "standard" red matching. It will be useful to isotope its endpoints to lie on a fixed circle in $\mathcal{A}$. To the bottom closure, we will assign the same standard matching but isotoped by the braid and colored blue. The braid moves are read from bottom to top.

2.1.2. To the Riemann surface $\mathcal{A}$ with marked points colored by pairs of conjugate, minuscule representations of ${}^L\mathfrak{g}$, we will associate a category. The category

$$\mathscr{D}_Y = D(\mathcal{F}S(Y, W))$$

is a category of equivariant A-branes on an exact symplectic manifold $Y = Y_{{}^L\mathfrak{g}}$ of $2c_1(K_Y) = 0$, equipped with a choice of top holomorphic form $\Omega = \Omega_{{}^L\mathfrak{g}}$ and with a superpotential $W = W_{{}^L\mathfrak{g}}$. The symplectic form must be compatible with Ω in the sense that both give rise to the same real volume form $\Omega \wedge \overline{\Omega}$. A comprehensive introduction to derived Fukaya categories is [**90**]; a brief review is in [**13**].

2.1.3. The target Y is a product of symmetric products, one for each node of the Dynkin diagram

$$Y = \prod_{a=1}^{\mathrm{rk}\,{}^L\mathfrak{g}} Sym^{d_a}(\mathcal{A}), \tag{2.1}$$

with potential W that couples them written in [**2**]. An important special feature of the theory is that W is not single-valued.

2.1.4. Objects of $\mathscr{D}_Y$ are Lagrangians L in Y, together with extra structures. Their grading is a choice of a lift of the phase of

$$\Omega e^{-W}|_L, \tag{2.2}$$

to a single-valued function on the brane. The choice of a lift of the phase of $\Omega|_L$ is the cohomological, or Maslov grading of the brane. Since W is not single-valued, the choice of a lift of the phase of $e^{-W}|_L$ leads to additional, equivariant, gradings in the theory. As always, the branes may be equipped with a local system and a choice of spin structure. We will see that both will play a role.

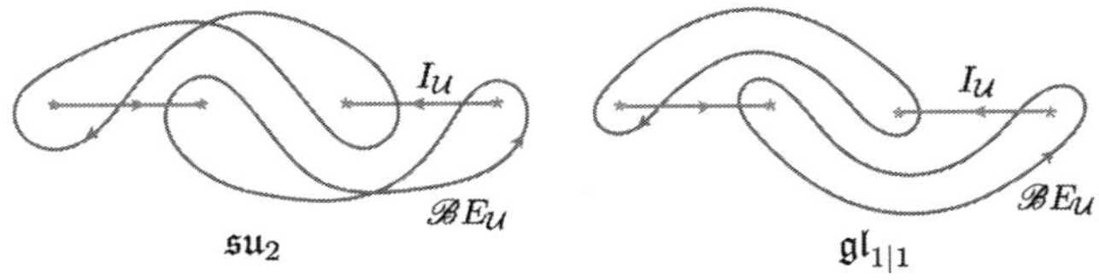

FIGURE 2. The branes corresponding to the unknot projection in Fig. 1 in the ${}^L\mathfrak{g} = \mathfrak{su}_2$ and $\mathfrak{gl}_{1|1}$ theories.

2.1.5. To the blue and red segments we will assign a pair of A-branes

$$I_{\mathcal{U}}, \quad \mathscr{B}E_{\mathcal{U}} \in \mathscr{D}_Y.$$

The branes depend on ${}^L\mathfrak{g}$ and on representations coloring the link components. The choice of branes is explained by equivariant mirror symmetry when ${}^L\mathfrak{g}$ is simply laced.

Homological $U_{\mathfrak{q}}({}^L\mathfrak{g})$ link invariants are the spaces of morphisms between the branes, viewed as objects of $\mathscr{D}_Y$:

$$Hom^{*,*}_{\mathscr{D}_Y}(\mathscr{B}E_{\mathcal{U}}, I_{\mathcal{U}}) = \bigoplus_{k,\vec{J}} Hom_{\mathscr{D}_Y}(\mathscr{B}E_{\mathcal{U}}, I_{\mathcal{U}}[k]\{\vec{J}\}). \tag{2.3}$$

Here, $[k]$ is $\mathbb{Z}$-valued cohomological, or Maslov degree. The vector $\vec{J}$ is the equivariant degree, whose J_0 component is half integral, and related to $\mathfrak{q}$. The $\mathrm{rk}^L\,\mathfrak{g}$ additional equivariant $\mathbb{Z}$-gradings couple to holonomies of Chern-Simons gauge fields on the S^1.

2.1.6. Spaces of morphisms between the branes in $\mathscr{D}_Y$ are Floer cohomology groups [**90**]:

$$Hom^{*,*}_{\mathscr{D}_Y}(\mathscr{B}E_{\mathcal{U}}, I_{\mathcal{U}}) = \mathrm{Ker}\,\delta_F/\mathrm{Im}\,\delta_F. \tag{2.4}$$

Floer theory, introduced in [**34**], generalizes the Morse theory approach to supersymmetric quantum mechanics, to a theory in one dimension up. The Floer complex is a vector space

$$hom^{*,*}_Y(\mathscr{B}E_{\mathcal{U}}, I_{\mathcal{U}}) \equiv CF^{*,*}(\mathscr{B}E_{\mathcal{U}}, I_{\mathcal{U}}) = \bigoplus_{\mathcal{P}\in \mathscr{B}E_{\mathcal{U}}\cap I_{\mathcal{U}}} \mathbb{C}\mathcal{P}, \tag{2.5}$$

spanned by the intersection points of the branes, keeping track of gradings, equipped with a differential δ_F

$$\delta_F : hom^{*,*}_Y(\mathscr{B}E_{\mathcal{U}}, I_{\mathcal{U}}) \to hom^{*+1,*}_Y(\mathscr{B}E_{\mathcal{U}}, I_{\mathcal{U}}). \tag{2.6}$$

The space $hom^{*,*}_Y(\mathscr{B}E_{\mathcal{U}}, I_{\mathcal{U}})$ of morphisms between the branes in the ordinary Fukaya category is the space of perturbative supersymmetric ground states of the $\mathcal{N}=2$ Landau-Ginsburg model on an interval, with the pair of A-branes as boundary conditions. The space $Hom^{*,*}_{\mathscr{D}_Y}(\mathscr{B}E_{\mathcal{U}}, I_{\mathcal{U}})$ of morphisms in the derived Fukaya category is spanned by the exact supersymmetric ground states.

2.1.7. The action of the Floer differential in (2.6) is defined by counting holomorphic maps $y : \mathrm{D} \to Y$ from the unit disk D with two marked points to Y. The unit disk with two marked points is biholomorphic to an infinite strip of unit width. The coefficient of $\mathcal{P}'$ in $\delta_F\mathcal{P}$ is the count of maps for which the two boundary components of D map to the branes $\mathscr{B}E_{\mathcal{U}}$ and $I_{\mathcal{U}}$ and the two marked points to $\mathcal{P}$ and $\mathcal{P}'$. Our orientation conventions are spelled out in [**2**] and are the same as in [**13**].

Holomorphic maps that contribute to the Floer differential, or the higher A_∞ maps, have finite symplectic area. In a physically sensible A-model, the potential W must pull back to a regular function on the disk D [**2**]. For theories with potentials which are regular this condition is vacuous, but not for us. It implies the equivariant degree of all Floer theory maps is 0. Maps that contribute to the differential δ_F have Maslov index 1. This is the index of the linearized Cauchy-Riemann operator for a fixed conformal structure on D. After dividing by reparameterizations of the unit disk keeping the two boundary marked points fixed, the moduli space of maps that contribute becomes zero dimensional.

2.1.8. There are no general means for computing the action of the Floer differential δ_F from its definition, rather, one has to approach the problem instanton by instanton. In Heegaard-Floer theory, for some very special link presentations [**66**], one can compute the differential by arranging that only a few different kinds of disks appear; however, there is nothing canonical about it.

In sufficiently nice derived Fukaya categories, there is an alternate approach. In our setting, it will enable us to compute the action of differential, and hence the Floer cohomology groups, algebraically.

In what follows, we will describe the category $\mathscr{D}_Y$ explicitly, and how to solve it. We will give an algorithm for computing $Hom^{*,*}_{\mathscr{D}_Y}(\mathscr{B}E_{\mathcal{U}}, I_{\mathcal{U}})$, and prove the resulting vector spaces are link invariants.

Remark. We will not prove that the algebraic and the geometric approaches to computing the action of the Floer differential agree. With current technology, proof can be given only in the complement of the divisor D_O in (3.1). For Lie algebras ${}^L\mathfrak{g}$ other than $\mathfrak{gl}_{1|1}$, general proof requires laying foundations of Fukaya categories with coefficients.

2.1.9. To keep the length of the paper contained, will specialize to the two simplest cases, when the Lie algebra is either $\mathfrak{su}_2$ or $\mathfrak{gl}_{1|1}$, with links colored by their two-dimensional defining representations. In both cases, Y is the symmetric product of d copies of $\mathcal{A}$

$$Y = Sym^d(A), \tag{2.7}$$

but with different Ω and W. A point in the symmetric product is d unordered points on $\mathcal{A}$. Correspondingly, an object in $\mathscr{D}_Y$, the category of A-branes on Y, is a product of d one-dimensional Lagrangians on $\mathcal{A}$, which we take to be disjoint.

In both cases, the $I_{\mathcal{U}}$ branes come from Lagrangians on $Y = Sym^d(\mathcal{A})$ that are simply the product of the corresponding d red segments on $\mathcal{A}$. The $\mathscr{B}E_{\mathcal{U}}$ branes are obtained by replacing the blue segments with products of d figure eights in the $\mathfrak{su}_2$ case, or ovals for $\mathfrak{gl}_{1|1}$, as in Fig. 2.

2.2. Algebra of thimbles. The category of A-branes on Y is generated by a finite set of branes. As the generating set, we will take the branes

$$T_{\mathcal{C}} = T_{i_1} \times \ldots \times T_{i_d},$$

which are products of d real line Lagrangians on $\mathcal{A}$, where d is the dimension of Y, colored by simple roots of the Lie algebra.

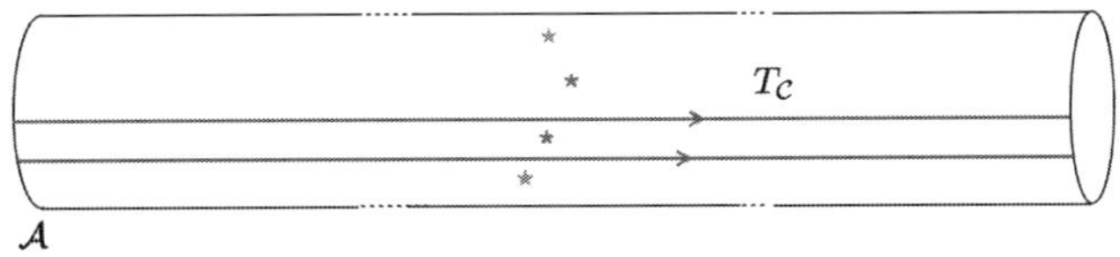

FIGURE 3. A $T_{\mathcal{C}}$-brane for ${}^L\mathfrak{g} = \mathfrak{gl}_{1|1}$ with $d = 2$.

There is a $T_{\mathcal{C}}$-brane for every inequivalent ordering of real line Lagrangians on $\mathcal{A}$. For Y as in (2.1), every $T_{\mathcal{C}}$ brane has d_a of its real line Lagrangians colored by a simple root ${}^Le_a$. Without changing the $T_{\mathcal{C}}$-brane, a one-dimensional component of it colored by a simple root ${}^Le_a$ can be isotoped through a marked point provided $\langle {}^Le_a, \mu_j\rangle = 0$, where μ_j is the highest weight labeling the puncture; it can pass through another real line Lagrangian colored by ${}^Le_b$ if $\langle {}^Le_a, {}^Le_b\rangle = 0$.

2.2.1. The T-branes should be thought of as thimbles [**44, 100**] of the potential W, which Y is equipped with, in a chamber of the parameter space [**2**]. The thimbles are the set of all initial conditions for upward gradient flows of $\mathrm{Re}(W)$, on which $\mathrm{Im}(W)$ is constant, and which asymptote to a critical point of the potential.

One subtlety is that, to generate the category of A-branes with the T-branes, one has include to additional branes which one can think of as thimbles associated to "critical points at infinity". Another subtlety, explained in section 5, is that when ${}^L\mathfrak{g} \neq \mathfrak{gl}_{1|1}$, the $T_{\mathcal{C}}$-branes used to generate $\mathscr{D}_Y$ get equipped with an additional structure, which is a local system of modules for a certain finite rank algebra.

2.2.2. The category $\mathscr{D}_Y$ of A-branes on Y being generated by the T-brane

$$T = \bigoplus\nolimits_{\mathcal{C}} T_{\mathcal{C}}, \tag{2.8}$$

implies the equivalence

$$\mathscr{D}_Y \cong \mathscr{D}_A \tag{2.9}$$

where $\mathscr{D}_A$ is the derived category of modules of an algebra A,

$$A = hom_Y^{*,*}(T, T), \tag{2.10}$$

which is the endomorphism algebra of the generator T. The sum in (2.8) is over all the thimbles, including those associated to critical points at infinity. A is the algebra of open strings that begin and end on the branes.

One of the virtues of this generating set of branes is that the algebra A is computable explicitly and in a sense, as simple as possible. When ${}^L\mathfrak{g}$ is a Lie superalgebra, the algebra A is an associative graded algebra with a differential. For $\mathfrak{gl}_{1|1}$, we will describe the algebra in section 4. When ${}^L\mathfrak{g}$ is an ordinary simply laced Lie algebra, the algebra A is even simpler, as it is an ordinary associative algebra in cohomological degree zero. For $\mathfrak{su}_2$, it is described in section 5.

2.2.3. One should think of the equivalence in (2.9) as one half of homological mirror symmetry,

$$\mathscr{D}_X \cong \mathscr{D}_A \cong \mathscr{D}_Y. \tag{2.11}$$

Here X is the mirror of Y, and $\mathscr{D}_X$ is the category of B-type branes on X, working equivariantly with respect to a torus $\mathbb{T}$. When ${}^L\mathfrak{g}$ is an ordinary Lie algebra, $\mathscr{D}_X$ is the derived category of $\mathbb{T}$-equivariant coherent sheaves on X; in Lie superalgebra cases, it is a category of matrix factorizations. The T-branes on Y map to vector bundles on X with the same label. Mirror to adding to Y its multi-valued potential

W is working equivariantly with respect to the torus $\mathbb{T}$. Conversely, turning off the $\mathbb{T}$-action sets W to zero.

2.2.4. Every $T_{\mathcal{C}}$-brane associated to an honest critical point of W has a dual $I_{\mathcal{C}}$-brane, which satisfies that the only non-vanishing Homs from the T-branes to the I-branes are

$$hom_Y(T_{\mathcal{C}}, I_{\mathcal{C}'}) = \delta_{\mathcal{C},\mathcal{C}'}\mathbb{C}, \tag{2.12}$$

in Maslov and equivariant degree zero. The $I_{\mathcal{C}}$-brane is its dual thimble, corresponding to the set of all initial conditions of downward gradient flows of $\mathrm{Re}\, W$. Correspondingly, a $T_{\mathcal{C}}$-brane intersects only the dual $I_{\mathcal{C}'}$-brane, and only once, at the critical point of the potential, leading to (2.12). Among the I-branes, one finds included the $I_{\mathcal{U}}$-branes that serve as caps. There is a Koszul dual way to generate $\mathscr{D}_Y$, in terms of I-branes instead of T-branes [**2**].

2.2.5. The equivalence of categories in (2.9) comes from the Yoneda functor

$$hom_Y(T, -) : \mathscr{D}_Y \to \mathscr{D}_A, \tag{2.13}$$

which maps A-branes on Y to complexes of modules of the algebra A. It maps $T_{\mathcal{C}}$-branes to projective modules of the algebra A, and the dual $I_{\mathcal{C}}$-branes to simple modules. A projective module is a direct summand of a free module. Since an algebra is always a free module for itself, the free module whose summand is $T_{\mathcal{C}}$ corresponds to the $T = \bigoplus_{\mathcal{C}} T_{\mathcal{C}}$ brane.

2.3. Floer complexes from T-brane algebra.

Like every brane in $\mathscr{D}_Y$, the braided cup branes have a description as a complex, every term of which is a direct sum of T-branes. The complex $(\mathscr{B}E(T), \delta)$ is a direct sum brane $\mathscr{B}E(T)$, together with a differential δ. The complex describes how to get the $\mathscr{B}E_{\mathcal{U}} \in \mathscr{D}_Y$ brane by starting with the direct sum of the generators

$$\mathscr{B}E(T) = \bigoplus_k \mathscr{B}E_k(T)[k], \tag{2.14}$$

and deforming differential $\delta \in A = hom_Y^{*,*}(T, T)$ away from zero. The differential δ acts on the brane as a cohomological degree one and equivariant degree zero operator

$$\delta : \mathscr{B}E(T) \xrightarrow{\delta} \mathscr{B}E(T)[1]$$

that squares to zero in the algebra A (in an appropriate sense).

2.3.1. Per definition, the direct sum brane $\mathscr{B}E(T)$ together with the differential δ, is equivalent to the brane we started with, as object of $\mathscr{D}_Y$

$$\mathscr{B}E_{\mathcal{U}} \cong (\mathscr{B}E(T), \delta). \tag{2.15}$$

Since the T-branes map to projective modules of the algebra A by the Yoneda functor (2.13), the complex $(\mathscr{B}E(T), \delta)$ is a projective resolution of the $\mathscr{B}E_{\mathcal{U}}$ brane. In particular, this means that all Homs of the $\mathscr{B}E_{\mathcal{U}}$-brane, viewed as a Lagrangian on Y are equivalent in $\mathscr{D}_Y$ to the Homs of the complex $(\mathscr{B}E(T), \delta)$ resolving the brane.

2.3.2. Because the $I_{\mathcal{U}}$-brane is one of the simples of the algebra A, from the projective resolution of the $\mathscr{B}E_{\mathcal{U}}$-brane we get for free a complex of vector spaces

$$(\mathrm{hom}^*_Y(\mathscr{B}E(T), I_{\mathcal{U}}\{\vec{J}\}), \delta_F), \tag{2.16}$$

one for each $\{\vec{J}\}$, whose cohomology is the link homology group we are after:

$$\mathrm{Hom}_{\mathscr{D}_Y}(\mathscr{B}E_{\mathcal{U}}, I_{\mathcal{U}}[k]\{\vec{J}\}) = H^k(\mathrm{hom}^*_Y(\mathscr{B}E_{\mathcal{U}}(T), I_{\mathcal{U}}\{\vec{J}\})). \tag{2.17}$$

The complex of vector spaces in (2.16) is obtained from the (twisted) complex of branes $(\mathscr{B}E(T), \delta)$ by applying the functor

$$hom^*_Y(-, I_{\mathcal{U}}\{\vec{J}\}) = \bigoplus_{k\in\mathbb{Z}} \mathrm{hom}_Y(-, I_{\mathcal{U}}[k]\{\vec{J}\}), \tag{2.18}$$

and making use of the duality between T-branes and the I-branes in (2.12). The fact that the k-th cohomology group of the complex on the right hand side in (2.17) agrees with the Hom space in $\mathscr{D}_Y$ on the left is a standard piece of homological algebra. We will explain this in section 3.5 in more detail.

2.3.3. The complex we get in (2.16) is the Floer complex. By construction of the direct sum brane $\mathscr{B}E(T)$ from section 3, the vector space we get at the k-th term of the complex, in equivariant degree $\vec{J}$, can be identified with the space spanned by the intersection points of the $\mathscr{B}E_{\mathcal{U}}$-brane and the $I_{\mathcal{U}}$-brane, in the same degree

$$hom_Y(\mathscr{B}E_{\mathcal{U}}, I_{\mathcal{U}}[k]\{\vec{J}\}) = \mathrm{hom}_Y(\mathscr{B}E_k(T), I_{\mathcal{U}}\{\vec{J}\}). \tag{2.19}$$

Since, in addition, the cohomology of the complex coincides with Floer cohomology, it follows that the functor in (2.18) maps the brane differential δ, to the Floer differential δ_F in (2.6). Thus, knowing δ, we obtain the Floer differential δ_F without counting holomorphic curves.

2.3.4. It follows that the link homology $\mathrm{Hom}^{*,*}_{\mathscr{D}_Y}(\mathscr{B}E_{\mathcal{U}}, I_{\mathcal{U}})$ captures only a small part of the geometry of $\mathscr{B}E_{\mathcal{U}}$, the braided cup brane. Because the T-branes are dual to the I-branes by (2.12), almost all terms in the complex (2.16) vanish. The cohomology (2.17) of the complex that remains is the $U_{\mathfrak{q}}({}^L\mathfrak{g})$ link homology.

2.4. Decategorification. The category of A-branes $\mathscr{D}_Y$ manifestly categorifies $U_{\mathfrak{q}}({}^L\mathfrak{g})$ link invariants [**1,2**].

2.4.1. A conformal block of $\widehat{{}^L\mathfrak{g}}$ on $\mathcal{A}$ corresponding to a brane $L \in \mathscr{D}_Y$ has an integral representation

$$\mathcal{V}_\alpha[L] = \int_L \Phi_\alpha \Omega e^{-W}, \tag{2.20}$$

developed by Feigin and Frenkel [**33**] and Schechtman and Varchenko [**88,89**] (see [**31**] for review) for simple Lie algebras ${}^L\mathfrak{g}$. $\mathcal{V}_\alpha[L]$ are solutions of the trigonometric version of the Knizhnik-Zamolodchikov [**57**] equation of $\widehat{{}^L\mathfrak{g}}$ because $\mathcal{A}$ is an infinite cylinder. It depends on the brane only through its K-theory class $[L]$.

The integral differs from the equivariant central charge function $\mathcal{Z} : \mathscr{D}_Y \to \mathbb{C}$ of the A-brane category

$$\mathcal{Z}[L] = \int_L \Omega e^{-W} \tag{2.21}$$

by insertions of chiral ring operators Φ_α [**2**]. These are rational functions on Y, so the actions of monodromies on (2.20) and (2.21) coincide. The equivariant central charge functions are also conformal blocks, this time of a W-algebra associated to the Lie algebra $\mathfrak{g}$, which is Langlands dual to ${}^L\mathfrak{g}$ [**6**].

Conformal blocks corresponding to collections of caps and cups come from K-theory classes of branes $E_{\mathcal{U}}$ and $I_{\mathcal{U}}$ in $\mathscr{D}_Y$, which are proportional, with constant of proportionality that is a $\mathfrak{q}$-number.

2.4.2. If the braiding functor takes L to $\mathscr{B}L \in \mathscr{D}_Y$, its action on K-theory classes of branes $[\mathscr{B}L] = \mathfrak{B}[L]$ is obtained from Picard-Lefshetz theory. In our case, $\mathfrak{B}$ represents the action of a braid group element on the space of conformal blocks,

$$\mathcal{V}_\alpha[\mathscr{B}L] = \mathfrak{B}\mathcal{V}_\alpha[L], \tag{2.22}$$

so by the Kohno-Drinfeld theorem [**30**], $\mathfrak{B}$ is an element of the $U_{\mathfrak{q}}({}^L\mathfrak{g})$ quantum group. It is a product of R-matrices, which describe exchanges of neighboring pairs of vertex operators associated with elementary braid moves.

2.4.3. The $U_{\mathfrak{q}}({}^L\mathfrak{g})$ invariant of a link described as plat closure of a braid is the matrix element of $\mathfrak{B}$ between conformal blocks $[E_{\mathcal{U}}]$ and $[I_{\mathcal{U}}]$ corresponding to caps and cups. This matrix element is, per construction, the equivariant intersection number of the branes:

$$\chi(\mathscr{B}E_{\mathcal{U}}, I_{\mathcal{U}}) = \bigoplus_{\mathcal{P}\in\mathscr{B}E_{\mathcal{U}}\cap I_{\mathcal{U}}} (-1)^{M(\mathcal{P})}\vec{\mathfrak{q}}^{\vec{J}(\mathcal{P})}. \tag{2.23}$$

In addition to the Maslov, or cohomological grading, we keep track of the equivariant gradings coming from the non-single-valued potential. Above, $M(\mathcal{P})$ and $\vec{J}(\mathcal{P})$ are the Maslov and equivariant gradings of the point $\mathcal{P}$. The equivariant intersection number is also, per definition, the graded Euler characteristic of the homology $Hom^{*,*}_{\mathscr{D}_Y}(\mathscr{B}E_{\mathcal{U}}, I_{\mathcal{U}})$ in (2.3),

$$\chi(\mathscr{B}E_{\mathcal{U}}, I_{\mathcal{U}}) = \bigoplus_{M,\vec{J}}(-1)^M\vec{\mathfrak{q}}^{\vec{J}}\dim_{\mathbb{C}}(Hom_{\mathscr{D}_Y}(\mathscr{B}E_{\mathcal{U}}, I_{\mathcal{U}}[M]\{\vec{J}\})) \tag{2.24}$$

where the dimension is that of homology group over $\mathbb{C}$, or over $\mathbb{Q}$, ignoring torsion.

2.4.4. The quantum link invariants depend on choice of normalization, which one can take to be the value of the invariant for the unknot. One common choice assigns to the unknot the quantum dimension of the representation coloring it. Another assigns to it 1. The link invariant that assigns 1 to the unknot is known as "reduced", as opposed to the "unreduced" invariant that assigns it the quantum dimension.

One way to get the reduced $U_{\mathfrak{q}}({}^L\mathfrak{g})$ invariant of a link K is to start with a presentation of the link as a closure of a (V, V^*) tangle, which has one strand open at the top and one at the bottom, colored by representations V and V^*. The reduced invariant of the link is the invariant of the tangle. The ratio of the original and reduced invariant is the quantum dimension of the representation V. In our framework, cutting a strand translates to removing a pair of blue and red segments that meet at a puncture. By construction, this leaves a puncture which one can take to infinity without crossing any of the remaining matchings [**67**]. For example, for the unknot and ${}^L\mathfrak{g} = \mathfrak{su}_2$ or $\mathfrak{gl}_{1|1}$, this changes the brane configurations from Fig. 2 to Fig. 4.

FIGURE 4. The branes for reduced unknot for $\mathfrak{su}_2$ and $\mathfrak{gl}_{1|1}$, respectively.

2.4.5. Representation theory sourced by KZ equations [**57**] and their quantum qKZ counterparts [**35**], based on bosonic Lie algebras, has played a central role in mathematical physics since their discoveries. The extension of the theory to Lie superalgebras is not as well developed, with the exception of the extension of Drinfeld's theorem to superalgebras from [**32**]. It should be only a matter of time before this is remedied. For bosonic simply laced Lie algebras, the contemporary understanding of the theory is due to Maulik and Okounkov [**69**], followed by [**8**, **9**, **75**–**78**] and building on work of Nekrasov and Shatashvili [**70**]. It is based on quantum geometry of certain holomorphic symplectic varieties that arise as moduli spaces of three-dimensional quiver gauge theories with $\mathcal{N}=4$ supersymmetry. The relation between representation theory of quantum groups and quantum geometry naturally extends to a larger set, coming from moduli spaces of certain very special three dimensional quiver gauge theories with only $\mathcal{N}=2$ supersymmetry. In one direction, this leads to extension of the theory to $\mathfrak{gl}(m|n)$, development of which commenced in [**86**]. These gauge theories can be shown to lead to equivariant central charge functions (2.21) coming from conformal blocks of the corresponding super W-algebra from [**65**]. The other direction of extension is to Verma modules [**10**].

While the conceptual explanation for why the theory decategorifies to $U_{\mathfrak{q}}({}^L\mathfrak{g})$ link invariants will ultimately be along the lines presented in this subsection for any ${}^L\mathfrak{g}$, there is an alternative, direct way to understand it, once one picks a specific Lie algebra.

2.5. Isotopy invariance and skein. It is manifest at the outset that the theory gives rise to homological invariants of braids. It is true, though not manifestly so, that it also gives rise to homological invariants of links, invariant under ambient isotopy:

THEOREM 1. *The homology groups $Hom^{*,*}_{\mathscr{D}_Y}(\mathscr{B}E_{\mathcal{U}}, I_{\mathcal{U}})$ are invariants of links.*

To prove the theorem, we need to verify homology groups $Hom^{*,*}_{\mathscr{D}_Y}(\mathscr{B}E_{\mathcal{U}}, I_{\mathcal{U}})$ satisfy Markov-type moves for plat closures due to Birman [**22**] and Bigelow [**18**]. In [**18**, **22**], it is proven that two oriented braids have isotopic plat closures if and only if they differ by the sequence of moves in Figs. 5, 6, and 7, and an analogous set with the tops and bottoms of Figs. 5 and 7 exchanged.

A proof of Thm. 1 was given in [**1**, **2**], for all simply laced Lie algebras. Using the results of this paper, we will give a more direct proof of the theorem for $\mathfrak{gl}_{1|1}$ and $\mathfrak{su}_2$ theories. We briefly sketch the strategy here. The detailed proofs are in section 4.6 for $\mathfrak{gl}_{1|1}$ and in section $\mathfrak{su}_2$ for 5.7. They rely on a technical result from appendix A.

2.5.1. The key fact used in the proof of Thm. 1 is that moves in Figs. 5, 6 and 7 hold as a consequence of equivalences satisfied by the $E_{\mathcal{U}}$-branes, viewed as objects of $\mathscr{D}_Y$, in other words, even before taking Homs. Equivalences of $E_{\mathcal{U}}$-branes under those moves

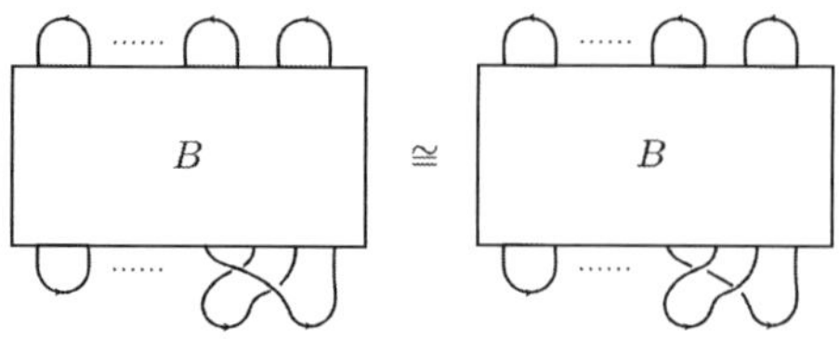

FIGURE 5. Markov-type move for plat closure of a braid B

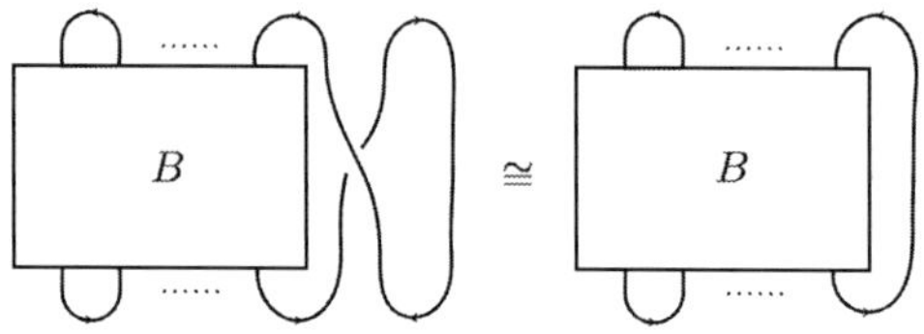

FIGURE 6. Stabilization, or Markov II-type move

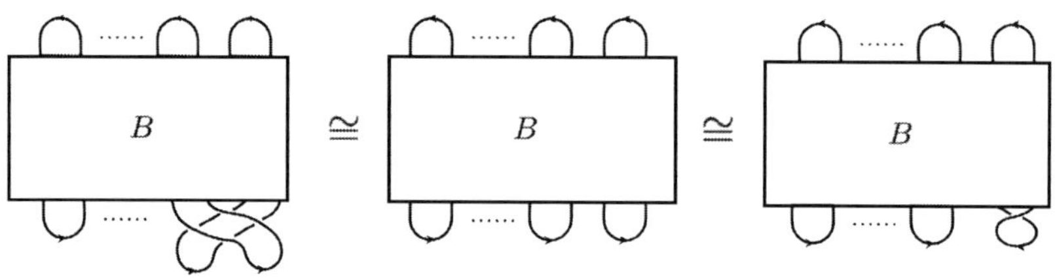

FIGURE 7. Further Markov-type moves.

imply that all Homs involving the branes are invariant, and $Hom^{*,*}_{\mathscr{D}_Y}(\mathscr{B}E_{\mathcal{U}}, I_{\mathcal{U}})$ in particular. We also learn that homology groups $Hom^{*,*}_{\mathscr{D}_Y}(\mathscr{B}E_{\mathcal{U}}, E_{\mathcal{U}})$ are invariant under the remaining Birman moves, obtained from moves in Figs. 5 and 7 by exchanging their top and bottom. The groups $Hom^{*,*}_{\mathscr{D}_Y}(\mathscr{B}E_{\mathcal{U}}, E_{\mathcal{U}})$ are not, however, invariant under the stabilization move in Fig. 6 – only $Hom^{*,*}_{\mathscr{D}_Y}(\mathscr{B}E_{\mathcal{U}}, I_{\mathcal{U}})$ are. Finally, we prove that $Hom^{*,*}_{\mathscr{D}_Y}(\mathscr{B}E_{\mathcal{U}}, E_{\mathcal{U}})$ coincides with $Hom^{*,*}_{\mathscr{D}_Y}(\mathscr{B}E_{\mathcal{U}}, I_{\mathcal{U}})$ times a fixed graded vector space of rank 2^d. It follows that $Hom^{*,*}_{\mathscr{D}_Y}(\mathscr{B}E_{\mathcal{U}}, I_{\mathcal{U}})$ are invariant under all Markov-type moves for plat closures, and hence that they are link invariants. This proves the theorem.

2.5.2. The Chern-Simons or $U_{\mathfrak{q}}({}^L\mathfrak{g})$ invariants are invariants of framed links [**85**, **99**]. Framing is a choice of a normal vector field to the knot, which makes the knot into a ribbon. Our category of branes, and conformal field theory, give all links a definite framing. This is the vertical or blackboard framing [**99**], where the vector fields are everywhere normal to the plane of the projection. In vertical framing, the second move in Fig. 7 does not hold as stated. Rather, the move adds a full twist to the ribbon, changing its framing by one unit. In $\mathscr{D}_Y$, change of framing is reflected in a shift of equivariant and homological grading of the $\mathscr{B}E_{\mathcal{U}}$ brane. An exception is the $\mathfrak{gl}_{1|1}$ theory for which the framing acts trivially.

We will change the framing of links from vertical to zero framing, so that the second relation in Fig. 7 holds on the nose. A link in zero framing differs from the vertical one by adding w twists to the corresponding ribbon. The writhe w is the number of positive ⤧ minus the number of negative ⤨ crossings in an oriented link. See section 5.7 for more detail.

2.5.3. The Euler characteristic of the theory is the corresponding quantum link invariant on general grounds from section 2.4.1, so in particular, we have:

THEOREM 2. *The equivariant Euler characteristics of* $Hom^{*,*}_{\mathscr{D}_Y}(\mathscr{B}E_{\mathcal{U}}, I_{\mathcal{U}})$ *satisfy the skein relations in eqn. (1.1), with* $n = 0$ *for* $\mathscr{D}_{Y_{\mathfrak{gl}_{1|1}}}$, *and* $n = 2$ *for* $\mathscr{D}_{Y_{\mathfrak{su}_2}}$.

Once we pick a specific Lie algebra, there is a simple, direct way to prove Thm. 2. The proof is in section 4.7 for the $\mathfrak{gl}_{1|1}$ theory and in section 5.8 for the $\mathfrak{su}_2$ theory.

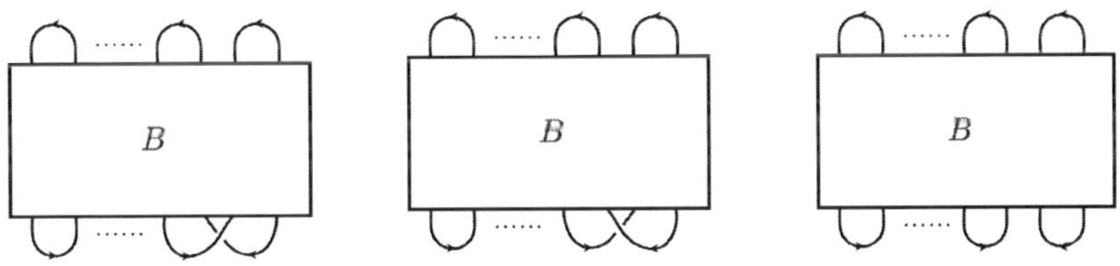

FIGURE 8. Plat closures for the skein relation in eqn. (1.1)

The $\mathfrak{q}$ variable used throughout the text is related to the $\mathbf{q}$ in eqn. (1.1) by a simple redefinitions spelled out in eqn. (4.59) for $\mathfrak{gl}_{1|1}$ and in eqn. (5.64) for $\mathfrak{su}_2$. Since the theory produces homological invariants of links, to prove the theorem, it suffices to consider only the very special link presentations in Fig. 8. The theorem is proven by showing that K-theory classes of the braided $E_{\mathcal{U}}$-branes associated to the very bottom of the diagrams in Fig. 8 satisfy the skein relation.

3. Resolutions of branes in $\mathscr{D}_{Y_{^L\mathfrak{g}}}$

A key result of the paper is an algorithm for computing projective resolutions, which describe branes $L \in \mathscr{D}_{Y_{^L\mathfrak{g}}}$ as complexes of T-branes in (2.15). Applied to the braided cap brane $L = \mathscr{B}E_{\mathcal{U}}$, this lets us compute the $U_{\mathfrak{q}}(^L\mathfrak{g})$ link homology, as described in section 2.3.

3.1. Finding resolutions. In general, finding a projective resolution of a brane $L \in \mathscr{D}_Y$ requires solving two difficult problems. Each is in principle solvable, though typically not in practice.

3.1.1. The first problem is to compute which right module of the algebra A the brane $L \in \mathscr{D}_Y$ gets mapped to by the Yoneda functor

$$hom_Y(T, -) : \mathscr{D}_Y \longrightarrow \mathscr{D}_A.$$

Apriori, computing the module requires counting disk instantons, which is a problem with no general solution. The second problem is to find the projective resolution of this module, as a complex of T branes $(L(T), \delta)$, as in (2.15). This problem is known to be solvable, however, only formally so, by infinite bar resolutions.

3.1.2. We will solve the two problems simultaneously for any $L \in \mathscr{D}_Y$, using features specific to our theory. Among the special features of our theory is the fact that Lagrangians of interest on Y are products $L = L_1 \times \ldots \times L_d$ of one-dimensional Lagrangians L_i on $\mathcal{A}$, where $d = \sum_{a=1}^{\mathrm{rk}^L \mathfrak{g}} d_a$, and that the thimble algebra $A = hom_Y^{*,*}(T,T)$ is explicitly known.

The further key feature is that the theory on Y has an even simpler cousin, corresponding to working on Y with a special divisor D_O deleted:

$$Y_O = Y \backslash D_O. \tag{3.1}$$

The divisor D_O contains a union of diagonals, one for each copy of symmetric product in $Y = \prod_{a=1}^{\mathrm{rk}^L \mathfrak{g}} Sym^{d_a}(\mathcal{A})$. Along a diagonal, a pair of points coincide on $\mathcal{A}$. By removing the diagonals one removes all aspects of the theory on Y that are difficult to understand. We will solve the problem on Y by first solving it on Y_O, and then describing the effect of filling D_O back in.

3.1.3. The category of A-branes on Y_O is derived equivalent to the category of modules of the algebra

$$A_O = hom_{Y_O}^{*,*}(T,T), \tag{3.2}$$

which is the endomorphism algebra of T-branes on Y_O,

$$\mathscr{D}_{A_O} \cong \mathscr{D}_{Y_O}. \tag{3.3}$$

The algebras A and A_O have the same elements, because T-branes are disjoint from D_O, but they have different A_∞ products - in our case, what differs is the associative product or the differential. As we will see, the algebra A_O is computable by elementary means, thanks to the very special features of Y_O.

3.1.4. Resolutions in terms of T-branes of branes L on Y_O, which are products of one-dimensional Lagrangians $L = L_1 \times \ldots \times L_d$, turn out to be computable by elementary means. The result is $(L(T), \delta_O)$, the Both the direct sum brane $L(T)$ together which is a direct sum of T-branes, together with the differential δ_O that squares to zero in the algebra A_O. Both $L(T)$ and δ_O are deduced simply from the d-tuple of one dimensional Lagrangians $L_1, \ldots, L_d$.

3.1.5. In going from Y_O to Y, we put D_O back in. This leaves the T-branes unchanged but replaces the algebra A_O by A, and the differential δ_O by δ, which are their deformations. Since algebra A is known exactly, finding the differential δ is an algebraic problem, with a straight-forward solution.

Below, we will explain this in a general setting, applicable for any ${}^L\mathfrak{g}$. In later sections, we will describe in detail the two special cases of ${}^L\mathfrak{g} = \mathfrak{gl}_{1|1}$ and $\mathfrak{su}_2$.

3.2. The geometric origin of the algorithm. Consider the theory in one complex dimension, for which $Y = \mathcal{A}$. A key simplification of the one-dimensional world is that cones over all morphisms are geometric: they correspond to taking connected sums of Lagrangians over their intersection points.

3.2.1. Suppose two graded one-dimensional Lagrangians L' and L'' on $\mathcal{A}$ intersect over a point $p \in L' \cap L''$. More precisely, let p be an element of the Floer complex

$$p \in hom_Y(L', L''). \tag{3.4}$$

We get a new one-dimensional Lagrangian L on $\mathcal{A}$, by taking the connected sum of $L'[1] \oplus L''$ over p. The resulting Lagrangian L is equivalent, as an object of $\mathscr{D}_Y$, to the complex

$$L \cong L' \xrightarrow{\ p\ } L''. \tag{3.5}$$

The complex is the direct sum of $L'[1] \oplus L''$ with differential deformed by p in (3.4).

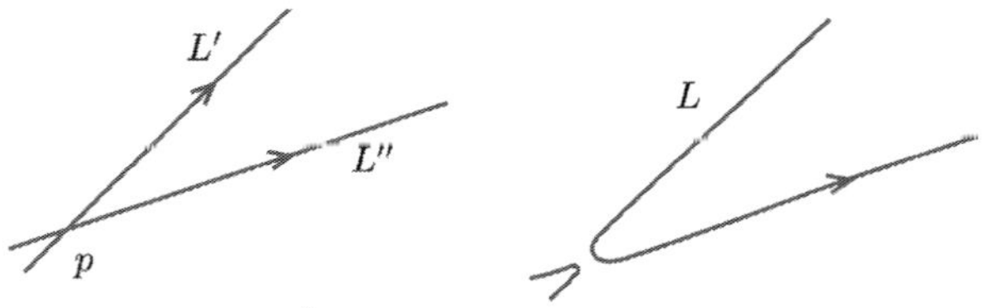

FIGURE 9. A cone $L' \xrightarrow{p} L''$ is a brane equivalent to L, the connected sum of $L'[1]$ and L''.

Conversely, any brane L which can be broken into $L'[1] \oplus L''$ intersecting over $p \in hom_Y(L', L'')$ is given by a complex of the form (3.5).

3.2.2. In general, our target Y is a (product of) symmetric products of $\mathcal{A}$, of dimension d. Take a pair of branes L', L'' on Y, each of which is a product of d one-dimensional Lagrangians on $\mathcal{A}$. Moreover, assume that L' and L'' coincide up to one of their factors. Take, for example,

$$L' = L_1' \times L_2 \times \ldots \times L_d, \qquad L'' = L_1'' \times L_2 \times \ldots \times L_d. \tag{3.6}$$

Suppose the one-dimensional Lagrangians L_1' and L_1'' intersect over a point p. If

$$p \in hom_Y(L_1', L_1''), \tag{3.7}$$

we can get a new Lagrangian L_1 on $\mathcal{A}$, by starting from $L_1'[1] \oplus L_1''$ and deforming the differential using p, as in (3.5). We also get a new Lagrangian L on Y given by

$$L = L_1 \times L_2 \times \ldots \times L_d, \tag{3.8}$$

obtained from $L'[1] \oplus L''$ by deforming the differential using

$$\mathcal{P} = (p, id_{L_2}, \ldots, id_{L_d}) \in hom_Y(L', L''), \tag{3.9}$$

where id_{L_j} is the identity morphism of the one-dimensional L_j brane. The result is the complex is

$$L \cong L' \xrightarrow{\mathcal{P}} L'', \tag{3.10}$$

with L' placed in degree [1], and L'' in degree [0].

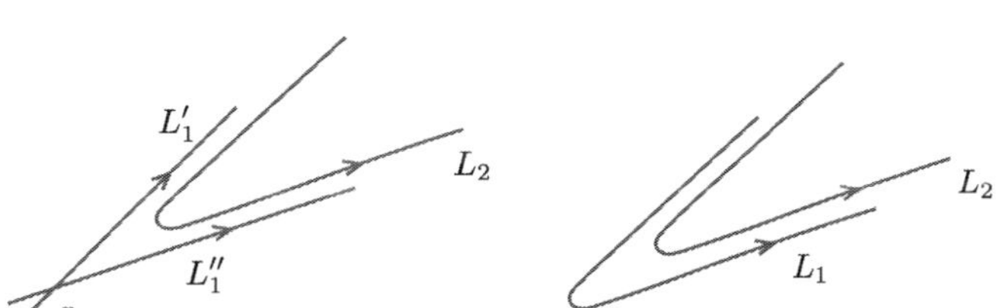

FIGURE 10. A cone over $\mathcal{P} = (p, id_{L_2})$, for $d = 2$.

Conversely, the $L \in \mathscr{D}_Y$ brane is equivalent to the complex, obtained from $L'[1] \oplus L''$ by deforming the differential, with the map coming from a one-dimensional intersection point in one of its factors as in (3.9).

3.3. The algorithm, part I. Let L be a product of d one-dimensional compact Lagrangians on Y, as in (3.8). By iterating the construction we just described, L can be broken up into $L(T)$, a direct sum of T-branes, each of which is a product of d real-line Lagrangians on $\mathcal{A}$.

3.3.1. The $L(T)$-brane and the induced differential can be represented as a d-dimensional rectangular grid. The grid is a "product" of d one-dimensional grids, each corresponding to resolution of a factor brane in L. The $L(T)$-brane is a sum of d-dimensional T-branes, one for each grid site. We will denote by δ_O the sum of terms associated to one-dimensional edges of the grid. Each edge corresponds to an intersection point of T-branes, and is oriented in the direction of the corresponding *hom*. Like in equation (3.9), the intersection points are identity in all but the one slot corresponding to the edge direction. Every intersection point translates to a specific element of the algebra.

3.3.2. The relative gradings of the T-brane summands in $L(T)$ are determined by the gradings of algebra elements of the grid's edges. Start by picking a T-brane at some vertex of the grid. Given this, the maps on the edges of the grid determine the relative grading of all the other T-branes in the complex. The difference in equivariant gradings of a T-brane an edge points to, and the T-brane the edge points from, is the equivariant degree of the element on the edge. The difference in their cohomological degrees is the cohomological degree of the edge element minus 1. The choice of absolute grading is equivalent to the choice of grading of L, viewed as an object of the category $L \in \mathscr{D}_Y$.

3.3.3. Starting from the grid, take the direct sum of T-branes on its sites to get

$$L(T) = \bigoplus_k L_k(T)[k], \tag{3.11}$$

where $L_k(T)[k]$ stands for the contribution of T-branes in cohomological degree k, and various equivariant degrees $\vec{J}$ which we suppress. Taking the sum of all the edge elements, we get a cohomological degree one operator

$$\delta_O : L(T) \to L(T)[1]. \tag{3.12}$$

Per construction, cones over all morphisms in δ_O are geometric; they describe taking connected sums of T-branes over their intersection points.

3.3.4. For $(L(T), \delta)$ to define a brane, an object of $\mathscr{D}_Y$, δ needs to be a differential. A differential is a degree one map $\delta : L(T) \to L(T)[1]$ which squares to zero in an appropriate sense. If A is an ordinary associative algebra, which is the case if ${}^L\mathfrak{g}$ is an ordinary Lie algebra like $\mathfrak{su}_2$, δ is a differential if

$$\delta \cdot \delta = 0, \tag{3.13}$$

so δ squares to zero in the algebra in the usual sense. If A is a differential graded algebra, as is the case when ${}^L\mathfrak{g}$ is a Lie superalgebra like $\mathfrak{gl}_{1|1}$, δ is a differential if it squares to zero in the sense of "twisted complexes":

$$\delta \cdot_{\mathrm{Tw}} \delta = 0, \tag{3.14}$$

the precise statement of which we will give in a later section.

3.3.5. In general, the operator δ_O we just found is not the full differential describing the brane $L \in \mathscr{D}_Y$ as a complex of T-branes. This is because δ_O may fail to square to zero in the sense of (3.13) or (3.14), as appropriate. The fact that δ_O in general fails to define a differential on Y is unsurprising since it is essentially a product of differentials on d copies of $\mathcal{A}$, but there are maps to Y that are not inherited from maps to the product. Such maps have non-zero intersection the divisor of the diagonal Δ in Y. (In the language of Coulomb branches, the diagonal Δ is not trivial if the original gauge theory is not abelian.)

3.3.6. While the operator δ_O may not correspond to a differential of a brane on Y, it defines a differential of a brane on $Y_O = Y \backslash D_O$. On Y_O, all maps that contribute to A_∞ relations are products of one dimensional maps, as a result of deleting Δ.

Thus, $L(T)$ together with the differential δ_O describes the resolution of the brane L viewed as an object of $\mathscr{D}_{Y_O}$, the category of A-branes on Y_O. As we will see, even though δ_O may not coincide with the true differential δ on Y, it determines it uniquely, up to gauge equivalence. Gauge equivalences are maps $e^b \in End(L(T))$ which take δ to $e^b\delta e^{-b}$. These preserve the cohomology of the differential, and take the brane $(L(T), \delta)$ to an equivalent one.

3.4. The algorithm, part II. The category of branes on Y is a deformation of the category of branes on Y_O corresponding to filling in D_O. Having found the resolution of the brane L on Y_O, we can determine a resolution on Y, as a solution to a deformation problem. This approach to understanding the categories of A-branes, namely solving the theory in the complement of a divisor and then filling it back in, was pioneered in [**91**] and is central to the subject.

The deformation that takes the theory on Y_O to the theory on Y is simple to describe since both the T-branes, which we use to generate the category, and the product Lagrangians L are disjoint from the divisor D_O we deleted. In particular, this means that the generators of the corresponding Floer complexes are the same both on Y and on Y_O, and only the A_∞ maps involving them may change.

3.4.1. For simplicity, assume we have deleted only the diagonal $\Delta \subset D_O$. Let $Y_0 = Y \backslash \Delta$, let δ_0 be the differential of the brane on Y_0, and denote by $\hbar$ be the parameter that keeps track of the intersection of the map to Y with the diagonal Δ. Every holomorphic map to Y intersects the diagonal some number $i \geq 0$ times, and gets a weight $\hbar^i$. The theory on Y_0 is obtained from the theory on Y by setting $\hbar$ to zero, since this is the same as restricting to maps that do not intersect the diagonal Δ. Conversely, the theory on Y is a deformation of the theory on Y_0, corresponding to setting $\hbar \neq 0$.

3.4.2. The algebra $A_\hbar$ on Y being a deformation of the algebra $A_{\hbar=0}$ on Y_0 means they coincide as vector spaces, and that, given any two of their elements a, b, the product $\cdot$ and differential ∂ are each polynomials in $\hbar$:

$$(a \cdot b)_\hbar = (a \cdot b)_0 + (a \cdot b)_1 \hbar + \ldots = \sum_{k \geq 0} (a \cdot b)_k \hbar^k, \tag{3.15}$$

where $(a \cdot b)_0$ is the product of a and b viewed as elements of A_0, and

$$(\partial a)_\hbar = (\partial a)_1 \hbar + \ldots = \sum_{k \geq 1} (\partial a)_k \hbar^k. \tag{3.16}$$

The algebra A_0 is always an ordinary associative algebra, with vanishing differential $(\partial a)_0 = 0$. The algebra $A_\hbar$ is known explicitly, as are all the terms in equations (3.15) and (3.16). If ${}^L\mathfrak{g}$ is the superalgebra $\mathfrak{gl}_{1|1}$, only the differential deforms. If ${}^L\mathfrak{g}$ is $\mathfrak{su}_2$, or more generally an ordinary Lie algebra, only the product deforms, since all the elements of the algebra have cohomological degree zero. For a general Lie superalgebra, both the product and the differential ∂ deform.

3.4.3. The fact that the series in (3.15) and (3.16) always truncate follows from the algebra $A_\hbar$, but it can be understood apriori. At $\hbar = 0$, the theory turns out to have an extra symmetry under which all algebra elements have non-negative degrees, while $\hbar$ has degree 1. Setting $\hbar \neq 0$, the symmetry is broken, as $\hbar$ transforms under it. The terms that can appear on the right hand sides of (3.15) and (3.16) are constrained by the symmetry, even a broken one, and finiteness follows.

3.4.4. Since the brane L does not intersect the diagonal, the set $L(T)$ of T-branes that resolve it are the same in Y and $Y_0 = Y\backslash\Delta$. In going from Y_0 to Y, only the brane differential may change.

The differential δ corresponding to the brane on Y is an $\hbar$ deformation of the differential δ_0 on Y_0. It is obtained from δ_0 by setting

$$\delta_\hbar = \delta_0 + \hbar\delta_1 + \ldots = \sum_{k=0} \hbar^k \delta_k, \tag{3.17}$$

and solving a set of equations which come from imposing either (3.13) or (3.14), depending on whether $A_\hbar$ is an ordinary associative algebra, or a differential graded one. At order $\hbar^k$, the equation solved by δ_k, given $\delta_0, \ldots, \delta_{k-1}$ is a linear inhomogenous equation, of the form

$$\{\delta_0, \delta_k\}_0 = D_k(\delta_0, \ldots, \delta_{k-1}), \tag{3.18}$$

where the left hand side is a product of δ_0 and δ_k in A_0, and $D_k(-)$ is an operator depending only on lower order terms $\delta_0, \ldots, \delta_{k-1}$.

3.4.5. In general, the deformation problem may not have any solutions, and if it does, the solution may not be unique, see e.g. [**91**].

Existence of solutions in our case follows from the fact that complexes $(L(T), \delta)$ and $(L(T), \delta_0)$ describe a resolution of the same product Lagrangian L on Y and on $Y_0 = Y\backslash\Delta$, which avoids the diagonal.

Uniqueness of the solution is a consequence of equivariance. On Y_0, equivariance forbids any deformation of δ_0 that does not simply scale its elements. This is because the graded algebra $A_0 = \bigoplus_{\mathcal{C},\mathcal{C}'} hom^{*,*}_{Y_0}(T_\mathcal{C}, T_{\mathcal{C}'})$ is non-degenerate: in every fixed degree, the $hom^{*,*}_{Y_0}(T_\mathcal{C}, T_{\mathcal{C}'})$ is at most one dimensional.

Since δ_0 is a differential of a product complex, the meaning of the remaining freedom to scale components of δ_0 is easily understood. All scaling actions of the form $\delta_0 \to e^b\delta_0 e^{-b}$, for some $b \in End(L(T))$, are gauge equivalences which do not change the brane. As we will see, any remaining freedom to scale components of δ_0, turns out to correspond to changing the complexified flat connection on a brane of topology of $(S^1)^d$. There is always a unique deformation of δ_0, the differential of the brane on Y_0, to the differential δ of the brane on Y that keeps the flat connection fixed.

3.4.6. The deformation idea is powerful and we will use it more than once if D_O has other components besides Δ. For example, while in the $\mathfrak{gl}_{1|1}$ theory $D_O = \Delta$, in the $\mathfrak{su}_2$ theory D_O is the union of Δ and the divisors D_{a_i} of punctures on $\mathcal{A}$.

The deformation theory works in the same way for all components of D_O, although practically computing the effect of the filling divisors of punctures is simpler than computing the effect of filling in the diagonal Δ, as the maps one ends up counting are simpler.

3.5. Floer differential from resolutions. The projective resolution of $L \in \mathscr{D}_Y$

$$L \cong (L(T), \delta) \tag{3.19}$$

encodes, with no further work, the Floer complex of L with any $I_{\mathcal{C}}$ brane, together with the Floer differential acting on it.

3.5.1. By isotoping the Lagrangian brane L to one obtained from $L(T)$ by taking connected sums over intersection points of T-branes at infinity on $\mathcal{A}$, we get a one to one correspondence between the set of graded intersection points of L with any $I_{\mathcal{C}}$-brane, and the set of graded intersection points of $L(T)$ with the same brane:

$$hom_Y(L, I_{\mathcal{C}}[k]\{\vec{J}\}) = \text{hom}_Y\big(L(T), I_{\mathcal{C}}[k]\{\vec{J}\}\big). \tag{3.20}$$

This follows because all the intersection points are in the interior of $\mathcal{A}$, and L differs from $L(T)$ only at its infinity. Moreover, the correspondence in (3.20) preserves the gradings by construction of $L(T)$: the relative degrees of summands in $L(T)$ are determined from the intersection points of T branes at infinity whose cones we take to get L.

3.5.2. Because complex $(L(T), \delta)$ is a projective resolution of the brane L in $\mathscr{D}_Y$, the k-th cohomology group of the complex obtained by applying the functor $hom^*_Y(-, I_{\mathcal{C}}\{\vec{J}\})$ to it is, per definition, the space $Hom_{\mathscr{D}_Y}(L, I_{\mathcal{C}}[k]\{\vec{J}\})$. The same functor sends the brane $L(T)$ to the set of intersection points of degree $[k]\{\vec{J}\}$, so it must send the brane differential δ to the Floer differential.

The functor whose action on $L \in \mathscr{D}_Y$ the algorithm computes, defined by finding a projective resolution of L, applying the $hom_Y(-, I_{\mathcal{C}}\{\vec{J}\})$ to that, is the derived functor usually denoted by the $\mathbf{L}hom(-, I_{\mathcal{C}}\{\vec{J}\})$. From this perspective, the identification of the resulting complex with the Floer complex is a tautological consequence of thinking of L as an object of the derived Fukaya category $\mathscr{D}_Y$, generated by the $T = \bigoplus_{\mathcal{C}} T_{\mathcal{C}}$ brane.

The problem we solved in $\mathscr{D}_Y$ arizes the B-side of mirror symmetry too. The $\mathbf{L}hom$ functor defines the Hom's in the derived category of coherent sheaves, see e.g. [**23**]. One needs to understand its action to compute the Hom's.

3.5.3. Computing δ_F given δ requires no work. Since the T-branes and the I-branes are dual in the sense of (2.12), the functor $hom^{*,*}_Y(-, I_{\mathcal{C}})$ sends all terms in the differential δ to zero, apart from those coming from the identity maps of its dual $T_{\mathcal{C}}$-brane.

The terms in the complex $(L(T), \delta)$ coming from the identity map of the $T_{\mathcal{C}}$-brane are of the form

$$T_{\mathcal{C}}\{\vec{J}\} \in L_k(T) \xrightarrow{c \cdot id_{T_{\mathcal{C}}}} T_{\mathcal{C}}\{\vec{J}\} \in L_{k-1}(T), \tag{3.21}$$

where $L(T) = \bigoplus_k L_k(T)[k]$ as in (3.11). The functor $hom^*_Y(-, I_{\mathcal{C}}\{\vec{J}\})$ sends (3.21) to a pair of intersection points in degrees $[k-1]\{\vec{J}\}$ and $[k]\{\vec{J}\}$,

$$\mathcal{P} \in hom_Y(L, I_{\mathcal{C}}[k-1]\{\vec{J}\}) \xrightarrow{\ c\ } \mathcal{P}' \in hom_Y(L, I_{\mathcal{C}}[k]\{\vec{J}\}), \tag{3.22}$$

and evaluates the differential between them. The direction of the differential changes from (3.21) to (3.22), as the functor $hom_Y^{*,*}(-, I_{\mathcal{C}})$ is contravariant. The coefficient c is count of holomorphic maps interpolating from $\mathcal{P}$ to $\mathcal{P}'$ and contributing to the Floer differential.

3.5.4. While the differential δ_F obtained algebraically should agree with the Floer differential on general grounds we explained here and in section 2.3, providing a direct proof of this is worthwhile, as it amounts to verifying the definition of Floer theory on Y in terms the maps that contribute to its amplitudes.

4. The $\mathfrak{gl}_{1|1}$ theory

We will now illustrate the general theory in the context of ${}^L\mathfrak{g} = \mathfrak{gl}_{1|1}$ and links colored by its two-dimensional representation. The result is a theory categorifying the Alexander polynomial. The theory should coincide with the familiar Heegaard-Floer theory, the flavor of it known as $\widehat{HFK}$. Details of our formulation of the theory are new however, as is our approach to solving it. It is distinct from previous approaches in [**64, 66**], which are based on genus ≥ 1 Heegaard surfaces, while ours is for $\mathcal{A} = \mathbb{C}^\times$, which is a genus zero surface with punctures.

4.1. $Y_{\mathfrak{gl}_{1|1}}$ and its potential. Take an oriented link K, colored by the defining representation V of $\mathfrak{gl}_{1|1}$. As in section 2, from a plat closure representation of the link with $d+1$ caps and cups at the top and the bottom, we get a configuration of $2(d+1)$ punctures on $\mathcal{A}$. We will call half of the punctures, those colored by the highest weight μ_+ of V, the even punctures and place them at $y = a_{2i}$, where i runs from 1 to d. The odd punctures are at $y = a_{2i-1}$ and are colored by the highest weight μ_- of the conjugate representation V^*. To get unreduced link homology, we would work with the target space which is the symmetric product of $d+1$ copies of $\mathcal{A}$, since the weight zero subspace of $(V \otimes V^*)^{\otimes(d+1)}$ is $d+1$ levels below the highest weight. To get reduced link homology we will delete a pair of matchings and the branes associated to them but not the corresponding punctures.

The target is, as in (2.7), the symmetric product of d copies

$$Y = Sym^d(\mathcal{A}).$$

of the Riemann surface $\mathcal{A}$ we started with. A point in Y is d unordered points on $\mathcal{A}$.

While one commonly introduces gradings in Heegaard-Floer theory "by hand", it is better to think of them as coming from a Landau-Ginsburg potential $W_{\mathfrak{gl}_{1|1}}$: $Y \to \mathbb{C}$ and the top holomorphic form $\Omega_{\mathfrak{gl}_{1|1}}$, described below. The result is a version of Heegaard-Floer theory known as "hat" or $\widehat{HF}$ for the way the punctures on $\mathcal{A}$ will end up being treated.

4.1.1. The top holomorphic form $\Omega = \Omega_{\mathfrak{gl}_{1|1}}$, which gives rise to the Maslov $\mathbb{Z}$-grading in the theory, coincides with the standard top holomorphic form on $Sym^d(\mathbb{C}^\times)$,

$$\Omega = \bigwedge_{\alpha=1}^{d} \frac{dy_\alpha}{y_\alpha} \cdot g(y_\alpha) \cdot \prod_{\alpha<\beta} (y_\alpha - y_\beta), \tag{4.1}$$

apart from the first order poles at all the odd, or "-" punctures:

$$g(y) = \prod_{i=1}^{d+1} \frac{1}{(1-y/a_{2i-1})} \cdot \prod_{i<j}(1-a_{2i-1}/a_{2j-1}). \tag{4.2}$$

Since one should regard $\Omega \wedge \overline{\Omega}$ as the volume form on Y, the odd punctures should be treated as coming from infinitely long cylindrical ends.

4.1.2. The potential

$$W = W_{\mathfrak{gl}_{1|1}} = \lambda_0 W^0 + \lambda_1 W^1, \tag{4.3}$$

is a sum of two multi-valued holomorphic functions on Y and gives rise to a pair of equivariant gradings in the theory. We will take λ_0, λ_1 to be complex parameters, and choose normalizations so that

$$c^0 = dW^0/2\pi i, \qquad c^1 = dW^1/2\pi i \in H^1(Y, \mathbb{Z}), \tag{4.4}$$

are one forms with integer periods. W^0 is associated to the $\mathfrak{q}$-grading, and W^1 to the equivariant grading coming from $\mathcal{A}$ being an infinite cylinder. The parameter λ_0 is related to $\mathfrak{q}$ by

$$\mathfrak{q} = e^{2\pi i\lambda_0}, \tag{4.5}$$

and λ_1 is associated to the holonomy around the S^1. Explicitly:

$$W^0 = \ln f, \qquad W^1 = \ln \prod_{\alpha=1}^{d} y_\alpha,$$

where

$$f(y) = \prod_{i=1}^{d+1} \prod_{\alpha=1}^{d} \frac{(1-a_{2i}/y_\alpha)}{(1-a_{2i-1}/y_\alpha)} \cdot \prod_{i<j} \frac{(1-a_{2i-1}/a_{2j-1})^{\frac{1}{2}}}{(1-a_{2i}/a_{2j})^{\frac{1}{2}}}. \tag{4.6}$$

The different contributions of the odd and the even punctures reflect the fact the even ones are associated to V and the odd ones to V^*. (An invariant way to write $f(y)$, in terms of highest weights and simple roots of Lie algebra ${}^L\mathfrak{g}$ is given in [**2**].)

Terms in W and Ω that depend only on the positions of punctures become relevant once we consider the action of braiding on $\mathscr{D}_Y$, where the a's start to vary.

4.2. The category of A-branes on $Y_{\mathfrak{gl}_{1|1}}$. The category of A-branes on $Y = Y_{\mathfrak{gl}_{1|1}}$ is an example of a Fukaya category with potential W, with a somewhat novel feature of equivariance, due to W being not single valued. Here, we briefly review the general features of the theory.

4.2.1. The objects of $\mathscr{D}_Y$ are based on graded Lagrangian branes in Y. Every Lagrangian L on Y is an unordered product of d one dimensional Lagrangians L_α on $\mathcal{A}$:

$$L = L_1 \times \ldots \times L_d, \tag{4.7}$$

since a point on Y is a d-tuple of unordered points on $\mathcal{A}$. The Maslov grading of L is the lift of the phase of $\Omega^{\otimes 2}$ to a real valued function $2\varphi : L \to \mathbb{R}$. The equivariant grading is the choice of a lift of the phase of e^{-W} to a function on L. For such a lift to exist, the one forms c^0 and c^1 in (4.4) must have vanishing periods restricted to L.

4.2.2. An example of an object in $\mathscr{D}_Y$ is the braided cup brane:

$$\mathscr{B}E_{\mathcal{U}} = \mathscr{B}E_1 \times \ldots \times \mathscr{B}E_d, \tag{4.8}$$

obtained by replacing every blue segment of a link, ending on a pair of + and − punctures, by an oval containing it. Another example of an object of $\mathscr{D}_Y$ is the cap brane

$$I_{\mathcal{U}} = I_1 \times \ldots \times I_d \tag{4.9}$$

which is a product of d non-intersecting intervals, where I_j ends on pair of + and − punctures at $y = a_{2j-1}$ and $y = a_{2j}$. Such cup and cap branes arise from a link K as the output of the construction from section 2.1.

4.2.3. The $\mathscr{B}E_{\mathcal{U}}$ brane is an object of $\mathscr{D}_Y$, even though it bounds disks on $\mathcal{A}$. The phase of Ω is well defined on the brane as winding of the phase of dy_i around the i-th factor brane $\mathscr{B}E_i$ gets canceled by the winding of the phase of $g(y_i)$ around the negative puncture inside it. By contrast, a Lagrangian which bounds a generic disk on $\mathcal{A}$ does not give rise to a brane in $\mathscr{D}_Y$, as it does not admit Maslov grading.

4.2.4. The $I_{\mathcal{U}}$ branes are allowed to end at the punctures on $\mathcal{A}$ due to the fact W has logarithmic singularities there. In general, Lagrangians L with boundaries on Y do not give rise to objects of $\mathscr{D}_Y$, since in particular, Floer complexes involving them would not be well defined. Lagrangians ending on the punctures of $\mathcal{A}$ avoid this as follows.

In Floer theory on Y with potential W, elements of the Floer complex $hom_Y^{*,*}(L, L')$ are really Reeb flows. They are solutions to equations for flows

$$\partial_s y = X_W,$$

that start at $s = 0$ on L and end at $s = 1$ on L'. Here, X_W is the vector field generated by the Hamiltonian $H_W = \frac{1}{2}\operatorname{Re} W$. Explicitly, $X_W^i = \frac{i}{2} g^{i\bar{j}} \partial_{\bar{y}^j} W$ where $g_{i\bar{j}}$ is the Kahler metric on Y [**39**]. One can set X_W to zero in the flow equation, thereby trading solving flow equations for finding intersection points of Lagrangians, at the expense of replacing L by a Lagrangian L^W obtained from L by time one flow of X_W. When W has logarithmic singularities, this introduces an infinite amount of wrapping of L^W at the punctures it ends on. As a result, Floer complexes of $I_{\mathcal{C}}$-branes are well defined, as any brane that intersects the $I_{\mathcal{C}}$-brane at the puncture, intersects it infinitely many times.

4.2.5. Given a pair of Lagrangians $L = L_1 \times \ldots \times L_d$ and $L' = L'_1 \times \ldots \times L'_d$, their Floer complex (2.5) is generated by intersection points

$$\mathcal{P} \subset L \cap L'.$$

The intersection points $\mathcal{P}$ are d-tuples of points on $\mathcal{A}$ that lie on intersection of the one dimensional factor branes, up to permutation:

$$\mathcal{P} = (p_1, \ldots, p_d), \qquad p_\alpha \in L_\alpha \cap L'_{\sigma(\alpha)}, \tag{4.10}$$

where σ is any element of S_d, the symmetric group of d elements. One needs to allow for an arbitrary permutation σ in (4.10) as, in going around any closed loop in Y, the corresponding d points on $\mathcal{A}$ come back to themselves only up to the action of the symmetric group. The action of the Floer differential δ_F (2.6) is best described using a formulation of Floer theory specialized to targets Y which are symmetric products or Hilbert schemes [**63**].

4.2.6. Floer theory with target Y that is symmetric product space (or a Hilbert scheme) has an alternative "cylindrical" formulation [**63**]. It relies on the fact that every holomorphic map to from a disk D to $Y = Sym^d(\mathcal{A})$ corresponds to Riemann surface that is embedded in the product of D and $\mathcal{A}$,

$$S \subset \mathrm{D} \times \mathcal{A},$$

and that has a projection to D as a d-fold cover. For any point $z \in \mathrm{D}$ on S, its preimage on S determines a d-tuple of points $y_1(z), \ldots, y_d(z)$ on $\mathcal{A}$, and hence a map from $z \in \mathrm{D}$ to a point $y(z) = (y_1(z), \ldots, y_d(z)) \in Y$, and vice versa. (The cylindrical formulation of Floer theory should be thought of as an open Gopakumar-Vafa theory [**41**]. Its existence follows from string theory realization of the theory [**3**].)

4.2.7. Projection of S to $\mathcal{A}$ defines a domain $A = A(y)$ on the Riemann surface with boundaries on one-dimensional Lagrangians. Many aspects of the map $y : \mathrm{D} \to Y$ can be reconstructed from a domain A, with little work. The Maslov index and the equivariant degree of the map $y : \mathrm{D} \to Y$ can be recovered from the 2-chain A, using (4.12), (4.16) and (4.18). Furthermore, any domain A that has both positive and negative multiplicities in some connected components cannot come from a holomorphic map, since holomorphic maps are orientation preserving.

4.2.8. In the $\mathfrak{gl}_{1|1}$ theory, the Floer differential δ_F and all other A_∞ products, count only maps that project to domains on $\mathcal{A}$ containing no punctures.

Apriori, as described in section 2.1.7, the A-model counts maps $y : \mathrm{D} \to Y$ which have finite symplectic area and pull the Landau-Ginzburg potential $W_{\mathfrak{gl}_{1|1}}$ back to a regular function on the disk D. In the current context, the first of these conditions implies that maps are not allowed to pass through the negative punctures on $\mathcal{A}$, as near every such puncture Ω has a simple pole per eqn. (4.1). The simple pole translates into the negative punctures being at infinity of Y, since $\Omega \wedge \overline{\Omega}$ is proportional to the real volume form on Y. Correspondingly, any map to Y passing through a negative puncture has an infinite area, and the e^{-Area} factor sets its contribution to any of the A_∞ amplitudes to zero. Every holomorphic map to Y which pulls the Landau-Ginzburg potential W back to a regular function on the disk D has equivariant degree zero. For this to be the case, the map to Y must project to a domain on $\mathcal{A}$ that contains equal numbers of positive and negative punctures. The two conditions together imply the domains the map projects to contain no punctures at all.

The version of Heegaard-Floer theory in which the only maps to Y that contribute to the Floer differential or to any other A_∞ maps are those that do not pass through any of the punctures is known as $\widehat{HF}$, of $\widehat{HFK}$.

4.2.9. Maslov index of any map $y : \mathrm{D} \to Y$ from a 2-pointed disk D to Y, which interpolates from $\mathcal{P}$ to $\mathcal{P}'$ viewed as elements of $hom_Y^{*,*}(L, L')$, is given by

$$ind(y) = M(\mathcal{P}') - M(\mathcal{P}) = \int_{\partial \mathrm{D}} y^* d\varphi/\pi, \tag{4.11}$$

where φ the phase of the holomorphic form Ω, and our orientation conventions are as in [**2**]. See section 4 of [**2**], for definition of the absolute grading $M(\mathcal{P})$.

The index $ind(y)$ be computed [**63, 87**] from the domain A on the Riemann surface $\mathcal{A}$ that the map to Y projects to:

$$ind(y) = 2e(A) + i(A) - 2n_-(A). \tag{4.12}$$

The three terms in the formula correspond to the three factors which contribute to Ω in (4.1). The Euler measure $e(A)$ in (4.12) is the contribution to the index coming from the $\prod_\alpha dy_\alpha/y_\alpha$ factor in Ω. It equals

$$e(A) = \chi(A) - \#\text{acute}/4 + \#\text{obtuse}/4, \tag{4.13}$$

where #acute is the number of acute and #obtuse the number of obtuse angles at vertices of A. The vertices are the intersection points of the one dimensional Lagrangians in L and L' which enter $\mathcal{P}$ and $\mathcal{P}'$. The second term in (4.12) is the contribution to the phase of Ω coming from the factor $\prod_{\alpha<\beta}(1 - y_\alpha/y_\beta)$ in (4.1). It equals the intersection $i(A)$ of the map to Y with the diagonal Δ

$$i(A) = n_{\mathcal{P}}(A) + n_{\mathcal{P}'}(A) - e(A), \tag{4.14}$$

where $n_{\mathcal{P}}(A) = \sum_{a=1}^d n_{p_\alpha}(A)$ and where $n_{p_\alpha}(A)$ is 1/4 of the sum of multiplicities of A in the four corners at p_α. The last contribution to (4.12) is the multiplicity of the domain A at the "−" punctures:

$$n_-(A) = \sum_{j=1}^{d+1} n_{a_{2j-1}}(A),$$

where $n_a(A)$ is the multiplicity of A at a point $y = a$ on $\mathcal{A}$. It comes from the factor $g(y)$ in Ω.

For a map to contribute to the Floer differential it has to have Maslov index $ind(y) = 1$, and $n_-(A)$ which vanishes per section 4.2.8. By contrast, for the purpose of computing relative degrees of a pair of intersection points, any topological disk interpolating between them will do.

4.2.10. The equivariant degree $\vec{J} = (J_0, J_1)$ of a map that interpolates from $\mathcal{P}$ to $\mathcal{P}'$ is given by

$$J_i(y) = J_i(\mathcal{P}') - J_i(\mathcal{P}) = -\int_{\partial \mathrm{D}} y^* c^i. \tag{4.15}$$

where c^i are the one forms in (4.4). While the absolute equivariant degree $J_i(\mathcal{P})$ of any one intersection point $\mathcal{P}$ depends on equivariant gradings of the Lagrangians $L, L' \in \mathscr{D}_Y$ [**2**], their relative degree (4.15) depends only on A. In terms of domain the A corresponding to y, the c^0 equivariant degree is given by

$$J_0(y) = n_-(A) - n_+(A) \tag{4.16}$$

Above, $n_+(A)$ is the multiplicity of A at the + punctures, $n_+(A) = \sum_{j=1}^{d+1} n_{a_{2j}}(A)$. The c^1 equivariant degree equals

$$J_1(y) = n_\infty(A) - n_0(A), \tag{4.17}$$

and automatically vanishes for all domains A with compact support. A map to Y that contributes to any A-model amplitude necessarily has all equivariant degrees zero, as it avoids all the punctures.

4.2.11. The Floer differential $\delta_F = m_1$ is the first in the sequence of A_∞-operations, obtained by counting holomorphic maps from a disk D to Y which take as an input ℓ incoming strings/intersection points, and output one. The corresponding index (the difference between the Maslov indices of the outgoing and incoming intersection points) equals

$$ind(y) = 2e(A) + i(A) - 2n_-(A) - (\ell - 1)d/2. \tag{4.18}$$

Maps that can contribute to m_ℓ are rigid holomorphic maps, for which $ind(y) = 2-\ell$, and which have $n_{a_j}(A) = 0$, for all j. All such maps have equivariant degrees zero.

4.3. The $\mathfrak{gl}_{1|1}$ T-branes and their algebra. The category of A-branes on $Y_{\mathfrak{gl}_{1|1}}$ with potential $W_{\mathfrak{gl}_{1|1}}$ is generated by a finite set of branes. For us, the relevant ones are the T-branes

$$T_{\mathcal{C}} = T_{i_1} \times \cdots \times T_{i_d} \in \mathscr{D}_Y,$$

which are products of d real lines on a cylinder (which one should regard as colored by the one simple root ${}^L e$ of $\mathfrak{gl}_{1|1}$). Passing a component of $T_{\mathcal{C}}$ across a puncture changes the brane, as its equivariant grading changes. The direct sum of T-branes

$$T = \bigoplus_{\mathcal{C}} T_{\mathcal{C}} \in \mathscr{D}_Y,$$

over all possible orderings, is the generator of $\mathscr{D}_Y$. We are including in the sum, some branes which are actually zero objects of $\mathscr{D}_Y$. This apparent redundancy is a crucial feature, resulting in a simple and explicit algorithm to obtain projective resolutions of branes in $\mathscr{D}_Y$. (The zero objects of $\mathscr{D}_Y$ correspond to $T_{\mathcal{C}}$ branes with an even number of components between a pair of punctures.)

4.3.1. The endomorphism algebra of the T-branes,

$$A = hom_Y^{*,*}(T,T),$$

which stands for:

$$A = \bigoplus_{\mathcal{C},\mathcal{C}'} \bigoplus_{M,\vec{J}} hom_Y(T_{\mathcal{C}}, T_{\mathcal{C}'}[M]\{\vec{J}\}),$$

is as a vector space, spanned by the generators of the Floer complex, which are the intersection points of T-branes. Explicitly

$$hom_Y(T_{\mathcal{C}}, T_{\mathcal{C}'}[M]\{\vec{J}\}) \tag{4.19}$$

is the span of intersection points of Maslov index M and equivariant degree $\vec{J}$.

4.3.2. The T-branes are not compact, so defining Floer complexes between them requires an additional step. The Floer complex

$$hom_Y(T_{\mathcal{C}}, T_{\mathcal{C}'}[M]\{\vec{J}\}) = CF^{M,\vec{J}}(T^{\zeta}_{\mathcal{C}}, T_{\mathcal{C}'}). \tag{4.20}$$

is obtained by replacing the $T_{\mathcal{C}}$ brane with $T^{\zeta}_{\mathcal{C}}$, in which each of the d real line Lagrangians in $T_{\mathcal{C}} = T_{i_1} \times \ldots \times T_{i_d}$ is taken to make an infinitesimal angle $+\zeta$ with the axis of the cylinder [**44**], as in Fig. 11. The $T^{\zeta}_{\mathcal{C}}$ brane winds infinitely may times around both asymptotic ends, and since ζ is infinitesimally small, all the intersections points in $T^{\zeta}_{\mathcal{C}} \cap T_{\mathcal{C}'}$ occur near the two infinities on $\mathcal{A}$. The deformation that replaces T by T^{ζ} has the same effect as turning on a Hamiltonian H_ζ which one takes to vanish in the interior and behave near the infinities as $H_\zeta \sim \zeta \sum_\alpha |Y_\alpha|^2$. Here, Y is the cylindrical coordinate on $\mathcal{A}$, $Y \sim Y + 2\pi i$, related to the coordinate y, $y \in \mathbb{C}^\times$, which we have been using so far by $y = e^Y$. On the cover where we open up the cylinder to a plane, the time 1 flow of H_ζ generates a rotation by an angle $+\zeta$. This leads to the definition of $\mathscr{D}_Y$ which is known as the (derived) wrapped Fukaya category of Y, described in [**13**] for example.

In general, one expects that [**39**], provided $|dW|^2 \gg |W|$ near an infinity of Y, the superpotential effectively compactifies Y, and the deformation we introduced here is not necessary. As is, to ensure the invariance of the theory under Hamiltonian symplectomorphisms, one needs to introduce the additional ζ-deformation, as we did above. In principle, one should further replace $T_{\mathcal{C}}^{\zeta}$ by a Lagrangian obtained from it by time one flow of $H_W = \frac{1}{2}\,\mathrm{Re}\, W$, as in section 4.2.4. This has no further effect however, since the definition of the ζ-deformation is designed to ensure invariance under Hamiltonian symplectomorphisms. (The need to perturb by H_ζ at infinity of $\mathcal{A}$, and at the odd punctures, but not near the even punctures, is due to different behavior of the Kahler form, which enters into $|dW|^2$. We have been assuming the Kahler form is the one compatible with the volume form coming from $\Omega \wedge \overline{\Omega}$.)

4.3.3. The A-model being a theory of open strings, a generator of the Floer complex corresponding to the intersection point

$$\mathcal{P} \in hom_Y(T_{\mathcal{C}}, T_{\mathcal{C}'}[M]\{\vec{J}\}), \tag{4.21}$$

is fundamentally a flow $\partial_s y = X^{\zeta}$ that begins on $T_{\mathcal{C}}$ at $s = 0$ and ends on $T_{\mathcal{C}'}$ at $s = 1$, generated by the vector field coming from the Hamiltonian H_ζ. Using the invariance of the theory under Hamiltonian symplectomorphisms of individual branes, we have been trading the flow for the intersection point

$$\mathcal{P} \in T_{\mathcal{C}}^{\zeta} \cap T_{\mathcal{C}},$$

where $T_{\mathcal{C}}^{\zeta}$ is the image of $T_{\mathcal{C}}$ under the time one flow of the Hamiltonian.

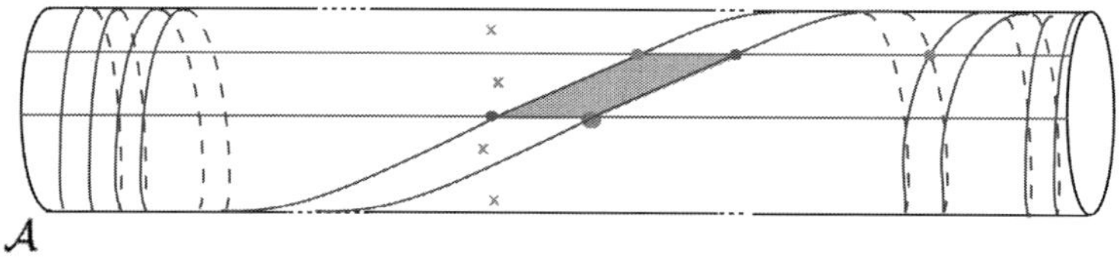

FIGURE 11. The Riemann surface $\mathcal{A}$ for $d = 2$. Three intersection points of $T_{\mathcal{C}}^{\zeta}$ and $T_{\mathcal{C}'}$ branes are shown, indicated as the pair of green, red and orange dots.

Now, we will reverse the above construction, to get an economical, graphical representation of the generators of A, as follows.

4.3.4. The intersection point $\mathcal{P}$, or rather a time-one flow corresponding to it, maps to a configuration of d blue strings on a cylinder whose vertical direction parameterizes the flow, and whose base corresponds to the S^1 in $\mathcal{A}$. The strings start at $s = 0$, from the positions of d one-dimensional components of $T_{\mathcal{C}}$ on the S^1 in $\mathcal{A}$, and end at $s = 1$, where the corresponding components of $T_{\mathcal{C}'}$ are. The path the strings take from $s = 0$ to $s = 1$ is determined uniquely by $\mathcal{P}$. The $+$ and $-$ punctures on the cylinder translate, respectively, to additional orange and green strings, whose positions on the S^1 are those of the corresponding punctures and independent of s.

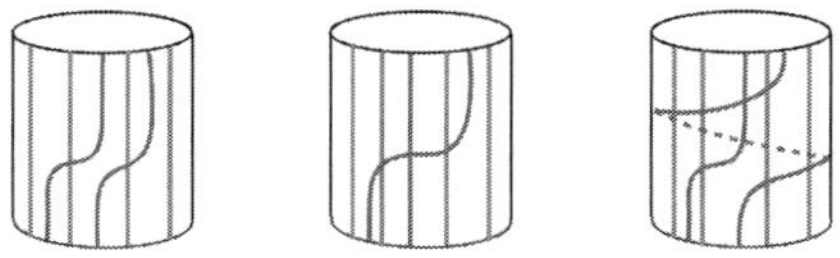

FIGURE 12. String diagrams corresponding to, respectively the green, red and orange intersection points, from Fig. 11.

4.3.5. The relative gradings of intersection points $\mathcal{P}$ and $\mathcal{P}'$ in $T_{\mathcal{C}}^{\zeta} \cap T_{\mathcal{C}'}$ can be computed from a domain A on the Riemann surface interpolates between them, using expressions (4.12) and (4.16). The absolute grading depend on the choice of relative grading of the $T_{\mathcal{C}}$ and $T_{\mathcal{C}'}$. We will use the same conventions as in [**2**].

The result is that the Maslov $M(\mathcal{P})$ and $J^0(\mathcal{P})$ equivariant degrees of any $\mathcal{P} \in T_{\mathcal{C}}^{\zeta} \cap T_{\mathcal{C}'}$ can be read of from the corresponding string diagram, as the sum of degrees of all the string bits, per Fig. 13. The $J^1(\mathcal{P})$ equivariant degree is the net winding number of the string diagram around the cylinder.

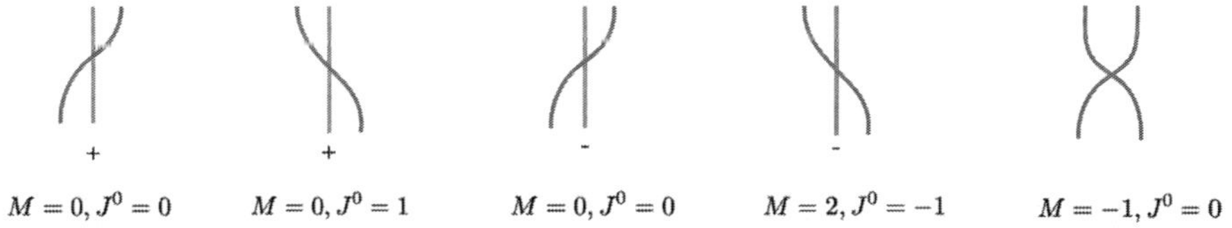

FIGURE 13. The string bits out of which any string diagram is composed. M and J^0 are their Maslov and $\mathfrak{q}$ equivariant degrees. The J^1 equivariant degree is given by the net winding number of the strands around the cylinder.

4.3.6. The Floer product, which reads

$$m_2 : hom_Y^{*,*}(T_{\mathcal{C}'}, T_{\mathcal{C}''}) \cdot hom_Y^{*,*}(T_{\mathcal{C}}, T_{\mathcal{C}'}) \to hom_Y^{*,*}(T_{\mathcal{C}}, T_{\mathcal{C}''}), \tag{4.22}$$

translates into stacking cylinders and rescaling, per construction. The composition of cylinders or $m_2(\mathcal{P}', \mathcal{P})$ leads to an algebra A with product $\cdot$ defined by [**90**]:

$$\mathcal{P}' \cdot \mathcal{P} = (-1)^{M(\mathcal{P})} m_2(\mathcal{P}', \mathcal{P}). \tag{4.23}$$

The product is defined by counting maps $y : \mathrm{D} \to Y$, from a three pointed disk to Y, with three boundary components mapping to $T_{\mathcal{C}}$, $T_{\mathcal{C}'}$ and $T_{\mathcal{C}''}$.

4.3.7. The algebra product can be computed explicitly, as follows [**14, 15**]. For a map to contribute to m_2, its Maslov index (obtained from (4.18) with $\ell = 2$):

$$ind(y) = 2e(A) + i(A) - 2n_-(A) - d/2, \tag{4.24}$$

must vanish, $ind(y) = 0$. Maps from a 3-pointed disk to Y with boundaries on the T-branes are shown in [**14, 15**] to have $e(A) = d/4$. They also have $n_-(A) = 0$, as explained in section 4.2.8. It follows that maps that contribute to m_2 in (4.22) have

$$i(A) = 0.$$

The quantity $i(A)$ is the branching index of the projection to D [**63, 87**], which vanishes for maps $y : \mathrm{D} \to Y$ which are disconnected d-tuples of maps from D to $\mathcal{A}$.

It follows that the algebra product is inherited from d copies of a one dimensional theory.

4.3.8. The algebra product A introduces relations between its elements. The relations say that the product of two string diagrams is either a taut diagram or it vanishes. A taut diagram has no "excess" intersection, one which can be removed by isotoping the diagrams keeping the endpoints fixed. Explicitly:

1. 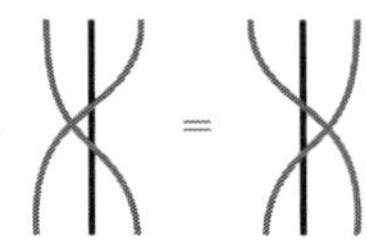

2. 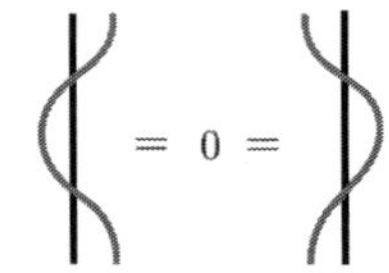

The diagrams above compute m_2, and hence the algebra product. The black string can stand for either a blue string, colored by the simple root, or the orange and green strings colored by the highest weights of V and V^*.

The products are based on counts of embedded triangles in Floer theory with target which is effectively d copies of $\mathcal{A}$, as described in [**14, 15**] and reviewed in [**2**]. Every such triangle contributes a factor of ± 1 to the amplitude, with the sign that depends on the orientations of the branes at the boundaries [**90**] (see also [**12**]). The orientations of T-branes are such that the triangles which contribute to $m_2(\mathcal{P}, \mathcal{P}')$ always contribute $+1$.

4.3.9. The Floer differential

$$m_1: \quad hom_Y^{*,*}(T_{\mathcal{C}}, T_{\mathcal{C}'}) \to hom_Y^{*+1,*}(T_{\mathcal{C}}, T_{\mathcal{C}'}), \tag{4.25}$$

turns the algebra A into a differential graded algebra. The algebra differential ∂ is defined by

$$\partial \mathcal{P} = (-1)^{M(\mathcal{P})} m_1(\mathcal{P}). \tag{4.26}$$

Maps that contribute to m_1 have Maslov index

$$ind(y) = 2e(A) + i(A) - 2n_-(A),$$

from (4.12), equal to $ind(y) = 1$. Every two-pointed map from D to Y with boundaries on T-branes necessarily projects to domains A that are (products of) rectangles and isolated points. The latter are images of constant maps from a disk d to $\mathcal{A}$. For such domains, the Euler measure from (4.13) vanishes $e(A) = 0$ (every rectangle has Euler characteristic one, and 4 acute angles). Maps also have $n_-(A) = 0$ as they are not allowed to pass through the punctures.

It follows that maps with Maslov index one all have

$$i(A) = 1,$$

so they intersects the diagonal in the symmetric product exactly once. Such maps have an image on $\mathcal{A}$ which consists of exactly one empty rectangle and the corresponding number of maps to points. Fortunately, the count of this kind of a map is known. See e.g. [**66**]. This translates into the differential ∂ which acts on any diagram by removing a crossing locally:

3. 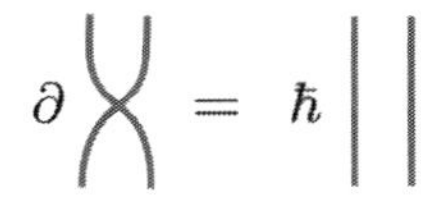

4.3.10. As in section 3, it is useful to introduce a parameter $\hbar$, which keeps count of the intersection of a map to Y with the diagonal Δ. The value of $\hbar$ is immaterial, as long as it is not zero, as it can be absorbed into the definition of the algebra element described by the crossing of two blue strands. The algebra

$$A_0 = hom^{*,*}_{Y_0}(T, T),$$

obtained from $A_\hbar$ by setting $\hbar$ to zero, is the endomorphism algebra of T-branes on $Y_0 = Y\backslash\Delta$. The algebra A_0 differs from $A_\hbar$ only in that its differential vanishes identically. The algebra A_0 plays a central role in the algorithm from section 3, applied to $\mathfrak{gl}_{1|1}$, which we will revisit below. The divisor D_O from (3.1) coincides with Δ in the present setting, so in particular $A_O = A_0$ and $Y_O = Y_0$.

4.3.11. All the higher A_∞ products m_ℓ with $\ell > 2$ incoming strings and one outgoing one string vanish. Maps that contribute to m_ℓ are rigid holomorphic maps for which the Maslov index equals $ind(y) = 2 - \ell$. From the formula for the index $ind(y)$ given in (4.18), together with the fact that for $\ell + 1$ pointed maps bounded by T-branes $e(A) = (\ell - 1)d/4$ [**14, 15**] and $i(A)$ is non-negative, it follows that to contribute to $m_{\ell>2}$, a map must have $n_-(A) > 0$. Such a map intersects the divisor of the negative punctures at least once. The divisor is deleted from Y, so the map vanishes.

4.3.12. The algebra $A_\hbar$ is an associative graded algebra with a differential. It is a version of the algebra discovered by Lipshitz, Osvath, and Thurston [**64**], adapted for the Heegaard surface that is a cylinder with punctures.

The signs in the relations between Floer operations and the algebra product (4.23) and the differential (4.26) are important. They are chosen so that the A_∞ relations of the Fukaya category translate into the standard relations of the differential graded algebra with a product that is associative. See appendix B for details. In particular, some products which are equal when computed by m_2 as described in section 4.3.8, are equal only up to a sign computed by the $\cdot$ product.

4.4. Projective resolutions. Because the algebra A has elements of non-zero Maslov degree, the complexes resolving the $\mathscr{B}E_{\mathcal{U}}$ branes (or any other brane) are really twisted complexes.

4.4.1. Resolution of the $\mathscr{B}E_{\mathcal{U}} \in \mathscr{D}_Y$ brane in terms of T-branes,

$$\mathscr{B}E_{\mathcal{U}} \cong (\mathscr{B}E(T), \delta)$$

is a direct sum brane

$$\mathscr{B}E(T) = \bigoplus_k \mathscr{B}E_k(T)[k] \tag{4.27}$$

together with a differential δ, which is a cohomological degree 1 and equivariant degree zero operator

$$\delta : \mathscr{B}E(T) \to \mathscr{B}E(T)[1] \tag{4.28}$$

that squares to zero. Informally, we get the direct sum brane by stretching the $\mathscr{B}E_{\mathcal{U}}$ brane straight along the cylinder $\mathcal{A}$ and breaking it at both ends at infinity to a direct sum of T-branes. The original brane $\mathscr{B}E_{\mathcal{U}}$ is recovered by turning on the differential δ. Writing this out explicitly, δ is a sum of algebra A elements

$$\delta = \sum_{k,\ell} d_{k,\ell} \in A \tag{4.29}$$

corresponding to Homs between T-branes in the direct sum brane:

$$d_{k,\ell} \in hom_Y(\mathscr{B}E_k(T)[k], \mathscr{B}E_\ell(T)[\ell+1]). \tag{4.30}$$

4.4.2. The twisted differential δ, viewed as acting on the brane (4.28) in $\mathscr{D}_Y$, squaring to zero means that it satisfies the Maurer-Cartan equation

$$\partial\delta + \delta \cdot \delta = 0. \tag{4.31}$$

This is true only up to signs, whose details are quite technical but crucial, unless one is satisfied with working over $\mathbb{Z}_2$ coefficients. The sign assignments in the category $\mathscr{D}_Y$ of twisted complexes are explained in appendix C. The upshot is that δ satisfies

$$m_1^p(\delta) + m_2^p(\delta, \delta) = 0, \tag{4.32}$$

where m_ℓ^p is a twist of the Floer theory m_ℓ from (4.23) and (4.26) by signs. The condition (4.32) is as simple as it is only because all $m_{\ell>2}$ maps vanish in our algebra. For a general A_∞ algebra, all higher morphisms would contribute and the condition would be $\sum_{\ell=1}^{\infty} m_\ell^p(\delta, \ldots, \delta) = 0$ instead. For simplicity, we will write (4.32) simply as (4.31).

4.4.3. The algorithm from section 3 starts by giving the projective resolution of any one component of the $\mathscr{B}E_{\mathcal{U}}$ brane in the $d = 1$ theory, as described in section 3.2. It recursively constructs the resolution of the brane in $\mathscr{D}_{Y_0}$, by taking a product with one factor brane at a time. The result of that section is the direct sum brane $\mathscr{B}E(T)$ and the differential

$$\delta_0 : \mathscr{B}E(T) \to \mathscr{B}E(T)[1],$$

which is a cohomological degree 1 and equivariant degree zero operator that squares to zero in algebra A_0. The differential δ_0, per construction, knows everything there is to know about the brane $\mathscr{B}E_{\mathcal{U}}$, viewed as a product of d Lagrangians on $\mathcal{A}$. Since A_0 is an ordinary associative algebra, the differential squaring to zero means that

$$\delta_0 \cdot \delta_0 = 0, \tag{4.33}$$

up to sign twists. The relation (4.33) holds both in A_0 and in $A_\hbar$, since A_0 and $A_\hbar$ differ only in the differential.

4.4.4. The complex $\mathscr{B}E(T)$ with differential δ_0 describes the resolution of the brane $\mathscr{B}E_{\mathcal{U}}$ in $\mathscr{D}_{Y_0}$. The full differential δ of the brane in $\mathscr{D}_Y$ is obtained by deforming δ_0 per section 3.4.4:

$$\delta = \delta_0 + \hbar\delta_1 + \ldots$$

and asking for the full differential to satisfy (4.32). For example, δ_1 is a solution to the linear equation

$$\partial\delta_0 + \delta_0 \cdot \hbar\delta_1 + \hbar\delta_1 \cdot \delta_0 = 0.$$

4.4.5. To get a Floer homology theory with $\mathbb{Z}$-coefficients we need to specify the spin structure on the $\mathscr{B}E_{\mathcal{U}}$ brane we started with [**37, 73**]. The choice of spin structure is necessary to orient the moduli spaces of holomorphic maps Floer theory is based on. For us, this translates into a $\mathbb{Z}_2$ worth of choices of sign for every component of the brane of S^1 topology. From perspective of its resolution, the $(\mathscr{B}E(T), \delta)$ brane, to get a $\mathbb{Z}$ graded theory, we need to specify signs of maps in the differential δ. The two ambiguities must match, so the inequivalent choices of signs in δ should correspond to the inequivalent choices of spin structure in the original $\mathscr{B}E_{\mathcal{U}}$ brane.

Since δ arises as a deformation of δ_0, to specify signs in δ it suffices to specify the signs in δ_0. Because δ_0 is a differential of a product complex, signs in it are determined from signs in its factor complexes, by a slight variant of the standard procedure of forming a product complex. There is, as expected, a $\mathbb{Z}_2$ worth of choices signs in a complex describing a one dimensional factor brane, as only the product of all signs turns out to be gauge invariant. The departure from the standard sign assignment on a product complex is needed because δ_0 is a differential of a twisted, not ordinary complex. The sign assignments in $\mathscr{D}_Y$ are explained thoroughly in appendix C. They follow verbatim the general prescription from [**90**].

For definiteness, out of 2^d possible sign choices, we choose the one for which the product of all the maps in the differentials of one dimensional factor branes is even.

4.4.6. The different choices of spin structure on the $\mathscr{B}E_{\mathcal{U}}$ brane of $(S^1)^d$ topology may be related by changing holonomies of the flat $U(1)$ connection around the S^1's by $e^{i\pi}$, as in [**26, 27**]. The moduli of flat connections on the $\mathscr{B}E_{\mathcal{U}}$ brane is complexified by the symplectic area of d annuli swept out by deforming its one-dimensional factor branes.

In terms of its resolution, changing the holonomy of the complexified flat $U(1)$ connection on the $\mathscr{B}E_{\mathcal{U}}$ brane should correspond to scaling the maps in δ_0 by $\mathbb{C}^{\times}$-valued parameters, modulo gauge transformations of $\mathscr{B}E(T)$, the direct sum brane, as in section 3.4.5. Gauge transformations send the differential δ_0 to $e^b\delta_0 e^{-b}$ where $e^b \in End(\mathscr{B}E(T))$. It is elementary to show that for every one dimensional factor brane, we get exactly one $\mathbb{C}^{\times}$-valued gauge invariant scaling parameter, which should be identified with the choice of holonomy.

The differential δ of the brane on Y is, per definition, the deformation of δ_0 keeping the flat connection fixed. Homology groups $Hom^{*,*}_{\mathscr{D}_Y}(\mathscr{B}E_{\mathcal{U}}, I_{\mathcal{U}})$ being invariant under the swiping move from section 4.6 and appendix A.1, should be independent of the choice of complexified flat connection one starts with, as long as one stays away from infinities in the moduli space, at which topology of the brane may change.

4.5. Simple examples. We will now illustrate the algorithm, starting with the simplest one-dimensional examples.

4.5.1. Consider a single cup brane in a theory with $d = 1$. The $E_{\mathcal{U}_1}$ brane in Fig. 14 is one of the four cup branes in the theory on $Y = \mathcal{A}$ with two even and two odd punctures. By isotoping the $E_{\mathcal{U}_1}$ brane, and allowing it to break at $y = 0$ and $y = \infty$ we get the direct sum brane:

$$E_{\mathcal{U}_1}(T) = T_0[1] \oplus T_2. \tag{4.34}$$

The cohomological degree shift comes from the flip of orientation, per Fig. 14.

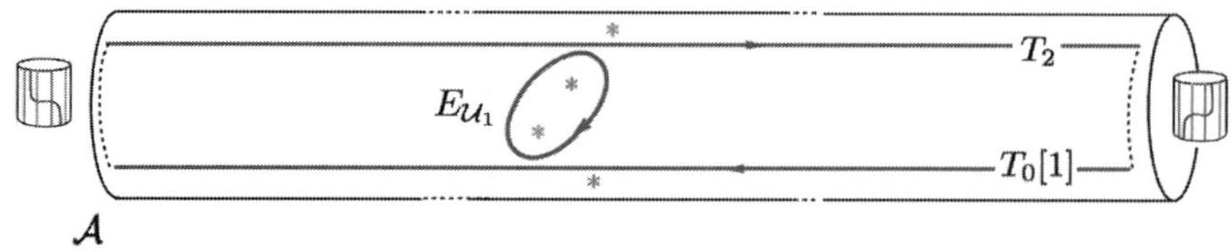

FIGURE 14. The $E_{\mathcal{U}_1}$ brane is the connected sum of the two T-branes over the intersection points corresponding to the open strings/Reeb chords shown.

In fact, the construction we are about to describe fixes all the degrees for us from the geometry of the brane. To recover the $E_{\mathcal{U}_1}$ brane, we take the connected sum of of $T_0[1]$ and T_2, over a specific pair of their intersection points at $y = \infty$ and $y = 0$. The intersection point at $y = \infty$ corresponds to the following element of the algebra A:

 $$\in hom_Y(T_0, T_2) = hom_Y(T_0[1], T_2[0+1])$$

which encodes the Reeb flow from T_0 to T_1 that we used in Fig. 14 to take the connected sum. The intersection point at $y = 0$, corresponds to a blue string going the other way:

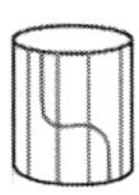 $$\in hom_Y(T_2, T_0[2]) = hom_Y(T_2, T_0[1+1])$$

The orientations of the maps in the complex follow from Fig. 14 and the definition, from section 4.3.2, of homs in the wrapped Fukaya category. The result is the twisted complex $(E_{\mathcal{U}_1}, \delta)$, with differential:

$$\delta = \begin{pmatrix} 0 & \cdot \\ \cdot & 0 \end{pmatrix} \tag{4.35}$$

The differential squares to zero thanks to the A-algebra relations:

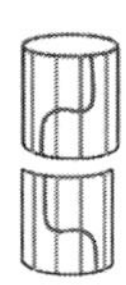 $= 0 =$ 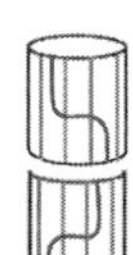

The complex is often written as

$E_{\mathcal{U}_1} \cong T_0[1]$ 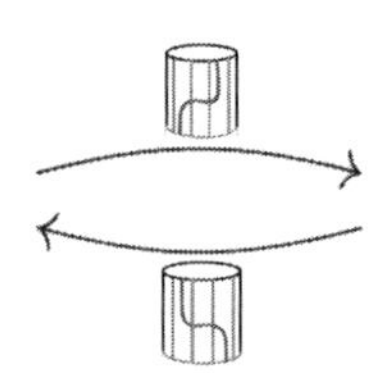 T_2

For clarity, for twisted complexes, we will write explicitly not only the equivariant degrees of all the terms in the complex, but also the homological ones. (An alternative is to write explicitly the degrees of all the maps. For ordinary complexes, degrees of all the maps are zero, so degrees of direct summand branes are fixed by knowing a degree of any one of its components. In a twisted complex, this is not the case.) When a degree is not written, it vanishes. For example, the equivariant degrees of both T-branes in (4.34) vanish.

The inequivalent choices of spin structure on a brane, viewed as a smooth Lagrangian on Y should, as in section 4.4.5, correspond to gauge inequivalent choices of signs in the differential. In this simple example, gauge transformations of the two T-branes in the complex can change signs of both maps in the differential simultaneously, leaving a $\mathbb{Z}_2$ choice for the sign of the product. In writing δ, we picked one.

The two choices of spin structure on the $E_{\mathcal{U}_1}$ brane of S^1 topology are related by changing holonomy of the flat $U(1)$ connection on the brane by $e^{i\pi}$. More generally, changing the holonomy of the complexified flat $U(1)$ connection, corresponds to a deformation of δ that scales the product of the two maps in the differential, as gauge transformations scale the two maps oppositely. The example at hand is in fact a variant of the following canonical example: replace Y by a cylinder without any punctures, and the $E_{\mathcal{U}_1}$ brane by a brane wrapping the S^1. That brane has a similar two-term resolution of the in terms of the real line Lagrangian which generates the category. The differential has an analogous deformation by scaling the product of terms in it, which corresponds to sliding the brane along the cylinder together with varying the holonomy of the $U(1)$ flat connection on the brane. Note that some definitions of twisted complexes [**90**] ask for the differential to be upper or lower triangular. In our setting this is not what one wants since it would eliminate this simple brane, which clearly exists as an object of $\mathscr{D}_Y$.

From the complex resolving the $E_{\mathcal{U}_1}$ brane, we can compute all its graded Homs to any of the I-branes with no further work. For example, it follows immediately that $Hom^{*,*}_{\mathscr{D}_\mathcal{Y}}(E_{\mathcal{U}_1}, I_1)$ vanishes, as applying $hom^*_Y(-, I_1\{\vec{J}\})$ to the complex sends all its terms to zero. This is because as no T_1 branes, which are dual to the I_1 brane, enter $E(T)$ in (4.34). This reproduces the fact the Floer complex $hom^{*,*}_Y(E_{\mathcal{U}_1}, I_1)$ vanishes, as there the $E_{\mathcal{U}_1}$ and the I_1 branes do not intersect. The homology

$$Hom^{*,*}_{\mathscr{D}_\mathcal{Y}}(E_{\mathcal{U}_1}, I_{\mathcal{U}_1}) = 0,$$

with $I_{\mathcal{U}_1} = I_1\{1/2\}$, is the reduced homology of a pair of unlinked unknots. By removing two of the four punctures, we would find that the unreduced homology of a single unknot also vanishes. We will explain the degree shift between the cup brane and the basic I_1 brane in section 4.6.

4.5.2. For another simple example, consider the left-handed trefoil. The knot has a representation as a plat closure of a braid of four strands, resulting is the pair of matchings shown in Fig. 15. To get the matchings for reduced trefoil, we remove a pair of blue and red segments, per section 2.4.4.

FIGURE 15. Left handed trefoil, and the corresponding matchings.

By replacing the one remaining blue segment by an oval we get a pair of branes $\mathscr{B}E_{\mathcal{U}}$ and $I_{\mathcal{U}} = I_1\{\frac{1}{2}\}$ on Y, which is the Riemann surface $\mathcal{A}$ with four punctures in Fig. 16. We have rotated the figures clockwise by 90° and suppressed the cylinder.

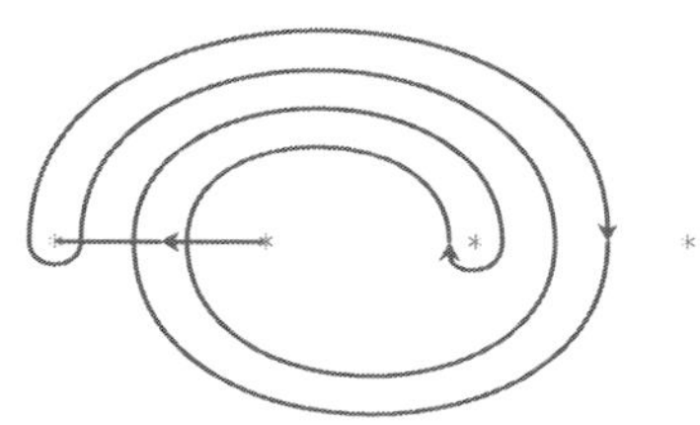

FIGURE 16. The branes for the reduced trefoil.

We can resolve the $\mathscr{B}E_{\mathcal{U}}$ brane into thimbles, as shown in Fig. 17. Starting with any one thimble, we can work our way around the cap brane, using Fig. 17 to write down the thimbles and the maps connecting them, which gives:

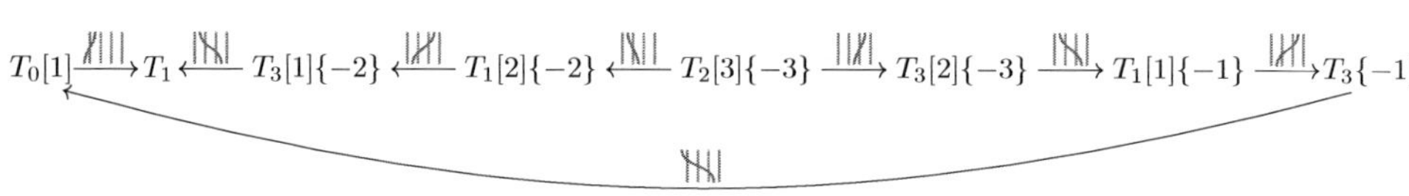

By construction, all relevant aspects of the Lagrangian $\mathscr{B}E_{\mathcal{U}}$ are encoded in complex. The relative degrees of all the T-branes in the complex follow from degrees of the maps between them, shown in Fig. 4, and general facts from section 3.2.1.

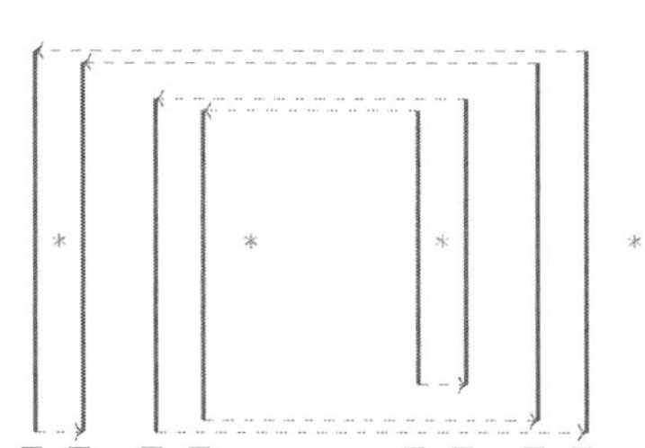

FIGURE 17. The resolution of the cap brane of the trefoil.

For example, all but one map in the complex have Maslov degree zero and so in the notation of (4.29), correspond to $d_{k,\ell}$ which sends a thimble in degree k to a thimble in degree $\ell = k-1$. The exception is the map  which has Maslov degree 2 and therefore sends a thimble in Maslov degree k to a thimble in Maslov degree $\ell = k+1$.

While the relative degrees of the T-branes in the complex depend only on the underlying Lagrangian, the absolute degree depends on the braid group element whose action $\mathscr{B}$ represents. We will explain how to find absolute degrees of the branes in section 4.6. For this, it suffices to determine the absolute degree of any one T-brane in the complex. In this example, the absolute degree of the T_1 brane is [1], unchanged from the $E_{\mathcal{U}}$ brane complex, as the functor $\mathscr{B}$ acts trivially on it.

There is an additional, non-geometric degree shift which accompanies the action of $\mathscr{B}$ on $\mathscr{D}_Y$, as explained in section 4.6. Since it is the same for all the branes in $\mathscr{D}_Y$, and in that sense non-geometric, we will not include it when we write the complex resolving the $\mathscr{B}E_{\mathcal{U}}$ brane. It is important, however, to get the link homology whose Euler characteristic is the Alexander polynomial in standard normalization, and we will need to include it in the very end. In this case, there is overall degree shift of $\Delta M = 0$, $\Delta J_0 = \frac{3}{2}$, coming from the fact $\mathscr{B}$ represents the element σ_2^{-3}, in the standard braid group notation, which braids a pair of even punctures.

The result is the direct sum brane

$$\mathscr{B}E_{\mathcal{U}}(T) = \bigoplus_k \mathscr{B}E_k(T)[k]$$

as in equation (4.27), which has

$$\mathscr{B}E_3 = T_2\{-3\},\ \mathscr{B}E_2 = \begin{pmatrix} T_3\{-3\} \\ T_1\{-2\} \end{pmatrix},\ \mathscr{B}E_1 = \begin{pmatrix} T_1\{-1\} \\ T_0 \\ T_3\{-2\} \end{pmatrix},\ \mathscr{B}E_0 = \begin{pmatrix} T_3\{-1\} \\ T_1 \end{pmatrix}.$$

Then, we can write

(4.36) $\delta =$

0	0	0	0	0	0	0	0
	0	0	0	0	0	0	0
	0	0	0	0	0	0	0
0		0	0	0	0	0	0
0	0	0	0	0	0		0
0	0		0	0	0	0	0
0	0	0	0	0	0	0	0
0	0	0		0	0	0	0
0	0	0	0			0	0

The $d = 1$ algebra is an ordinary associative algebra whose differential ∂ vanishes identically. Indeed, the differential squares to zero: $\delta \cdot \delta = 0$.

The differential δ would still square to zero if we rescaled the maps in the complex by arbitrary $\mathbb{C}^\times$-valued parameters. Complexified gauge transformations of the direct sum brane reduce this to one parameter's worth, and correspond to the choice of complexified flat connection on the $\mathscr{B}E_{\mathcal{U}}$ brane of S^1 topology. Varying the spin structure is equivalent to tensoring with an additional $\mathbb{Z}_2$ flat connection.

The link homology $Hom^{*,*}_{\mathscr{D}_Y}(\mathscr{B}E_{\mathcal{U}}, I_{\mathcal{U}})$ is, up to overall degree shifts, the cohomology of the complex of vector spaces obtained by applying the $hom^{*,*}_Y(-, I_1)$ functor to the complex resolving the $\mathscr{B}E_{\mathcal{U}}$ brane. Non-vanishing contributions come only from the terms in the complex involving the T_1 brane, since it is the dual to the I_1 brane. The complex that results is

$$\mathbb{C}[0] \xrightarrow{0} \mathbb{C}\{-1\}[1] \xrightarrow{0} \mathbb{C}\{-2\}[2] \to 0, \tag{4.37}$$

with the differential δ that gets mapped to zero. (The complex is really extended by 0's on both sides, as is the brane we obtained it from.)

The three copies of $\mathbb{C}$ in the complex (4.37) correspond, respectively, to the three intersection points which we labeled p_1, p_2, and p_3 in Fig. 16, which generate the Floer complex. As in section 3.5, one establishes this by isotoping the $\mathscr{B}E_{\mathcal{U}}$ brane to the connected sum of T-branes. One can verify the identification by considering disks interpolating between the three intersection points. For example, the disk in Fig. 18 interpolates from intersection point p_2 to p_3, and has $J_0(A) = -1$ and $M(A) = 1$ using (4.16) and (4.12), reproducing their relative degrees from (4.37).

The absolutely graded homology differs from (4.37) by an overall grading shift. It comes from two places, as explained in section 4.6. The first is the overall degree shift of branes that accompanies the action of $\mathscr{B}$, mentioned above, which in this case adds $+\frac{3}{2}$ to J^0, and nothing to M. The second is the fact the cap brane is related to the I_1 brane by a canonical grading shift: $I_1 = I_{\mathcal{U}}\{-\frac{1}{2}\}$. The net result is a shift of equivariant grading by $\{1\}$. The result is the reduced $\mathfrak{gl}_{1|1}$ homology of the trefoil:

$$Hom^{*,*}_{\mathscr{D}_Y}(\mathscr{B}E_{\mathcal{U}}, I_{\mathcal{U}}) = \mathbb{C}\{1\} \oplus \mathbb{C}[1] \oplus \mathbb{C}\{-1\}[2]. \tag{4.38}$$

Our J^0 grading is, as we will explain in section 4.7, related to the usual one in (1.1) by $J^0 \to -J^0$. The graded Euler characteristic recovers the Alexander polynomial $\mathfrak{q} - 1 + \mathfrak{q}^{-1}$.

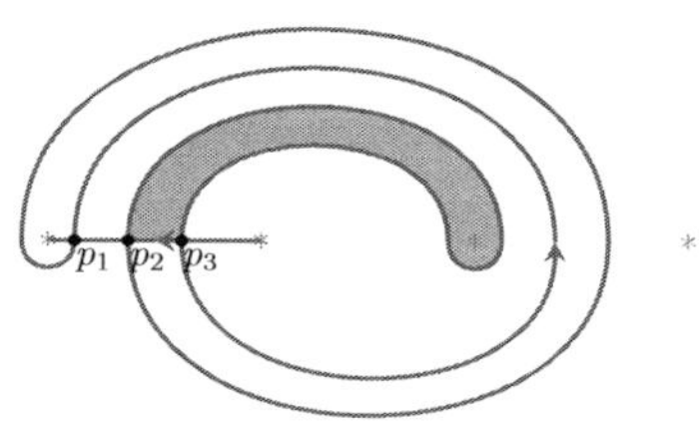

FIGURE 18. A disk interpolating from the intersection point p_2 to p_3 when the latter are viewed as elements of $Hom^{*,*}_{\mathscr{D}_Y}(\mathscr{B}E_{\mathcal{U}}, I_{\mathcal{U}})$.

4.5.3. Computing the reduced $U_{\mathfrak{q}}(\mathfrak{gl}_{1|1})$ link homology of any two-bridge link is as simple as the current trefoil example, because the target Y is always one-dimensional. Stretching the $\mathscr{B}E_{\mathcal{U}}$ brane along the cylinder, breaking it up into T-branes and recording which maps one needs to turn on to glue the brane back together, results in the full differential δ from the outset, as there is no diagonal to deform it. It is also manifest that the brane differential δ computes the Floer differential δ_F which in this case is easy to compute directly. The Floer differential either trivial, or receives contributions from one dimensional disks connecting pairs of intersection points which can cancel out in pairs by isotoping the branes.

In what follows, we will explain how to compute the reduced $U_{\mathfrak{q}}(\mathfrak{gl}_{1|1})$ link homology for arbitrary links K, and any corresponding d. We will do so without relying on any special link presentations; the resulting theory is completely general.

4.5.4. To begin with, we will start with a very simple example which, even though not relevant for knot theory, is a good starting point to illustrate how to find projective resolutions and how they compute the Floer differential without counting solutions to instanton equations.

We can recover the brane $L_1 \times L_2$ below

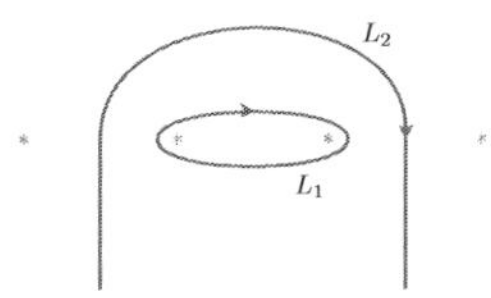

by starting with a direct sum of T-branes, and taking connected sums.

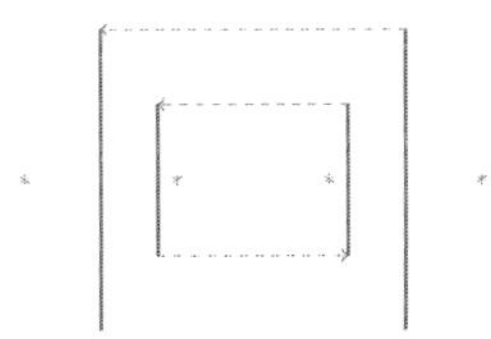

FIGURE 19. The resolution of the $L_1 \times L_2$ brane in terms of T-branes.

To start, pick one of the two factor branes, say the L_1 brane, and write down its resolution. Keep track of the morphisms themselves, but ignore any degree shifts until all the factor branes have been incorporated. The resolution of the L_1 brane is

$$L_1 \quad \cong \quad T_2[1] \rightleftarrows T_4 \tag{4.39}$$

Then, take the product of L_1 with each of the two T-brane that L_2 branes breaks into, namely T_2 and T_4. The product with T_2 gives the complex:

$$L_1 \times T_2 \quad \cong \quad T_{22}[1] \rightleftarrows T_{24}\,. \tag{4.40}$$

The specific morphisms follow from Fig. 19. In writing down the maps, we keep track of the positions of T-branes relative to each other and to the punctures.

Similarly, the product of L_1 with T_4 gives:

$$L_1 \times T_4 \quad \cong \quad T_{24}[1] \rightleftarrows T_{44}\,. \tag{4.41}$$

The complexes encodes all aspects of the brane relevant to the A-model. For example, because the T_2 brane is to the left of the L_1 brane, and hence of both of its T-branes, the complex resolving $T_2 \times L_1$ does not contain blue strands that cross.

To get $L_1 \times L_2$, we take a connected sum of $L_1 \times T_2$ and $L_1 \times T_4$, using the specific maps that follow from Fig. 19. The maps are identity morphisms of the summands of L_1, times the morphism that makes up L_2. The corresponding string diagrams now involve blue strings that cross, since the open strings connecting the two T-branes in L_2 cross the open strings corresponding to identity of L_1. This gives:

(4.42)

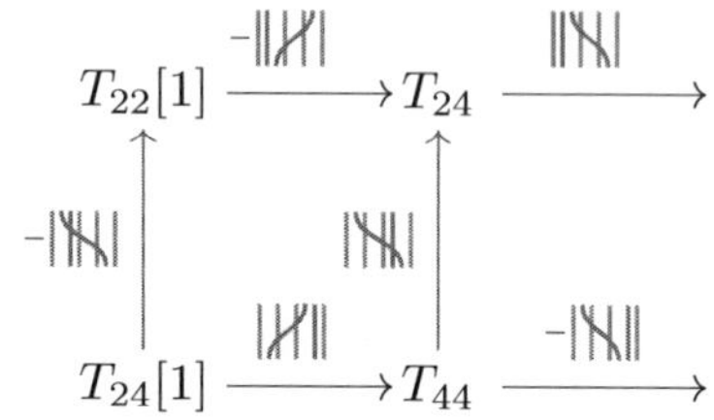

To avoid cluttering, we now present the complexes as periodic d-dimensional grids. The top and the bottom row of (4.42) are rewriting of the complexes in (4.40) and in (4.41). In this specific example, the complex "lives" on $S^1 \times \mathbb{R}$, where the periodic direction is the horizontal one. Correspondingly, rightmost arrows implicitly point to the beginning of the complex on the left. The new maps we have turned on run vertically, and glue $L_1 \times T_2$ and $L_1 \times T_4$ to get $L_1 \times L_2$. All the relative Maslov and equivariant degrees of the T-branes in the complex are determined from the maps in the complex. For example, the morphism that takes the T_{44} brane to lower left T_{24} has Maslov degree 2, and equivariant degree zero. Consequently if we were to place T_{44} brane in Maslov degree k, the lower left T_{24} brane would be in degree $k + 2 - 1 = k + 1$ and the same equivariant degree. We chose the overall cohomological degree to be $k = 0$; a different choice gives a resolution of a brane obtained from $L_1 \times L_2$ by a degree shift.

The resulting complex has the direct sum brane

$$L(T) = \bigoplus_k L_k(T)[k] = \begin{pmatrix} T_{22} \\ T_{24} \end{pmatrix} [1] \oplus \begin{pmatrix} T_{24} \\ T_{44} \end{pmatrix}$$

and differential

(4.43) $\delta_0 =$

The differential δ_0 squares to zero in the algebra $A_{\hbar=0}$ and gives a projective resolution of the brane on $Y_0 = Y \backslash \Delta$. The brane has one complex parameter worth of deformations, corresponding to turning on a generic flat connection on the L_1 brane, of which we picked one.

The signs in δ_0 are assigned according to the discussion in appendix C. (On the first pass, it is instructive to ignore the signs and work with $\mathbb{Z}_2$ coefficients instead.) Specializing the signs in the appendix to the current $d = 2$ case, we start with assigning the $+1$ sign to all the arrows in the first row as well as to all the arrows in the left-most column in (4.42). We then assign the same $+1$ sign to all columns, but alternate the overall signs in subsequent rows. Finally, we twist the sign of map going from a T-brane of odd Maslov degree by an additional -1. In checking that $\delta_0 \cdot \delta_0 = 0$, one has to recall that the composition of strand-algebra elements, viewed as the Homs in a category of twisted complexes of A-modules, come with extra signs. These signs, explained in appendix B, amount to replacing the Floer theory m_2, with its sign-twisted version m_2^p.

The differential δ that squares to zero in the full $A = A_\hbar$ algebra satisfies $\partial\delta + \delta\cdot\delta = 0$, or more precisely eqn. (4.32). As explained in section 4.3, for $\hbar \neq 0$ the algebra product stays the same, but the algebra differential ∂ deforms. We have that $\partial\delta_0 + \delta_0 \cdot \delta_0 = \partial\delta_0$, where

(4.44)

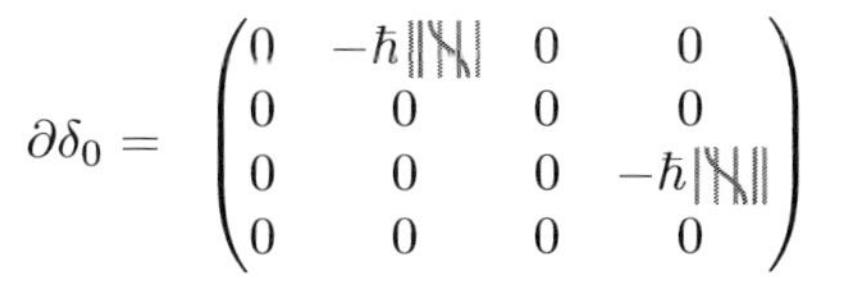

or, written as a complex:

(4.45)

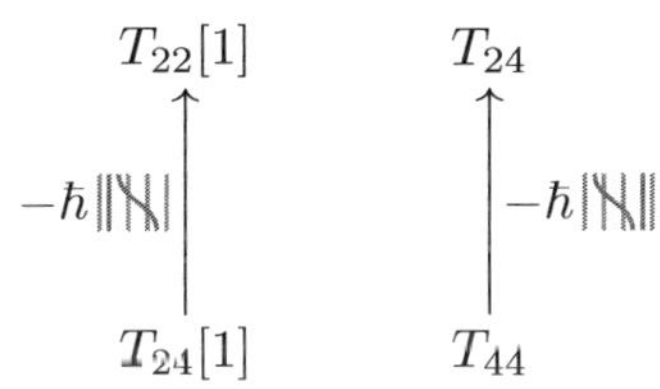

Since $\partial\delta_0$ does not vanish, the actual differential δ must be a non-trivial deformation of δ_0. In fact, we will show momentarily that the exact differential δ, one which satisfies (4.32) in the algebra $A_\hbar$, is a first order deformation of δ_0.

On equivariant and cohomological degree grounds, there are only two new maps that one could turn in δ. Solving for $\partial\delta + \delta\cdot\delta = 0$ fixes their coefficients. One coefficient gets set to zero and the other to $-\hbar$, so that the exact brane differential δ equals

(4.46)

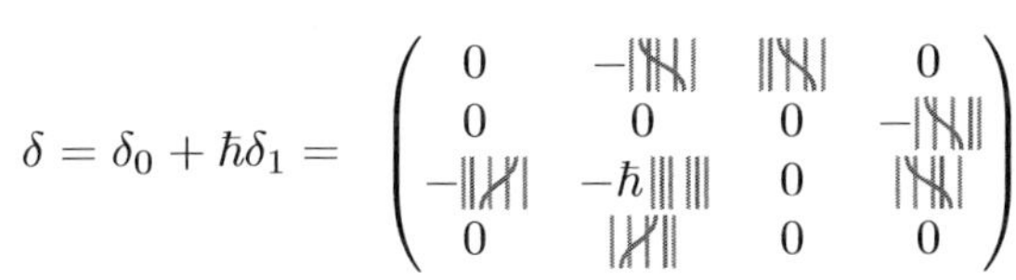

Written as a complex, the $\hbar$ deformation corresponds adding the orange arrow below:

(4.47)

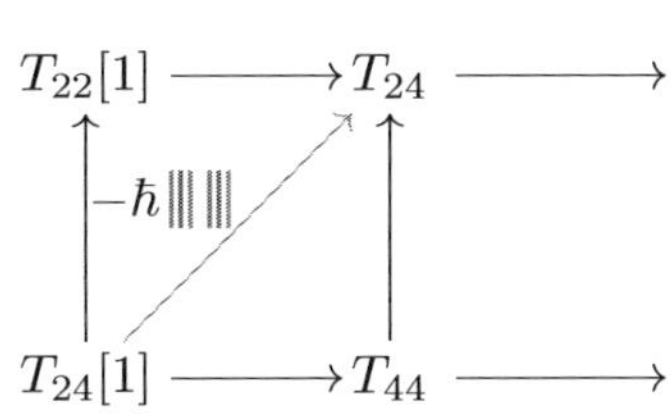

Per discussion in section 3.4.5, in looking for the full differential δ, we excluded deformations of δ_0 by terms already in it. They are either pure gauge, so they do not change the brane, or they change the complexified flat connection. In deforming the $\mathscr{B}E_{\mathcal{U}}$ brane from Y_0 to Y, we want to keep the flat connection fixed.

The term in the differential (4.47) proportional to $\hbar$ encodes a non-trivial disk instanton on Y. To compute the space $Hom^{*,*}_{\mathscr{D}_Y}(L, I_{24})$ of morphisms from the brane L to $I_{24} = I_2 \times I_4$, start by applying $hom^*_Y(-, I_{24}\{\vec{J}\})$ to the complex resolving L. The result is a complex of vector spaces one for each $\{\vec{J}\}$, to which only the terms with $T_{24}\{\vec{J}\}$ contribute. The k-th cohomology of this complex is $Hom_{\mathscr{D}_Y}(L, I_{24}[k]\{\vec{J}\})$. Recall the functor is contravariant, so reverses the direction of maps relative to the complex we started with. The only non-zero contribution is a two-term complex in degree $\vec{J} = 0$

$$\delta_F : \qquad \mathbb{C} \xrightarrow{-\hbar} \mathbb{C}[1] \tag{4.48}$$

with vanishing cohomology. This complex coincides with the Floer complex.

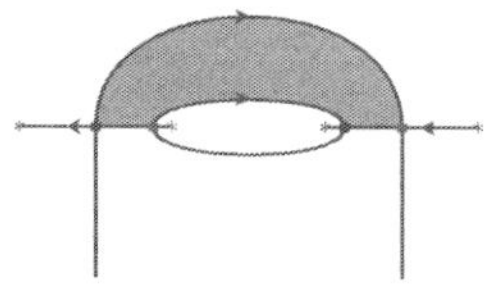

FIGURE 20. A domain which supports a holomorphic disk that generates the Floer differential acting on $hom_Y^{*;*}(L, I_{24})$. The red intersection point corresponds to the $\mathbb{C}$ factor in (4.48) and the green one to $\mathbb{C}[1]$.

The two terms in the complex (4.48) are the two intersection points of the I_{24}-brane with L. The differential in the complex is proportional to $\hbar$ and so counts maps to Y that intersect the diagonal once. It accounts for the disk which projects to the shaded domain A shown in Fig. 20. Every such "empty rectangle" has $M(A) = 1$ and $J_0(A) = 0$ and is known [**87**] to contribute ± 1 to the Floer differential, by an explicit calculation. Here have re-derived this, together with the sign, from the resolution of the brane $L \cong (L(T), \delta)$.

4.5.5. Here is a simple example relevant to knot theory.

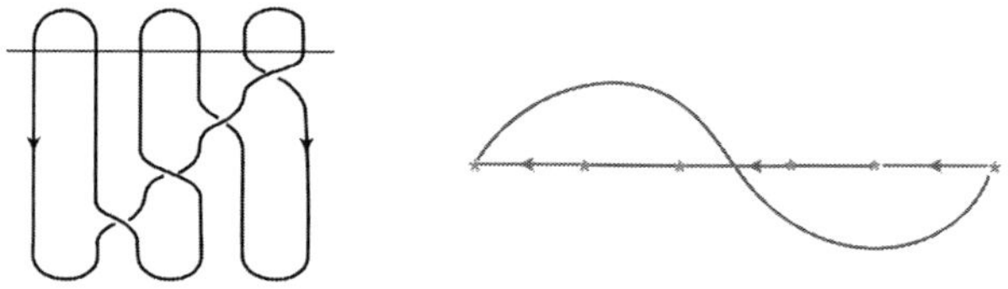

FIGURE 21. A $d = 3$ presentation of the unknot as a plat closure of a braid, and the corresponding matchings.

Reduced homology of unknot presented as in Fig. 21 comes from the brane configuration in Fig. 22, which is obtained from by removing a pair of red/blue matchings, and replacing the remaining blue matchings with ovals.

The cap and the cup branes have 3 intersection points: p_1q_1, p_2q_2, and p_2q_3. The disk shown in Fig. 22 is an empty rectangle, which therefore has $M = 1$ and $J_0 = 0$, and interpolates from p_2q_2 to p_1q_1. As we explained in our toy example above, such empty rectangles are known to contribute ± 1 to the differential in Heegaard-Floer theory. So we expect to find that

$$\delta_F p_2 q_2 = \pm \hbar p_1 q_1 \tag{4.49}$$

and homology that is one dimensional, as expected for the unknot.

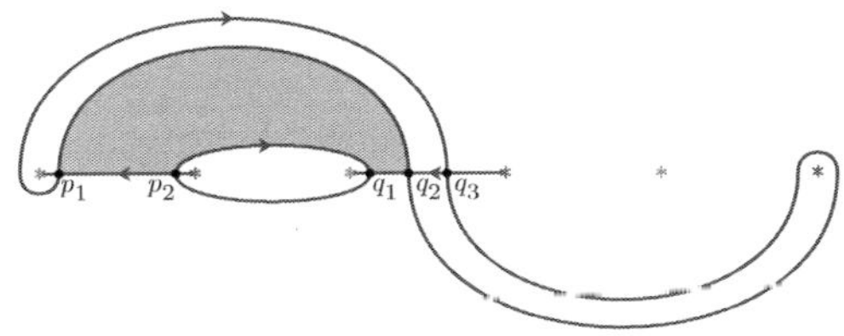

FIGURE 22. Branes for the unknot, and a disk which generates the Floer differential.

This kind of counting of disks "by hand" is possible only for very special examples because in general, maps of disks to symmetric product can be arbitrarily complicated. We will now show how the algorithm from section 3, which is completely general, recovers this disk and moreover fixes the sign of the disk contribution in (4.49).

Pick one of the two factor branes in $\mathscr{B}E_{\mathcal{U}} = \mathscr{B}E_1 \times E_2$, and find its resolution. For example, the resolution of $\mathscr{B}E_1$ is

$$T_1[1] \longrightarrow T_2 \longleftarrow T_4[-1] \longrightarrow T_7[-2] \longrightarrow T_6[-3]\{1\} \longleftarrow T_4[-2]\{1\} \longrightarrow .$$

Since the E_2 brane breaks into T_2 and T_4, to find the resolution of $\mathscr{B}E_1 \times E_2$, begin by reading off the complex for the products $\mathscr{B}E_1 \times T_2$ and $\mathscr{B}E_1 \times T_4$ using Fig. 23. As before, it is important to keep track of the placement of the T_2 and T_4 branes relative to the branes in the complex for $\mathscr{B}E_1$. Just as picture of the brane encodes the actual Lagrangian, so does the resolution read off from it. Taking these products will change some of the cohomological degrees of $\mathscr{B}E_1$ because some of the maps end up containing crossings, each of which changes the Maslov degree by -1.

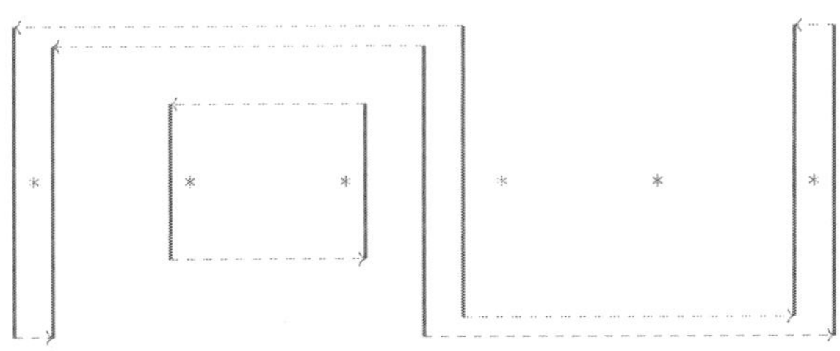

FIGURE 23. The cap brane for the unknot, broken up into thimbles.

Taking these shifts into account, we find that the resolution of $\mathscr{B}E_1 \times T_2$ is

$$T_{12}[2] \longrightarrow T_{22}[1] \longleftarrow T_{24}[1] \longrightarrow T_{27} \longrightarrow T_{26}[-1]\{1\} \longleftarrow T_{24}\{1\} \longrightarrow .$$

and the resolution of $\mathscr{B}E_1 \times T_4$ is

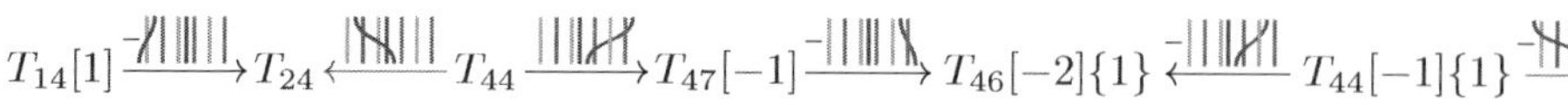

Next, we take the direct sum of $\mathscr{B}E_1 \times T_2$ and $\mathscr{B}E_1 \times T_4$ and turn on the geometric maps between T_2 and T_4 to combine $\mathscr{B}E_1 \times T_2$ and $\mathscr{B}E_1 \times T_4$ into $\mathscr{B}E_1 \times E_2$. Unfolding the complex into a toric grid, the result is

(4.50)

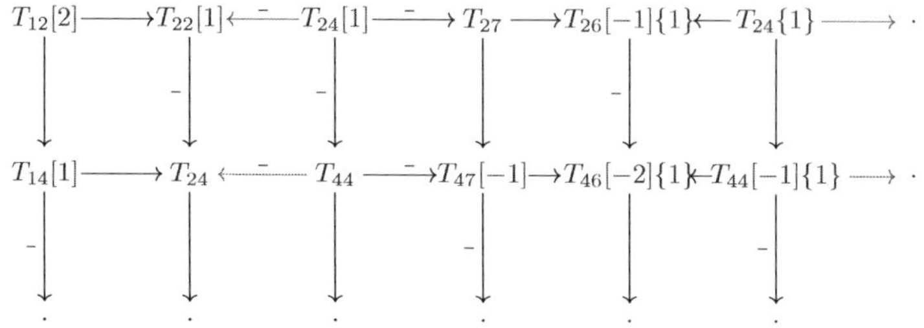

For the sake of readability, we have suppressed the maps, but they are fixed uniquely by the graded thimbles they map between. Additionally, we colored purple the maps containing strands that cross.

The signs are assigned as described in section 4.4.5 and appendix C. First, we start by fixing signs of the first row and all columns to $+1$. Next, we alternate signs of the horizontal maps according to the row. Finally, we multiply all maps starting at a site of odd Maslov degree by an additional factor of -1.

One can easily check that δ_0, the collection of maps we found so far, defines a differential in the algebra A_0, which simply requires $\delta_0^2 = 0$ (taking all the signs from the twisted A_∞ products into the account). Turning on $\hbar \neq 0$, δ_0 still squares to zero, but it no longer defines a differential, as $\partial\delta_0$ does not vanish in $A_\hbar$. It consists of:

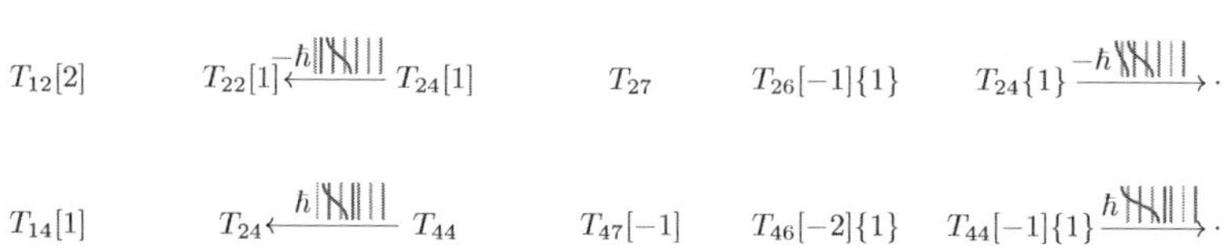

The differential δ that satisfies $\partial\delta + \delta \cdot \delta = 0$ in the algebra $A_\hbar$ is a deformation of δ_0. In principle, on degree grounds, there are 16 new maps which may be turned on in δ shown below:

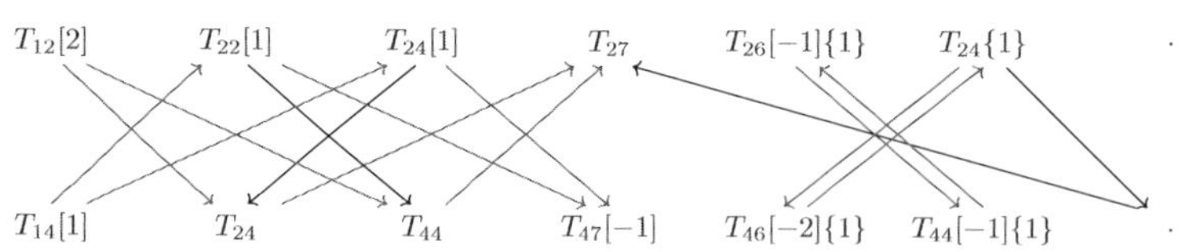

The ansatz for δ_1, the first order deformation of δ_0 thus contains 16 coefficients of these maps. Solving the first-order equation $\partial\delta_0 + \delta_0 \cdot \hbar\delta_1 + \hbar\delta_1 \cdot \delta_0 = 0$ reduces the 16-dimensional space to a 12-dimensional one. One can find the exact solution, without introducing higher order terms in δ, by asking that $\hbar\partial\delta_1 + \hbar\delta_1 \cdot \hbar\delta_1 = 0$. This uniquely determines the remaining 12 coefficients in δ_1, leading to $\hbar\delta_1$ composed of only two maps:

$$\begin{array}{cccccc} T_{12}[2] & T_{22}[1] & T_{24}[1] & T_{27} & T_{26}[-1]\{1\} & T_{24}\{1\} \\ & \swarrow{\scriptstyle -\hbar} & & & & \searrow{\scriptstyle -\hbar} \\ T_{14}[1] & T_{24} & T_{44} & T_{47}[-1] & T_{46}[-2]\{1\} & T_{44}[-1]\{1\} \end{array}.$$

One of these maps, the one from $T_{24}[1]$ to T_{24}, corresponds to the disk in Fig. 22. The complex resolving the $\mathscr{B}E_1 \times E_2$ brane is thus the direct sum brane:

$$T_{12}[2] \oplus \begin{pmatrix} T_{14} \\ T_{22} \\ T_{24} \end{pmatrix} [1] \oplus \begin{pmatrix} T_{24} \\ T_{24}\{1\} \\ T_{27} \\ T_{44} \end{pmatrix} \oplus \begin{pmatrix} T_{26}\{1\} \\ T_{44}\{1\} \\ T_{47} \end{pmatrix} [-1] \oplus T_{46}[-2]$$

with differential $\delta = \delta_0 + \hbar\delta_1$, which we just computed.

To calculate homology, we take $hom_Y^{*,*}(-, I_{24})$ of the complex, to find

$$0 \longrightarrow 0 \longrightarrow \begin{pmatrix} \mathbb{C} \\ \mathbb{C}\{1\} \end{pmatrix} \xrightarrow{\begin{pmatrix} -\hbar & 0 \end{pmatrix}} \mathbb{C}[1] \longrightarrow 0$$

The three terms above are the three intersection points of the $\mathscr{B}E_1 \times E_2$ brane with the I_{24} brane in Fig. 22.

Taking cohomology, the the final result is

$$Hom_{\mathscr{D}_Y}(\mathscr{B}E_{\mathcal{U}}, I_{\mathcal{U}}) = \mathbb{C}, \tag{4.51}$$

which is what we expect for the unknot. We used that the cap brane differs from the I_{24} brane by $I_{\mathcal{U}} = I_{24}\{1\}$, and that the constant shifts ΔM, ΔJ_0 from section 4.6 vanish for the braid in Fig. 21.

4.5.6. As the final example, consider the 8_{19} knot. The link homology is comes from the following branes:

FIGURE 24. A presentation of 8_{19} as a braid closure and the corresponding branes

Compared to previous examples, finding the differential δ requires a second-order correction in $\hbar$ to δ_0. Intersecting with the I_{24} Lagrangian leads to the complex

$$\mathbb{C}\{5\} \xrightarrow{\begin{pmatrix} 0 \\ 0 \end{pmatrix}} \begin{matrix} \mathbb{C}\{3\} \\ \mathbb{C}\{4\} \end{matrix} \xrightarrow{\begin{pmatrix} -\hbar & 0 \\ 0 & 0 \\ 0 & 0 \\ 0 & 0 \end{pmatrix}} \begin{matrix} \mathbb{C}\{3\} \\ \mathbb{C}\{2\} \\ \mathbb{C}\{2\} \\ \mathbb{C}\{2\} \end{matrix} \xrightarrow{\begin{pmatrix} 0 & 0 & \hbar & 0 \\ 0 & \hbar & 0 & -\hbar \\ 0 & 0 & 0 & 0 \\ 0 & 0 & 0 & 0 \end{pmatrix}} \begin{matrix} \mathbb{C}\{2\} \\ \mathbb{C}\{2\} \\ \mathbb{C}\{1\} \\ \mathbb{C}\{1\} \end{matrix} \xrightarrow{\begin{pmatrix} 0 & 0 & 0 & 0 \\ 0 & 0 & -\hbar & 0 \\ 0 & 0 & -\hbar & \hbar \end{pmatrix}}$$

$$\begin{matrix} \mathbb{C} \\ \mathbb{C}\{1\} \\ \mathbb{C}\{1\} \end{matrix} \xrightarrow{\begin{pmatrix} 0 & 0 & 0 \\ 0 & 0 & 0 \\ -\hbar & 0 & 0 \end{pmatrix}} \begin{matrix} \mathbb{C}\{-1\} \\ \mathbb{C} \\ \mathbb{C} \end{matrix} \xrightarrow{\begin{pmatrix} 0 & 0 & 0 \\ \hbar & 0 & 0 \end{pmatrix}} \begin{matrix} \mathbb{C}\{-1\} \\ \mathbb{C}\{-1\} \end{matrix} \xrightarrow{\begin{pmatrix} 0 & 0 \end{pmatrix}} \mathbb{C}\{-2\} \xrightarrow{\begin{pmatrix} -\hbar \end{pmatrix}} \mathbb{C}\{-2\}$$

which has homology

$$\mathbb{C}[-2]\{5\} \oplus \mathbb{C}[-1]\{4\} \oplus \mathbb{C}\{2\} \oplus \mathbb{C}[3] \oplus \mathbb{C}[4]\{-1\}.$$

The degree shifts that come from the braid, which has $e_{++} = 0$ and $e_{--} = -2$, are $\Delta M = 2$ and $\Delta J_0 = -1$, by eqn. (4.54). Accounting for $I_{\mathcal{U}} = I_{24}\{1\}$, we find that the regraded homology is

$$Hom^{*,*}_{\mathscr{D}_Y}(\mathscr{B}E_{\mathcal{U}}, I_{\mathcal{U}}) = \mathbb{C}\{3\} \oplus \mathbb{C}[1]\{2\} \oplus \mathbb{C}[2] \oplus \mathbb{C}[5]\{-2\} \oplus \mathbb{C}[6]\{-3\}, \tag{4.52}$$

which agrees with $\widehat{HFK}$ after replacing J_0 by $-J_0$, as expected from section 4.7.

4.6. Homological $U_{\mathfrak{q}}(\mathfrak{gl}_{1|1})$ link invariants from $\mathscr{D}_{Y_{\mathfrak{gl}_{1|1}}}$. In this section we will prove that $Hom^{*,*}_{\mathscr{D}_Y}(\mathscr{B}E_{\mathcal{U}}, I_{\mathcal{U}})$ are homological link invariants, which is Thm. 1 for $\mathfrak{gl}_{1|1}$. Along the way, we will explain how to assign absolute gradings to the branes and their homology groups.

The key ingredient in the proofs is the following relation, which says that the branes on the left and the right hand sides of:

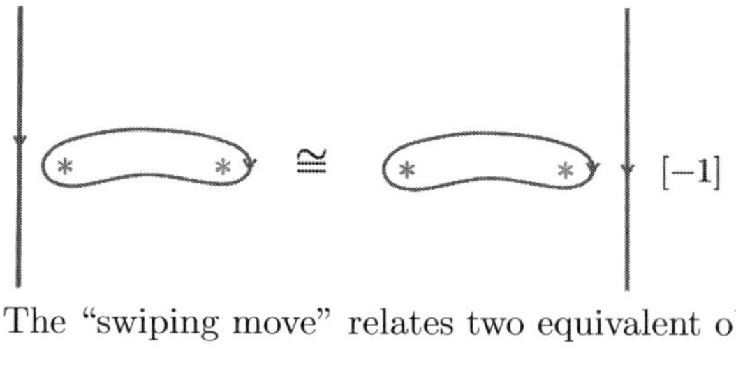

FIGURE 25. The "swiping move" relates two equivalent objects of $\mathscr{D}_Y$

are equivalent objects of $\mathscr{D}_Y$. We prove this in appendix A.1 where we show that the resolutions of the two branes are homotopy equivalent for any $\hbar \neq 0$.

The fact that the swiping move holds for any $\hbar \neq 0$, but not for $\hbar = 0$ is why one must work in the A-model on Y, instead of the simpler theory on $Y_0 = Y \backslash \Delta$. While the Euler characteristics of the homology theories coming from Y and Y_0 are the same, only the theory on Y will give homological link invariants.

4.6.1. Once we know the swiping move in Fig. 25 holds, all four moves in Figs. 5, 6 and 7 follow, up to possible overall Maslov and equivariant degree shifts. They in fact hold with gradings included, as we will explain below.

The moves in Fig. 6, up to possible grading shifts, hold by isotopy invariance alone. The Markov I type move from Fig. 5, translated to cup branes, reads

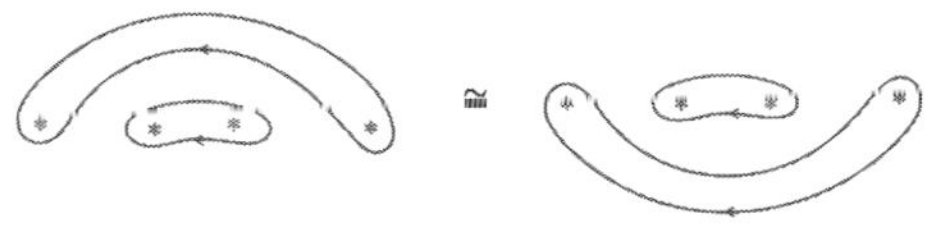

FIGURE 26. Markov I type move, in terms of branes

To show the relation holds, we first show the brane on the left is equivalent, using the swiping move, to the following brane:

FIGURE 27. Swiping move simplifies the brane.

To show this, start with the $E_{\mathcal{U}}$ brane which is a product of two basic oval branes. Next, use the swiping move in Fig. 25 to get an equivalent description of the brane in terms of one where the left brane "swallows" the right. Finally, braid both branes to get the equivalence in Fig. 27. The resulting brane is based on the product Lagrangian pictured, with the overall degree that needs to be determined. The degree will depend on the braid used to obtain the brane.

The brane on the right is similarly equivalent to:

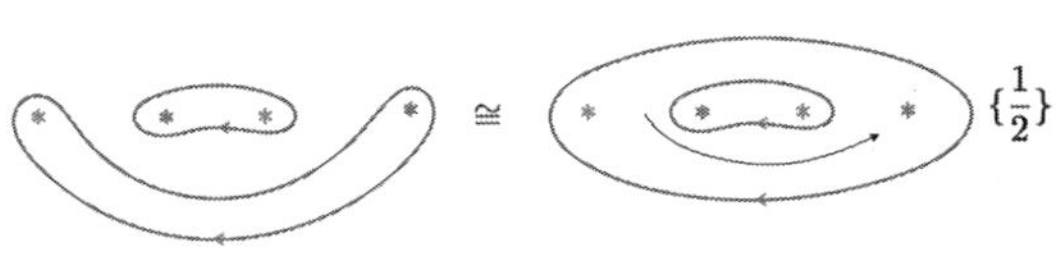

FIGURE 28. The grading of the resulting brane in principle depends on the path of the green puncture.

The two branes that result are manifestly equivalent objects of $\mathscr{D}_Y$, up to possible overall Maslov and equivariant degree shifts which come from the fact we obtained them by braiding in two different ways. We will show below they are equivalent on the nose, with the grading included. There are analogous versions of these relations for different choices of strand orientations, which can be similarly shown to hold exactly.

The stabilization move from Fig. 7 reads:

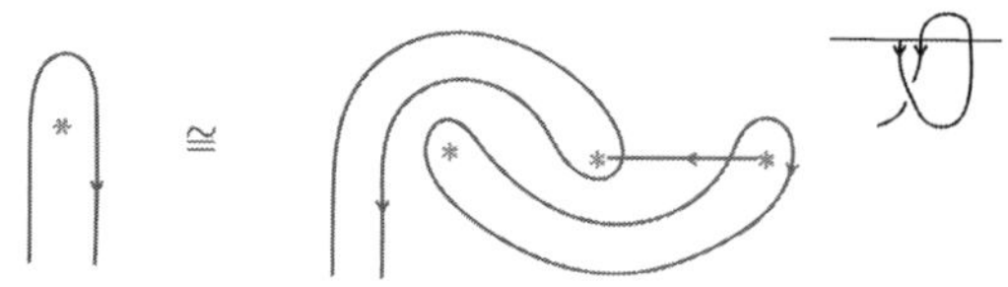

FIGURE 29. Stabilization move in terms of branes

By first applying the swiping move to the $E_{\mathcal{U}}$ brane and then braiding, one gets an equivalent presentation of the brane on the right hand side as:

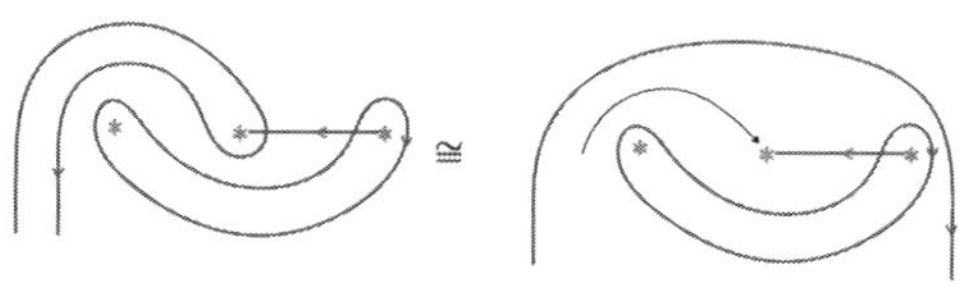

Since the I-brane intersects the only the interior component of the E-brane, and only once, the stabilization move in Fig. 29 follows up to a grading shift.

4.6.2. The moves in Figs. 5, 6 and 7 in fact hold exactly, with Maslov and equivariant gradings of the branes included. To explain this, we need to explain how absolute gradings of the branes are assigned.

Absolute equivariant gradings, as well as cohomological gradings mod 2, are visible already in K-theory, so they are the "easy part" of the theory, that was not central to us so far. Now, we will fill the gap.

4.6.3. The grading of the $\mathscr{B}E_{\mathcal{U}}$ brane is determined uniquely from the grading of the $E_{\mathcal{U}}$ brane we started with and the braid group element whose action on $\mathscr{D}_Y$ the functor $\mathscr{B}$ represents. We will take the $E_{\mathcal{U}}$ brane to be the product of d one dimensional E-branes from section 4.5.4. The degrees of the T-branes in the resolution of the brane range from $[0]$ to $[d]$, where d is the dimension of Y; the equivariant degrees all vanish.

Because the relative degrees of the T-branes in the complex resolving the $\mathscr{B}E_{\mathcal{U}}$ branes are fixed from geometry of the underlying Lagrangian alone, knowing the absolute degree of the $\mathscr{B}E_{\mathcal{U}}$-brane is the same as knowing absolute degree of any one T-brane in its resolution.

The absolute grading of the $\mathscr{B}E_{\mathcal{U}}$ brane receives two contributions. The first contribution is the obvious one, due to the geometric braid group action on $\mathscr{D}_Y$, whose objects are graded branes: Pick any $T_{\mathcal{C}}$-brane in the direct sum brane resolving the $E_{\mathcal{U}}$ brane. Consider its connected sum with a "spectator" T-brane on which $\mathscr{B}$ acts trivially. Since braiding leaves the spectator brane fixed, the relative gradings of the connected sum T-brane determine the absolute gradings of all branes in $\mathscr{B}T_{\mathcal{C}}$. As an example, the braid in Fig. 27 gives a brane with no geometric degree shift at all. The braid in Fig. 28 results in a geometric degree shift of $\{1\}$.

4.6.4. There is a second contribution to the grading of the $\mathscr{B}E_{\mathcal{U}}$ brane, which originates from terms in Ω and W, in (4.2) and (4.6), which are constant on Y, but depend on the positions of the punctures in a non-single-valued way. The constant terms in Ω and W have a conformal field theory origin, recalling the relation to $\widehat{{}^L\mathfrak{g}}$ chiral algebra from section 2.4.1. They become important if we want the quantum group to have the canonical action on K-theory and its categorification in $\mathscr{D}_Y$.

As a result of these terms, the geometric braid group action on $\mathscr{D}_Y$ is accompanied with an action of a "constant" degree shift functor. The action of this functor is constant in that it depends only on the braid group element that $\mathscr{B}$ represents, and is the same for all objects in $\mathscr{D}_Y$.

Recall that every braid group element is a word in the generators $\sigma_i^{\pm 1}$, where σ_i exchanges i-th and $i+1$-st puncture counterclockwise. The generators satisfy $\sigma_i \sigma_{i+1} \sigma_i = \sigma_{i+1} \sigma_i \sigma_{i+1}$, and $\sigma_i \sigma_j = \sigma_j \sigma_i$, for $i > j$. Let e_{++} and e_{--} be the sums of exponents of braid group generators that braid a pair of even and odd punctures, respectively. Then terms in (4.2) and (4.6) result in shift functors that act on every object of $L \in \mathscr{D}_Y$ by a shift

$$L \to L[\Delta M]\{\Delta J_0\}, \tag{4.53}$$

in cohomological and equivariant degrees, given by

$$\begin{aligned} \Delta M &= -e_{--} \\ \Delta J_0 &= -\frac{e_{++}}{2} + \frac{e_{--}}{2}. \end{aligned} \tag{4.54}$$

For example, the Markov I move from Fig. 26 involves a pair of braids that have $e_{++} = -1$ and $e_{++} = 1$; both have $e_{--} = 0$. The resulting constant shifts of $\{\frac{1}{2}\}$ and $\{-\frac{1}{2}\}$ combine with the geometric degree shifts of zero and $\{1\}$, respectively, to result in both branes having degree $\{\frac{1}{2}\}$. Consequently, the Markov I move holds exactly, with gradings included.

4.6.5. The absolute grading on the homology groups $Hom^{*,*}_{\mathscr{D}_Y}(\mathscr{B}E_{\mathcal{U}}, I_{\mathcal{U}})$ depends also on the absolute grading of the $I_{\mathcal{U}}$ branes.

The grading is determined by asking that the stabilization move in Fig. 29 holds, as well as its variants obtained by changing the orientations of the strands. The result is that the cap brane $I_{\mathcal{U}}$ is not simply the product of basic intervals oriented right to left in degree zero, as is the $I_{\mathcal{C}}$ brane, a brane canonically dual to one of the T-branes, as in (2.12). Rather, it differs by degree shifts which depend on orientation of the caps:

FIGURE 30. The cap brane $I_{\mathcal{U}}$ is related to the basic corresponding I-brane by degree shift, which is either $\{\frac{1}{2}\}$ or $\{-\frac{1}{2}\}[1]$, depending on the orientation.

The fact the stabilization move holds comes from three degree shifts canceling out, two shifts coming from the E-brane and one from the I-brane. There is an overall shift of $\mathscr{B}E_{\mathcal{U}}$ brane degree $[1]\{-\frac{1}{2}\}$ coming from a pair of odd punctures braided an element whose exponent is -1. The geometric degree of the intersection point of the $\mathscr{B}E_{\mathcal{U}}$ brane with the basic I-brane has degree $[-1]\{1\}$. The two shifts combined give the intersection point of the $\mathscr{B}E_{\mathcal{U}}$ brane with a basic I-brane net degree $\{\frac{1}{2}\}$. Due to the degree shift between the $I_{\mathcal{U}}$ brane and the basic I-brane from Fig. 30, this is equivalent to an intersection point of the $\mathscr{B}E_{\mathcal{U}}$ brane with an $I_{\mathcal{U}}$ brane of Maslov and equivariant degree zero. It is easy to check that, varying orientations of the strands, we get net zero degree shifts in all other cases as well.

4.6.6. The remaining two relations in Fig. 6 follow similarly. The first of the relations Fig. 6 is an easy exercise, which amounts to showing the geometric degree shift of the $E_{\mathcal{U}}$ brane gets canceled by the constant overall degree shift from (4.53). The second relation in Fig. 6, corresponding to freedom to add a twist to a cup, holds manifestly for $E_{\mathcal{U}}$-branes, as it translates to a freedom to reorder the odd and even puncture within it without changing the brane. It also holds for the $I_{\mathcal{U}}$ branes, but only after including the shifts from Fig. 30. These categorify the statement that the change of framing acts trivially on $U_{\mathfrak{q}}(\mathfrak{gl}_{1|1})$ group invariants.

4.6.7. To complete the proof of Thm. 1 for $\mathfrak{gl}_{1|1}$, it remains to show the remaining Markov type moves. These are moves obtained from those in Figs. 5 and 7 by exchanging tops and bottoms, also hold. As sketched in section 2.5.1, the proof proceeds in two steps. First, we observe that $Hom^{*,*}_{\mathscr{D}_Y}(\mathscr{B}E_{\mathcal{U}}, E_{\mathcal{U}})$ is invariant under the two moves since the $E_{\mathcal{U}}$ branes are. Next, we compute how $Hom^{*,*}_{\mathscr{D}_Y}(\mathscr{B}E_{\mathcal{U}}, E_{\mathcal{U}})$ relates to $Hom^{*,*}_{\mathscr{D}_Y}(\mathscr{B}E_{\mathcal{U}}, I_{\mathcal{U}})$.

Per definition, $Hom^{*,*}_{\mathscr{D}_Y}(\mathscr{B}E_{\mathcal{U}}, E_{\mathcal{U}})$ is the cohomology of a double complex obtained from $hom^{*,*}_{Y}(\mathscr{B}E_{\mathcal{U}}, E_{\mathcal{U}})$ by replacing the $\mathscr{B}E_{\mathcal{U}}$ brane its projective resolution, and the $E_{\mathcal{U}}$ brane by its *injective* resolution. The injective resolution is a resolution in terms of I-branes, instead of the T-branes. Each one-dimensional E_i-brane in

$$E_{\mathcal{U}} = E_1 \times \ldots \times E_d$$

has a resolution $E_i \cong (E_i(I), \delta_i)$ in terms of a pair of corresponding I_i branes, each with a relative shift in degree by $[1]\{-1\}$. Taking the product of the resulting one-dimensional complexes, we get the injective resolution of the $E_{\mathcal{U}}$ brane as $E_{\mathcal{U}} = (E(I), \delta_E)$ where the $E(I)$ is a direct sum $I_{\mathcal{U}}$ branes in different degrees:

$$E(I) = I_{\mathcal{U}}\Big(\{1/2\} \oplus [1]\{-1/2\}\Big)^{\bigotimes d} \equiv I_{\mathcal{U}} \otimes W. \tag{4.55}$$

We introduced a 2^d dimensional graded vector space W to implement the degree shifts for us. Analogously to section 3.5.3, the only terms in the corresponding brane differential δ_E that survive to the double complex involve identity maps between $I_{\mathcal{U}}$-branes. These do not exist on degree grounds. Correspondingly, it follows

$$Hom^{*,*}_{\mathscr{D}_Y}(\mathscr{B}E_{\mathcal{U}}, E_{\mathcal{U}}) = Hom^{*,*}_{\mathscr{D}_Y}(\mathscr{B}E_{\mathcal{U}}, I_{\mathcal{U}}) \otimes W. \tag{4.56}$$

Since W is fixed and independent of the braid $\mathscr{B}$, $Hom^{*,*}_{\mathscr{D}_Y}(\mathscr{B}E_{\mathcal{U}}, I_{\mathcal{U}})$ must transform in the same way under the outstanding Markov-type moves, as $Hom^{*,*}_{\mathscr{D}_Y}(\mathscr{B}E_{\mathcal{U}}, E_{\mathcal{U}})$ does. Since the latter is invariant, so is $Hom^{*,*}_{\mathscr{D}_Y}(\mathscr{B}E_{\mathcal{U}}, I_{\mathcal{U}})$. This completes the proof of Thm. 1 for $\mathfrak{gl}_{1|1}$.

4.7. $U_{\mathfrak{q}}(\mathfrak{gl}_{1|1})$ **skein from branes.** We will now prove Thm. 2, which says the Euler characteristic of $Hom^{*,*}_{\mathscr{D}_Y}(\mathscr{B}E_{\mathcal{U}}, I_{\mathcal{U}})$ satisfies the skein of the Alexander polynomial. Since homology groups are independent of the choice of the link projection by Thm. 1, it suffices to show that Euler characteristic of homological invariants obtained by the very special link presentation in Fig. 8 satisfy the skein relation.

The Euler characters of $Hom^{*,*}_{\mathscr{D}_Y}(\mathscr{B}E_{\mathcal{U}}, I_{\mathcal{U}})$ will satisfy the skein relation if K-theory classes of the three branes which represent the braided cups at the bottoms of Fig. 8, satisfy the skein relation. The branes whose K-theory classes we need to consider are given in Fig. 31. Using the swiping move, for each of three branes in Fig. 31 can be simplified by letting the rightmost brane "swallow" the left.

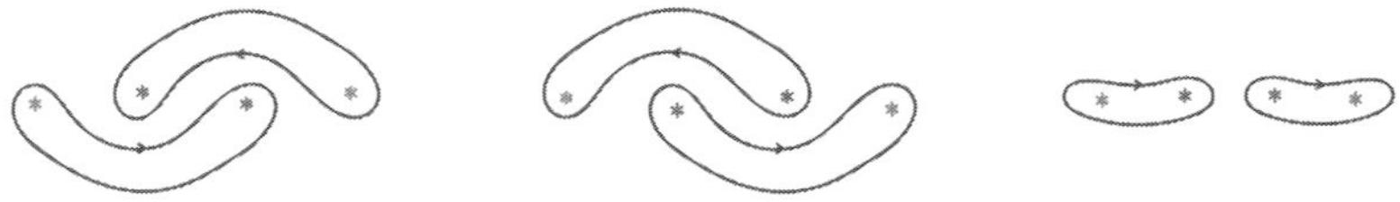

FIGURE 31. The three branes that the skein relates.

Having done so, it suffices to show that K-theory classes of the three branes in Figs. 32, 33 and 34 satisfy the skein relation.

The first of the three branes, given in Fig. 32, is obtained from the simple $E_{\mathcal{U}}$ brane by braiding two positive punctures by σ_2. Its resolution is given by

$$T_0[1] \longrightarrow T_3 \longrightarrow T_2[-1]\{1\} \longleftarrow T_1\{1\} \longrightarrow . \tag{4.57}$$

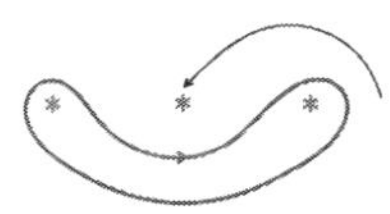

FIGURE 32. The reduced brane corresponding to crossing

The overall degree of the brane, before any non-geometric shifts, follows from the fact that braiding acted trivially on the T_0 brane, hence its degree could not have shifted. With $e_{++} = 1$, $e_{--} = 0$, there is an additional degree shift of the brane by $\{-1/2\}$. The final K-theory class of the brane is $\mathfrak{q}^{-1/2}(\mathfrak{q}[T_1] - \mathfrak{q}[T_2] + [T_3] - [T_0])$.

FIGURE 33. The reduced brane corresponding to crossing

The second brane, in Fig. 33, is obtained from the simple $E_{\mathcal{U}}$ brane by braiding a pair of positive punctures by σ_2^{-1}. The resolution of the brane

$$T_0[1] \longrightarrow T_1 \longleftarrow T_2[1]\{-1\} \longrightarrow T_3\{-1\} \longrightarrow . \tag{4.58}$$

gets accompanied by an additional degree shift of $\{1/2\}$, since $e_{++} = -1$, $e_{--} = 0$, so final K-theory class of the brane is $\mathfrak{q}^{1/2}([T_1] - \mathfrak{q}^{-1}[T_2] + \mathfrak{q}^{-1}[T_3] - [T_0])$.

FIGURE 34. The reduced brane corresponding to

The last brane, in Fig. 34, is the simple $E_{\mathcal{U}}$ brane from section 4.5.1, whose K-theory class is $[T_2] - [T_0]$. K-theory classes of the three branes in Figs. 32, 33 and 34 satisfy one relation:

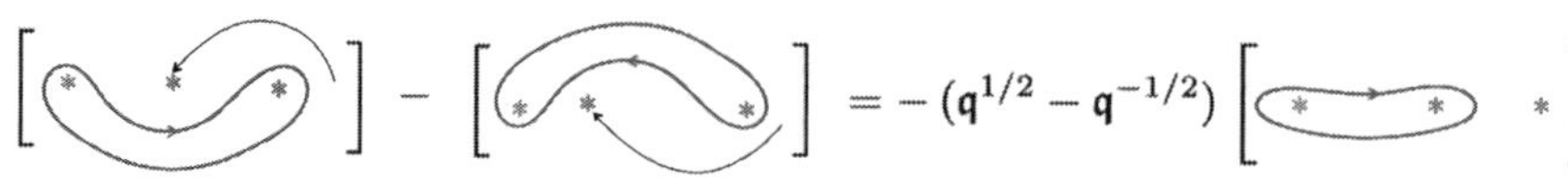

This is the skein relation satisfied by the $U_{\mathfrak{q}}(\mathfrak{gl}_{1|1})$ R-matrices from e.g. [**74**]. It coincides with the $n = 0$ skein relation in eqn. (1.1) after identifying

$$\mathbf{q} = \mathfrak{q}^{-1}. \tag{4.59}$$

The fact that our $\mathfrak{q}$, and $\mathbf{q}$ in terms of which the Alexander polynomial and its skein relation are written, are related in this way is responsible for the flip of J_0 grading between our homology theory and the conventions of $\widehat{HFK}$ homology theory which we have seen in all our examples.

5. The $\mathfrak{su}_2$ theory

In this section, we describe the generalization of Heegaard-Floer theory, from previous section, to ${}^L\mathfrak{g} = \mathfrak{su}_2$, categorifying the Jones polynomial.

5.1. The A-model on $Y_{\mathfrak{su}_2}$. We take the Riemann surface $\mathcal{A}$, with $n = 2d$ punctures colored by the two-dimensional defining representation V of $\mathfrak{su}_2$. The representation is self-conjugate, so V and V^* coincide, and we have only one kind of a puncture, located at $y = a_i$, where i runs from 1 to $2d$. The target $Y = Y_{\mathfrak{su}_2}$,

$$Y = Sym^d(\mathcal{A}) \tag{5.1}$$

is again the symmetric product of d points on $\mathcal{A}$, but equipped with a different top holomorphic form $\Omega_{\mathfrak{su}_2}$ and a different potential $W_{\mathfrak{su}_2}$ than in the $\mathfrak{gl}_{1|1}$ case.

5.1.1. The Maslov grading comes from the top holomorphic form $\Omega = \Omega_{\mathfrak{su}_2}$,

$$\Omega = \bigwedge_{\alpha=1}^{d} \frac{dy_\alpha}{y_\alpha}, \tag{5.2}$$

which coincides with the holomorphic volume form on the Cartesian product of d copies of $\mathbb{C}^\times$, with coordinate $y \neq 0, \infty$. The theory admits a $\mathbb{Z}$-valued Maslov grading since $\Omega^{\otimes 2}$ is a globally well defined section of $K_Y^{\otimes 2}$, and hence

$$2c_1(K_Y) = 0. \tag{5.3}$$

The form Ω differs from the standard top holomorphic form on $Sym^d(\mathbb{C}^\times)$ by having a first order pole along the diagonal Δ in Y. If $\sigma_1, \dots, \sigma_d$ are elementary symmetric functions of $y_1, \dots, y_d$, the latter equals $\bigwedge_{\alpha=1}^d \frac{dy_\alpha}{y_\alpha} \prod_{\alpha<\beta}(y_\alpha - y_\beta) = (\bigwedge_{\alpha=1}^d d\sigma_\alpha)/\sigma_d$. Unlike Ω, this canonical top holomorphic form is invariant under the action of the permutation group S_d on $y_1, \dots, y_d$.

The fact that $\Omega_{\mathfrak{su}_2}$ differs from $\Omega_{\mathfrak{gl}_{1|1}}$ in 4.1.1 means that the curves one counts in the two theories are not the same, even though they both map to $Y = Sym^d(\mathcal{A})$.

5.1.2. The equivariant gradings come from a potential $W = W_{\mathfrak{su}_2}$,

$$W = \lambda_0 W^0 + \lambda_1 W^1, \tag{5.4}$$

which is a multi-valued holomorphic function on Y. We take λ_0 to be related to $\mathfrak{q}$ as in (4.5), and λ_1 is associated to the holonomy around the S^1, as in the $\mathfrak{gl}_{1|1}$ case. The functions W^0 and W^1 are given by:

$$W^0 = \ln f, \qquad W^1 = \ln \prod_{\alpha=1}^{d} y_\alpha,$$

for

$$f(y) = \frac{\prod_{i=1}^{2d}\prod_{\alpha=1}^{d}(1 - a_i/y_\alpha)}{\prod_{1\leq\beta\neq\alpha\leq d}(1 - y_\beta/y_\alpha)} \prod_{1\leq i<j\leq 2d} (1 - a_i/a_j)^{1/2}. \tag{5.5}$$

For comparison to the general simple Lie algebra ${}^L\mathfrak{g}$ case, if μ is the highest weight of the fundamental representation coloring the punctures and ${}^Le$ the positive simple root, the exponents in the numerator are $1 = \langle {}^Le, \mu\rangle$ and $\frac{1}{2} = \langle \mu, \mu\rangle$, the exponent in the denominator is $1 - \langle {}^Lc, {}^L c\rangle/2$.

The factors of (5.5) that depend only on the positions of punctures on $\mathcal{A}$ play no role in any local considerations. They become relevant when we consider the action of braiding on $\mathscr{D}_Y$, as explained in section 5.7.

5.1.3. A proposal that a similar A-model should categorify the Jones polynomial was made by Gaiotto and Witten [**38**] and pursued in [**39, 40**]. It was meant to be a finite dimensional effective theory for Witten's five dimensional gauge theory [**101**]. The theory in [**101**] can also be interpreted as an A-model, except coming from a Landau-Ginsburg model with an infinite dimensional target space [**39**].

5.2. The category of A-branes on $Y_{\mathfrak{su}_2}$. The categories of A-branes on $Y_{\mathfrak{su}_2}$ and on $Y_{\mathfrak{gl}_{1|1}}$ closely parallel each other. In particular, for both theories, objects of $\mathscr{D}_Y$ are product Lagrangians of the form (4.7), which are graded by Maslov and equivariant gradings, where the grading is a choice of a lift of the phase of Ω and of e^{-W} to real valued functions on L. One novelty of the present case is that immersed Lagrangians play a key role.

5.2.1. As in the $\mathfrak{gl}_{1|1}$ theory, the branes are allowed to end at punctures of $\mathcal{A}$, due to the fact potential has logarithmic poles there. An example is a cap brane $I_{\mathcal{U}}$ on Y, which is a product of d non-intersecting intervals,

$$I_{\mathcal{U}} = I_1 \times \ldots \times I_d,$$

where I_j begins at $y = a_{2j-1}$ and ends at $y = a_{2j}$.

The cup brane $E_{\mathcal{U}}$ on Y,

$$E_{\mathcal{U}} = E_1 \times \ldots \times E_d,$$

is a product obtained by replacing each I_j brane by an immersed, figure eight Lagrangian, encircling the corresponding punctures. The brane has well-defined Maslov grading, since the phase of dy_j comes back to itself around a figure eight. It also has a well defined equivariant grading, as it winds around the two punctures in opposite directions.

5.2.2. For a holomorphic map to contribute to A-model amplitudes, as discussed in section 2.1.7, its image in Y should have finite area, computed using any symplectic form compatible with Ω, and moreover W should pull back to a regular function on the disk. As always, the latter condition guarantees that the equivariant degree of the map is zero, but in the $\mathfrak{su}_2$ theory, it does not preclude maps from passing through punctures on $\mathcal{A}$.

As shown in [**2**], for W to pull back to a regular function on the disk D under the map $y : \mathrm{D} \to Y$, every order one branch point of the projection $S \to \mathrm{D}$ must map to a puncture on $\mathcal{A}$ with multiplicity 1, and every node of the projection to D must map to a puncture with multiplicity 2. Here, S is a Riemann surface in $\mathrm{D} \times \mathcal{A}$, which the map $y : \mathrm{D} \to Y$ corresponds to in the cylindrical formulation [**63**] of Floer theory, as in section 4.2.6.

5.2.3. As in Heegaard-Floer theory, there is a simpler cousin of Y, the manifold

$$Y_O = Y \backslash D_O, \tag{5.6}$$

obtained from Y by removing the divisor

$$D_O = \Delta + D_a. \tag{5.7}$$

Along the diagonal Δ, at least two out of d points coincide on $\mathcal{A}$; along D_a, at least one of them coincides with a finite puncture. The theory on Y_O is simple to solve [**2**]; in particular, removing Δ eliminates all but the maps without branching in $S \to \mathrm{D}$, as the branching index coincides the intersection of the image with the diagonal Δ.

The theory on $Y_{\mathfrak{su}_2}$ is obtained from the theory on Y_O by carefully understanding the effect of filling in D_O. Direct A-model analysis appears in [**5**]. Equivalently, we can use homological mirror symmetry as a guide, as [**5**] proves it.

5.2.4. In the theory on $Y = Y_{\mathfrak{su}_2}$, the Maslov index of a map $y : \mathrm{D} \to Y$, where D is a disk with two marked points on the boundary, is given by [**2**]

$$ind(y) = 2e(A), \tag{5.8}$$

where $e(A)$ is the Euler measure defined in (4.13). The index of the $\mathfrak{gl}_{1|1}$ theory in (4.12) has two additional contributions, corresponding to the two additional contributions to the top holomorphic form of that theory in (4.1) relative to (5.2). (The indices of $\ell+1$ pointed disks for other values of ℓ are related similarly as well.)

The c^0-equivariant degree (corresponding to the $\mathfrak{q}$ grading) is given by

$$J_0(y) = i(A) - 2n(A). \tag{5.9}$$

Above, $n(A)$ is the multiplicity of A at all the finite a-punctures, and $i(A)$ is the intersection of the map with the diagonal, defined in (4.14). The factor $i(A)$ in the equivariant degree originates from the factor of $\Delta(y) = \prod_{\alpha<\beta}(y_\alpha - y_\beta)$ in the potential W^0, through (5.5). The c^1 equivariant degree is the same as in (4.17).

5.2.5. The fact that the Euler characteristic of $Hom^{*,*}_{\mathscr{D}_Y}(\mathscr{B}E_{\mathcal{U}}, I_{\mathcal{U}})$, which is the graded count of intersection points of the $\mathscr{B}E_{\mathcal{U}}$ and $I_{\mathcal{U}}$ branes in (2.23), reproduces the Jones polynomial of the link

$$\chi(\mathscr{B}E_{\mathcal{U}}, I_{\mathcal{U}}) = \sum_{\mathcal{P}} (-1)^{M(\mathcal{P})} \mathfrak{q}^{J(\mathcal{P})} = J_K(\mathfrak{q}).$$

is a theorem of Bigelow [**18**], building on the work of Lawrence [**60–62**]. In section 5.8, we will give a different, direct proof of this. Our theory explains the homological origin of Bigelow's peculiar construction and generalizes it to other $U_{\mathfrak{q}}({}^L\mathfrak{g})$ link invariants. A relation of Bigelow's representation of the Jones polynomial to symplectic Khovanov homology of [**92**] was first observed in [**67**].

5.3. Equivariant mirror symmetry. Mirror symmetry relating Y to its mirror X fits into a larger framework of mirror symmetries. The theory on Y, and its mirror X, are defined by the rectangle

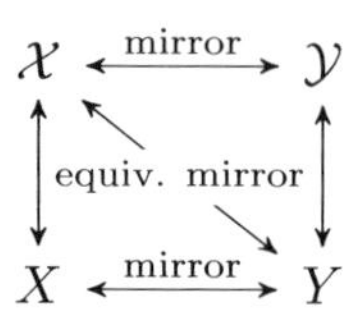

Here, $\mathcal{Y}$ is a holomorphic symplectic manifold which fibers over Y with holomorphic Lagrangian $(\mathbb{C}^\times)^d$ fibers. Its mirror $\mathcal{X}$ is also holomorphic symplectic, and X embeds into it as a holomorphic Lagrangian called the "core". We will call Y the equivariant mirror of $\mathcal{X}$. The $\mathfrak{gl}_{1|1}$ theory is exceptional in this regard, in that there is no "upstairs".

The category of branes on $\mathcal{Y}$ is $\mathscr{D}_\mathcal{Y}$, the derived Fukaya-Seidel category of $\mathcal{Y}$ with potential $\mathcal{W}$, described in **[2, 5]**. $\mathscr{D}_\mathcal{X}$ is the derived category of $\mathbb{T}$-equivariant coherent sheaves on $\mathcal{X}$ from **[1]**.

The equivariant mirror symmetry framework, which provides two descriptions of the theory, one upstairs and one downstairs, is key for proper understanding of many features of the theory, starting with the asymmetry between cup and cap branes. It is also crucial for proper understanding of how the divisor D_O gets filled in.

5.3.1. In our current example, the upstairs $\mathcal{Y} = \mathcal{Y}_{\mathfrak{su}_2}$ is **[2, 5]** an open subset of the Hilbert scheme of d points on (a deformation of) the multiplicative A_{2d-1} surface singularity,

$$\mathcal{C}: \qquad u^+u^- = \prod_{i=1}^{2d}(1 - y/a_i), \tag{5.10}$$

which is a $\mathbb{C}^\times$ fibration over $\mathcal{A}$. The fibration degenerates at $2d$ points where the punctures are.

A Hilbert scheme of points on the surface is a blow-up of the locus in the symmetric product of d copies of the surface $\mathcal{C}$ in (5.10) where a pair of points comes together. Explicitly, take $(y_\alpha, u_\alpha^+, u_\alpha^-)$ for α running from 1 to d, modulo permutations, to be coordinates on the symmetric product of d copies of the surface in (5.10). The Hilbert scheme is a resolution of it, by remembering the direction $\frac{u_\alpha^+ - u_\beta^+}{y_\alpha - y_\beta} \in \mathbb{P}^1$, in which pairs of points come together, for any two $\alpha \neq \beta$ (and similarly with $+$ and $-$ exchanged, modulo obvious relations).

To get $\mathcal{Y}$, one starts with the "horizontal Hilbert scheme" of d points on the surface $\mathcal{C}$ in (5.10), which is an open subset of the Hilbert scheme, where one deletes the locus where a pair of points comes together on $\mathcal{A} = \mathbb{C}_y^\times$, without coming together on the fiber. This corresponds to asking that $\frac{u_\alpha^- - u_\beta^-}{y_\alpha - y_\beta} \neq \infty$, and similarly with $+$ and $-$ exchanged. This space is simply the moduli space of d $SU(2)/\mathbb{Z}_2$ monopoles on $\mathbb{R}^2 \times S^1$, in the presence of $2d$ singular ones, or a Coulomb branch of a four-dimensional $\mathcal{N} = 2$ $U(d)$ gauge theory with $2d$ hypermultiplets.

5.3.2. The upstairs $\mathcal{Y}$ is the horizontal Hilbert scheme of d points on the surface $\mathcal{C}$, equipped with the potential

$$\mathcal{W} = \lambda_0 \mathcal{W}_0 + \lambda_1 \mathcal{W}_1 + \mathcal{W}_2 \tag{5.11}$$

where

$$\mathcal{W}_0 = \sum_{\alpha=1}^{d} \log u_\alpha^+, \qquad \mathcal{W}_1 = \sum_{\alpha=1}^{d} \log y_\alpha, \qquad \mathcal{W}_2 = \sum_{\alpha=1}^{d} \frac{u_\alpha^-}{\prod_{\beta\neq\alpha}(1 - y_\beta/y_\alpha)}.$$

The $\mathcal{W}_0$ and $\mathcal{W}_1$ terms are mirror to equivariant action on $\mathcal{X}$. The $\mathcal{W}_2$ term is a holomorphic function on the horizontal Hilbert scheme. For $\mathcal{W}_0$ to be defined on $\mathcal{Y}$, we must further remove from it the points where $u_\alpha^+ = 0$. So, $\mathcal{Y}$ is simply the horizontal Hilbert scheme of d points on $\mathbb{C}^\times_{u+} \times \mathbb{C}^\times_y$, or

$$\mathcal{Y} = \mathrm{Hilb}^d(\mathbb{C}^\times_{u+} \times \mathbb{C}^\times_y \longrightarrow \mathbb{C}^\times_y).$$

5.3.3. The upstairs $\mathcal{Y}$ fibers over the downstairs Y, with generic fiber

$$Y_F = (\mathbb{C}^\times_u)^d.$$

It is helpful to introduce new local coordinates $u_1, \ldots, u_d \in \mathbb{C}^\times$, related to the old global ones by

$$u_\alpha = u_\alpha^+ \frac{\prod_{\beta\neq\alpha}(1 - y_\beta/y_\alpha)}{\prod_{i=1}^{2d}(1 - y_\alpha/a_i)}. \tag{5.12}$$

In terms of u's and y's, the potential $\mathcal{W}$ on $\mathcal{Y}$ splits as

$$\mathcal{W} = W + W_F. \tag{5.13}$$

The first term is the downstairs potential W on Y, from (5.4). The remainder is the potential on the fiber Y_F over Y:

$$W_F = \sum_{\alpha=1}^{d}(\lambda_0 \log u_\alpha + u_\alpha^{-1}). \tag{5.14}$$

The category $\mathscr{D}_{Y_F}$ of A-branes on Y_F with potential W_F is familiar. It is mirror to the category of B-branes on X_F, which is a product of d copies of $\mathbb{C}_z$, each with a $\mathbb{C}^\times_{\lambda_0}$ action on it.

This has an important consequence. It means that the upstairs category of A-branes $\mathscr{D}_{\mathcal{Y}}$, has a description purely from the downstairs perspective, as the Fukaya category of Y "with coefficients in $\mathscr{D}_{Y_F}$". We will make explicit below what this means.

5.3.4. The mirror of $\mathcal{Y}$ with potential $\mathcal{W}$ is

$$\mathcal{X} = \mathrm{Hilb}^d(\mathcal{C}^\vee \longrightarrow \mathbb{C}_z),$$

together with an equivariant T-action. Here, $\mathcal{C}^\vee$ is the resolution of the ordinary A_{2d-1} surface singularity

$$\mathcal{C}^\vee : \qquad x_+x_- = z^{2d}.$$

The equivariant T-action on $\mathcal{X}$ is a product $\mathbb{C}^\times_{\lambda_0} \times \mathbb{C}^\times_{\lambda_1}$, where the first factor comes from scaling z with weight $\mathfrak{q} = e^{2\pi i \lambda_0}$, and the second scales x_+ and x_- oppositely, preserving the holomorphic symplectic form on $\mathcal{C}^\vee$ and hence on $\mathcal{X}$. These two actions mirror the $\mathcal{W}_0$ and $\mathcal{W}_1$ terms in the potential, respectively.

The core X, the mirror of Y, is the fiber of $\mathcal{X}$ over the point in the base $Hilb^d(\mathbb{C}_z)$, where d points come together at the origin of the z plane.

5.3.5. The upstairs $\mathcal{X}$ is the moduli space of d $SU(2)/\mathbb{Z}_2$ monopoles on $\mathbb{R}^3$, in the presence of $2d$ singular ones, or a Coulomb branch of a three-dimensional $\mathcal{N} = 2$ $U(d)$ gauge theory with $2d$ hypermultiplets. It is based on $\mathcal{C}^\vee$, which is the ordinary A_{2d-1} surface singularity and not the multiplicative one, due to the $\mathcal{W}_2$ term in the potential on $\mathcal{Y}$. This is an example of the fact that filling in a divisor in $\mathcal{X}$ is mirror to adding a term in the potential on $\mathcal{Y}$. This interplay between removing/filling in the divisors on one side, and changing the superpotential is first described in [**45**]. It is further developed in [**16**] and many other works.

5.3.6. The upstairs and the downstairs categories $\mathscr{D}_{\mathcal{X}}$ and $\mathscr{D}_Y$ are related by a pair of adjoint functors

$$\begin{array}{ccc} \mathscr{D}_{\mathcal{X}} & & \\ & \underset{h^*}{\overset{h_*}{\rightleftarrows}} & \\ & & \mathscr{D}_Y \end{array}$$

which come from composing the downstairs mirror symmetry with a pair of functors $f_* : \mathscr{D}_X \to \mathscr{D}_{\mathcal{X}}$, which interprets a sheaf "downstairs" on X as a sheaf "upstairs" on $\mathcal{X}$, and $f^* : \mathscr{D}_{\mathcal{X}} \to \mathscr{D}_X$ which corresponds to tensoring with the structure sheaf $\otimes\mathcal{O}_X$, and restricting.

Alternatively, h^* and h_* come by composing the upstairs mirror symmetry with a pair of functors k^* and k_*

$$\begin{array}{c} \mathscr{D}_{\mathcal{Y}} \\ k^* \big\downarrow \big\uparrow k_* \\ \mathscr{D}_Y \end{array}$$

which come from a Lagrangian correspondence on $\mathcal{Y} \times Y_-$, where Y_- is Y with sign of symplectic form reversed [**7**]. The functors are intersections with the Lagrangian $\mathcal{K}$ on $\mathcal{Y} \times Y_-$. The Lagrangian $\mathcal{K}$ is a product of Y with a real d–dimensional fiber over it, constructed in [**5**].

5.3.7. Upstairs, a cup and a cap are the same object. From perspective of $\mathcal{X}$, the brane $\mathcal{U} \in \mathscr{D}_{\mathcal{X}}$ is the structure sheaf $\mathcal{U} = \mathcal{O}_U$ of vanishing cycle. Continuing with the $\mathfrak{su}_2$ example, the vanishing cycle is a product of d non-intersecting $\mathbb{P}^1$'s in $\mathcal{C}^\vee$, $U = (\mathbb{P}^1)^d$. In $\mathscr{D}_{\mathcal{X}}$ the homological link invariants are simply

$$Hom^{*,*}_{\mathscr{D}_{\mathcal{X}}}(\mathscr{B}\mathcal{U}, \mathcal{U}),$$

where $\mathscr{B}$ is the action of the braiding functor. Equivariant mirror symmetry relates $\mathscr{D}_{\mathcal{X}}$ and $\mathscr{D}_Y$ and dictates what the cup and cap branes are in $\mathscr{D}_Y$.

The $\mathcal{U} \in \mathscr{D}_{\mathcal{X}}$ brane comes from the cap branes $I_{\mathcal{U}}$ on Y, as the image of the $h_* : \mathscr{D}_Y \to \mathscr{D}_{\mathcal{X}}$ functor,

$$h_* : \quad I_{\mathcal{U}} \in \mathscr{D}_Y \longrightarrow \mathcal{U} = h_* I_{\mathcal{U}} \in \mathscr{D}_{\mathcal{X}}.$$

The functor h^* that sends the upstairs $\mathcal{U}$-brane from $\mathcal{X}$ back down to Y gives the $E_{\mathcal{U}}$ branes, as well as their images under the braiding functor:

$$h^* : \quad \mathscr{B}\mathcal{U} \in \mathscr{D}_{\mathcal{X}} \longrightarrow \mathscr{B}E_{\mathcal{U}} = h^*\mathscr{B}\mathcal{U} \in \mathscr{D}_Y.$$

An explicit computation of this may be found in [**2**]. The adjointness of the functors h^* and h_* implies that the Homs between the branes upstairs are equal to the Homs between the branes downstairs

$$Hom^{*,*}_{\mathscr{D}_Y}(\mathscr{B}E_{\mathcal{U}}, I_{\mathcal{U}}) = Hom^{*,*}_{\mathscr{D}_{\mathcal{X}}}(\mathscr{B}\mathcal{U}, \mathcal{U}).$$

The proof that the right hand side gives homological link invariants (for any simply-laced Lie algebra ${}^L\mathfrak{g}$) is in [**1**], and homological mirror symmetry proven in [**5**] implies the same is true of the left hand side. We will give a second proof that the left hand side is a link invariant later in this section.

5.3.8. In general, the functor h^*, or equivalently $k^* : \mathscr{D}_{\mathcal{Y}} \to \mathscr{D}_Y$, equips the branes on Y that come from $\mathcal{Y}$ as the images of this functor with an extra structure of a local system. The familiar case corresponds equipping Lagrangian branes with a unitary bundle of rank one or higher. Here, we find a generalization of this [**5**].

The functor h^* equips the branes on $\mathscr{D}_Y$ that come from $\mathscr{D}_{\mathcal{Y}}$ with a local system of, possibly trivial, modules for a graded algebra $\mathcal{B}$. Examples of branes that come from upstairs as images of h^* are E-branes and T-branes. By contrast, the I-branes in $\mathscr{D}_Y$ do not arise in this way.

In our case, the algebra $\mathcal{B}$ turns out to be the quotient of the algebra of polynomials in d variables $z_1, \ldots, z_d$, by an ideal $\mathcal{I}$ which sets their symmetric functions to zero. The z's have equivariant $\mathfrak{q}$-degree equal to one. We will next explain where this structure comes from, following [**5**].

5.4. $\mathfrak{su}_2$ T-branes and their algebra. The branes generating $\mathscr{D}_Y$ are products of d real lines on a cylinder,

$$T_{\mathcal{C}} = T_{i_1} \times \cdots \times T_{i_d} \in \mathscr{D}_Y. \tag{5.15}$$

colored by the simple root of $\mathfrak{su}_2$. We get one such brane for each inequivalent ordering of T-branes, where passing a component across a puncture changes the brane. Unlike in Heegaard-Floer theory, there is only one kind of a puncture, colored by the fundamental representation. The branes with more than one T_i brane between a pair of punctures are associated to critical points of W at infinity; unlike in the $\mathfrak{gl}_{1|1}$ case, all these branes are non-zero objects of the category. The generator of $\mathscr{D}_Y$ is the direct sum of $T_{\mathcal{C}}$'s,

$$T = \bigoplus_{\mathcal{C}} T_{\mathcal{C}} \in \mathscr{D}_Y,$$

over all possible orderings. Moreover, as an example of the structure in section 5.3.8, the T-branes get equipped with a local system of modules for the algebra $\mathcal{B}$, equal to $\mathcal{B}$ itself. To explain where this extra structure comes from, we will in parallel consider the generators of $\mathscr{D}_{\mathcal{Y}}$, the upstairs category of A-branes.

5.4.1. The upstairs category $\mathscr{D}_{\mathcal{Y}}$ also has a single generator:

$$\mathcal{T} = \bigoplus_{\mathcal{C}} \mathcal{T}_{\mathcal{C}}$$

which is a direct sum of

$$\mathcal{T}_{\mathcal{C}} = \mathcal{T}_{i_1} \times \cdots \times \mathcal{T}_{i_d} \in \mathscr{D}_{\mathcal{Y}}.$$

Each upstairs $\mathcal{T}_{\mathcal{C}}$ brane fibers over the downstairs $T_{\mathcal{C}} \in \mathscr{D}_{\mathcal{Y}}$ brane with fibers which are a product of d real line Lagrangians, one for each copy of $\mathbb{C}^\times$.

The real line Lagrangians in the fiber $Y_F = (\mathbb{C}^\times_u)^d$ with potential W_F generate the fiber Fukaya category $\mathscr{D}_{Y_F}$. The fiber category is in fact a product of d one dimensional categories, one for each copy of $\mathbb{C}^\times_u$.

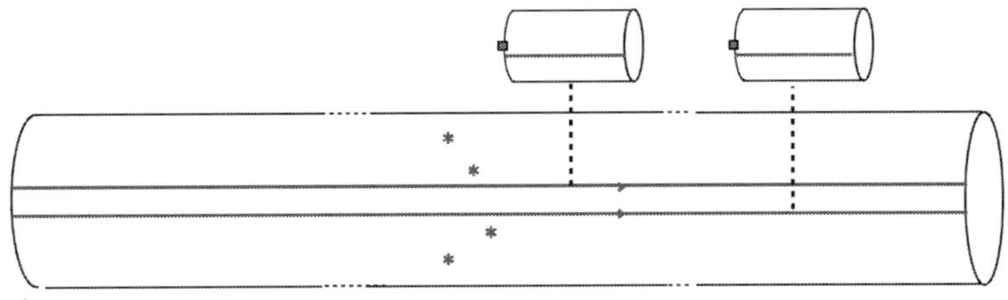

FIGURE 35. An upstairs $\mathcal{T}_{\mathcal{C}}$ brane pairs every point on the downstairs $T_{\mathcal{C}}$ brane in Y (which is a d-tuple of points on $\mathcal{A}$) with a product of d real-line Lagrangians, in the fiber Y_F.

5.4.2. The endomorphism algebra of the upstairs generators

$$\mathscr{A} = Hom^*_{\mathscr{D}_{\mathcal{Y}}}(\mathcal{T}, \mathcal{T}) \tag{5.16}$$

is generated by intersection points of $\mathcal{T}_{\mathcal{C}}$ branes in $\mathcal{Y}$, defined by the wrapped Fukaya category. The kind of wrapping one needs, in the base and in the fiber, is determined by the potential W on Y and W_F on Y_F, which add up to the potential $\mathcal{W}$ on the total space $\mathcal{Y}$, per (5.13).

Since the upstairs branes fiber over the downstairs, their intersection points are obtained by choosing an intersection point of the downstairs generators,

$$\mathcal{P} \in T^{\zeta}_{\mathcal{C}} \cap T_{\mathcal{C}'}, \tag{5.17}$$

and having chosen $\mathcal{P}$, picking an intersection point of the branes in the fiber over that point, as in Fig. 36.

5.4.3. The A-model on the fiber Y_F with potential W_F from (5.14), is very well known. The corresponding category of A-branes $\mathscr{D}_{Y_F}$ is generated by a single brane T_F, which is the product of d real line Lagrangians, each on a copy of $\mathbb{C}^{\times}_{u}$.

As we recalled in section 5.3.3, mirror to Y_F with potential W_F is $X_F = \mathbb{C}^d$, with the equivariant action that scales each copy of $\mathbb{C}$ by $\mathfrak{q}$. Mirror symmetry relates the generator T_F of $\mathscr{D}_{Y_F}$ to the generator of $\mathscr{D}_{X_F}$, which we will also call T_F, which is the trivial line bundle on X_F.

5.4.4. The endomorphism algebra of the real-line Lagrangian $T_F \in \mathscr{D}_{Y_F}$, defined by the wrapped Fukaya category, is shown in [**13**] to be $\mathcal{B}_{\infty} = \mathbb{C}[z_1, \ldots, z_d]$ the algebra of polynomials in d variables, $z_1, \ldots, z_d$, with a single grading coming from the equivariant action $z_{\alpha} \to \mathfrak{q} z_{\alpha}$. The endomorphism algebra of the line bundle T_F generating $\mathscr{D}_{X_F}$ is the algebra of homolorphic functions on X_F, with grading coming from the $\mathbb{C}^{\times}_{\mathfrak{q}}$ action, in other words the $\mathcal{B}_{\infty}$ algebra. The fact the two algebras

$$Hom^*_{\mathscr{D}_{X_F}}(T_F, T_F) = \mathcal{B}_{\infty} \cong Hom^*_{\mathscr{D}_{Y_F}}(T_F, T_F), \tag{5.18}$$

are the same leads to homological mirror symmetry:

$$\mathscr{D}_{X_F} \cong \mathscr{D}_{\mathcal{B}_{\infty}} \cong \mathscr{D}_{Y_F}, \tag{5.19}$$

Any brane in $\mathscr{D}_{X_F}$ or in $\mathscr{D}_{Y_F}$, gets mapped, by the Yoneda functor $Hom^*(T_F, -)$ as in (2.13), to a complex of modules of $\mathcal{B}_{\infty}$. Here T_F is viewed as the brane generating the corresponding category. This identifies as mirror the branes that give rise to the same object of $\mathscr{D}_{\mathcal{B}_\infty}$, the derived category of $\mathcal{B}_\infty$ modules. In particular, the Yoneda functors map the T_F branes to the module for $\mathcal{B}_\infty$ equal to itself.

5.4.5. Let $\mathcal{L}$ be any Lagrangian on $\mathcal{Y}$. Since $\mathcal{Y}$ fibers over Y with fibers Y_F, $\mathcal{L}$ is a fibration over a Lagrangian L in the base. Let $\Lambda_{\mathcal{P}}$ be the Lagrangian which is the fiber of $\mathcal{L}$ over a point $\mathcal{P} \in L$. The Yoneda functor maps it to a module $M_{\mathcal{P}}$ for $\mathcal{B}_\infty$,

$$Hom^*_{\mathscr{D}_{Y_F}}(T_F, \Lambda_{\mathcal{P}}) = M_{\mathcal{P}}.$$

In this way, we identify every Lagrangian $\mathcal{L} \in \mathscr{D}_{\mathcal{Y}}$ upstairs on $\mathcal{Y}$ with a Lagrangian $L \in \mathscr{D}_Y$ downstairs, equipped with a local system M of modules for $\mathcal{B}_\infty$,

$$\mathcal{L} \cong (L, M). \tag{5.20}$$

The local system is a $\mathcal{B}$-module $M_{\mathcal{P}}$ over each point $\mathcal{P} \in L$ on the Lagrangian $L \in Y$, together with a flat connection that identifies the fibers over different points.

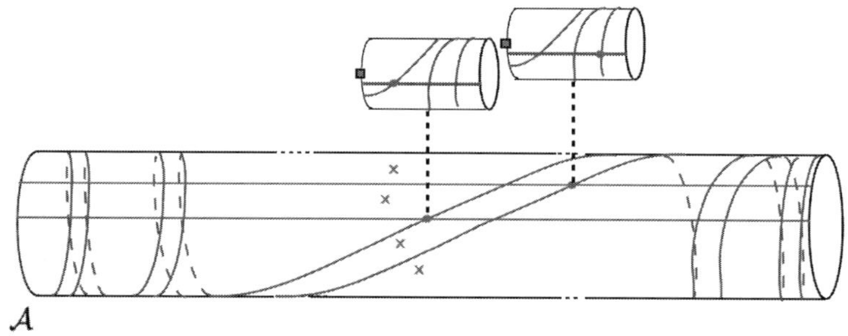

FIGURE 36. An intersection point of the downstairs $T_{\mathcal{C}}$-branes, together with the choice of the intersection in the fiber over it.

The Floer complex of $\mathcal{L}$ and $\mathcal{L}' \in \mathscr{D}_{\mathcal{Y}}$ gets a description as a complex generated by intersection points $\mathcal{P} \in L \cap L'$ each tensored by the vector space $hom_{\mathscr{D}_{\mathcal{B}_\infty}}(M_{\mathcal{P}}, M'_{\mathcal{P}})$

$$hom^{*,*}_{\mathcal{Y}}(\mathcal{L}, \mathcal{L}') = hom^{*,*}((L, M), (L', M')) = \bigoplus_{\mathcal{P} \in L \cap L'} hom_{\mathcal{B}_\infty}(M_{\mathcal{P}}, M'_{\mathcal{P}}). \tag{5.21}$$

of homomorphisms of the corresponding $\mathcal{B}_\infty$-modules, restricted to $\mathcal{P}$.

5.4.6. As a special case, every upstairs $\mathcal{T}_{\mathcal{C}}$ brane is the downstairs $T_{\mathcal{C}}$ brane, equipped with a local system with $M = \mathcal{B}_\infty$,

$$\mathcal{T}_{\mathcal{C}} \cong (T_{\mathcal{C}}, \mathcal{B}_\infty). \tag{5.22}$$

the latter viewed as a module over itself, using $\mathcal{B}_\infty = hom_{\mathcal{B}_\infty}(\mathcal{B}_\infty, \mathcal{B}_\infty)$.

Furthermore, the Floer complex corresponding to the upstairs $\mathcal{T}_{\mathcal{C}}$ and $\mathcal{T}_{\mathcal{C}'}$ branes is generated by intersection points $\mathcal{P} \in T^{\zeta}_{\mathcal{C}} \cap T_{\mathcal{C}'}$ of the downstairs T-branes, tensored with the vector space $\mathcal{B}_\infty$,

$$Hom^*_{\mathscr{D}_{\mathcal{Y}}}(\mathcal{T}_{\mathcal{C}}, \mathcal{T}_{\mathcal{C}'}) = \bigoplus_{\mathcal{P} \in T^{\zeta}_{\mathcal{C}} \cap T_{\mathcal{C}'}} \mathcal{B}_\infty \mathcal{P} \tag{5.23}$$

We equated the Floer complex $hom^{*,*}_{\mathcal{Y}}(\mathcal{T}_{\mathcal{C}}, \mathcal{T}_{\mathcal{C}'})$ with $Hom^*_{\mathscr{D}_{\mathcal{Y}}}(\mathcal{T}_{\mathcal{C}}, \mathcal{T}_{\mathcal{C}'})$, as the space is concentrated in cohomological degree zero with necessarily trivial differential, as we will see momentarily, so the complex and its cohomology coincide.

5.4.7. This translates to a graphical representation of the algebra $\mathscr{A}$. By the same argument as for the $\mathfrak{gl}_{1|1}$ theory, the intersection points $\mathcal{P}$ of the downstairs $T_{\mathcal{C}}$-branes in (5.17) translate to diagrams on the cylinder. The choice of an intersection point in the fiber over $\mathcal{P}$, which corresponds to an element of $\mathcal{B}_\infty$ per section 5.4.3, can be represented graphically as well. To represent an element $z_1^{k_1}\cdots z_d^{k_d}$ of $\mathcal{B}_\infty$, post-compose the first blue string in the diagram with k_1 dots, and so on, the last with k_d. For example, the intersection point in Fig. 36 corresponds to the string diagram:

FIGURE 37. The intersection of the downstairs T-branes determines the string diagram; the intersection of the $\mathcal{T}$-branes in the fiber determines the dots the string diagram gets post-composed with. This string diagram corresponds to the intersection point in Fig. 36; the intersection in the fiber is $z_1^0 z_2^2 \in \mathcal{B}_\infty$.

5.4.8. The relative equivariant gradings of intersection points $\mathcal{P}$ and $\mathcal{P}'$ in $T_{\mathcal{C}}^{\zeta} \cap T_{\mathcal{C}'}$ can be computed [**2**] from a domain A on the Riemann surface that interpolates between them. The gradings are a sum of gradings of string bits, which the string diagram decomposes into. This is as in Heegaard-Floer theory; except the gradings differ.

FIGURE 38. The Maslov degree of every element of $hom(T_{\mathcal{C}}, T_{\mathcal{C}'})$ vanishes. The J^0 equivariant degree is determined locally as the sum of the degrees of the string bits in the figure. The J^1 equivariant degree is the winding number of strings in the string diagram, around the cylinder.

The Maslov degrees of all the morphisms of $\mathcal{T}$-branes vanish [**5**]. This follows from the fact the holomorphic volume form on $\mathcal{Y}$ is the Cartesian product of holomorphic forms on d copies of $\mathbb{C}^\times \times \mathbb{C}^\times$, with branes that are products of real line Lagrangians in the two copies of $\mathbb{C}^\times$.

5.4.9. The algebra $\mathscr{A}$ has a product coming from the product on Floer cohomology,

$$m_2 : Hom^*_{\mathscr{D}_\mathcal{Y}}(\mathcal{T}_{\mathcal{C}'}, \mathcal{T}_{\mathcal{C}''}) \cdot Hom^*_{\mathscr{D}_\mathcal{Y}}(\mathcal{T}_{\mathcal{C}}, \mathcal{T}_{\mathcal{C}'}) \to Hom^*_{\mathscr{D}_\mathcal{Y}}(\mathcal{T}_{\mathcal{C}}, \mathcal{T}_{\mathcal{C}''}), \tag{5.24}$$

defined using the wrapped Fukaya category of $\mathcal{Y}$ with potential $\mathcal{W}$. As in the $\mathfrak{gl}_{1|1}$ case, the product corresponds to stacking cylinders and rescaling.

Because all elements of $\mathscr{A}$ are in degree zero, only the product in (5.24) needs computing as all other A_∞ morphisms vanish. For m_ℓ products to not vanish the difference between Maslov indices of outgoing intersection point and the ℓ incoming ones must be $2-\ell$, and all algebra elements have degree zero.

5.4.10. The associative product on the algebra $\mathscr{A}$ is computed in [**5**]. The algebra product translates into local relations of strands elements, given in #1 to #7 below.

1. 2. 3.

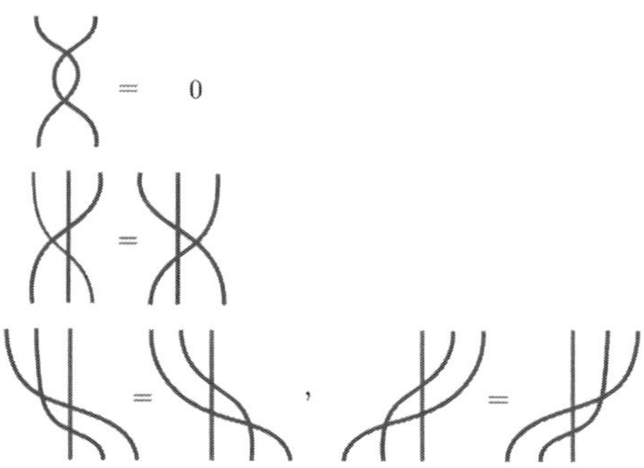

To compute the algebra $\mathscr{A}$, [**5**] first consider the theory in the complement of the divisor D_O, which is the union of the diagonal divisor Δ and the divisor of all the punctures $D_a = \sum_i D_{a_i}$. More precisely, in the context of $\mathcal{Y}$, the divisor D_O is really the pullback of the divisor D_O on Y from (5.7) by the map that projects $\mathcal{Y}$ to Y.

Deleting the D_O from $\mathcal{Y}$ does not affect the generators of the Floer complex, since the $\mathcal{T}$-branes are disjoint from it. As a result, the algebras as vector spaces do not change, only the relations between their elements may. In the complement of D_O, the theory is extremely simple. The only maps that contribute to the algebra come from counts of one-dimensional triangles, which are easy to evaluate. The result is the algebra

$$\mathscr{A}_O = \mathscr{A}_{u=0,\hbar=0} \tag{5.25}$$

given by relations #1 − #7, with both u and $\hbar$ set to zero.

Filling in D_a corresponds to deforming the algebra to $u \neq 0$. The deformation comes from including disks in $\mathcal{Y}$ that project to $\mathcal{A}$ as products of one-dimensional triangles passing through the punctures. For the superpotential $\mathcal{W}$ to remain regular on such a disk, which is the same as saying the disk does not run off to infinity on $\mathcal{Y}$, the point on the domain that maps to a puncture must also map to $u=\infty$ in the $\mathbb{C}^\times$ fiber over it, in such a way that u_+ remains finite by (5.12). The result was shown in [**5**] to be the deformation of the product in #5.

4.

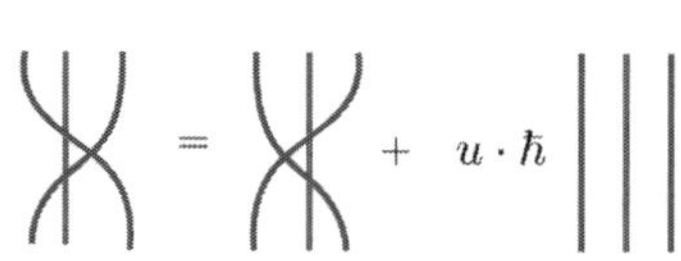

Filling in the diagonal Δ corresponds to deforming to $\hbar \neq 0$. On equivariant degree grounds, the only possible correction is add a single term to relations #4, #6 and #7, with coefficient to be computed. The corrections to those relations come from counting disks in the symmetric product of two copies of $\mathscr{A}$. In fact, it suffices to count a single such disk, as knowing the correction to either of the three relations fixes the other two by associativity of the algebra. Finding the $\hbar$ deformation of the algebra thus amounts to a single, but difficult calculation, performed in [**5**].

5. 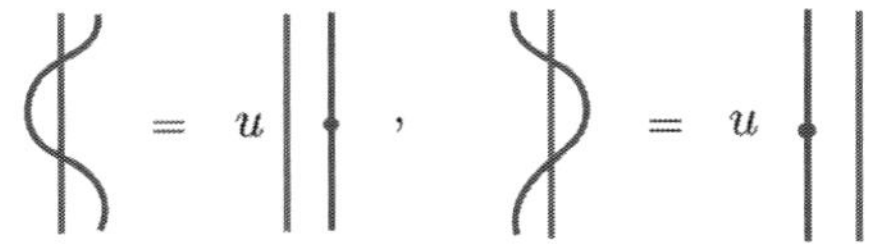

Both parameters u and $\hbar$ may be set to 1 by rescaling the algebra generators; if we keep them generic, they are useful as book-keeping parameters.

6. 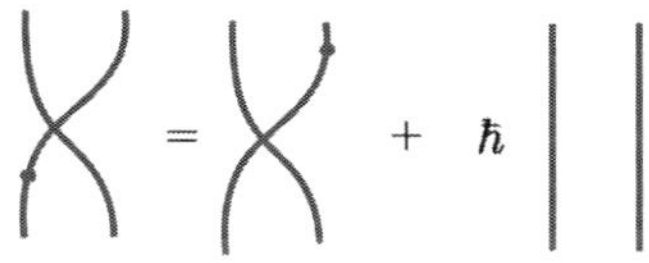

7. 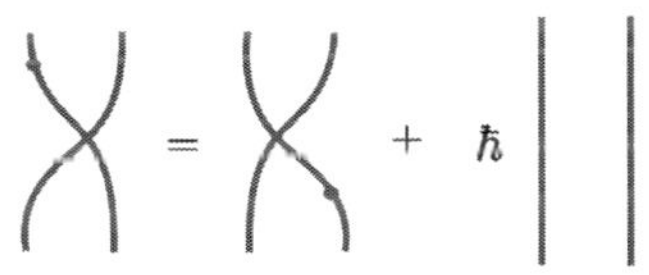

5.4.11. The algebra relations should be interpreting as stating that the difference between the products of pairs of generators on the left and the right hand sides of relation #4, for example, is not zero, thanks to the contribution of a map D to $\mathcal{Y}$ which intersects the diagonal Δ and the divisor of the punctures D_a once, and which computes the coefficient of the corresponding identity morphism. To show that relations #6 and #7 follow from relations #4 and #5 by associativity of the algebra, pre- or post-compose relation #4 with an element that, in one of the compositions, creates a crossing of the kind appearing on the left hand side of #5.

Viewing the $\mathcal{T}$-brane as the downstairs T-brane equipped with the local system $\mathcal{B}_\infty$, the relation #5 says that the product of the two intersection points in Y on the left hand side of the relation does not vanish (as it would without the bundle), but rather defines a non-trivial endomorphism of the bundle over the brane.

5.4.12. The relations #1, #2, #6 and #7 are the basic relations of the $\mathfrak{su}_2$ KLR algebra from [**52, 87**]. The resulting upstairs algebra $\mathscr{A} = Hom^*_{\mathscr{D}_\mathcal{Y}}(\mathcal{T}, \mathcal{T})$ is the cylindrical version of the KLRW algebra of [**52, 87**], which [**95**] generalized to adding red strands. The KLRW algebras are originally formulated as algebras of strands on a plane, not a cylinder, but the local relations are the same.

This shows that the string diagrams, in terms of which the algebras of [**52, 53**] and [**95**] are phrased, are nothing but Reeb chords associated to T-branes that generate the wrapped Fukaya categories of $\mathcal{Y}$.

5.4.13. The $\mathcal{T}$-brane, which generates $\mathscr{D}_\mathcal{Y}$, is related by mirror symmetry to a very special generator of $\mathscr{D}_\mathcal{X}$, which we will also label $\mathcal{T}$ and which is known as the tilting vector bundle. The tilting property of the generator $\mathcal{T}$ is the statement that its endomorphism algebra is vanishes outside of cohomological degree zero, as it does in $\mathscr{A}$.

The construction of tilting bundle $\mathcal{T}$ on a holomorphic symplectic $\mathcal{X}$ is due to Bezrukavnikov and Kaledin [**20**]. It uses methods of characteristic p. The "exotic

t-structure" which enters the construction of [**21**] corresponds to choosing the positions of the punctures on $\mathcal{A}$ along a fixed S^1, as in our setup from section 2.1.1. The fact that for this choice of t-structure, the $\mathcal{T}$ brane is a good generator of the category is manifest on the mirror $\mathcal{Y}$.

In [**96,97**], Webster computed the endomorphism algebra of the Bezrukavnikov-Kaledin tilting generator $\mathcal{T}$, corresponding to ${}^L\mathfrak{g}$ which is the ordinary simply laced Lie algebra, and showed it equals the cylindrical KLRW algebra $\mathscr{A}$. This, together with the computation of the algebra $\mathscr{A}$ as the endomorphism algebra of $\mathcal{T}$-branes on $\mathcal{Y}$ proves [**5**] the homological mirror symmetry:

$$\mathscr{D}_{\mathcal{X}} \cong \mathscr{D}_{\mathscr{A}} \cong \mathscr{D}_{\mathcal{Y}}. \tag{5.26}$$

The two equivalences separately reflect the fact that $\mathcal{T} = \bigoplus_{\mathcal{C}} \mathcal{T}_{\mathcal{C}}$ generates the category and that $\mathscr{A}$ is its endomorphism algebra. Homological mirror symmetry relating $\mathscr{D}_{\mathcal{X}}$ and $\mathscr{D}_{\mathcal{Y}}$ follows from the fact one gets the same algebra from both $\mathcal{X}$ and $\mathcal{Y}$.

5.4.14. The functor k^*, which takes the branes from upstairs down, comes from a Lagrangian correspondence. The correspondence acts by intersecting the brane $\mathcal{L}$ on $\mathcal{Y}$ with a Lagrangian $\mathcal{K} \in \mathcal{Y} \times Y_-$ to get a brane L on Y.

As any A-brane in Y_F, the fiber of the brane $\mathcal{K}$ in Y_F is a module for the algebra $\mathcal{B}_\infty$. The module corresponding to it turns out to not be a trivial module of the algebra $\mathcal{B}_\infty$ [**5**]. The module is rather a finite rank quotient of $\mathcal{B}_\infty$

$$\mathcal{B} = \mathcal{B}_\infty/\mathcal{I}, \tag{5.27}$$

in which the symmetric functions of z's are set to zero. The symmetric functions of z's generate an ideal, which we denoted by $\mathcal{I}$. The algebra $\mathcal{B}$ has rank $d!$, so is non-trivial for $d > 1$.

5.4.15. An immediate implication is that the functor k^* maps the upstairs $\mathcal{T}_{\mathcal{C}}$ to the downstairs $T_{\mathcal{C}}$ brane but equipped with the extra structure

$$k^*\mathcal{T}_{\mathcal{C}} \cong (T_{\mathcal{C}}, \mathcal{B}), \tag{5.28}$$

of a $\mathcal{B}$-local system of finite rank, instead of the infinite one in (5.22). Inherited from the upstairs theory, the local system is not based on $\mathcal{B}$ simply as a vector space, but rather as a module over itself [**5**].

5.4.16. The relation between the upstairs and downstairs generators from (5.28) implies that the downstairs algebra $A = Hom^*_{\mathscr{D}_Y}(T, T)$ has the same elements and the same relations as $\mathscr{A} = Hom^*_{\mathscr{D}_{\mathcal{X}}}(\mathcal{T}, \mathcal{T})$, the latter given by $\#1 - \#7$ above, except one quotients by the ideal

$$A = \mathscr{A}/\mathcal{I}, \tag{5.29}$$

$\mathcal{I}$ since $\mathcal{B} = \mathcal{B}_\infty/\mathcal{I}$. One can show the ideal $\mathcal{I}$ is a two sided ideal not only of $\mathcal{B}_\infty$, but also of $\mathscr{A}$, so the quotient makes sense.

5.4.17. Along with the proof of upstairs homological mirror symmetry (5.26), we get the proof of the downstairs one [**5**],

$$\mathscr{D}_X \cong \mathscr{D}_A \cong \mathscr{D}_Y. \tag{5.30}$$

by applying the functor f^* and its mirror k^*.

From the mirror perspective, the first equivalence in (5.29) is a consequence of the fact X embeds into $\mathcal{X}$, not as a locus where z's are set to zero, but as a locus where one sets their symmetric functions to zero. The action of the symmetric group makes only the latter good variables on $\mathcal{X}$. The fact that $\mathcal{I}$ is a two-sided ideal of $\mathscr{A}$ follows [**5**] from the fact that, since $\mathcal{T}$ is a vector bundle on $\mathcal{X}$, the center of its endomorphism algebra is the algebra of holomorphic functions on $\mathcal{X}$. The two-sided ideal is the subalgebra of it, consisting of those functions that vanish on X.

5.4.18. Every Lagrangian $\mathcal{L}$ of $\mathscr{D}_{\mathcal{Y}}$ gets mapped by the functor k^* to a Lagrangian L in $\mathscr{D}_Y$, but equipped with a possibly non-trivial local system of modules for the algebra $\mathcal{B}$, as in the case of the $T_{\mathcal{C}}$ branes that arise as images of $\mathcal{T}_{\mathcal{C}}$ branes under the functor k^*. The vanishing cycle branes $E_{\mathcal{U}}$ and $I_{\mathcal{U}}$ come equipped with trivial modules for $\mathcal{B}$. For the $I_{\mathcal{U}}$ brane, this is obvious, as it does not come from the upstairs. The $E_{\mathcal{U}}$ brane does come from the upstairs; however, it comes from a brane $\mathcal{U} = \mathscr{O}_U \in \mathscr{D}_X$ supported on the core X. The vanishing cycle U, whose structure sheaf is $\mathcal{U}$, is in X. Such a brane is a trivial module for the algebra $\mathcal{B}_\infty$ and hence for $\mathcal{B}$ as well.

5.4.19. Since the upstairs and the downstairs generators have direct summands

$$\mathcal{T}_{\mathcal{C}} \cong (T_{\mathcal{C}}, \mathcal{B}_\infty) \in \mathscr{D}_{\mathcal{Y}}, \qquad k^*\mathcal{T}_{\mathcal{C}} \cong (T_{\mathcal{C}}, \mathcal{B}) \in \mathscr{D}_Y, \tag{5.31}$$

corresponding to same Lagrangian $T_{\mathcal{C}}$ on Y, just equipped with the local systems $\mathcal{B}_\infty$ and $\mathcal{B}$ respectively, from now on we will simply denote both by $T_{\mathcal{C}}$, leaving implicit the corresponding local system. Most of the time, it is clear from the context which we are referring to; we will spell it out when not.

5.4.20. The structure we found above means that there should be a description of the entire upstairs category $\mathscr{D}_{\mathcal{Y}}$ as a category of A-branes on Y "with coefficients" in $\mathscr{D}_{Y_F}$ or, equivalently, in $\mathscr{D}_{\mathcal{B}_\infty}$. The objects of $\mathscr{D}_{\mathcal{Y}}$ can be described as A-branes on Y, equipped with a local system of $\mathcal{B}_\infty$-modules, or complexes thereof. Morphisms between its objects are intersection points of the corresponding Lagrangians on Y, dressed with homorphisms between the corresponding $\mathcal{B}_\infty$ modules.

Just as the upstairs category $\mathscr{D}_Y$ should have a downstairs description based on a category of A-branes on Y with coefficients in $\mathscr{D}_{\mathcal{B}_\infty}$, the downstairs category is a category of A-branes on Y with coefficients in $\mathscr{D}_{\mathcal{B}}$.

5.5. Projective resolutions. In this section, we will explain how to find the complex $(\mathscr{B}(T), \delta)$ resolving the $\mathscr{B}E_{\mathcal{U}}$ brane in $\mathscr{D}_Y$ in terms of its T-branes. Since all elements of the algebra A have zero cohomological degree, the complexes resolving the brane are ordinary, rather than twisted, complexes.

A remarkable fact is that we will be able to find the resolution not only of the downstairs brane $\mathscr{B}E_{\mathcal{U}} \in \mathscr{D}_Y$ but also its upstairs parent $\mathscr{B}\mathcal{U} \in \mathscr{D}_{\mathcal{Y}}$. The two branes are related by

$$k^*\mathscr{B}\mathcal{U} = \mathscr{B}E_{\mathcal{U}}. \tag{5.32}$$

We will first explain how to find the resolution of the downstairs brane on Y, and then how to get the resolution of the upstairs brane on $\mathcal{Y}$ that it came from as the image of the k^* functor.

5.5.1. The projective resolution of $\mathscr{B}E_{\mathcal{U}} \in \mathscr{D}_Y$ in terms of T-branes is based on the direct sum brane

$$\mathscr{B}E(T) = \bigoplus_k \mathscr{B}E_k(T)[k], \tag{5.33}$$

together with the differential δ. The direct sum brane is obtained by stretching the $\mathscr{B}E_{\mathcal{U}}$ brane straight along the cylinder $\mathcal{A}$ and allowing it to break at both ends.

The differential δ, which we turn on to recover the brane we started with, is a cohomological degree 1 and equivariant degree zero operator

$$\delta : \mathscr{B}E(T) \to \mathscr{B}E(T)[1]$$

that squares to zero

$$\delta \cdot \delta = 0 \tag{5.34}$$

in the algebra A. Because all elements of algebra A have cohomological degree zero, the product in (5.34) is an ordinary product, and the ordinarily twisted complex becomes an ordinary complex. One commonly writes the complex as

$$\mathscr{B}E_{\mathcal{U}} \cong \ldots \xrightarrow{t_1} \mathscr{B}E_1(T) \xrightarrow{t_0} \mathscr{B}E_0(T), \tag{5.35}$$

where $t_k \in Hom^*_{\mathscr{D}_Y}(\mathscr{B}E_{k+1}(T), \mathscr{B}E_k(T))$, and $\delta = \sum_k t_k \in A$. Unlike for a twisted complex, is not necessary to explicitly write the cohomological degrees of all the terms in the complex, as they are fixed by knowing any one.

To find the differential δ, we will proceed in steps, as in Sec. 3.

5.5.2. The algorithm from section 3 starts by finding the resolution of the $\mathscr{B}E_{\mathcal{U}}$-brane (or any other product brane) with the divisor D_O in (5.7) deleted. The result is a direct sum brane $\mathscr{B}E_{\mathcal{U}}(T)$, together with the product differential δ_O

$$(\mathscr{B}E_{\mathcal{U}}(T), \delta_O), \tag{5.36}$$

both of which can be read off from geometry of the brane. As we explained in section 3, the direct sum brane and the product differential form a rectangular toric grid in dimension d.

5.5.3. The differential δ of the brane in $\mathscr{D}_Y$ is found iteratively, as in section 3, by deforming $\delta_O = \delta_{0,0}$ to a differential $\delta = \delta_{u,\hbar}$ that squares to zero in the full algebra A. The deformation corresponds to filling the divisor $D_O = \Delta + D_a$ back in.

Filling D_a back in corresponds to deforming the differential $\delta_O = \delta_{0,0}$ to $\delta_{u,0}$ which squares to zero in the algebra $A_{u,0}$, in which $\hbar$, but not u is set to zero:

$$\delta_{u,0} = \sum_{\ell=0} u^\ell \delta_{\ell,0} = \delta_{0,0} + u\delta_{1,0} + \ldots, \tag{5.37}$$

analogous to the expansion in (3.17). The terms in $\delta_{\ell,0}$ are corrections involving ℓ dots, since at $\hbar = 0$ the theory has a symmetry that scales the dots and u oppositely. There are at most finitely many terms in (5.37) that may be non-vanishing, on degree grounds, since $\delta_{u,0}$ must define a differential on the same direct sum brane we started with. For the deformation problem to make sense, we need to regard the T-branes as equipped with a $\mathcal{B}$-local system. This simply means that their endomorphism algebra is $A_{u,0}$, which however enriches each geometric intersection point $\mathcal{P} \in T^\zeta \cap T$ with a copy of $\mathcal{B}$.

Next, we fill the diagonal Δ back in. This corresponds to deforming the differential $\delta_{u,0}$ to $\delta_{u,\hbar}$

$$\delta_{u,\hbar} = \sum_{k=0} \hbar^k \delta_{u,k} = \delta_{u,0} + \hbar\delta_{u,1} + \ldots, \tag{5.38}$$

which squares to to zero in the algebra full algebra $A = A_{u,\hbar}$, to get the resolution of the $\mathscr{B}E_{\mathcal{U}}$ brane in $\mathscr{D}_Y$.

5.5.4. Along with the resolution of the downstairs brane $\mathscr{B}E_{\mathcal{U}} \in \mathscr{D}_Y$, we can also obtain the resolution of its parent, the upstairs brane $\mathscr{B}\mathcal{U} \in \mathscr{D}_{\mathcal{Y}}$.

This is a consequence of two facts. The first is that the same downstairs T-branes, just equipped with a different the local system, as in eqn. (5.31), describe both the upstairs and the downstairs generators.

The second is that our starting point, the direct sum brane $\mathscr{B}E(T)$ and the product differential δ_O in (5.36) describes not only a brane downstairs in $\mathscr{D}_{Y_O}$, but also a brane upstairs, in $\mathscr{D}_{\mathcal{Y}_O}$. The differential δ_O, based on the intersection points of $T_{\mathcal{C}}$ branes on the Riemann surface $\mathcal{A}$ and containing no contributions from algebra elements that come from either $\mathcal{B}$ or from $\mathcal{B}_\infty$, is easily proven to square to zero both in the algebra A_O and in $\mathscr{A}_O$.

5.5.5. The resolution of the upstairs brane $\mathscr{B}\mathcal{U} \in \mathscr{D}_{\mathcal{Y}}$ is the same the direct sum brane $\mathscr{B}E(T)$ as for the downstairs one, with differential $\delta = \delta_{u,\hbar}$ that squares to zero in the upstairs algebra $\mathscr{A} = \mathscr{A}_{u,\hbar}$, and reduces to δ_O uppon setting u and $\hbar$ to zero.

Finding δ is a problem entirely analogous to finding the differential that squares to zero in the downstairs algebra $A = A_{u,\hbar}$ from section 5.5.3, just the algebra itself is different. For the $u \neq 0$ deformation problem to make sense, the T-branes on $\mathcal{A}$ as need to be equipped with a copy of the infinite rank algebra $\mathcal{B}_\infty$, which makes them be equivalent to the upstairs $\mathcal{T}$-branes, as in (5.22).

5.5.6. Since the algebra A is a quotient of the upstairs algebra, $A = \mathscr{A}/\mathcal{I}$, the differential $\delta : \mathscr{B}E(T) \to \mathscr{B}E(T)[1]$ that squares to zero in the upstairs algebra $\mathscr{A}$ also always squares to zero in the downstairs algebra A, by simply setting any terms in $\mathcal{I}$ to zero. The converse is of course not true.

The brane $(\mathscr{B}E(T), \delta)$ with differential δ that squares to zero in the upstairs algebra $\mathscr{A}$ thus describes *both* the resolution of a brane in $\mathscr{D}_{\mathcal{Y}}$ and, after dividing by $\mathcal{I}$, downstairs in $\mathscr{D}_Y$.

5.5.7. In the remainder of this section, when we write the differential $\delta = \delta_{u,\hbar}$ it will always be the differential of the upstairs complex, since it has more information than the downstairs one, and we can find it just as easily. This is particularly pertinent because, taking the hom with the $I_{\mathcal{U}}$-branes, the downstairs or the upstairs ones, all the dots in δ get set to zero anyhow. This follows because the $I_{\mathcal{U}}$ branes are simples of the corresponding upstairs and downstairs algebras [**2**].

5.5.8. To get a homology theory over $\mathbb{Z}$, we need to assign signs to maps in δ_O, which then determine the signs in δ.

Up to gauge equivalence, there is a unique choice of signs in the product differential δ_O that deforms to a theory with $u \neq 0$. This is natural, as sign choices in the complex should reflect the choices of spin structure on the brane that it describes, and both the upstairs and the downstairs $\mathscr{B}E_{\mathcal{U}}$ branes admit a unique spin structure once the divisor D_a of the punctures is filled in. The spin structure is the one inherited from the unique spin structure on the product of d disks in Y which the brane bounds.

Like in the $\mathfrak{gl}_{1|1}$ theory whose signs are discussed in 4.4.5, the differential δ_O is a differential on a product complex whose factors resolve the one-dimensional branes, with the signs derived from signs from the one-dimensional factor branes. Signs of the individual maps in one-dimensional complexes resolving the factor branes may be changed by gauge transformations on the T-branes they run between, so they are not meaningful. For the figure eight E-brane, only the product of all the signs in the complex is gauge invariant. The choice of signs that does deform to $u \neq 0$ can be shown to be $(-1)^{1+n_j/2}$ where n_j is the number, always even, of T-branes in the j-th one-dimensional complex.

In a product complex, one gains an additional standard sign needed to ensure the product differential squares to zero if its factors do. This sign is determined recursively. Assume we are given the signs in a product differential on a $k-1$ dimensional grid, which describing a brane which is the product of $k-1$ one-dimensional $\mathscr{B}E_i$-branes. Taking the product with $\mathscr{B}E_k$, the signs of the maps in the first $k-1$ directions of the grid stay the same, whereas signs in the k'th direction get multiplied by an overall alternating sign. For thorough discussion of how signs are assigned in $delta_O$, see appendix C.

5.6. Simple examples. We will now illustrate the $\mathfrak{su}_2$ algorithm in some simple examples.

5.6.1. Consider Y which is a copy of $\mathcal{A}$, the Riemann surface in Fig. 39. Y is the equivariant mirror to $\mathcal{X}$, the resolved A_3 surface. There are four T_j-branes generating $\mathscr{D}_Y$, which are images of the h^* functor of four line bundles which generate $\mathscr{D}_\mathcal{X}$ [**2**]. They are real line Lagrangians passing between consecutive punctures on the cylinder.

A cap/cup brane on $\mathscr{D}_\mathcal{X}$ is a structure sheaf of a $\mathbb{P}^1$. We will pick one of them, say $\mathcal{U}_2 \in \mathscr{D}_\mathcal{X}$, and let $E_{\mathcal{U}_2} \in \mathscr{D}_Y$ brane be its equivariant mirror, the image of the h^* functor. (More precisely, only three of four cup branes are structure sheaves of the three vanishing $\mathbb{P}^1$'s in $\mathcal{X}$ and the fourth one is described in [**2**].)

By isotoping the $E_{\mathcal{U}_2}$ brane and allowing it to break at $y=0$ and $y=\infty$ we get the direct sum brane:

$$E_{\mathcal{U}_2}(T) = T_2\{-1\}[2] \oplus \begin{pmatrix} T_1 \\ T_3\{-1\} \end{pmatrix} [1] \oplus T_2, \tag{5.39}$$

where the cohomological degree shift comes from the flip of orientation, per Fig. 39.

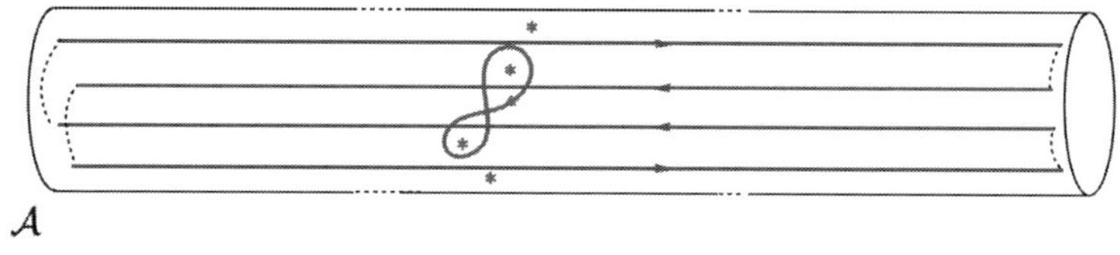

FIGURE 39. By isotoping the $E_{\mathcal{U}_2}$ brane, it becomes a connected sum of T-branes, which enter the direct sum brane $E_{\mathcal{U}_2}(T)$.

To recover the $E_{\mathcal{U}_2}$ brane, take the connected sum of the T-branes, over their intersection points at $y=\infty$ and $y=0$. The intersection points at $y=\infty$ correspond to the following elements of algebra A:

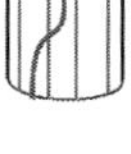 $\in Hom_{\mathscr{D}_Y}(T_1, T_2)$

 $\in Hom_{\mathscr{D}_Y}(T_2\{-1\}, T_3\{-1\}) = Hom_{\mathscr{D}_Y}(T_2, T_3)$

The intersection points at $y = 0$, correspond to the string diagrams going the other way, by the choice of wrapping in Fig. 39:

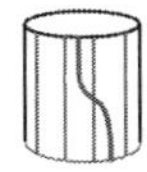 $\in Hom_{\mathscr{D}_Y}(T_3\{-1\}, T_2)$

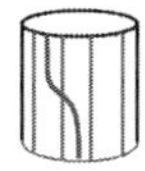 $\in Hom_{\mathscr{D}_Y}(T_2\{-1\}, T_1)$

The resulting complex has differential

As before, we suppressed the cylinder to simplify the notation. Thus, the $E_{\mathcal{U}}$ brane has a description as the following complex of T-branes:

(5.40)

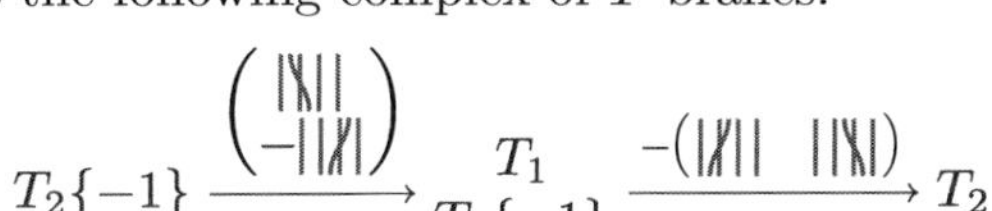

Unless stated otherwise, the last term in the complex is placed cohomological degree zero. Since A is an ordinary associative algebra, every map in the complex has degree zero and thus reduces the cohomological degree by 1. Since cohomological degrees of all terms in the complex are determined from the degree of the first, we do not need to write them.

In this example, the differential δ in (5.40) squares to zero not only downstairs in A, but also upstairs in $\mathscr{A}$. Thus the complex which describes the downstairs brane $E_{\mathcal{U}_2} = k^*\mathcal{U}_2 \in \mathscr{D}_Y$ also describes its upstairs parent $\mathcal{U}_2 \in \mathscr{D}_{\mathcal{Y}}$. We get one or the other, by simply reinterpreting the downstairs T-branes as carrying a trivial local system – as $\mathcal{B}$ is trivial in one dimension – or $\mathcal{B}_\infty$ which is not, as explained in section (5.5). The complex is the standard Koszul resolution of the structure sheaf of a $\mathbb{P}^1$ in $\mathcal{X}$.

5.6.2. The case of one dimensional $Y = \mathcal{A}$, is simple but essential, as it is the basis of the algorithm to solve the theory on $Y = Sym^d(\mathcal{A})$, for d arbitrary.

In one dimension, in general, the geometric differential $\delta_O = \delta_0$, which we simply read off from the $\mathscr{B}E_{\mathcal{U}}$ Lagrangian, is easily proven to square to zero in the full downstairs algebra A. Because there is no diagonal, maps passing through the punctures do not contribute to the A-model on $Y = \mathcal{A}$, and the $u = 0$ theory is the full answer. The differential $\delta_{u=0}$ thus describes the resolution of the brane in $\mathscr{D}_Y$

In general, the differential which squares to zero in $\mathscr{A} = \mathscr{A}_u$, is a one parameter deformation δ_u, of the geometric differential $\delta_{u=0}$. From δ_u, we get the downstairs differential $\delta_{u=0}$ by either taking the quotient by the ideal $\mathcal{I}$, which kills the dots,

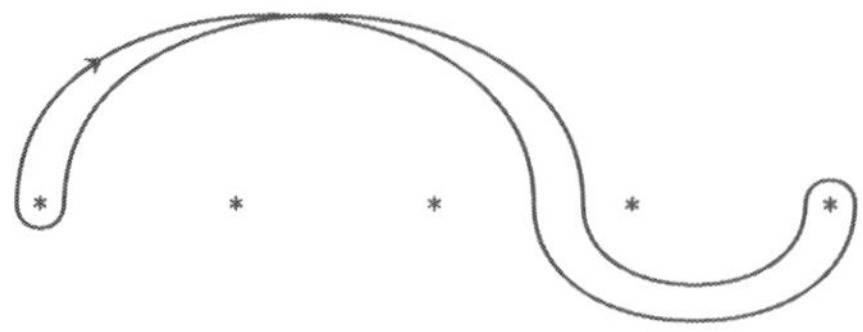

FIGURE 40. An $\mathscr{B}E_{\mathcal{U}}$-brane in the $\mathfrak{su}_2$ theory.

or by setting $u = 0$. The example in Fig. 40 illustrates that. The figure has been rotated clockwise by 90 degrees, and we have also omitted drawing the cylinder. The corresponding direct sum brane is

$$\mathscr{B}E_{\mathcal{U}}(T) = T_4\{-3\}[2] \oplus \begin{pmatrix} T_6\{-3\} \\ T_4\{-2\} \\ T_1 \end{pmatrix} [1] \oplus \begin{pmatrix} T_5\{-2\} \\ T_2 \end{pmatrix}. \tag{5.41}$$

The geometric differential δ_0 is

(5.42) $\delta_{u=0} =$

The differential squares to zero in the downstairs algebra A and describes the resolution of the brane on $\mathscr{D}_Y$.

The same direct sum brane from (5.41), together with the differential $\delta_{u=0}$, describes the resolution of the parent brane upstairs on $\mathcal{Y}_0 = \mathcal{Y}/D_a$ with the divisor D_a of punctures removed. The corresponding algebra is $\mathscr{A}_{u=0}$, with the parameter u from relation #5 set to zero. As explained in section 5.5, to get $(\mathscr{B}E(T), \delta_0)$ to describe a resolution of the brane on $\mathcal{Y}_0$, we merely need to reinterpet the T-branes as the upstairs ones. This just changes the local system the branes are implicitly equipped, in this case, from a trivial one to $\mathcal{B}_\infty$, allowing arbitrary numbers of dots in their encomorphism algebra, rather than none.

The full upstairs differential on $\mathcal{Y}$, after filling in the divisor D_a, is the $u \neq 0$ deformation of $\delta_{u=0}$, as in (5.37). The correction of order u^ℓ corresponds to introducing maps with ℓ dots. In our simple example, on degree grounds, there is only one such map we may introduce:

$\delta =$

A short calculation shows that δ squares to zero in the $\mathscr{A}_u$ algebra if we set $x = -u$, as its square equals to

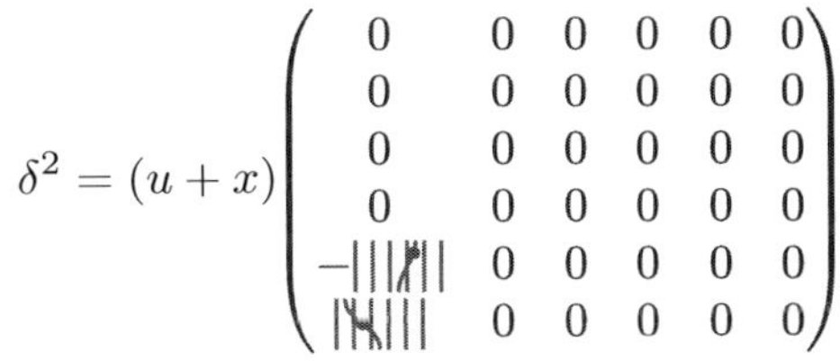

Thus, the complex resolving the upstairs $\mathscr{B}\mathcal{U} \in \mathscr{D}_{\mathcal{Y}}$ brane is

(5.43)

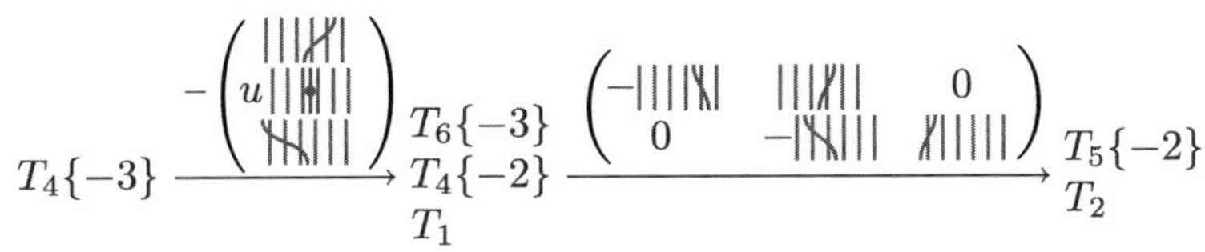

The complex resolving the downstairs brane $\mathscr{B}E_{\mathcal{U}} \in \mathscr{D}_Y$ is obtained from (5.43) as the image of the $k^* : \mathscr{D}_{\mathcal{Y}} \to \mathscr{D}_Y$ functor. This amounts to thinking of T's as the downstairs generators and setting the dots to zero. Its differential coincides with the $\delta_{u=0}$ in (5.42).

5.6.3. Projective resolutions encode counts of disk instantons. We will start with a very simple example demonstrating that; later we will turn to examples interesting for knot theory. Consider a Lagrangian in $d = 2$ theory in Fig. 41.

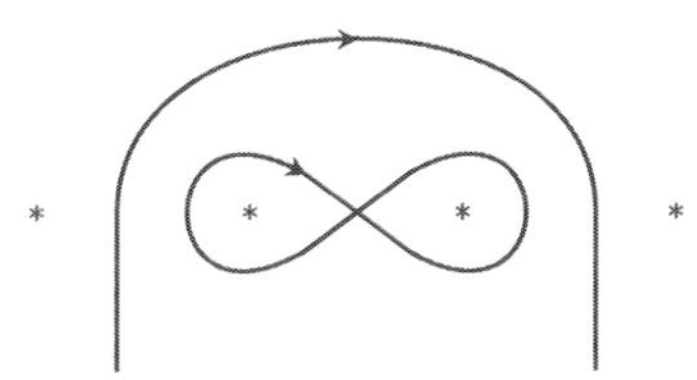

FIGURE 41. The simplest example of $d = 2$ in the $\mathfrak{su}_2$ theory.

We found the resolution of the figure-eight brane in section 5.6.1. The resolution of the outer brane is the complex

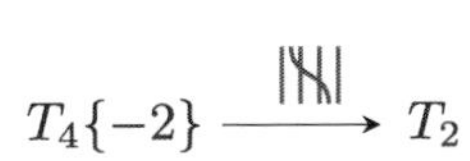

The product of the two complexes can be represented as a 2-dimensional grid:

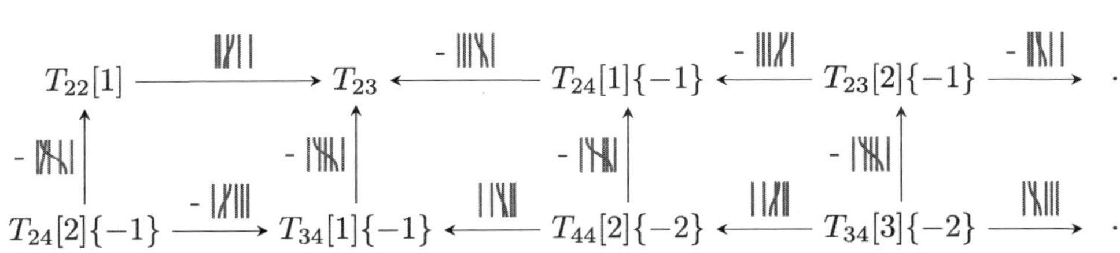

The maps in the complex encode the relative positions of branes in Fig. 41, since they represent open strings stretching between them. A map which is a product of

the identity map of the inner brane times a non-trivial morphism from the outer brane is a pair of blue strands, one constant, the other not, which cross. Their crossing reflects the fact that, in this brane configuration, an open string from the differential of the outer brane must cross a thimble of the inner. Similarly, a map which is a product of a non-trivial morphism from the inner brane complex times the identity map of an outer T-brane, is pair of blue strands, one constant and one not, which do not cross.

It is straightforward to read off the Maslov and the equivariant degrees of all the branes in the complex from the degrees of the maps. The signs, which are unique up to gauge transformations, are assigned per section 5.5.8 and appendix C.2. In particular, the horizontal maps in each line multiply to $(-1)^{4/2+1}$. The sign of the leftmost vertical map gets assigned to all other vertical maps, and the signs in the top row gets inherited by other rows, but multiplied with an overall alternating sign.

Organizing the T-branes by their cohomological degrees, we get the direct sum brane

$$T_{34}\{-1\}[3] \oplus \begin{pmatrix} T_{44}\{-1\} \\ T_{23} \\ T_{24} \end{pmatrix} [2] \oplus \begin{pmatrix} T_{24} \\ T_{34} \\ T_{22} \end{pmatrix} [1] \oplus T_{23}, \tag{5.44}$$

with differential

(5.45) $\delta_{u,0} =$ 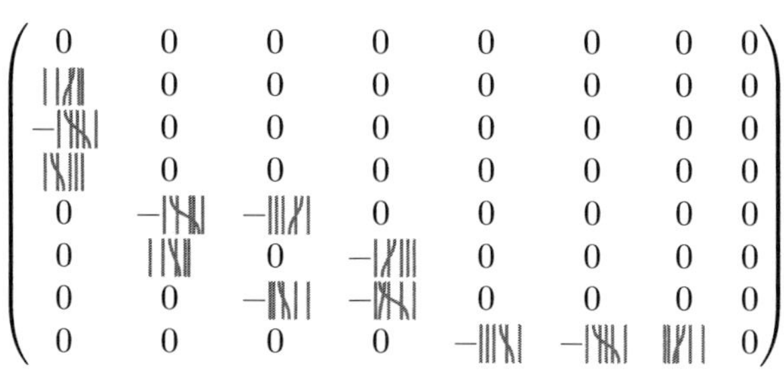 .

In this simple example, the geometric differential happens to also square to zero in $\mathscr{A}_{u,0}$. This follows from the fact that the two one-dimensional complexes we started had geometric differentials that squared to zero for any u, as in our first example in this section.

In general, all $u \neq 0$ corrections in $\delta_{u,0}$ should come from the corrections from one-dimensional complexes, except that not all of them survive to the product on degree grounds. This reflects the fact that, as long as $\hbar = 0$, the theory effectively behaves as a product theory.

It is straightforward to check that $\delta_{u,0}$ does not square to zero in the full $\mathscr{A}_{u,\hbar}$ algebra. Thus, we need to look for a deformation of it by terms of order $\hbar$. In fact, the first order deformation $\delta = \delta_{u,0} + \hbar\delta_{u,1}$ turns out to suffice. The deformation $\delta_{u,1}$ may contain all possible corrections allowed on degree grounds that are not already included in $\delta_{u,0}$, as those correspond to gauge transformations. By uniqueness of δ, discussed in section 3.4.5, any two solutions to the deformation problem should be equivalent, so we may try to find the solution in a subspace of the deformation space. In this case, the following simple deformation suffices:

(5.46) $\delta =$

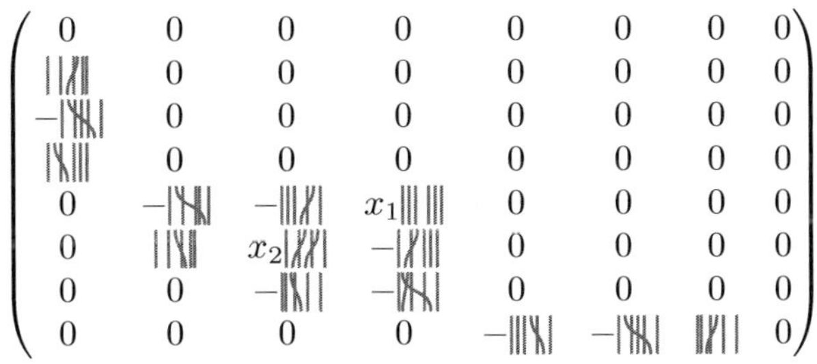

Computing its square

(5.47) $\delta^2 =$

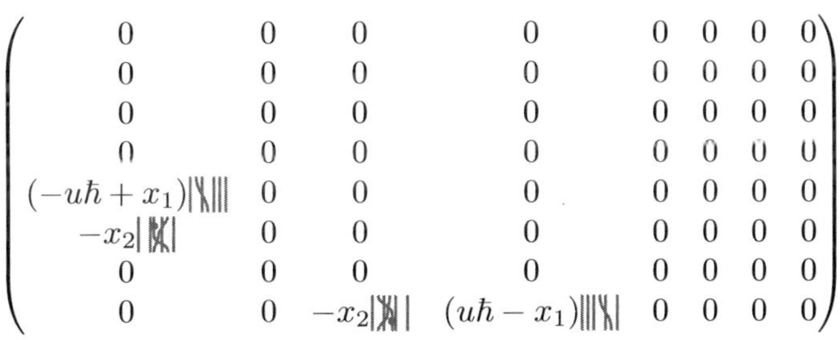

we can see that setting $x_1 = u\hbar$ and $x_2 = 0$ leads to a differential which, when written as a differential of a standard complex, reads

(5.48)

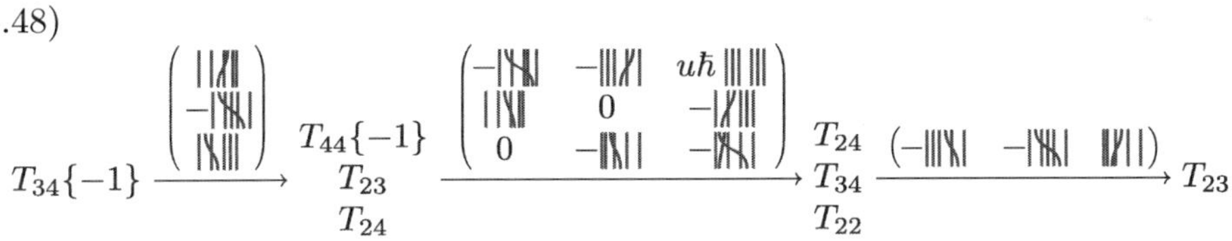

The above purely algebraic construction of the brane differential δ accounts for disks contributing to the Floer differential of the brane with any I-brane.

For example, the Floer complex of the brane configuration in Fig. 42 should be obtained by applying $hom_Y^{*,*}(-, I_{24})$ to the complex in (5.48). The only terms in the complex that give a non-trivial contribution contain T_{24} branes. The result is the complex

$$\delta_F : \qquad \mathbb{C} \xrightarrow{u\hbar} \mathbb{C} \tag{5.49}$$

The right term is placed in degree [2], the left in degree [1], as the functor is contravariant. Two copies of $\mathbb{C}$ correspond to the two intersection points in Fig. 42. The map between them is the Floer differential, together with its sign. The differential is proportional to $u\hbar$ since the map in Fig. 42 intersects a puncture and the diagonal Δ once. In particular, the complex has trivial homology whenever $u\hbar \neq 0$.

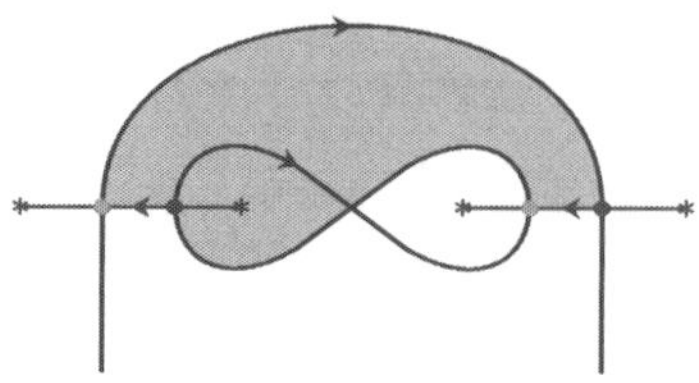

FIGURE 42. The holomorphic disk counted by the differential in (5.49).

One could in principle do the corresponding disk count by hand. The count of holomorphic maps $y : \mathrm{D} \to Y$ may be defined using the cylindrical approach to Floer theory [**63**]. The maps that the theory should count are described in section 5.2.2. We will merely check that the disk is not forbidden on degree grounds. From the shaded domain A which the disk projects to in Fig. 42, we can read off its Maslov index and equivariant degree. Using (5.8), its Maslov index is $ind(A) = 2e(A) = 1$ since $e(A) = 1 - 3 \times \frac{1}{4} + \frac{1}{4} = \frac{1}{2}$ per (4.13). It has equivariant degree $J^0(A) = 0$ per (5.9) because intersects a puncture once so $n_a(A) = 1$, and it also intersects the diagonal once $i(A) = 3 \times \frac{1}{4} + \frac{3}{4} - e(A) = 1$ per (4.14). Thus, the disk has Maslov index 1 and equivariant degree 0, as needed for it to contribute to δ_F. The fact that the disk intersects a puncture on $\mathcal{A}$ and the diagonal Δ once is in agreement with the fact that its contribution to the differential is at order $u\hbar$.

5.6.4. Here is another example, now more relevant to knot theory.

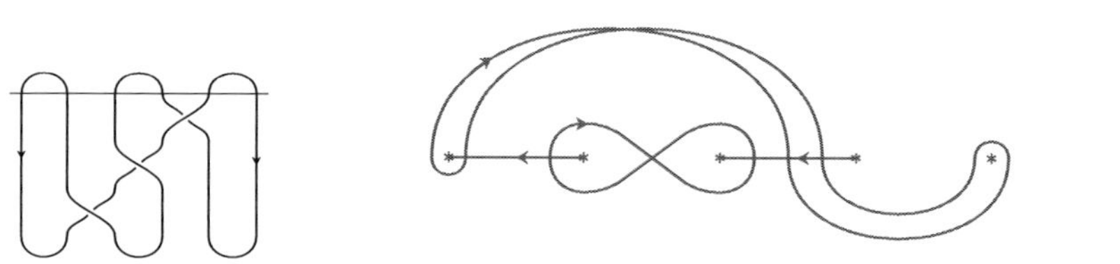

FIGURE 43. Branes for a presentation of the reduced unknot.

It corresponds to a presentation of the unknot similar to that in section 4.5.5. The brane configuration in Fig. 43 corresponds to the reduced unknot.

Resolutions of the two one-dimensional factor branes were worked out in sections 5.6.1 and 5.6.2. Taking their product, since the two one-dimensional branes have the topology of S^1, we get a grid on a torus of size 6×4, in (5.50). We suppressed the maps in the complex, since they are uniquely determined from the two initial one dimensional complexes. We merely indicated the maps in which blue strands cross by purple, instead of black, arrows. They come from maps $T_4 \to T_2$ and $T_4 \to T_1$ of the longer complex, which intersect the thimbles T_2, T_3, T_4, T_3 of the shorter complex.

(5.50)

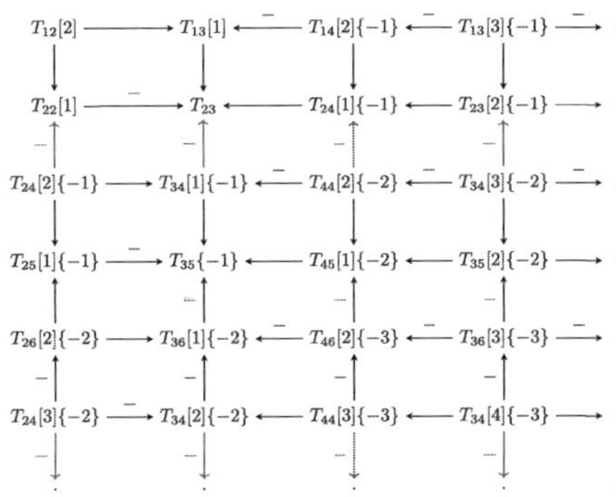

The relative Maslov and equivariant degrees of the thimbles in the complex are the degrees of the maps between them. Per section C.2, we took signs of the vertical maps to be those of the original one-dimensional complex, but then signs of the horizontal maps get multiplied by an overall $\pm$ sign, alternating by row. This is the standard sign on the tensor product of chain complexes.

The differential $\delta_Q = \delta_{u=0,\hbar=0}$ of the complex in (5.50) is the geometric one; it squares to zero in the algebra $\mathscr{A}_{u=0,\hbar=0}$, which has u and $\hbar$ both set to zero.

The differential that deforms to $u \neq 0$ is simply the product differential coming from the two one-dimensional $u \neq 0$ complexes. It is easy to check that the product differential squares to zero in the $\mathscr{A}_{u,\hbar=0}$ algebra. It gives the complex in (5.51). In a general, only a subset of one-dimensional $u \neq 0$ corrections lift to the product complex because the equivariant degrees of the thimbles in the product complex are not product of their degrees in the one-dimensional complexes. The maps that end up with wrong equivariant degrees get set to zero, as we will see in the later examples. In all cases we checked, this construction leads to a differential that squares to zero in the $\hbar = 0$ algebra.

(5.51)

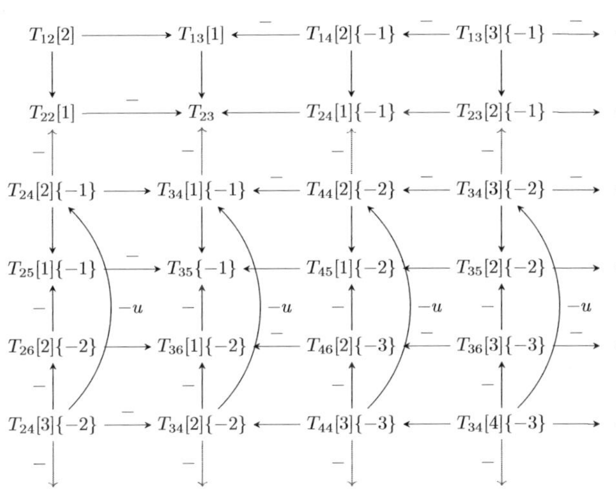

Finally, we look for a deformation of the differential that squares to zero in the full $\mathscr{A} = \mathscr{A}_{u,\hbar}$ algebra. To find the first order in $\hbar$ deformation $\delta = \delta_{u,0} + \hbar\delta_{u,1} + \ldots$, we could start by including in $\delta_{u,1}$ all possible maps that have consistent equivariant degree and that are not already present in $\delta_{u,0}$, with some unknown coefficients. We solve for the coefficients by asking that $\delta^2 = 0$ to order $\hbar$, in the algebra $\mathscr{A}_{u,\hbar}$.

Since any two solutions to the deformation problem are equivalent, one can start by including in $\delta_{u,1}$ only the simplest terms, without crossings or dots. In the example at hand, there are 19 such possible entries in $\delta_{u,1}$. Solving the first-order equation, we find a one-parameter family of solutions. The first-order $\hbar$ deformation we get in this way turns out to be exact: all these solutions turn out to automatically satisfy the higher-order equations as well. One can easily check explicitly that solutions within this family are gauge equivalent, so we pick one. This leads to 4 non-vanishing coefficients in $\delta_{u,1}$, given in (5.52). If we were not not able to find a solution with the initial ansatz, we would relax the assumption about which of the finite number of possible corrections have non-zero coefficients, as we generally do in more complicated examples.

(5.52)

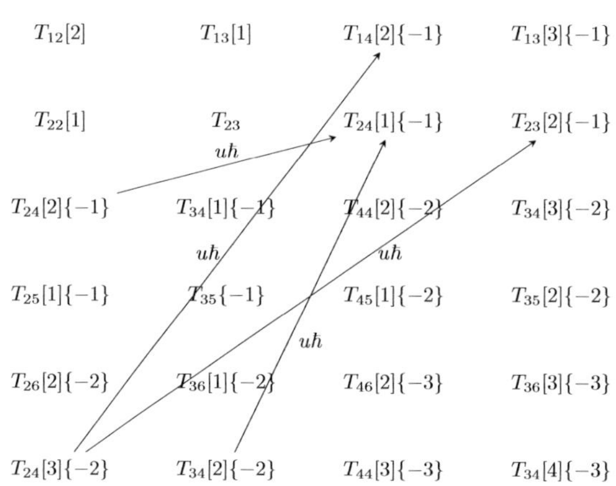

The complex, written in standard form, reads:

$$T_{34}\{-3\} \xrightarrow{d_4} \begin{array}{c} T_{36}\{-3\} \\ T_{44}\{-3\} \\ T_{34}\{-2\} \\ T_{24}\{-2\} \\ T_{13}\{-1\} \end{array} \xrightarrow{d_3} \begin{array}{c} T_{46}\{-3\} \\ T_{44}\{-2\} \\ T_{35}\{-2\} \\ T_{26}\{-2\} \\ T_{34}\{-2\} \\ T_{14}\{-1\} \\ T_{23}\{-1\} \\ T_{24}\{-1\} \\ T_{12} \end{array} \xrightarrow{d_2} \begin{array}{c} T_{45}\{-2\} \\ T_{36}\{-2\} \\ T_{24}\{-1\} \\ T_{34}\{-1\} \\ T_{25}\{-1\} \\ T_{13} \\ T_{22} \end{array} \xrightarrow{d_1} \begin{array}{c} T_{35}\{-1\} \\ T_{23} \end{array} \tag{5.53}$$

with differential given by

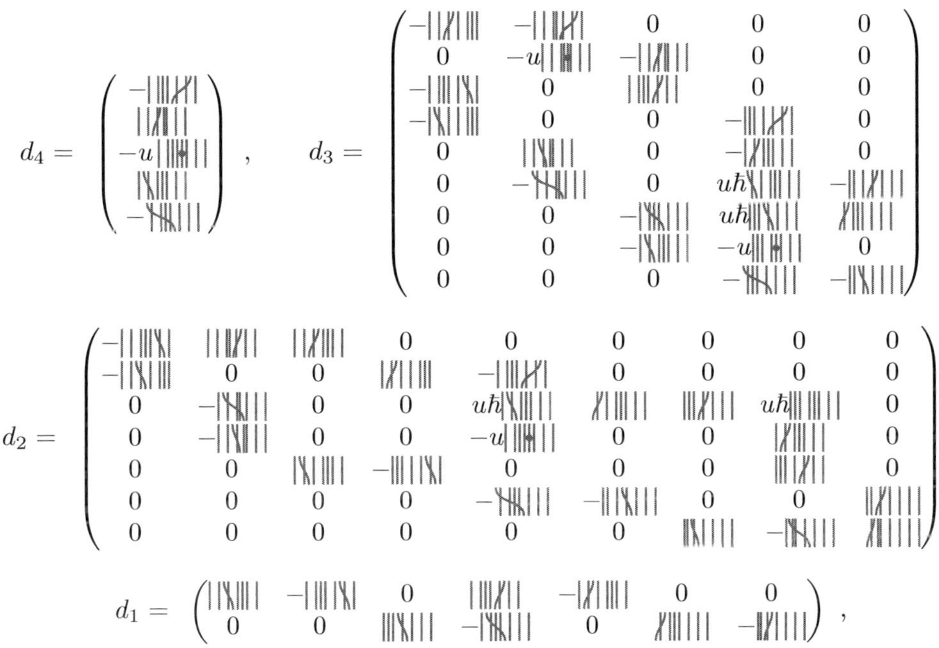

Applying the $hom^{*,*}_Y(-, I_{24})$ functor to (5.53) results in a three-dimensional complex

$$0 \to \mathbb{C}\{-1\} \xrightarrow{u\hbar} \mathbb{C}\{-1\} \xrightarrow{0} \mathbb{C}\{-2\} \to 0 \tag{5.54}$$

with the rightmost term in homological degree 3. The three terms in the complex correspond to, respectively, the three intersection points p_1q_1, p_2q_3 and p_2q_2 in Fig. 44. The maps in the complex should compute the action of the Floer differential.

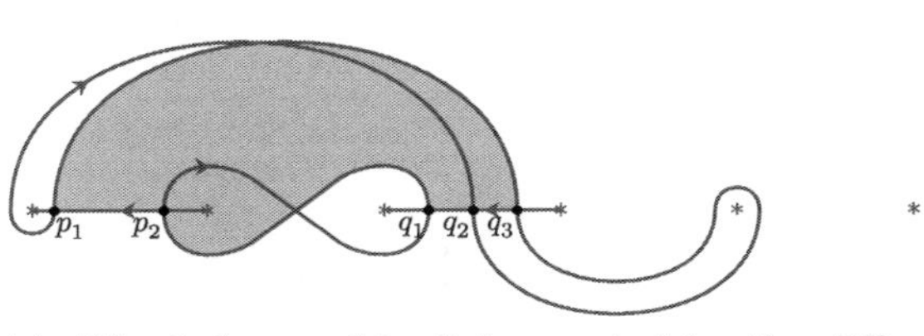

FIGURE 44. The holomorphic disk counted by the differential δ_F.

A disk that could contribute to coefficient of p_2q_3 in $\delta_F p_1q_1$, on degree grounds, projects to the shaded domain in Fig. 44. The corresponding disk would have Maslov index 1 and equivariant degree zero per (5.8) and (5.9). From the complex, we deduce that the count of such disks is 1. The fact that the map is proportional to $u\hbar$ corresponds to the fact the disk intersects the diagonal and a puncture once, as the one in Fig. 42. The homology of the complex is $\mathbb{C}[3]\{-2\}$.

To get the absolute grading of the link homology, we need to include the additional, overall grading shifts explained in section 5.7. The total shift is the sum of the braid dependent shift in (5.62) and the shift associated with grading of the cap branes relative to the canonical I-branes, which depends only on the dimension of Y. From link diagram in Fig. 43 we read off that the writhe of the braid (the difference in the number of positive and negative crossings) is $w = 1$, the exponent sum of the braid generators is $e = -1$, so (5.62) contributes $\Delta M = -1$ and $\Delta J_0 = 1$ to the shift. Since the dimension is $d = 2$, the $I_{\mathcal{U}}$ brane is related to the I_{24} brane by the grading shift which is twice that in Fig. 51, so $I_{\mathcal{U}} = I_{24}[2]\{-1\}$. The net shift is $M \to M - 3$ and $J^0 \to J^0 + 2$. The result is

$$Hom^{*,*}(\mathscr{B}E_{\mathcal{U}}, I_{\mathcal{U}}) = \mathbb{C},$$

which is the reduced homology of the unknot.

5.6.5. As our final explicit example, consider the Hopf link as shown in Fig. 45.

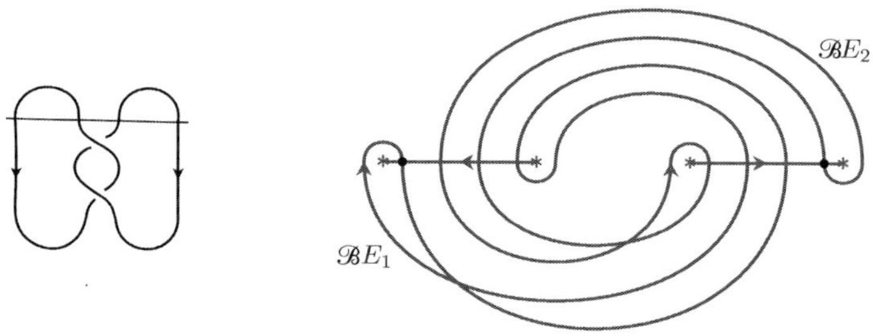

FIGURE 45. The Hopf link and the corresponding Lagrangians

We first write down the resolutions of the one dimensional Lagrangians $\mathscr{B}E_1$ and $\mathscr{B}E_2$. Including only the geometric terms, the resolution of $\mathscr{B}E_1$ is:

$$T_1[1] \longrightarrow T_4 \longrightarrow T_3[-1]\{1\} \longleftarrow T_2\{1\} \longleftarrow T_4[1]\{-1\} \longleftarrow T_2[2]\{-1\} \longrightarrow .$$

For $\mathscr{B}E_2$ we get:

$$T_2\{2\} \longrightarrow T_3[-1]\{2\} \longleftarrow T_4\{1\} \longleftarrow T_2[1]\{1\} \longleftarrow T_4[2]\{-1\} \longrightarrow T_5[1]\{-1\} \longrightarrow .$$

The product of these gives is a toric grid whose maps encode the differential $\delta_O = \delta_{0,0}$:

$$
\begin{array}{ccccccccccccc}
T_{12}[1]\{2\} & \xrightarrow{-} & T_{13}\{2\} & \longleftarrow & T_{14}[1]\{1\} & \longleftarrow & T_{12}[2]\{1\} & \xleftarrow{-} & T_{14}[3]\{-1\} & \longrightarrow & T_{15}[2]\{-1\} & \longrightarrow & \cdot \\
{\scriptstyle -}\downarrow & & {\scriptstyle -}\downarrow & & {\scriptstyle -}\downarrow & & {\scriptstyle -}\downarrow & & {\scriptstyle -}\downarrow & & {\scriptstyle -}\downarrow & & \\
T_{24}\{1\} & \longrightarrow & T_{34}[-1]\{1\} & \xleftarrow{-} & T_{44} & \xleftarrow{-} & T_{24}[1] & \longleftarrow & T_{44}[2]\{-1\} & \xrightarrow{-} & T_{45}[1]\{-1\} & \xrightarrow{-} & \cdot \\
{\scriptstyle -}\downarrow & & {\scriptstyle -}\downarrow & & {\scriptstyle -}\downarrow & & {\scriptstyle -}\downarrow & & {\scriptstyle -}\downarrow & & {\scriptstyle -}\downarrow & & \\
T_{23}[-1]\{2\} & \xrightarrow{-} & T_{33}[-2]\{1\} & \longleftarrow & T_{34}[-1] & \longleftarrow & T_{23}\{1\} & \xleftarrow{-} & T_{34}[1] & \longrightarrow & T_{35} & \longrightarrow & \cdot \\
\uparrow & & \uparrow & & \uparrow & & \uparrow & & \uparrow & & \uparrow & & \\
T_{22}\{2\} & \longrightarrow & T_{23}[-1]\{1\} & \xleftarrow{-} & T_{24} & \xleftarrow{-} & T_{22}[1]\{1\} & \longleftarrow & T_{24}[2] & \xrightarrow{-} & T_{25}[1] & \xrightarrow{-} & \cdot \\
\uparrow & & \uparrow & & \uparrow & & \uparrow & & \uparrow & & \uparrow & & \\
T_{24}[1] & \xrightarrow{-} & T_{34} & \longleftarrow & T_{44}[1]\{-1\} & \longleftarrow & T_{24}[2]\{-1\} & \xleftarrow{-} & T_{44}[3]\{-2\} & \longrightarrow & T_{45}[2]\{-2\} & \longrightarrow & \cdot \\
\uparrow & & \uparrow & & \uparrow & & \uparrow & & \uparrow & & \uparrow & & \\
T_{22}[2]\{1\} & \longrightarrow & T_{23}[1]\{1\} & \xleftarrow{-} & T_{24}[2] & \xleftarrow{-} & T_{22}[3] & \longleftarrow & T_{24}[4]\{-2\} & \xrightarrow{-} & T_{25}[3]\{-2\} & \xrightarrow{-} & \cdot \\
\downarrow & & \downarrow & & \downarrow & & \downarrow & & \downarrow & & \downarrow & & \\
\cdot & & \cdot & & \cdot & & \cdot & & \cdot & & \cdot & & \cdot
\end{array}
$$

We determined the absolute degrees of the branes in the complex, using the fact that the marked intersection point in Fig. 45 is fixed under braiding. Therefore, the thimble T_{24} corresponding to this point must have the same degrees before and after braiding. It has $M = 4$ and $J^0 = -2$.

The grid can be rearranged into the standard chain complex

$$
T_{24}\{-2\} \xrightarrow{d_4} \begin{matrix} T_{44}\{-2\} \\ T_{25}\{-2\} \\ T_{14}\{-1\} \\ T_{22} \end{matrix} \xrightarrow{d_3} \begin{matrix} T_{45}\{-2\} \\ T_{15}\{-1\} \\ T_{44}\{-1\} \\ T_{24}\{-1\} \\ T_{24} \\ T_{24} \\ T_{12}\{1\} \\ T_{22}\{1\} \end{matrix} \xrightarrow{d_2} \begin{matrix} T_{45}\{-1\} \\ T_{44}\{-1\} \\ T_{24} \\ T_{34} \\ T_{25} \\ T_{24} \\ T_{14}\{1\} \\ T_{22}\{1\} \\ T_{23}\{1\} \\ T_{12} \end{matrix} \xrightarrow{d_1} \begin{matrix} T_{44} \\ T_{35} \\ T_{24} \\ T_{34} \\ T_{24}\{1\} \\ T_{23}\{1\} \\ T_{13} \\ T_{22} \end{matrix} \xrightarrow{d_0} \begin{matrix} T_{34} \\ T_{34}\{1\} \\ T_{23}\{1\} \\ T_{23} \end{matrix} \xrightarrow{d_{-1}} T_{33}\{1\}
$$

with the rightmost term in homological degree -2 and where, for example, the first two differentials are

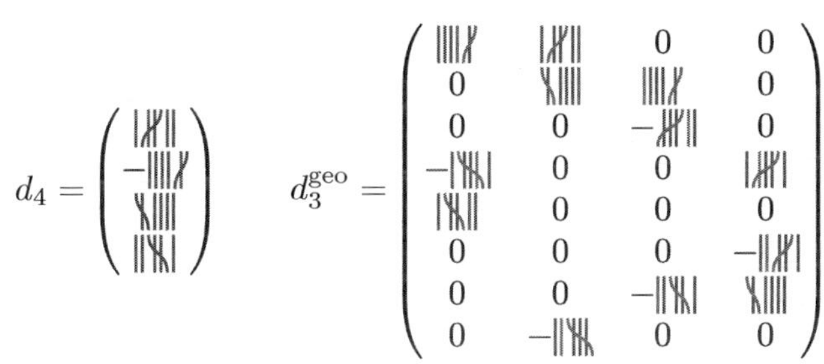

The next step is to find the u-deformation of the geometric complex. It is a product of one-dimensional complexes resolving $\mathscr{B}E_1$ and $\mathscr{B}E_2$ for $u \neq 0$, except that not all of its terms survive to the product complex, as some may have degrees which are inconsistent with the degrees of thimbles they are to map between. The terms that do survive are

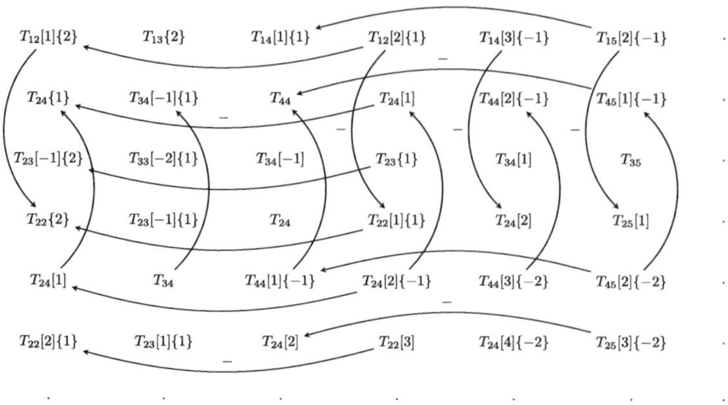

where each map is multiplied by an overall factor u. Continuing with the example of d_4 and d_3, we find that d_4 is unchanged and

$d_3^{\hbar=0} =$

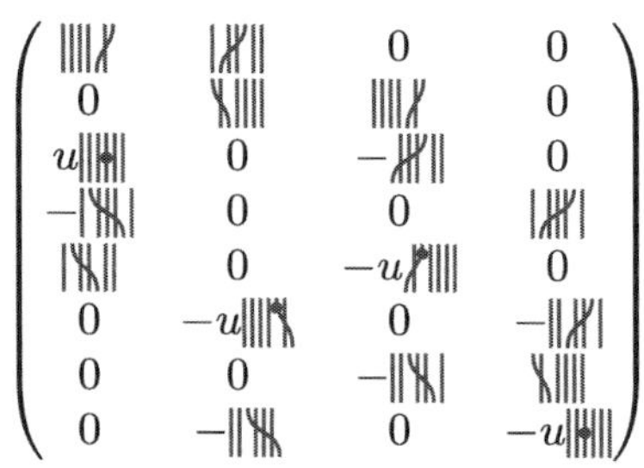

The result is a complex that satisfies $\delta_{u,0}^2 = 0$ at $\hbar = 0$.

Finally, we solve for the corrections necessary for δ to satisfy $\delta^2 = 0$ with $\hbar \neq 0$. It proves sufficient to turn on only the first order in $\hbar$ deformation, of the form $\delta = \delta_{u,0} + \hbar\delta_{u,1}$, and further restrict the possible terms in $\delta_{u,1}$ to maps without dots so $\delta_{u,1} = \delta_{u=0,1}$. For example, d_3 with the possible deformations terms added is given by

$d_3 =$

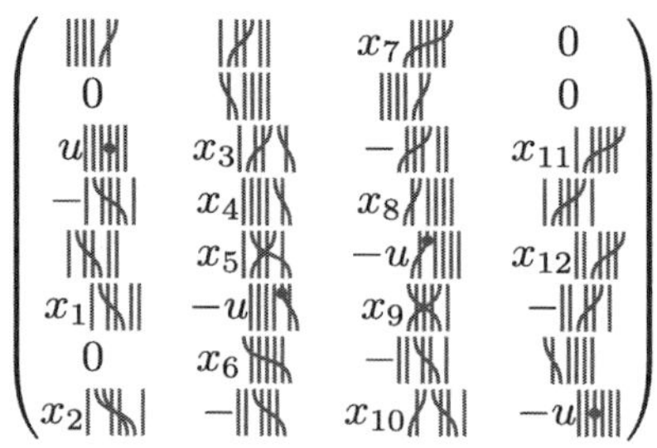

with coefficients x_i, which we want to solve for. Asking that $d_3 \cdot d_4 = 0$ leads to a system of equations. For example, the fourth row of d_3 gives the equation

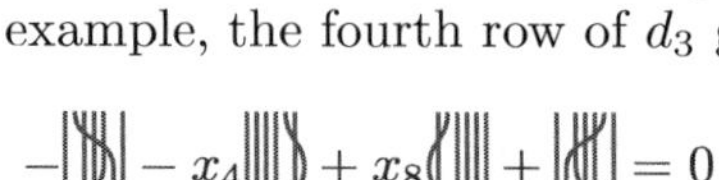

which has the solution

$$x_4 = -x_8 = -u\hbar.$$

The other coefficients in d_3 turn out to vanish, and altogether we find

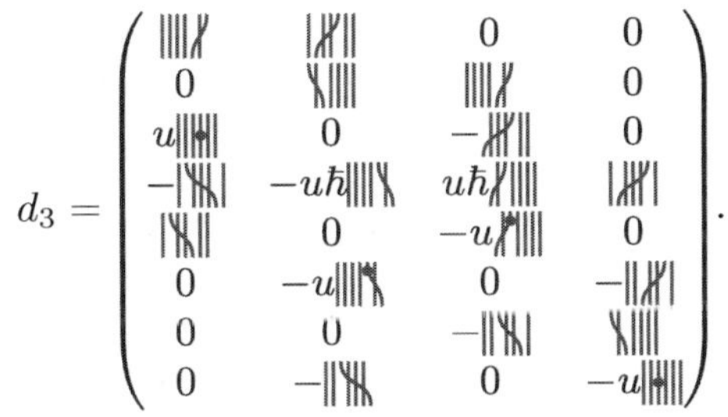

Proceeding further, the d_3 deformation we just found will determine the $\delta_{u,1}$ deformation of d_2, and so on. Altogether, the maps in $\delta_{u,1}$ are

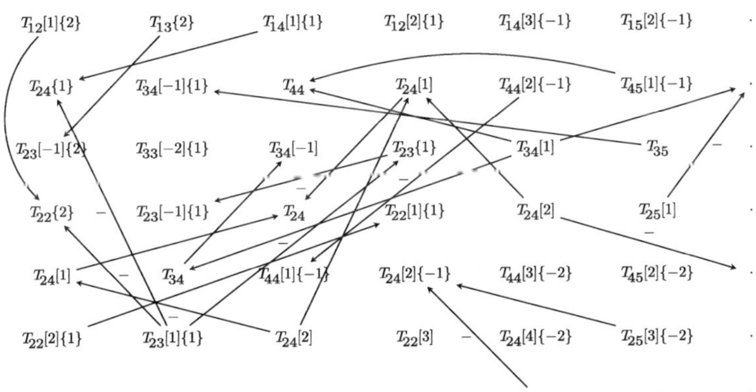

To calculate the $\mathfrak{su}_2$ link homology, we apply the $hom^*_Y(-, I_{24}\{J\})$ functor to the complex $(\mathscr{B}E(T), \delta)$. The functor sends all components the direct sum brane $\mathscr{B}E(T)$ to zero except for the 8 copies of T_{24}, for a suitable choice of $\{J\}$. The result is a complex whose differential comes from identity maps in δ. It is given by:

$$\begin{matrix}\mathbb{C}\\ \mathbb{C}\{1\}\end{matrix} \xrightarrow{u\hbar\cdot\begin{pmatrix}1&0\\1&0\end{pmatrix}} \begin{matrix}\mathbb{C}\\ \mathbb{C}\end{matrix} \xrightarrow{u\hbar\cdot\begin{pmatrix}0&0\\1&-1\\1&-1\end{pmatrix}} \begin{matrix}\mathbb{C}\{-1\}\\ \mathbb{C}\\ \mathbb{C}\end{matrix} \to 0 \to \mathbb{C}\{-2\} \tag{5.55}$$

with the rightmost term in homological degree 4.

The 8 terms of the complex in (5.55) correspond, per construction of the direct sum brane $\mathscr{B}E(T)$, to the 8 intersection points of the $\mathscr{B}E_{\mathcal{U}}$ brane with the I_{24}-brane in Fig. 46. The cohomology of the complex coincides with Floer cohomology, as we explained in section 3.5. It follows that the differential in (5.55) is the Floer differential δ_F, and that its terms count disk instantons.

The differential δ_F, obtained algebraically, squares to zero by construction. From the geometric perspective, it should square to zero because the cancelling contributions to δ_F^2 correspond to two different ways a Maslov index 2 map can degenerate into a pair of Maslov index 1 maps (all with equivariant degree zero). The moduli space of Maslov index 2 maps is a one real dimensional manifold, an interval, whose two boundary points correspond to the two ways the map can degenerate, and come with opposite orientations. Fig. 46 gives an example.

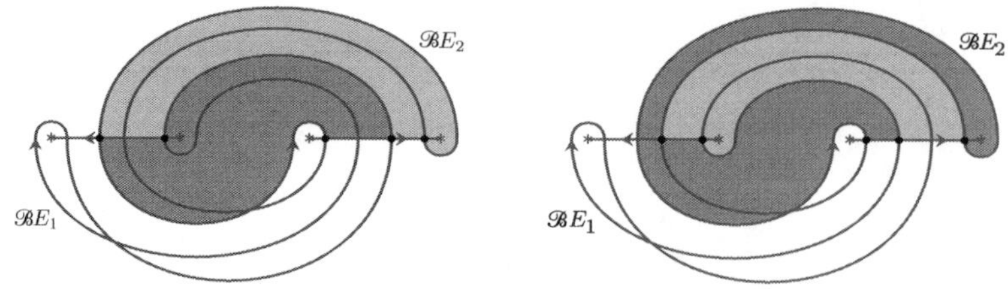

FIGURE 46. The shaded domains correspond to four terms in the differential δ_F whose pairwise product cancels in computing δ_F^2.

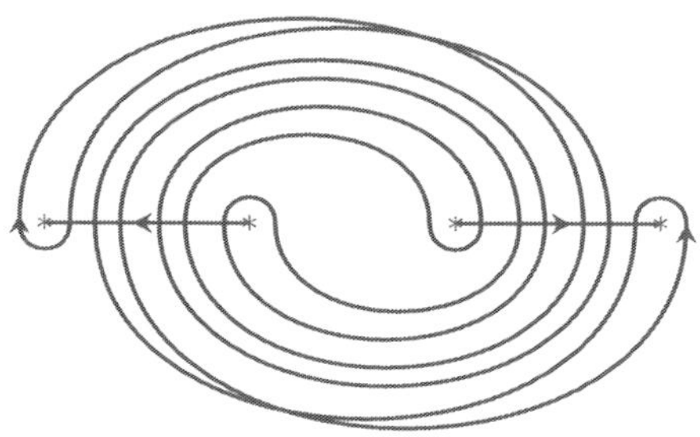

FIGURE 47. Lagrangians corresponding to left handed trefoil in $\mathfrak{su}_2$ trefoil theory.

The cohomology of the complex in (5.55) is

$$\mathbb{C}[4]\{-2\} \oplus \mathbb{C}[2]\{-1\} \oplus \mathbb{C}[2] \oplus \mathbb{C}\{1\}. \tag{5.56}$$

To get the absolute grading of link homology we need to include the additional, overall grading shifts from section 5.7. They depend only on the braid, so can be read off from the link diagram in Fig. 45. The net shift is by $M \to M - 4$ and $J^0 \to J^0 + 2$ from $w = 2$, $e = 2$ and $d = 2$, giving

$$\mathbb{C} \oplus \mathbb{C}[-2]\{1\} \oplus \mathbb{C}[-2]\{2\} \oplus \mathbb{C}[-4]\{3\}. \tag{5.57}$$

This agrees with the known Khovanov homology of the Hopf link after regrading $i = M + 2J^0$, $j = 2J^0$.

5.6.6. The same procedure outlined above can be used for any knot. Let us briefly sketch the analysis for the left-handed trefoil from Fig. 15. The Lagrangian corresponding to the unreduced trefoil is depicted in Fig. 47. Each cycle separately can be resolved in terms of a complex consisting of 8 one-dimensional T-branes. The product complex is a toric grid of size 8×8, whose vertices are the 64 two-dimensional T-branes. The geometric differential $\delta_{0,0}$ obtained in this way squares to zero in the $\mathscr{A}_{u=0,\hbar=0}$ algebra.

The differential deforms to $\delta_{u,0}$, which squares to zero in the $u \neq 0$ $\mathscr{A}_{u=0,\hbar=0}$, by including all the u-corrections inherited from the product of one-dimensional complexes, provided they are allowed on degree grounds. In this case, each of the two one-dimensional complexes has four u-corrections. Out of possible $8\times4+8\times4 =$

64 u-corrections, 12 end up having wrong equivariant degrees and get set to zero. The deformation is exact at first order in u, so $\delta_{u,0} = \delta_{0,0} + u\delta_{1,0}$.

The differential $\delta = \delta_{u,\hbar}$, which squares to zero in the full algebra $\mathscr{A}_{u,\hbar}$ is the first order in $\hbar$ deformation of $\delta_{u,0}$. There are 366 in total maps that can contribute to $\delta_{0,1}$ in $\delta = \delta_{u,0} + \hbar\delta_{0,1}$, which we include with some arbitrary coefficients x_i. The equation $\delta^2 = 0$ is a linear equation for x's which has a unique solution, exact to all orders in both u and $\hbar$. Note that, compared to the previous examples where none of the maps in $\delta_{0,1}$ contained any crossings of strands, here they do.

Applying the $hom_Y^{*,*}(-, I_{24})$ functor, the original 64-dimensional complex becomes the 18-dimensional complex:

$$\mathbb{C} \to \emptyset \to \begin{array}{c}\mathbb{C}\{-2\}\\ \mathbb{C}\{-2\}\\ \mathbb{C}\{-1\}\end{array} \xrightarrow{u\hbar\cdot\begin{pmatrix}0&0&0\\0&0&0\\1&1&0\\-1&-1&0\end{pmatrix}} \begin{array}{c}\mathbb{C}\{-3\}\\ \mathbb{C}\{-3\}\\ \mathbb{C}\{-2\}\\ \mathbb{C}\{-2\}\end{array} \xrightarrow{u\hbar\cdot\begin{pmatrix}1&-1&0&0\\-1&1&0&0\\-1&-1&0&0\\0&0&-1&-1\end{pmatrix}}$$

$$\begin{array}{c}\mathbb{C}\{-3\}\\ \mathbb{C}\{-3\}\\ \mathbb{C}\{-3\}\\ \mathbb{C}\{-2\}\end{array} \xrightarrow{u\hbar\cdot\begin{pmatrix}0&0&0&0\\0&0&0&0\\-1&-1&0&0\\1&1&0&0\end{pmatrix}} \begin{array}{c}\mathbb{C}\{-4\}\\ \mathbb{C}\{-4\}\\ \mathbb{C}\{-3\}\\ \mathbb{C}\{-3\}\end{array} \xrightarrow{u\hbar\cdot\begin{pmatrix}1&1&0&0\\0&0&1&1\end{pmatrix}} \begin{array}{c}\mathbb{C}\{-4\}\\ \mathbb{C}\{-3\}\end{array}$$

with the rightmost term in homological degree 6.

The cohomology of the complex as written is

$$\mathbb{C} \oplus \mathbb{C}[2]\{-1\} \oplus \mathbb{C}[2]\{-2\} \oplus \mathbb{C}[5]\{-4\}, \tag{5.58}$$

As a final step, to obtain the link homology groups $Hom_{\mathscr{D}_Y}^{*,*}(\mathscr{B}E_{\mathcal{U}}, I_{\mathcal{U}})$ we need to include in (5.58) the non-geometric, overall shift of grading from section 5.7. The net shift is by $M \to M+1$ and $J^0 \to J^0 - \frac{1}{2}$, because the braid has $w = -3$, $e = -3$ and $d = 2$. With proper gradings included, the link homology is

$$Hom_{\mathscr{D}_Y}^{*,*}(\mathscr{B}E_{\mathcal{U}}, I_{\mathcal{U}}) = \mathbb{C}[1]\left\{-\tfrac{1}{2}\right\} \oplus \mathbb{C}[3]\left\{-\tfrac{3}{2}\right\} \oplus \mathbb{C}[3]\left\{-\tfrac{5}{2}\right\} \oplus \mathbb{C}[6]\left\{-\tfrac{9}{2}\right\}. \tag{5.59}$$

Taking cohomology with coefficients in $\mathbb{Z}$ instead of $\mathbb{C}$, we discover that it has $\mathbb{Z}_2$ torsion:

$$\mathbb{Z}[1]\left\{-\tfrac{1}{2}\right\} \oplus \mathbb{Z}[3]\left\{-\tfrac{3}{2}\right\} \oplus \mathbb{Z}[3]\left\{-\tfrac{5}{2}\right\} \oplus \mathbb{Z}_2[5]\left\{-\tfrac{7}{2}\right\} \oplus \mathbb{Z}[6]\left\{-\tfrac{9}{2}\right\}, \tag{5.60}$$

which comes from the quotient

$$\mathbb{Z}_2 = \left\langle \begin{pmatrix}1\\-1\\0\\0\end{pmatrix}, \begin{pmatrix}0\\0\\1\\0\end{pmatrix} \right\rangle \Big/ \left\langle \begin{pmatrix}1\\-1\\-1\\0\end{pmatrix}, \begin{pmatrix}-1\\1\\-1\\0\end{pmatrix} \right\rangle.$$

The cohomologies in (5.59) and (5.60) agree with the Khovanov homology of the left-handed trefoil, when re-written in terms of the (i, j) gradings of [**50**], which are related to M and J^0 by $i = M + 2J^0$ and $j = 2J^0$. While the cohomologies are the same, the complexes are not. The complex defined by Khovanov in [**50**] (see also [**17**]), is 30-dimensional in this case, as opposed to ours which is 18-dimensional.

5.6.7. One can often vastly simplify the $\mathscr{B}E_{\mathcal{U}} \in \mathscr{D}_Y$ branes by using the swiping move to map the brane to an equivalent but simpler object of $\mathscr{D}_Y$. The swiping move plays the key role in proving the theory produces homological link invariants. It is described in section 5.7 in Fig. 49. Simplicity is measured by the number of T-branes needed to describe the brane, or rather, its resolution.

Having simplified the brane, we get a formulation of $Hom^{*,*}_{\mathscr{D}_Y}(\mathscr{B}E_{\mathcal{U}}, I_{\mathcal{U}})$ as the cohomology of a much simpler complex. The move should have a version any for Lie algebra; in the $\mathfrak{gl}_{1|1}$ case, it was described in section 4.6.

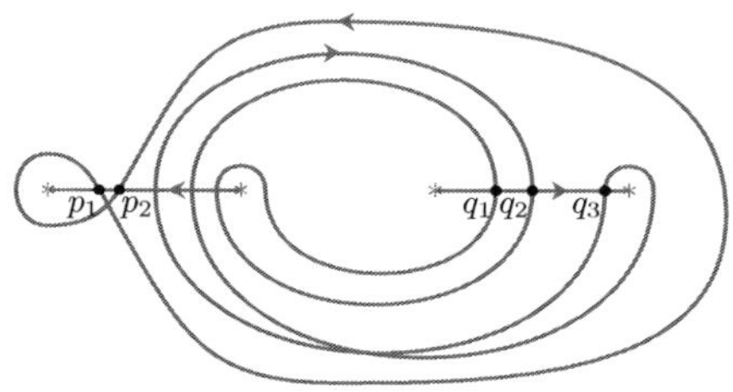

FIGURE 48. A simplification, using the swiping move, of the branes associated with left-handed trefoil

For example, to apply it for the branes associated with the left-handed trefoil, start with an unbraided $E_{\mathcal{U}}$ brane, which is a product of two simple figure eights. Using the move, one of the two branes can "swallow" the other, as in the proof of Thm. 1 in section 5.7. After that, we braid the brane to get the $\mathscr{B}E_{\mathcal{U}}$ brane. For the left handed trefoil, the result is shown in Fig. 48. The brane has a resolution consisting of only 32 thimbles, instead of 64 we started with:

$$T_{23}\{-4\} \xrightarrow{d_6} \begin{matrix} T_{13}\{-3\} \\ T_{35}\{-5\} \\ T_{24}\{-4\} \\ T_{22}\{-3\} \end{matrix} \xrightarrow{d_5} \begin{matrix} T_{14}\{-3\} \\ T_{12}\{-2\} \\ T_{23}\{-3\} \\ T_{45}\{-5\} \\ T_{25}\{-4\} \\ T_{22}\{-2\} \\ T_{24}\{-3\} \end{matrix} \xrightarrow{d_4} \begin{matrix} T_{12}\{-1\} \\ T_{14}\{-2\} \\ T_{24}\{-3\} \\ T_{22}\{-2\} \\ T_{25}\{-3\} \\ T_{45}\{-4\} \\ T_{22}\{-1\} \\ T_{25}\{-2\} \end{matrix} \xrightarrow{d_3} \begin{matrix} T_{12} \\ T_{15}\{-1\} \\ T_{22}\{-1\} \\ T_{24}\{-2\} \\ T_{25}\{-2\} \\ T_{55}\{-3\} \\ T_{24}\{-1\} \end{matrix} \xrightarrow{d_2} \begin{matrix} T_{14} \\ T_{22} \\ T_{25}\{-1\} \\ T_{45}\{-2\} \end{matrix} \xrightarrow{d_1} T_{24}$$

To find the differential, we proceed as in the case without swallowing. We start by writing down the differential $\delta_{0,0}$ that squares to zero in $A_{u=0,\hbar=0}$. The price to pay for the simplification of the brane gained by the swallowing move, is that the differential $\delta_{u,\hbar}$ that squares to zero in the full algebra, is a deformation not of the naive geometric differential $\delta_{0,0}$, but of its gauge transform $\delta_{0,0} \to e^b \delta_{0,0} e^{-b}$ where b depends on $u\hbar$. Proceeding order by order in $\hbar$ and u, we find the final differential δ, as a deformation of $\delta_{0,0}$ by four new terms.

After applying the $hom^{*,*}_Y(-, I_{24})$ functor, the result is a complex which is just 6 dimensional, spanned by the intersection points $p_i q_j$ in Fig. 48:

$$\mathbb{C} \to 0 \to \begin{matrix} \mathbb{C}\{-2\} \\ \mathbb{C}\{-1\} \end{matrix} \xrightarrow{0} \mathbb{C}\{-3\} \xrightarrow{-2(u\hbar)^2} \mathbb{C}\{-3\} \xrightarrow{0} \mathbb{C}\{-4\} \tag{5.61}$$

where the rightmost term is in homological degree 5. After introducing the same degree shifts as in section 5.6.6, the cohomology of the complex reproduces the homology trefoil homology (5.59) over $\mathbb{C}$, and after shifting to $\mathbb{Z}$ coefficients, the homology with torsion in (5.60). The simpler complex in (5.61) also makes it manifest where the torsion comes from: it is due to a holomorphic disk instanton that connects a pair of intersection points which has multiplicity -2. It generates the coefficient of p_1q_2 in $\delta_F p_2 q_1$. The fact the homology is unchanged is a consequence of the fact, proven in appendix A.2, that the swiping move leaves the brane invariant as an object of $\mathscr{D}_Y$, and hence all its morphisms as well.

5.6.8. Reduced link homology for all two-bridge knots involves a one-dimensional Y and does not require $\hbar$ deformations, so its computation may be done by hand. Our current computer code, implemented in Mathematica and specialized to the two-strand algebra, allows us to compute the reduced homology of 3-bridge links and the full homology (including the $\mathbb{Z}_2$ torsion) of 2-bridge links. Generalizing it to multiple strands and arbitrary links is straightforward, in principle.

We used the code to compute the unreduced homology for all knots up to 6 crossings, and for the first 6 knots with 7 crossings, as well as the reduced homology of several three-bridge knots including 8_5, 8_{19} and 9_{42}. In all the examples we checked, solving the system of linear equations coming from the first order in $\hbar$ and setting all the remaining coefficients to zero sufficed to find the full differential δ. Furthermore, for two-bridge knots we checked, δ can be made to square to zero at first-order in $\hbar$ and u which further speeds up the algorithm.

Our homology theory agrees with Khovanov's [**50**], including torsion, in all cases we checked. More precisely, the theories agree after a regrading: the cohomological index i and the Jones grading j of Khovanov homology are related to ours by $i = M+2J^0$ and $j = 2J^0$. The fact our homology theory coincides with Khovanov's to be expected by homological mirror symmetry proven in [**5**] and abstract theorems of [**95–97**].

5.7. Homological $U_\mathfrak{q}(\mathfrak{su}_2)$ link invariants from $\mathscr{D}_{Y_{\mathfrak{su}_2}}$. In this section we will prove Theorems 1 and 2 for $\mathfrak{su}_2$. The proof is analogous to the $\mathfrak{gl}_{1|1}$ case.

The key ingredient is the swiping move, shown for $\mathfrak{su}_2$ in Fig. 49. Just as in case of $\mathfrak{gl}_{1|1}$, the move holds for any $u\hbar \neq 0$, so for it to hold and for the theory to give homological link invariants, we need to work on Y without divisors deleted. We prove this move for $\mathfrak{su}_2$ in appendix A.2.

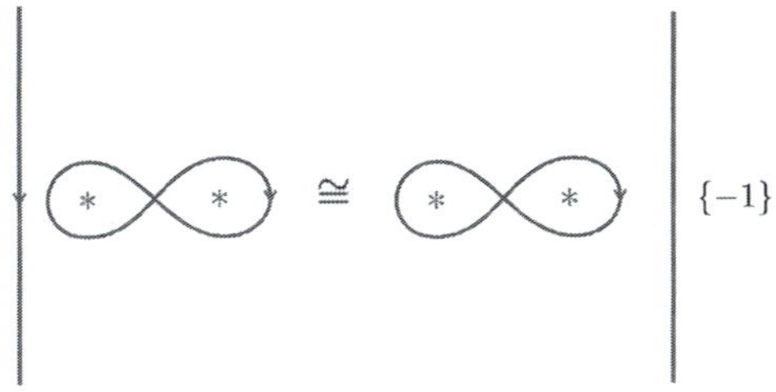

FIGURE 49. These two brane are equivalent objects of $\mathscr{D}_Y$

The swiping move holding in $\mathscr{D}_Y$ suffices to ensure that all Markov moves from Figs. 5, 6 and 7 hold, up to overall degree shifts. To show that the moves hold exactly, we need to keep track of the gradings of the $\mathscr{B}E_{\mathcal{U}}$ brane through the sequence of braid moves.

The principle for keeping track of degree shifts due to the geometric braid group action on an object of $\mathscr{D}_Y$ is described in section 4.6.3; it is independent of the choice of a Lie algebra. The other degree shifts depend on the Lie algebra.

5.7.1. As in the $\mathfrak{gl}_{1|1}$ case, the geometric action of $\mathscr{B}$ on the $E_{\mathcal{U}}$ brane gets combined with a constant degree shift functor. The functor acts on all objects of $\mathscr{D}_Y$ in the same way, as in (4.53). In the $\mathfrak{su}_2$ case, only the potential W, and not Ω, contains terms that depend on the positions of the punctures. These terms are introduced by (5.5). Their contribution to ΔJ_0 is equal to $-\frac{1}{4}e$, where e is the sum of the exponents of braid group generators $\sigma_i^{\pm}$ in the word representing the braid.

The one novel ingredient for $\mathfrak{su}_2$ case is that framing no longer acts trivially on link invariants in vertical framing. Vertical framing is the framing which the geometry picks, as discussed in section 2.5.2. As a consequence, the second relation in Fig. 7 is really true only up to a degree shift that depends on the writhe w of the braid. (The framing factor in general depends on the writhe of the link, but in our case, for links represented as plat closures of braids, this is the same as the writhe of the braid. The writhe is number of positive minus ⤧ minus the number of negative ⤨ crossings.) The Jones polynomial is an invariant of links in zero framing and satisfies the skein relation with no framing factor. To cancel the framing factor and get the theory whose Euler characteristic is the Jones polynomial, we will introduce an additional, w dependent degree shift which shifts both the Maslov and equivariant degrees. The net constant shift, due to to sources combined, is thus:

$$\begin{aligned} \Delta M &= -w, \\ \Delta J_0 &= \frac{3}{4}w - \frac{1}{4}e \end{aligned} \tag{5.62}$$

5.7.2. The Markov I move from Fig. 5 reads

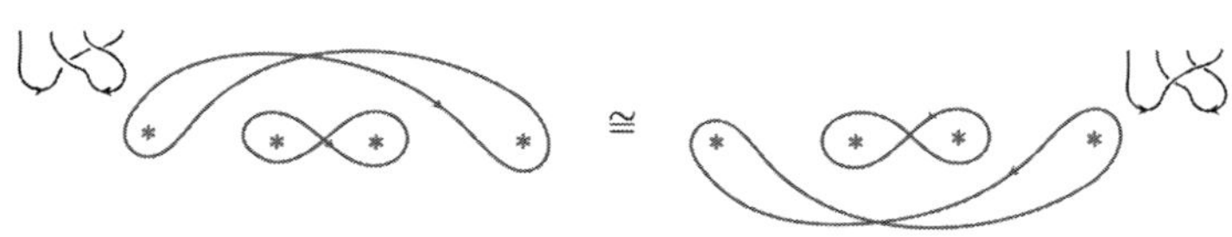

FIGURE 50. These two brane are equivalent objects of $\mathscr{D}_Y$

To show that it holds, simplify the two branes, using the swiping move, as follows:

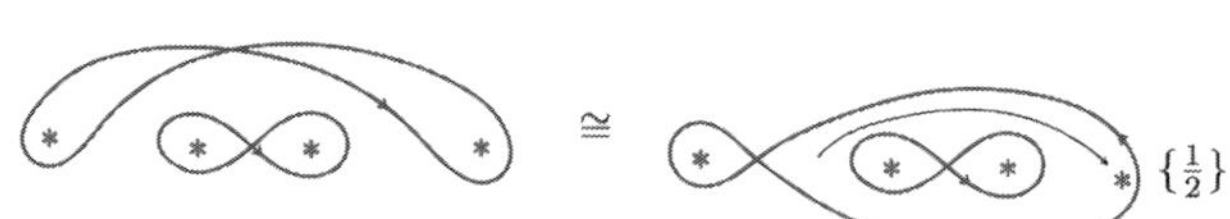

and

.

To compare the branes, we can ignore the outside figure eight, as it is the same for both. The two branes differ by the path the puncture took around the inner figure eight brane; in particular, they resulting branes are the same modulo shifts of gradings. Using resolutions of the two branes, it is easy to show that the top brane receives no geometric shift due to braiding the puncture, and in particular simply equals the brane on the right hand side, forgetting the path. The bottom brane, by contrast, gets a geometric shift of $\{1\}$. (The particular shifts depend on a placement of cuts used to define a single-valued branch of W_0. This is implicit throughout the theory, in particular in the definition of T-branes and their algebra.) The final degree of brane has to be supplemented with contributions of the constant degree shift functors in (4.53) which are, in the $\mathfrak{su}_2$ case, given in (5.62), and which depend only on the braid and not on the brane. The path at the top has $w = 0$ and $e = -2$, the path at the bottom has $w = 0$ and $e = 2$, resulting in shifts of $\{\frac{1}{2}\}$ and $\{-\frac{1}{2}\}$. These shifts combine with the geometric ones to ensure the two branes are equivalent objects of $\mathscr{D}_Y$. The move in Fig. 5 follows.

It is not difficult to show that with degree shifts in (5.62), both of the moves in Fig. 7 hold as well. For example, braid group generator σ that generates the twist in second move Fig. 7 has a geometric action on the $E_{\mathcal{U}}$ brane, giving a degree shift of $[-1]\{1\}$. This is canceled by the constant shift in (5.62), since the element has $w = -1$ and $e = 1$. Similarly, the constant shift functor of the braid in the first move, which has $w = 0$ and $e = 4$, cancels the geometric degree shift of the brane, which equals $\{1\}$. This proves both moves in Fig. 7.

Next, consider the stabilization move in Fig. 6. As in the $\mathfrak{gl}_{1|1}$ theory case, asking for the stabilization move to hold as stated should determine the grading of the $I_{\mathcal{U}}$ brane, the cap, relative to the basic I-branes, which are defined to be dual to T-branes via (2.12). The stabilization move reads:

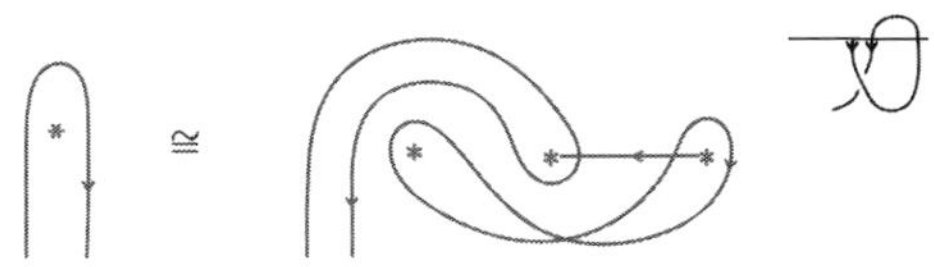

in terms of branes. By starting with the trivial braid, applying the sliding move, and then braiding, the brane on the right is equivalent to,

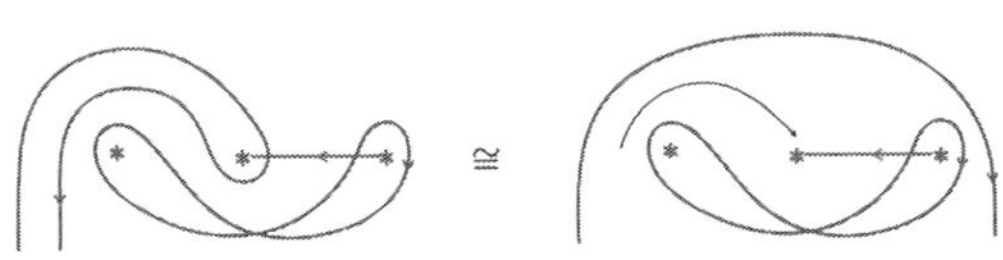

The single intersection point of the figure eight brane with the basic I-brane has the geometric degree 0, both equivariant and cohomological. The constant shift functor that accompanies the action of $\mathscr{B}$ is a degree shift of $\{-\frac{1}{2}\}[1]$, since the braid has $w = -1$ and $e = -1$. Asking for the stabilization move to hold as stated, with no accompanying degree shifts, determines the cap brane $I_{\mathcal{U}}$ to be equal to the basic I-brane, shifted in degree by $\{-\frac{d}{2}\}[d]$, where d is the dimension d of Y.

As an aside, the $I_{\mathcal{U}}$ brane is invariant under the second move in Fig. 7, just like the corresponding $E_{\mathcal{U}}$-brane, and for the same reason: the geometric degree shift, which is the same for both branes, gets canceled by the constant one from (5.62). The overall degree shift of $\{-\frac{d}{2}\}[d]$, goes along for the ride.

FIGURE 51. The $\mathfrak{su}_2$ cap brane $I_{\mathcal{U}}$ equals the corresponding canonical I-brane shifted in degree by $\{-\frac{1}{2}\}[1]$, for each interval it contains.

5.7.3. To complete the proof of Thm. 1 for $\mathfrak{su}_2$ we need to show the remaining Birman moves, obtained from those in Figs. 5 and 7 by exchanging tops and bottoms, also hold. The proof is analogous to that for $\mathfrak{gl}_{1|1}$ in section 4.6.7.

Since the $E_{\mathcal{U}}$ branes are invariant under moves in Figs. 5 and 7, the homology groups $Hom^{*,*}_{\mathscr{D}_Y}(\mathscr{B}E_{\mathcal{U}}, E_{\mathcal{U}})$ are invariant under those moves applied to the second entry in the Hom, as well as the first. Similarly to section 4.6.7, the homology groups $Hom^{*,*}_{\mathscr{D}_Y}(\mathscr{B}E_{\mathcal{U}}, E_{\mathcal{U}})$ are related to $Hom^{*,*}_{\mathscr{D}_Y}(\mathscr{B}E_{\mathcal{U}}, I_{\mathcal{U}})$ by (4.56), with W which is a fixed graded vector space of rank 2^d. This comes about because each one dimensional figure eight brane is equivalent to complex $E_i \cong (E_i(I), \delta_i)$ where $E_i(I) = I_i \oplus I_i[1]\{-1\}$. It follows that $Hom^{*,*}_{\mathscr{D}_Y}(\mathscr{B}E_{\mathcal{U}}, I_{\mathcal{U}})$ is invariant under the moves in Figs. 5 and 7 applied to the second entry rather than the first. This proves the invariance under the remaining moves, and hence proves Thm. 1.

5.8. $U_\mathfrak{q}(\mathfrak{su}_2)$ skein from branes. Finally, we prove Thm. 2 which says that the Euler characteristics of $Hom^{*,*}_{\mathscr{D}_Y}(\mathscr{B}E_{\mathcal{U}}, I_{\mathcal{U}})$ satisfy the $\mathfrak{su}_2$ skein relation in eqn. (1.1), with $n = 2$. The strategy of the proof is the same as for $\mathfrak{gl}_{1|1}$: Since $Hom^{*,*}_{\mathscr{D}_Y}(\mathscr{B}E_{\mathcal{U}}, I_{\mathcal{U}})$ are link invariants, to show their Euler characteristics satisfies the skein relation, it suffices to consider the special link presentations in Fig. 8. Proving the theorem amounts to showing K-theory classes of the three branes corresponding to the bottom of that figure satisfy the relation.

The three branes corresponding to the bottom of Fig. 8 are:

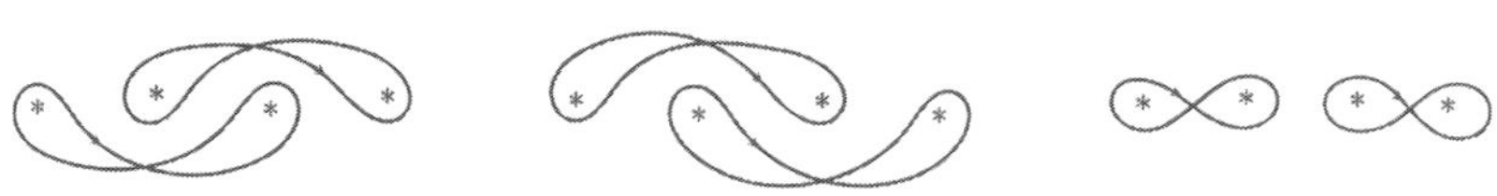

FIGURE 52. The three branes entering the Skein relation

As in the $\mathfrak{gl}_{1|1}$ case, we can simplify the branes using the swiping move. The swiping move lets the right figure eight swallow the left, in each of the three branes, without changing the corresponding object of $\mathscr{D}_Y$. Having done so, we learn that to compare K-theory classes of the three branes in Fig. 52, it suffices to compare K-theory classes of three simpler branes in Figs. 53, 54, and 55.

The first of the three branes, the brane in Fig. 53, is obtained from the simple figure eight $E_{\mathcal{U}}$ brane by braiding the second and third punctures by σ_2.

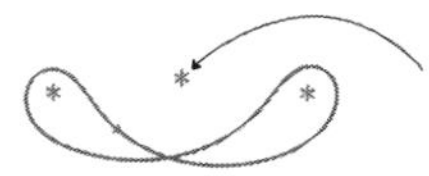

FIGURE 53. The reduced brane corresponding to positive crossing.

The brane has resolution:

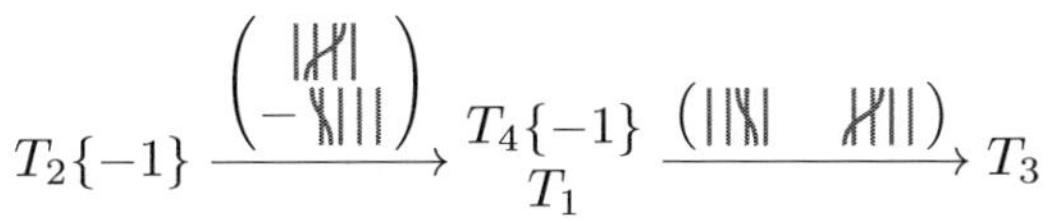

The overall degree of the brane, before any non-geometric shifts, follows from the fact that braiding acted trivially on the T_1 brane, hence its degree remained fixed. The degrees of the other branes in the complex are then fixed by the degrees of the maps in it. K-theory class of the brane is $(-\mathfrak{q}^{1/2})([T_3] - [T_1] - \mathfrak{q}^{-1}[T_4] + \mathfrak{q}^{-1}[T_2])$, where the prefactor comes from the constant degree shift of the braid that has a single positive crossing, with $e = 1$.

The second brane, from Fig. 54, has resolution:

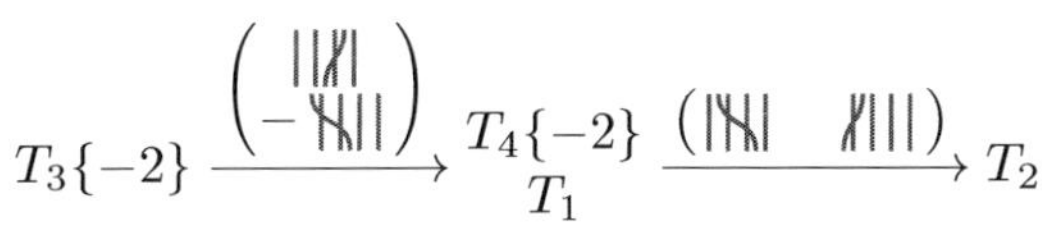

It is obtained from the $E_{\mathcal{U}}$ brane by braiding corresponding to σ_2^{-1}. The braid has has $w = -1$ and $\sigma = -1$,

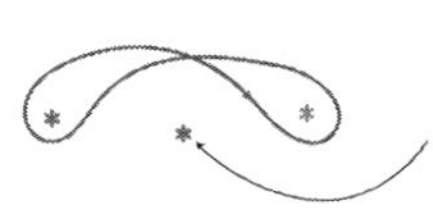

FIGURE 54. The reduced brane corresponding to the negative crossing.

so K-theory class of the brane equals $(-\mathfrak{q}^{-1/2})([T_2] - [T_1] - \mathfrak{q}^{-2}[T_4] + \mathfrak{q}^{-2}[T_3])$.

The third brane, from Fig. 55, is the $E_{\mathcal{U}}$ brane itself, with no braiding applied. It has resolution:

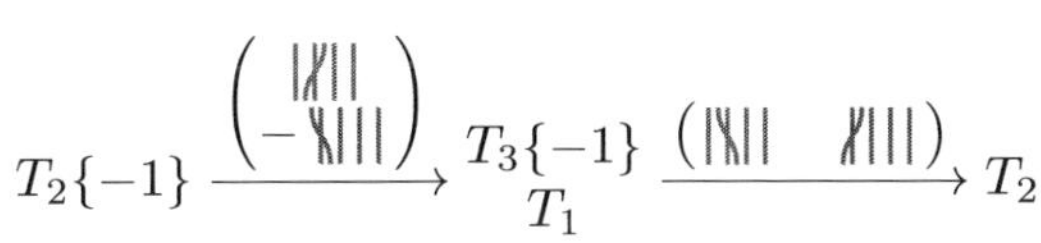

and K-theory class $(1 + \mathfrak{q}^{-1})[T_2] - [T_1] - \mathfrak{q}^{-1}[T_3]$.

FIGURE 55. The reduced brane corresponding to crossing.

K-theory classes of the three branes satisfy one linear relation:

(5.63)

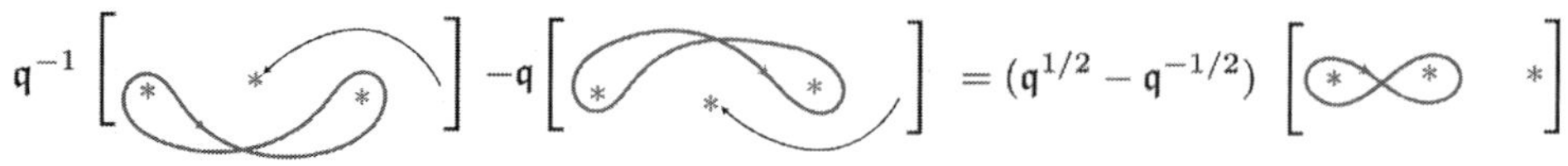

The relation (5.63) is the skein relation satisfied by the R matrices of $U_{\mathfrak{q}}(\mathfrak{su}_2)$, see e.g. [**74**]. It coincides with the $n = 2$ skein relation in eqn. (1.1) after setting

$$\mathbf{q}^{-1/2} = -\mathfrak{q}^{1/2}. \tag{5.64}$$

This proves Thm. 2 for $\mathfrak{su}_2$.

Appendix A. The swiping move

The swiping move plays the key role in proving the homology groups are invariants of links. In this appendix we prove that the branes related by the swiping moves are equivalent objects of $\mathscr{D}_Y$ by proving their resolutions in terms of T-branes are homotopy equivalent. The swiping move is shown in Fig. 25 for $\mathfrak{gl}_{1|1}$ and in Fig. 49 for $\mathfrak{su}_2$.

In the derived Fuykaya category a brane $L \in \mathscr{D}_Y$ is equivalent to the complex $(L(T), \delta)$ resolving it. Two complexes $(L_1(T), \delta_1)$ and $(L_2(T), \delta_2)$ describe the resolution of the same brane, and are hence equivalent, if they are homotopy equivalent. This means there exist chain maps $f : L_1 \to L_2$ and $f' : L_2 \to L_1$ such that $f' \circ f$ and $f \circ f'$ are homotopic to the identity on $(L_1(T), \delta_1)$ and $(L_2(T), \delta_2)$, respectively. The idea is that, if such maps f and f' exist, they are represent the identity maps from L to itself, in terms of the two resolutions of L, see e.g. [**43, 46**]. Once we specialize to $\mathfrak{su}_2$ and $\mathfrak{gl}_{1|1}$, respectively, we will explain in more detail how the homotopies are defined, in the case of ordinary and twisted complexes. Ordinary complexes are special cases of the twisted ones.

A.1. The move for $\mathfrak{gl}_{1|1}$. The swiping move for $\mathfrak{gl}_{1|1}$ is the equivalence of the branes $T_1 \times E_1$ and $E_1 \times T_3[-1]$ on the left and the right hand sides of Fig. 25, repeated here for convenience:

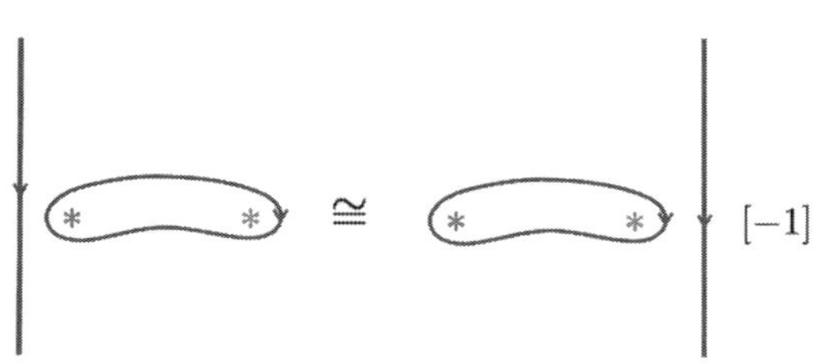

A.1.1. Given a pair of branes in L, $L' \in \mathscr{D}_Y$ and the twisted complexes of T-branes resolving them, $L \cong (L(T), \delta)$ and $L' \cong (L'(T), \delta')$, a morphism

$$\mathcal{P} \in Hom_{\mathscr{D}_Y}(L, L'),$$

can be represented as a map of the corresponding complexes, which satisfies [**90**]:

$$0 = m_1^{\mathrm{Tw}}(\mathcal{P}) = m_1^{\mathcal{P}}(\mathcal{P}) + m_2^{\mathcal{P}}(\delta', \mathcal{P}) + m_2^{\mathcal{P}}(\mathcal{P}, \delta). \tag{A.1}$$

A morphism $\mathcal{P}$ satisfying (A.1) plays a role of a chain map for ordinary complexes. Twisted A_∞ products are sign twists of the ordinary ones, as we recall in appendix

B. In writing (A.1), we take the higher A_∞ maps $m^{p}_{\ell>2}$ to vanish, as they do in our theory, for otherwise they would contribute to the definition of $m_1^{\mathrm{Tw}}(\mathcal{P})$.

Next, a given pair of morphisms are equivalent in $\mathscr{D}_Y$

$$\mathcal{P} \cong \mathcal{P}' \in Hom_{\mathscr{D}_Y}(L, L')$$

if they differ by an exact morphism $m_1^{Tw}(\mathcal{Q})$:

$$\mathcal{P} = \mathcal{P}' + m_1^{Tw}(\mathcal{Q}) \tag{A.2}$$

where $\mathcal{Q} \in Hom_{\mathscr{D}_Y}(L, L'[-1])$ is analogous to a chain homotopy of ordinary complexes.

A.1.2. We will show the two branes $L_1 = E_1 \times T_3$ and $L_2 = T_1 \times E_1[1]$ are equivalent as objects of $\mathscr{D}_Y$, by showing that there is a map $f \in Hom_{\mathscr{D}_Y}(L_1, L_2)$ with an inverse $f' \in Hom_{\mathscr{D}_Y}(L_2, L_1)$ which satisfy

$$f' \circ f \cong id_{L_1}, \qquad f \circ f' \cong id_{L_2}. \tag{A.3}$$

where the equivalence is a homotopy in the sense of (A.2).

A.1.3. The resolutions of the two branes are:

$$L_1 \quad \cong \quad T_{13}[1] \underset{\delta'_{10}}{\overset{\delta'_{01}}{\rightleftarrows}} T_{33} \tag{A.4}$$

where $\delta'_{01} =$  and $\delta'_{10} =$ , and:

$$L_2 \quad \cong \quad T_{11}[2] \underset{\delta_{10}}{\overset{\delta_{01}}{\rightleftarrows}} T_{13}[1] \tag{A.5}$$

where $\delta_{01} =$  and $\delta_{10} =$ . The signs in the differentials come from the factor of $(-1)^{M(T_C)}$ in taking the product of T_i with E_1 combining with the factor of $(-1)^{(n-1)k_0}$ from the individual oval branes. As explained in section 4.5.1, we have the freedom to twist the differential by a sign in one of the maps, corresponding to a different choice of spin structure on the brane we started with, or more generally, by a complex parameter which corresponds to choosing a generic holonomy on the brane of S^1 topology which the complex resolves. The discussion that follows readily generalizes, as is simple to verify.

One can easily find the morphisms f and f' we need. The morphism $f \in Hom_{\mathscr{D}_Y}(L_1, L_2)$, going from the resolution of L_1 to the resolution of L_2, is given by following diagram:

$$\begin{array}{ccccc} & & T_{13}[1] & \underset{\delta'_{10}}{\overset{\delta'_{01}}{\rightleftarrows}} & T_{33} \\ & \swarrow^{f_{11}} & \downarrow^{f_{01}} & \swarrow_{f_{00}} & \\ T_{11}[2] & \underset{\delta_{10}}{\overset{\delta_{01}}{\rightleftarrows}} & T_{13}[1] & & \end{array}$$

where $f_{11} =$ , $f_{01} = \hbar$ and $f_{00} =$ . The morphism $f' \in Hom_{\mathscr{D}_Y}(L_2, L_1)$ is given by:

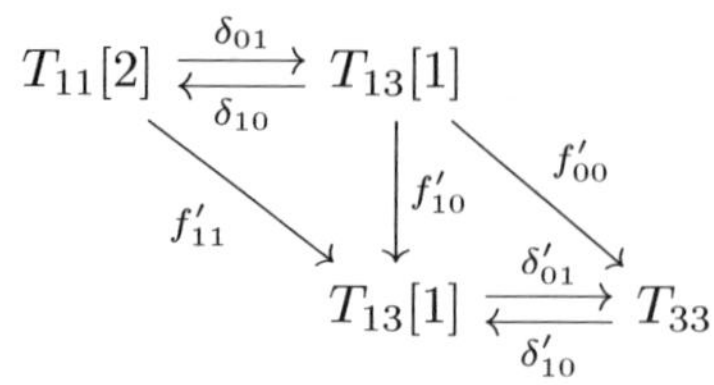

where $f'_{11} =$  , $f'_{10} = \hbar$  and $f'_{00} =$  . One easily checks that

$$m_1^{Tw}(f) = 0 = m_1^{Tw}(f').$$

A.1.4. To show that the compositions $f \circ f'$ and $f' \circ f$ are each homotopic to the corresponding identity, let

$$g = f \circ f' - \hbar^2 \cdot id_{L_2}, \qquad g' = f' \circ f - \hbar^2 \cdot id_{L_1}. \tag{A.6}$$

We will now show that g and g' are exact in the sense of (A.2). The power of $\hbar$ multiplying the identity elements in (A.6) is why the homomorphisms in (A.3) only hold for $\hbar \neq 0$. For any non-zero $\hbar$, its value is in (A.6) immaterial: and we can set it to 1 by appropriately rescaling f, f' and g, g'. For $\hbar = 0$, we cannot.

A.1.5. The maps g and g' can be simplified, as shown:

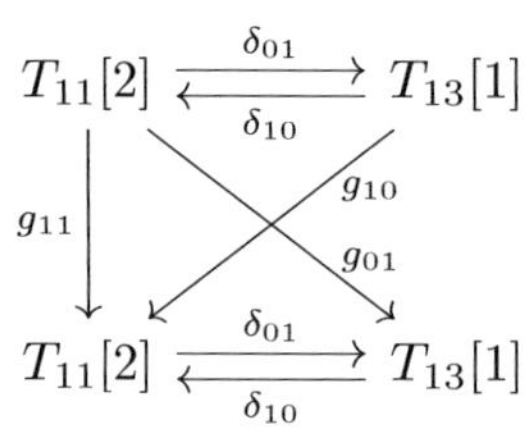

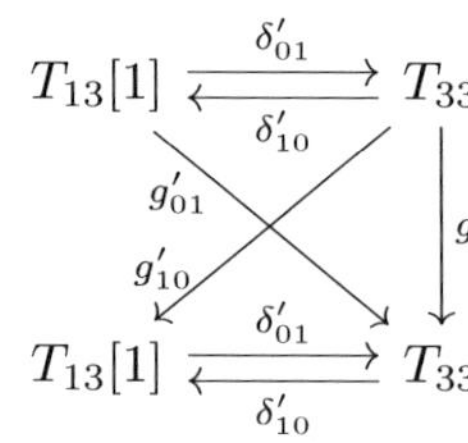

with components

(A.7) $g_{11} = -\hbar^2$  $\qquad g_{10} = -\hbar$  $\qquad g_{01} = -\hbar$ 

(A.8) $g'_{10} = -\hbar$ 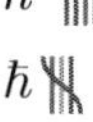 $\qquad g'_{01} = -\hbar$  $\qquad g'_{00} = -\hbar^2$ 

To show that the maps g and g' are exact, we show that there exist s and s' such that

$$g = m_1^{\mathrm{Tw}}(s), \qquad g' = m_1^{\mathrm{Tw}}(s'), \tag{A.9}$$

Explicitly, the maps s and s' are given by

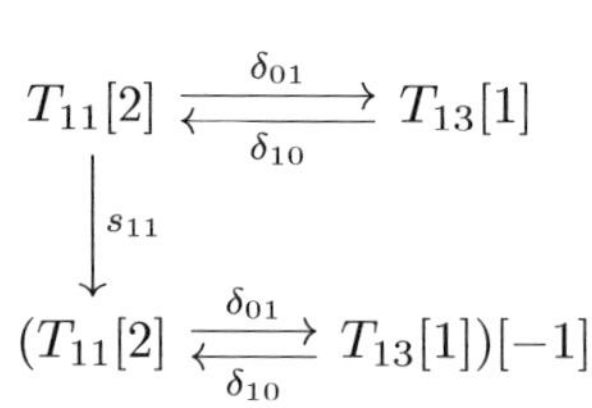

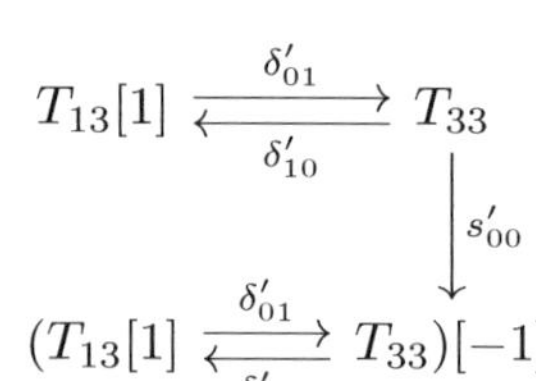

where $s_{11} = \hbar$ and $s'_{00} = \hbar$. One can then verify (A.9) by explicit computation.

A.2. The move for $\mathfrak{su}_2$. The swiping move for $\mathfrak{su}_2$ is the equivalence of branes $T_1 \times E_2$ and $E_2 \times T_3\{-1\}$ branes on the left and right hand sides of Fig. 49, repeated in Fig. 56 for convenience, as objects of $\mathscr{D}_Y$.

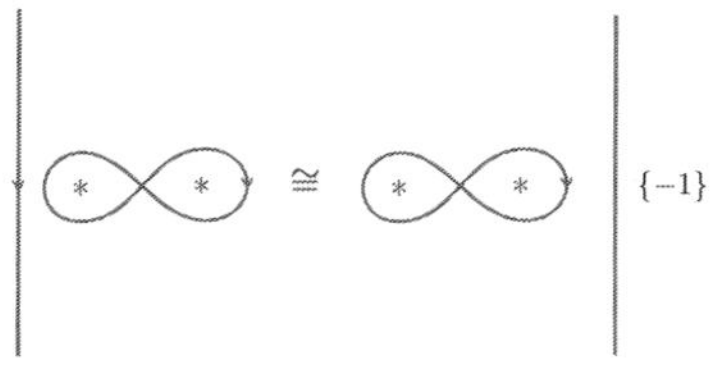

FIGURE 56. The "swiping move" for $\mathfrak{su}_2$

A.2.1. The resolution of a brane $L \in \mathscr{D}_Y$ in terms of T-branes is an ordinary complex $L \cong (L(T), \delta)$, as we have seen in section 5, since the T-brane endomorphism algebra is concentrated in degree zero. Specializing the section A.1.1 to this case, we get the familiar statements, as follows.

Given a pair of branes in L, $L' \in \mathscr{D}_Y$ and their projective resolutions $(L(T), \delta)$ and $(L'(T), \delta')$, a morphism $\mathcal{P} \in hom_Y(L, L')$, is represented by chain map of the resolving complexes. This is a map that satisfies

$$\delta' \cdot \mathcal{P} - \mathcal{P} \cdot \delta = 0,$$

where $\cdot$ coincides with m_2 product. Further, a pair of morphisms $\mathcal{P}, \mathcal{P}' \in hom_Y(L, L')$, are equivalent in $\mathscr{D}_Y$ if the difference between them

$$\mathcal{P} = \mathcal{P}' + \delta' \cdot \mathcal{Q} + \mathcal{Q} \cdot \delta$$

is a null homotopic chain map, for some $\mathcal{Q} \in hom_Y(L, L'[-1])$.

A.2.2. The resolutions for the $L_1 = T_1 \times E_2$ and $L_2 = E_2 \times T_3\{-1\}$ branes on the left and right hand sides of Fig. 49, respectively are

$$\begin{array}{ccc} L_1 & \cong & T_{12}\{-1\} \xrightarrow{d_2} \begin{matrix} T_{13}\{-1\} \\ T_{11} \end{matrix} \xrightarrow{d_1} T_{12} \\ \\ L_2 & \cong & T_{23}\{-2\} \xrightarrow{d_2'} \begin{matrix} T_{33}\{-2\} \\ T_{13}\{-1\} \end{matrix} \xrightarrow{d_1'} T_{23}\{-1\} \end{array}$$

where $d_2 =$  , $d_1 =$  , and $d_2' =$  , $d_1' =$  .

To show homotopy equivalence, we must find chains maps f and f', satisfying that $f' \circ f$ is homotopic to the identity on L_1 and $f \circ f'$ is homotopic to the identity on L_2.

The chain map f is found by asking for the maps in the following diagram to commute:

$$\begin{array}{ccccc} T_{12}\{-1\} & \xrightarrow{d_2} & \begin{matrix} T_{13}\{-1\} \\ T_{11} \end{matrix} & \xrightarrow{d_1} & T_{12} \\ \downarrow{\scriptstyle f_2} & & \downarrow{\scriptstyle f_1} & & \downarrow{\scriptstyle f_0} \\ T_{23}\{-2\} & \xrightarrow{d_2'} & \begin{matrix} T_{33}\{-2\} \\ T_{13}\{-1\} \end{matrix} & \xrightarrow{d_1'} & T_{23}\{-1\} \end{array}$$

One finds easily that the components of f are $f_2 =$ , $f_1 =$  and $f_0 =$ . The chain map f' is found from a similar commutative diagram and has components $f'_2 = -$ , $f'_1 =$ 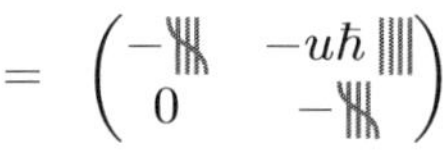, and $f'_0 = -$ .

A.2.3. To prove that $f \circ f'$ and $f' \circ f$ are each homotopic to the identify on corresponding brane, let

$$g = f \circ f' - (u\hbar)^2 \cdot id_{L_2}, \qquad g' = f' \circ f - (u\hbar)^2 \cdot id_{L_1}. \tag{A.10}$$

We want to find chain homotopies s and s' such that

$$g = \delta' \circ s + s \circ \delta', \qquad g' = \delta \circ s' + s' \circ \delta, \tag{A.11}$$

where δ and δ' are the differentials in the resolutions of the two branes. The first of these chain homotopies is given by the following diagram:

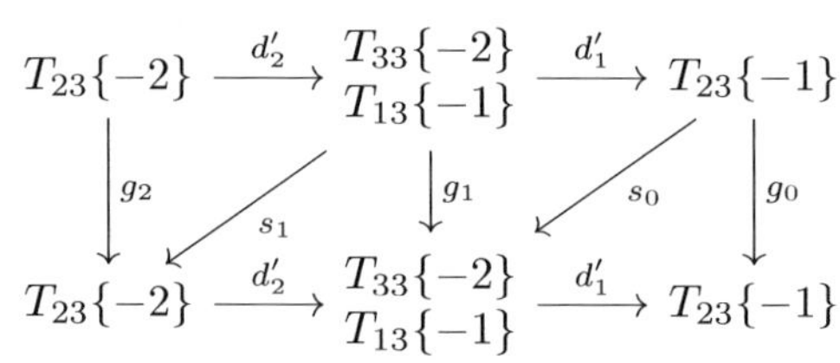

where g is given by

$g_2 = -u\hbar$  $\qquad g_1 =$ 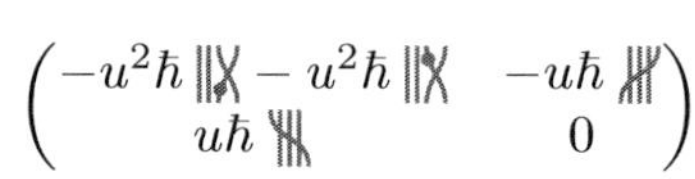 $\qquad g_0 = -u\hbar$ .

The needed chain homotopy s is given by

$s_1 =$  $\qquad s_0 =$ 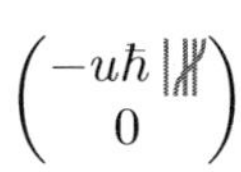.

One can then explicitly verify that the first of two equalities in (A.10) holds. Similarly, we find that s' is given by $s'_1 =$ 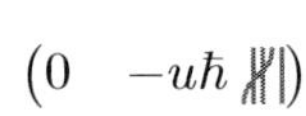 and $s'_0 =$ 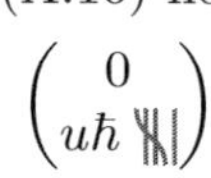.

Provided $u\hbar \neq 0$, one can always rescale f, f', g, and g' to set the factor $(u\hbar)^2$ to 1. This thus completes the proof that the swiping move holds in the $\mathfrak{su}_2$ theory. The move does not hold if either of the two components of the divisor D_O is deleted, corresponding to setting either u or $\hbar$ to zero.

Appendix B. Signs in the category of twisted complexes

The most technical aspect of the construction of a consistent assignment of signs, which allows us to obtain link homologies over $\mathbb{Z}$, and not merely over $\mathbb{Z}_2$ coefficients. The appropriate way to do this follows the familiar construction from standard treatments of derived Fukaya categories, for example in [**36, 90**]. In our setting, there is a slight simplification in the construction, since the algebra A is at most a differential graded algebra, and not a general A_∞ algebra. Because the sign assignments are so crucial for us, we will give a self contained summary. We will suppress equivariant gradings, since they play no role for this aspect of the story.

This section is aimed at the $\mathfrak{gl}_{1|1}$ theory, whose strands algebra is a differential graded algebra. The $\mathfrak{su}_2$ algebra is an ordinary associative algebra, with trivial differential, all of whose elements have Maslov degree 0 so the discussion of signs simplifies vastly.

B.1. Differential-graded algebra as an A_∞ algebra. A differential-graded algebra with multiplication $a \cdot b$ and a differential ∂ satisfying

$$\partial^2 = 0, \qquad \partial(a \cdot b) = \partial(a) \cdot b + (-1)^{M(a)} a \cdot \partial(b), \qquad (a \cdot b) \cdot c = a \cdot (b \cdot c), \tag{B.1}$$

where $M(a)$ is the Maslov degree of a.

It originates from the A_∞ algebra underlying the Fukaya category, after we identify

$$m_1(a) = (-1)^{M(a)} \partial(a), \qquad m_2(a, b) = (-1)^{M(b)} a \cdot b, \tag{B.2}$$

where m_1 and m_2 are respectively, the Floer differential and product, assuming the higher A_∞ operations $m_{\ell>2} = 0$ vanish.

Let us check, as an exercise, that the differential graded algebra relations in (B.1) coincide with the A_∞ relations. We obviously have

$$m_1(m_1(a)) = -\partial^2(a) = 0, \tag{B.3}$$

and

$$\begin{aligned} m_1(m_2(a_2, a_1)) &= (-1)^{M(a_2)+2M(a_1)} \partial(a_2 \cdot a_1) \\ &= (-1)^{M(a_2)} (\partial(a_2) \cdot a_1 + (-1)^{M(a_2)} a_2 \cdot \partial(a_1)) \\ &= (-1)^{M(a_1)} m_2(m_1(a_2), a_1) - m_2(a_2, m_1(a_1)), \end{aligned} \tag{B.4}$$

Finally, the associativity of the algebra is equivalent to

$$\begin{aligned} m_2(m_2(a_3, a_2), a_1) &= (-1)^{M(a_1)+M(a_2)} (a_3 \cdot a_2) \cdot a_1 \\ &= (-1)^{M(a_1)+M(a_2)} a_3 \cdot (a_2 \cdot a_1) \\ &= (-1)^{M(a_1)} m_2(a_3, m_2(a_2, a_1)). \end{aligned} \tag{B.5}$$

These are indeed the A_∞ relations from [**90**], restricted to a theory with $m_{p>2} = 0$.

B.2. Pre-twisted complexes. A pre-twisted complex (V, δ) is

$$(V, \delta) = \left(\bigoplus_i V_i[k_i], \sum_{ij} d_{i,j} \right) \tag{B.6}$$

where V_i are objects of an A_∞ category, and $\delta : V \to V[1]$ is a collection of maps

$$d_{i,j} : V_i[k_i] \to V_j[k_j + 1]. \tag{B.7}$$

$d_{i,j}$ are morphisms between the objects of the A_∞ category.

A morphism between a pair of pre-twisted complexes $(V_\alpha, \delta_\alpha)$ and (V_β, δ_β), of degree n, $A_{\alpha,\beta} : V_\alpha \to V_\beta[n]$ is simply a collection of maps $A_{\alpha,\beta} = \bigoplus_{i,j} a_{\alpha i,\beta j}$ between the summands of the form

$$a_{\alpha i,\beta j} : V_{\alpha i}[k_i] \to V_{\beta j}[k_j + n]. \tag{B.8}$$

The sign assiggnments on the category of pre-twisted complexes are determined in [**90**]. (What we call the category of pre-twisted complexes is called the additive enlargment of the A_∞ category in [**90**].) First, one modifies the A_∞ morphisms by replacing

$$a_{0,1} \to (-1)^{(n-1)k_0} a_{0,1}, \tag{B.9}$$

whenever $a_{0,1} : V_0[k_0] \to V_1[k_1 + n]$. One further modifies the A_∞ operations by introducing an extra sign, sending

$$m_i(a_{i-1,i}, \dots, a_{0,1}) \to (-1)^{k_0} m_i(a_{i-1,i}, \dots, a_{0,1}). \tag{B.10}$$

The maps $m_i^p(a_{i-1,i}, \dots, a_{0,1})$ will be understood as modifications of the standard A_∞ maps of the Fukaya category by the two additional signs from (B.9) and (B.10). The sign twisted maps satisfy the standard A_∞ relations [**90**].

B.3. Twisted objects. The category of twisted complexes, denoted by TwA in [**90**], is itself an A_∞ category, whose operations we will denote by m_ℓ^{Tw}.

Objects in the category are pre-twisted complexes (V, δ) introduced above, satisfying

$$m_1^{Tw}(\delta) \equiv m_1^p(\delta) + m_2^p(\delta, \delta) = 0. \tag{B.11}$$

A morphism a of degree n from (V_1, δ_1) to (V_2, δ_2) in the category of twisted complexes, an element of $hom_{\mathrm{Tw}(A)}(V_1, V_2[n])$, is a morphism from V_1 to $V_2[n]$ in the category of pre-twisted complexes, on which the twisted differential of TwA acts by

$$m_1^{Tw}(a) = m_1^p(a) + m_2^p(a, \delta_1) + m_2^p(\delta_2, a). \tag{B.12}$$

Further, if b is a map from (V_0, δ_0) to (V_1, δ_1) of any degree,

$$m_2^{Tw}(a, b) = m_2^p(a, b), \tag{B.13}$$

because we assumed A is a dga, with only m_1 and m_2 non-vanishing. If A were a general A_∞ algebra all its higher A_∞ operations would also contribute.

The derived category of the A_∞ category, or the derived Fukaya category in the text, is the cohomology category $H^0(\mathrm{Tw}A)$, whose objects are the same as in TwA, but whose morphisms are cohomology classes of m_1^{Tw} [**90**]. Rather than a complication, signs introduced in this section are necessary for the category of twisted complexes TwA to have the standard sign assignments. With the added signs, the pre-twisted maps $A_{\alpha,\beta} : V_\alpha \to V_\beta[n]$ become honest cohomological degree n maps between objects in the category of twisted complexes TwA.

Appendix C. Signs in the differential

In this section, we explain how to assign signs to maps in δ for a brane in $\mathscr{D}_Y$. Having a consistent sign assignment is crucial as we want a homology theory with a $\mathbb{Z}$-valued homological grading. In our case, it suffices to explain how to assign signs to the differential δ_O of a brane on Y_O, which is Y with the divisor D_O deleted. Having understood how to assign signs in δ_O, the deformation corresponding to filling in D_O determines the signs in the full differential δ as well.

A special feature of the differential δ_O is that the corresponding complex is a product of one dimensional complexes, as explained in section 3. Assigning signs to a product complex is essentially a standard problem which is easily solved. The only subtlety is that, when the complex is a twisted, rather than an ordinary complex, there are additional signs that enter, sourced by signs from section B.

C.1. Signs in one dimension. We will start our discussion on Y_O which is one dimensional. As explained in section 3, a compact brane L of topology of S^1 has the resolution $(L(T), \delta_O)$ which may be described in terms of a one-dimensional periodic toric grid. The sites of the grid are the direct summands in $L(T) = \bigoplus_i T_i$, and δ_O is the collection of maps between them:

$$T_1 \longleftrightarrow T_2 \longleftrightarrow \ldots \longleftrightarrow T_n \longleftrightarrow \cdot \tag{C.1}$$

We do not specify directions of arrows since they are example-dependent and irrelevant in our present discussion. The one dimensional geometry of Y_O ensures that the complex (C.1) is a linear chain of maps.

A special feature of the theory on Y_O is that the differential δ_O squares to zero for every choice of signs of the n maps in the grid. Since gauge transformations of the form $T_i \to -T_i$ do not change the brane, there is a single choice of sign that is meaningful, given by the product of all signs in the grid. As discussed in section 4.4.5, the $\mathbb{Z}_2$ worth of choices of differentials for a brane of S^1 topology should correspond to the choice of the spin structure of the Lagrangian brane L which the complex $(L(T), \delta_O)$ describes.

C.1.1. If the Lie algebra is $\mathfrak{su}_2$ only one sign choice deforms from $Y_O = Y_{0,0}$ with both the diagonal and the punctures deleted, to $Y_{u,0}$, where only the diagonal is deleted, or to $\mathcal{Y}_{u,0}$, the upstairs space. The choice that deforms to $u \neq 0$ is has a product of all the signs given by $(-1)^{1+n/2}$ where n is number of T-branes in the complex. One shows this by starting with an elementary figure eight brane, for which n is four, and considering the effect of braiding. This is in line with the identification of the $\mathbb{Z}_2$ choice of signs with that of the spin structures on L. The deformation $u \neq 0$, corresponds from perspective of L, to filling in the disk which the brane bounds and which picks out a unique spin strucutre that extends to the disk.

C.1.2. If the Lie algebra is $\mathfrak{gl}_{1|1}$ and Y is one dimensional, the corresponding a deformation does not exist. The $\mathbb{Z}_2$ worth of choices of signs in the full theory reflects the $\mathbb{Z}_2$ worth of choices spin structure on the brane whose topology is S^1, which is not contractible in Y.

C.2. Signs in $\mathfrak{su}_2$ complexes. Since the complex describing the brane $(L(T), \delta_O)$ on Y_O is a product of ordinary complexes when the Lie algebra is $\mathfrak{su}_2$, there is a simple, standard construction for assigning signs so that δ_O squares to zero, which works as follows.

C.2.1. The best way to describe the construction is recursive. We know how to assign signs if the brane is one dimensional. Suppose we are given a d dimensional brane $(L_d(T), \delta_d)$. The brane has summands which are labeled by sites of a d-dimensional toric grid

$$L_d(T) = \bigoplus_{i_1,\ldots,i_d} T_{i_1\ldots i_d},$$

with a product differential δ_d which squares to zero, $\delta_d^2 = 0$ (δ_d stands for δ_O for the d dimensional brane). We suppressed the gradings of the T-branes.

We will explain how to assign the signs on the product $L_{d+1} = L_d \times L_1$ of $(L_d(T), \delta_d)$ with another one-dimensional Lagrangian $(L_1(T), \delta_1)$, where

$$L_1(T) = \bigoplus_{i_{d+1}} T_{i_{d+1}}.$$

The product brane $L_{d+1}(T)$ lives on a $d+1$-dimensional grid. We claim its differential is given by

$$\delta_{d+1} = \sum_{i_1,\dots,i_d,i_{d+1}} (-1)^{i_1+\dots+i_d} \mathbb{1}_{i_1,\dots,i_d} \otimes \delta_{i_{d+1}} + \sum_{i_{d+1}} \delta_d \otimes \mathbb{1}_{i_{d+1}} \tag{C.2}$$

The sign twist in (C.2) preserves the $\mathbb{Z}_2$ sign choice we made for the $d+1$-th brane, because for a brane of S^1 topology, each one dimensional complex has an even number T-branes, and of maps connecting them, and we changed the overall sign only.

To check that the differential squares to zero, we need to show it squares to zero around every square in the grid. Per assumption, it does so for all squares that lie entirely in the first d directions, those associated with L_d. Due to the alternating sign we introduced in (C.2), it also squares to zero for squares with vertices

$$T_{i_1,\dots,i_j,\dots,i_d}\times T_{i_d},\quad T_{i_1,\dots,i_j+1,\dots,i_d}\times T_{i_d},\quad T_{i_1,\dots,i_j,.,i_d}\times T_{i_d+1},\quad T_{i_1,\dots,i_j+1,.,i_d}\times T_{i_d+1}. \tag{C.3}$$

which involve the $d+1$-st direction associated to L_1.

C.3. Signs in $\mathfrak{gl}_{1|1}$ complexes. Twisted complexes introduce additional signs which we reviewed in appendix B. For ordinary complexes, the additional sign we introduced in (C.2) suffices to ensure the differential squares to zero, because products the of elements, going two ways around a square with vertices in (C.3), commute. Thinking of an ordinary complex as a special case of a twisted one, the products commute because all elements of differential have Maslov index 0.

For a twisted complex, differential consists of maps of arbitrary Maslov index. In taking products, one gets additional sources of signs which we summarized in appendix B. Due to these signs, there are additional signs that enter in forming the product complex, to get a differential δ_O that squares to zero.

C.3.1. The first source of signs comes from (B.2) which says that $a \cdot b = (-1)^{M(b)} m_2(a,b)$. For the differential δ_O of the product complex to square to zero, we need to cancel this sign.

Consider one of the new squares formed in the $d+1$ dimensional grid with vertices in (C.3). If Maslov degrees of branes $T_{i_1,\dots,i_j+1,\dots,i_d} \times T_{i_d}$ and $T_{i_1,\dots,i_j,.,i_d} \times T_{i_d+1}$, located at a pair of opposite vertices of the square, are both odd or both even, then the sign in (B.2) does not introduce any new signs, and the differential squares to zero as before. If however one is odd, the other even, the product of maps around the square anti-commutes, instead of commuting, and we have a sign that needs cancelling.

To cancel the sign, it suffices to multiply every map in the complex by $(-1)^{M(T)}$ where $M(T)$ is the Maslov index of the thimble T at the grid site where the map lands at. If Maslov degrees of $T_{i_1,\dots,i_j+1,\dots,i_d} \times T_{i_d}$ and $T_{i_1,\dots,i_j,.,i_d} \times T_{i_d+1}$ branes are both odd, or both even, the new sign introduces an even number of sign changes around the square, so nothing changes. If Maslov index of one brane is even and the other odd, we get an extra minus sign which cancels the sign from (B.2) and which ensures the new differential squares to zero. The signs in the differental are no longer of the simple form in (C.2), but they are a simple modification of it.

C.3.2. There are two more sources of signs in the differential of the twisted complex, due to the fact the ordinary A_∞ products get replaced with the twisted ones from section B.2. One can show that both of these additional sign contributes equally to the two paths around the square with vertices (C.3). As a result, it does not lead to any further modification of the differential δ_O from previous subsection.

Acknowledgments

The first author is grateful to Vivek Shende and Peng Zhou for many discussions about Floer theory, and for collaboration on closely related work [**5**]. We thank the organizers and participants of the AIM workshop on "Floer theory of symmetric products and Hilbert schemes", in December 2022, for a stimulating workshop which centered, in part, on this work. The first author especially benefited from discussions with Mohammed Abouzaid, Robert Lipshitz and Ivan Smith during the workshop.

References

[1] M. Aganagic, *Knot Categorification from Mirror Symmetry,* Part I: Coherent Sheaves, arXiv:2004.14518.

[2] M. Aganagic, *Knot Categorification from Mirror Symmetry,* Part II: Lagrangians, arXiv:2105.06039.

[3] M. Aganagic, *Knot Categorification from Mirror Symmetry,* Part III: String theory origins, to appear.

[4] Mina Aganagic, *Homological knot invariants from mirror symmetry*, ICM—International Congress of Mathematicians. Vol. 3. Sections 1–4, EMS Press, Berlin, 2023, pp. 2108–2144. MR4680311

[5] M. Aganagic, I. Danilenko, Y. Li, P. Zhou, and V. Shende, *Hecke algebras from Floer homology in Couloumb branches*, to appear.

[6] M. Aganagic, E. Frenkel, and A. Okounkov, *Quantum q-Langlands correspondence*, Trans. Moscow Math. Soc. **79** (2018), 1–83, DOI 10.1090/mosc/278. MR3881458

[7] M. Aganagic, M. McBreen, V. Shende, and P. Zhou, to appear.

[8] Mina Aganagic and Andrei Okounkov, *Elliptic stable envelopes*, J. Amer. Math. Soc. **34** (2021), no. 1, 79–133, DOI 10.1090/jams/954. MR4188815

[9] Mina Aganagic and Andrei Okounkov, *Quasimap counts and Bethe eigenfunctions*, Mosc. Math. J. **17** (2017), no. 4, 565–600, DOI 10.17323/1609-4514-2016-16-4-565-600. MR3734654

[10] M. Aganagic and S. Tamagni, *Stable envelopes for Verma modules from vortex moduli spaces*, to appear.

[11] J. W. Alexander, *Topological invariants of knots and links*, Trans. Amer. Math. Soc. **30** (1928), no. 2, 275–306, DOI 10.2307/1989123. MR1501429

[12] Paul S. Aspinwall, Tom Bridgeland, Alastair Craw, Michael R. Douglas, Mark Gross, Anton Kapustin, Gregory W. Moore, Graeme Segal, Balázs Szendrői, and P. M. H. Wilson, *Dirichlet branes and mirror symmetry*, Clay Mathematics Monographs, vol. 4, American Mathematical Society, Providence, RI; Clay Mathematics Institute, Cambridge, MA, 2009. MR2567952

[13] Denis Auroux, *A beginner's introduction to Fukaya categories*, Contact and symplectic topology, Bolyai Soc. Math. Stud., vol. 26, János Bolyai Math. Soc., Budapest, 2014, pp. 85–136, DOI 10.1007/978-3-319-02036-5_3. MR3220941

[14] Denis Auroux, *Fukaya categories of symmetric products and bordered Heegaard-Floer homology*, J. Gökova Geom. Topol. GGT **4** (2010), 1–54. MR2755992

[15] Denis Auroux, *Fukaya categories and bordered Heegaard-Floer homology*, Proceedings of the International Congress of Mathematicians. Volume II, Hindustan Book Agency, New Delhi, 2010, pp. 917–941. MR2827825

[16] Denis Auroux, *Special Lagrangian fibrations, wall-crossing, and mirror symmetry*, Surveys in differential geometry. Vol. XIII. Geometry, analysis, and algebraic geometry: forty years of the Journal of Differential Geometry, Surv. Differ. Geom., vol. 13, Int. Press, Somerville, MA, 2009, pp. 1–47, DOI 10.4310/SDG.2008.v13.n1.a1. MR2537081

[17] Dror Bar-Natan, *On Khovanov's categorification of the Jones polynomial*, Algebr. Geom. Topol. **2** (2002), 337–370, DOI 10.2140/agt.2002.2.337. MR1917056

[18] Stephen Bigelow, *A homological definition of the Jones polynomial*, Invariants of knots and 3-manifolds (Kyoto, 2001), Geom. Topol. Monogr., vol. 4, Geom. Topol. Publ., Coventry, 2002, pp. 29–41, DOI 10.2140/gtm.2002.4.29. MR2002601

[19] Stephen Bigelow, *A homological definition of the HOMFLY polynomial*, Algebr. Geom. Topol. **7** (2007), 1409–1440, DOI 10.2140/agt.2007.7.1409. MR2350288

[20] R. Bezrukavnikov and D. Kaledin, *Fedosov quantization in positive characteristic*, J. Amer. Math. Soc. **21** (2008), no. 2, 409–438, DOI 10.1090/S0894-0347-07-00585-1. MR2373355

[21] R. Bezrukavnikov, *Noncommutative Counterparts of the Springer Resolution*, arXiv:math/0604445.

[22] Joan S. Birman, *On the stable equivalence of plat representations of knots and links*, Canadian J. Math. **28** (1976), no. 2, 264–290, DOI 10.4153/CJM-1976-030-1. MR402715

[23] Andrei Căldăraru, *Derived categories of sheaves: a skimming*, Snowbird lectures in algebraic geometry, Contemp. Math., vol. 388, Amer. Math. Soc., Providence, RI, 2005, pp. 43–75, DOI 10.1090/conm/388/07256. MR2182889

[24] Sabin Cautis and Joel Kamnitzer, *Knot homology via derived categories of coherent sheaves. I. The $\mathfrak{sl}(2)$-case*, Duke Math. J. **142** (2008), no. 3, 511–588, DOI 10.1215/00127094-2008-012. MR2411561

[25] Sabin Cautis and Joel Kamnitzer, *Knot homology via derived categories of coherent sheaves. II. $\mathfrak{sl}_m$ case*, Invent. Math. **174** (2008), no. 1, 165–232, DOI 10.1007/s00222-008-0138-6. MR2430980

[26] Cheol-Hyun Cho, *Holomorphic discs, spin structures and Floer cohomology of the Clifford torus*, ProQuest LLC, Ann Arbor, MI, 2003. Thesis (Ph.D.)–The University of Wisconsin - Madison. MR2705021

[27] Cheol-Hyun Cho and Yong-Geun Oh, *Floer cohomology and disc instantons of Lagrangian torus fibers in Fano toric manifolds*, Asian J. Math. **10** (2006), no. 4, 773–814, DOI 10.4310/AJM.2006.v10.n4.a10. MR2282365

[28] Louis Crane and Igor B. Frenkel, *Four-dimensional topological quantum field theory, Hopf categories, and the canonical bases*, J. Math. Phys. **35** (1994), no. 10, 5136–5154, DOI 10.1063/1.530746. Topology and physics. MR1295461

[29] I. Danilenko, *Slices of the Affine Grassmannian and Quantum Cohomology*, PhD thesis, Columbia University.

[30] V. G. Drinfel′d, *Quasi-Hopf algebras and Knizhnik-Zamolodchikov equations*, Problems of modern quantum field theory (Alushta, 1989), Res. Rep. Phys., Springer, Berlin, 1989, pp. 1–13. MR1091757

[31] Pavel I. Etingof, Igor B. Frenkel, and Alexander A. Kirillov Jr., *Lectures on representation theory and Knizhnik-Zamolodchikov equations*, Mathematical Surveys and Monographs, vol. 58, American Mathematical Society, Providence, RI, 1998, DOI 10.1090/surv/058. MR1629472

[32] Pavel Etingof and Nathan Geer, *Monodromy of trigonometric KZ equations*, Int. Math. Res. Not. IMRN **24** (2007), Art. ID rnm123, 15, DOI 10.1093/imrn/rnm123. MR2377010

[33] B. L. Feĭgin and È. V. Frenkel′, *A family of representations of affine Lie algebras* (Russian), Uspekhi Mat. Nauk **43** (1988), no. 5(263), 227–228, DOI 10.1070/RM1988v043n05ABEH001935; English transl., Russian Math. Surveys **43** (1988), no. 5, 221–222. MR971497

[34] Andreas Floer, *Morse theory for Lagrangian intersections*, J. Differential Geom. **28** (1988), no. 3, 513–547. MR965228

[35] I. B. Frenkel and N. Yu. Reshetikhin, *Quantum affine algebras and holonomic difference equations*, Comm. Math. Phys. **146** (1992), no. 1, 1–60. MR1163666

[36] Kenji Fukaya, *Floer homology and mirror symmetry. II*, Minimal surfaces, geometric analysis and symplectic geometry (Baltimore, MD, 1999), Adv. Stud. Pure Math., vol. 34, Math. Soc. Japan, Tokyo, 2002, pp. 31–127, DOI 10.2969/aspm/03410031. MR1925734

[37] Kenji Fukaya, Yong-Geun Oh, Hiroshi Ohta, and Kaoru Ono, *Lagrangian intersection Floer theory: anomaly and obstruction. Part I*, AMS/IP Studies in Advanced Mathematics, vol. 46, American Mathematical Society, Providence, RI; International Press, Somerville, MA, 2009, DOI 10.1090/amsip/046.1. MR2553465

[38] Davide Gaiotto and Edward Witten, *Knot invariants from four-dimensional gauge theory*, Adv. Theor. Math. Phys. **16** (2012), no. 3, 935–1086. MR3024278

[39] D. Gaiotto, G. W. Moore, and E. Witten, *Algebra of the Infrared: String Field Theoretic Structures in Massive $\mathcal{N} = (2,2)$ Field Theory In Two Dimensions*, `arXiv:1506.04087`.

[40] D. Galakhov and G. W. Moore, *Comments On The Two-Dimensional Landau-Ginzburg Approach To Link Homology*, `arXiv:1607.04222`.

[41] R. Gopakumar and C. Vafa, *M theory and topological strings. 1. and 2.*, `arXiv:hep-th/9809187` and `arXiv:hep-th/9812127`.

[42] Sergei Gukov, Albert Schwarz, and Cumrun Vafa, *Khovanov-Rozansky homology and topological strings*, Lett. Math. Phys. **74** (2005), no. 1, 53–74, DOI 10.1007/s11005-005-0008-8. MR2193547

[43] Peter John Hilton and Urs Stammbach, *A course in homological algebra*, Graduate Texts in Mathematics, Vol. 4, Springer-Verlag, New York-Berlin, 1971. MR346025

[44] K. Hori, A. Iqbal, and C. Vafa, *D-branes and mirror symmetry*, `arXiv:hep-th/0005247`.

[45] Kentaro Hori, *Mirror symmetry and quantum geometry*, Proceedings of the International Congress of Mathematicians, Vol. III (Beijing, 2002), Higher Ed. Press, Beijing, 2002, pp. 431–443. MR1957551

[46] M. Herbst, K. Hori, and D. Page, *Phases Of N=2 Theories In 1+1 Dimensions With Boundary*, `arXiv:0803.2045`

[47] Nafiz Ishtiaque, Seyed F. Moosavian, Surya Raghavendran, and Junya Yagi, *Superspin chains from superstring theory*, SciPost Phys. **13** (2022), no. 4, Paper No. 083, 60, DOI 10.21468/scipostphys.13.4.083. MR4503298

[48] Vaughan F. R. Jones, *A polynomial invariant for knots via von Neumann algebras*, Bull. Amer. Math. Soc. (N.S.) **12** (1985), no. 1, 103–111, DOI 10.1090/S0273-0979-1985-15304-2. MR766964

[49] V. G. Kac, *Lie superalgebras*, Advances in Math. **26** (1977), no. 1, 8–96, DOI 10.1016/0001-8708(77)90017-2. MR486011

[50] Mikhail Khovanov, *A categorification of the Jones polynomial*, Duke Math. J. **101** (2000), no. 3, 359–426, DOI 10.1215/S0012-7094-00-10131-7. MR1740682

[51] Mikhail Khovanov, *Link homology and categorification*, International Congress of Mathematicians. Vol. II, Eur. Math. Soc., Zürich, 2006, pp. 989–999. MR2275632

[52] M. Khovanov and A. D. Lauda, *A diagrammatic approach to categorification of quantum groups I*, `arXiv:0803.4121`.

[53] M. Khovanov and A. D. Lauda, *A diagrammatic approach to categorification of quantum groups II*, `arXiv:0804.2080`.

[54] Mikhail Khovanov and Lev Rozansky, *Matrix factorizations and link homology*, Fund. Math. **199** (2008), no. 1, 1–91, DOI 10.4064/fm199-1-1. MR2391017

[55] Mikhail Khovanov and Lev Rozansky, *Matrix factorizations and link homology. II*, Geom. Topol. **12** (2008), no. 3, 1387–1425, DOI 10.2140/gt.2008.12.1387. MR2421131

[56] Mikhail Khovanov and Paul Seidel, *Quivers, Floer cohomology, and braid group actions*, J. Amer. Math. Soc. **15** (2002), no. 1, 203–271, DOI 10.1090/S0894-0347-01-00374-5. MR1862802

[57] V. G. Knizhnik and A. B. Zamolodchikov, *Current algebra and Wess-Zumino model in two dimensions*, Nuclear Phys. B **247** (1984), no. 1, 83–103, DOI 10.1016/0550-3213(84)90374-2. MR853258

[58] Toshitake Kohno, *Monodromy representations of braid groups and Yang-Baxter equations* (English, with French summary), Ann. Inst. Fourier (Grenoble) **37** (1987), no. 4, 139–160, DOI 10.5802/aif.1114. MR927394

[59] Maxim Kontsevich, *Homological algebra of mirror symmetry*, Proceedings of the International Congress of Mathematicians, Vol. 1, 2 (Zürich, 1994), Birkhäuser, Basel, 1995, pp. 120–139. MR1403918

[60] R. J. Lawrence, *Homology representations of braid groups*, D.Phil. Thesis, University of Oxford (June 1989).

[61] R. J. Lawrence, *A functorial approach to the one-variable Jones polynomial*, J. Differential Geom. **37** (1993), no. 3, 689–710. MR1217166

[62] R. J. Lawrence, *The homological approach applied to higher representations*, Harvard preprint (1990), available at `http://www.ma.huji.ac.il/~ruthel/`

[63] Robert Lipshitz, *A cylindrical reformulation of Heegaard Floer homology*, Geom. Topol. **10** (2006), 955–1096, DOI 10.2140/gt.2006.10.955. [Paging previously given as 955–1097]. MR2240908

[64] R. Lipshitz, P. Ozsvath and D. Thurston, *Bordered Heegaard Floer homology: invariance and pairing*, arXiv:0810.0687.

[65] Alexey Litvinov and Lev Spodyneiko, *On W algebras commuting with a set of screenings*, J. High Energy Phys. **11** (2016), 138, front matter+17, DOI 10.1007/JHEP11(2016)138. MR3594801

[66] Ciprian Manolescu, Peter Ozsváth, and Sucharit Sarkar, *A combinatorial description of knot Floer homology*, Ann. of Math. (2) **169** (2009), no. 2, 633–660, DOI 10.4007/annals.2009.169.633. MR2480614

[67] Ciprian Manolescu, *Nilpotent slices, Hilbert schemes, and the Jones polynomial*, Duke Math. J. **132** (2006), no. 2, 311–369, DOI 10.1215/S0012-7094-06-13224-6. MR2219260

[68] Ciprian Manolescu, *An introduction to knot Floer homology*, Physics and mathematics of link homology, Contemp. Math., vol. 680, Amer. Math. Soc., Providence, RI, 2016, pp. 99–135, DOI 10.1090/conm/680. MR3591644

[69] Davesh Maulik and Andrei Okounkov, *Quantum groups and quantum cohomology* (English, with English and French summaries), Astérisque **408** (2019), ix+209, DOI 10.24033/ast. MR3951025

[70] Nikita A. Nekrasov and Samson L. Shatashvili, *Supersymmetric vacua and Bethe ansatz*, Nuclear Phys. B Proc. Suppl. **192/193** (2009), 91–112, DOI 10.1016/j.nuclphysbps.2009.07.047. MR2570974

[71] Nikita Nekrasov, *Superspin chains and supersymmetric gauge theories*, J. High Energy Phys. **3** (2019), 102, 16, DOI 10.1007/jhep03(2019)102. MR3951164

[72] Hirosi Ooguri and Cumrun Vafa, *Knot invariants and topological strings*, Nuclear Phys. B **577** (2000), no. 3, 419–438, DOI 10.1016/S0550-3213(00)00118-8. MR1765411

[73] Yong-Geun Oh and Kenji Fukaya, *Floer homology in symplectic geometry and in mirror symmetry*, International Congress of Mathematicians. Vol. II, Eur. Math. Soc., Zürich, 2006, pp. 879–905. MR2275627

[74] Tomotada Ohtsuki, *Quantum invariants*, Series on Knots and Everything, vol. 29, World Scientific Publishing Co., Inc., River Edge, NJ, 2002. A study of knots, 3-manifolds, and their sets. MR1881401

[75] Andrei Okounkov, *Lectures on K-theoretic computations in enumerative geometry*, Geometry of moduli spaces and representation theory, IAS/Park City Math. Ser., vol. 24, Amer. Math. Soc., Providence, RI, 2017, pp. 251–380, DOI 10.1090/pcms/024. MR3752463

[76] A. Okounkov and A. Smirnov, *Quantum difference equation for Nakajima varieties*, Invent. Math. **229** (2022), no. 3, 1203–1299, DOI 10.1007/s00222-022-01125-w. MR4462626

[77] Andrei Okounkov, *Enumerative geometry and geometric representation theory*, Algebraic geometry: Salt Lake City 2015, Proc. Sympos. Pure Math., vol. 97, Amer. Math. Soc., Providence, RI, 2018, pp. 419–457, DOI 10.1090/pspum/097.1/01681. MR3821158

[78] A. Okounkov, *Nonabelian stable envelopes, vertex functions with descendents, and integral solutions of q-difference equations*, arXiv:2010.13217.

[79] Peter Ozsváth and Zoltán Szabó, *Holomorphic disks and topological invariants for closed three-manifolds*, Ann. of Math. (2) **159** (2004), no. 3, 1027–1158, DOI 10.4007/annals.2004.159.1027. MR2113019

[80] Peter Ozsváth and Zoltán Szabó, *Holomorphic disks and three-manifold invariants: properties and applications*, Ann. of Math. (2) **159** (2004), no. 3, 1159–1245, DOI 10.4007/annals.2004.159.1159. MR2113020

[81] Peter Ozsváth and Zoltán Szabó, *Holomorphic disks, link invariants and the multi-variable Alexander polynomial*, Algebr. Geom. Topol. **8** (2008), no. 2, 615–692, DOI 10.2140/agt.2008.8.615. MR2443092

[82] Peter Ozsváth and Zoltán Szabó, *An overview of knot Floer homology*, Modern geometry: a celebration of the work of Simon Donaldson, Proc. Sympos. Pure Math., vol. 99, Amer. Math. Soc., Providence, RI, 2018, pp. 213–249, DOI 10.1090/pspum/099/08. MR3838884

[83] P. S. Ozsvath and Z. Szabo, *Algebras with matchings and link Floer homology*, arXiv:2004.07309.

[84] Jacob Andrew Rasmussen, *Floer homology and knot complements*, ProQuest LLC, Ann Arbor, MI, 2003. Thesis (Ph.D.)–Harvard University. MR2704683

[85] N. Yu. Reshetikhin and V. G. Turaev, *Ribbon graphs and their invariants derived from quantum groups*, Comm. Math. Phys. **127** (1990), no. 1, 1–26. MR1036112

[86] R. Rimányi and L. Rozansky, *New quiver-like varieties and Lie superalgebras*, Comm. Math. Phys. **400** (2023), no. 1, 341–370, DOI 10.1007/s00220-022-04608-2. MR4581477

[87] R. Rouquier, *2-Kac-Moody algebras*, `arXiv:0812.5023`.

[88] Vadim V. Schechtman and Alexander N. Varchenko, *Quantum groups and homology of local systems*, Algebraic geometry and analytic geometry (Tokyo, 1990), ICM-90 Satell. Conf. Proc., Springer, Tokyo, 1991, pp. 182–197. MR1260946

[89] Vadim V. Schechtman and Alexander N. Varchenko, *Arrangements of hyperplanes and Lie algebra homology*, Invent. Math. **106** (1991), no. 1, 139–194, DOI 10.1007/BF01243909. MR1123378

[90] Paul Seidel, *Fukaya categories and Picard-Lefschetz theory*, Zurich Lectures in Advanced Mathematics, European Mathematical Society (EMS), Zürich, 2008, DOI 10.4171/063. MR2441780

[91] Paul Seidel, *Fukaya categories and deformations*, Proceedings of the International Congress of Mathematicians, Vol. II (Beijing, 2002), Higher Ed. Press, Beijing, 2002, pp. 351–360. MR1957046

[92] P. Seidel and I. Smith, *A link invariant from the symplectic geometry of nilpotent slices*, `arXiv:math/0405089`.

[93] Paul Seidel and Jake P. Solomon, *Symplectic cohomology and q-intersection numbers*, Geom. Funct. Anal. **22** (2012), no. 2, 443–477, DOI 10.1007/s00039-012-0159-6. MR2929070

[94] Richard P. Thomas, *An exercise in mirror symmetry*, Proceedings of the International Congress of Mathematicians. Volume II, Hindustan Book Agency, New Delhi, 2010, pp. 624–651. MR2827812

[95] Ben Webster, *Knot invariants and higher representation theory*, Mem. Amer. Math. Soc. **250** (2017), no. 1191, v+141, DOI 10.1090/memo/1191. MR3709726

[96] B. Webster, *Coherent sheaves and quantum Coulomb branches I: tilting bundles from integrable systems*, `arXiv:1905.04623`.

[97] B. Webster, *Coherent sheaves and quantum Coulomb branches II: quiver gauge theories and knot homology*, to appear.

[98] Edward Witten, *Supersymmetry and Morse theory*, J. Differential Geometry **17** (1982), no. 4, 661–692 (1983). MR683171

[99] Edward Witten, *Quantum field theory and the Jones polynomial*, Comm. Math. Phys. **121** (1989), no. 3, 351–399. MR990772

[100] Edward Witten, *Analytic continuation of Chern-Simons theory*, Chern-Simons gauge theory: 20 years after, AMS/IP Stud. Adv. Math., vol. 50, Amer. Math. Soc., Providence, RI, 2011, pp. 347–446, DOI 10.1090/amsip/050/19. MR2809462

[101] Edward Witten, *Fivebranes and knots*, Quantum Topol. **3** (2012), no. 1, 1–137, DOI 10.4171/QT/26. MR2852941

Physics Department, University of California, Berkeley, California 94720; and Departments of Mathematics, University of California, Berkeley, California 94720

Physics Department, University of California, Berkeley, California 94720

Theory Department, CERN, Geneva, Switzerland

Contemporary Mathematics
Volume **817**, 2025
https://doi.org/10.1090/conm/817/16323

A braid group action on an A_∞-category for zigzag algebras

Benjamin Cooper, You Qi, and Joshua Sussan

Dedicated to Igor Frenkel on the occasion of his seventieth birthday

Abstract. We construct differential graded enhancements of the zigzag algebras which were used by Khovanov, Seidel and Thomas to produce categorical braid group actions. These enhancements are related to p-differential graded structures by a version of Koszul duality. We prove that the minimal model A_∞-structure on the zigzag algebras is *not* formal. We construct a braid group action in this setting and suggest a symplectic interpretation.

Contents

1. Introduction

In the 1990s, Crane and Frenkel conjectured that the 3-dimensional topological field theories of Witten, Reshetikhin and Turaev could be rewritten in the new categorical language that was appearing in geometric representation theory at the time [**CF94**]. Over the years, these ideas have revealed a vast mathematical landscape. Although the conjecture remains open, many parts of what we know today are expected to play a role in the final answer. Among these parts, the zigzag algebras C_{n-1}, where $n \geq 2$ is a natural number, introduced by Khovanov and Seidel [**KS02**], also [**RZ03**], serve as a kind-of simplest non-trivial example: sophisticated enough for its bounded homotopy category $K^b(C_{n-1})$ to support a faithful categorical representation of the braid group

$$\Psi_n : \mathrm{Br}_n \to \mathrm{Aut}(K^b(C_{n-1})) \qquad\qquad \mathrm{Ker}(\Psi_n) = \{e\},$$

The second author was partially supported by a Simons Foundation Collaboration Grant for Mathematicians. The first author's participation was facilitated by Simons Award #638089.

yet simple enough to appear in many places. There are interpretations of C_{n-1} in terms of the category $\mathcal{O}(\mathfrak{sl}_n)$ and perverse sheaves on $\mathbb{P}^{n-1}$. The zigzag algebra also arises naturally in homological realization of Nakajima quiver varieties [**FKS05**], as well as the endomorphism algebra of a system of Lagrangian spheres [**KS02**]. This latter observation was exploited by Seidel and Thomas to construct categorical braid group actions on many different examples [**ST01**].

In a slightly different direction, Chen, Khovanov, and independently Stroppel, studied, as a special case of their constructions, the Grothendieck group $K_0(A_n)\otimes\mathbb{C}$ of graded modules over a certain algebra A_n related to C_{n-1}, proving that it can be identified with the second highest weight space in the representation $(\mathbb{C}(q)^2)^{\otimes n}$ of $U_q(\mathfrak{sl}_2)$ [**CK14, Str09**]. Under this correspondence, the homomorphism Ψ_n becomes the Burau representation, well-studied in low-dimensional topology [**Bir74**].

In [**Kho16**], Khovanov suggested to categorify the operation of setting q to be a pth root of unity by introducing a p-nilpotent derivation

$$\partial : A \to A \qquad \text{such that} \qquad \partial^p = 0$$

and passing from the category of graded A-modules to a triangulated category of (A, ∂)-modules, (see Section 2). This led Khovanov and the second author to find p-nilpotent operators on nilHecke algebras, Khovanov-Lauda-Rouquier algebras and Webster algebras underlying many categorifications of quantum groups and their representations [**KQ15, EQ16, KQS17, QS18**]. The observation that the Koszul dual $A_n^!$ of the algebra A_n admits such a p-derivation $\partial \in \mathrm{End}(A_n^!)$ compatible with its identification as the Webster algebra associated to the second highest weight of the $U_q(\mathfrak{sl}_2)$ representation $(\mathbb{C}(q)^2)^{\otimes n}$ led to a detailed study of $(A_n^!, \partial)$, ultimately constructing an analogue of the categorical representation Ψ_n above evaluated at a pth root of unity [**QS16, QS17**].

This paper continues the study of zigzag algebras by examining what happens to this p-differential graded structure $\partial : A_n^! \to A_n^!$ under Koszul duality. In short,

$$\text{what is } (A_n^!, \partial)^!?$$

We answer this question when $p = 2$. In this case, $(A_n^!, \partial)$ becomes a differential graded algebra and we compute its Koszul dual A_∞-algebra. This is done in stages. There is a natural bijection between the simple modules $\{L_i\}$ of $A_n^!$ and those of $(A_n^!, \partial)$. We observe that the cofibrant resolutions $\mathbb{P}(L_i) \to L_i$ in Lemma 3.3 of the simple modules L_i over $(A_n^!, \partial)$ are closely related to, but not the same as projective resolutions of corresponding simple modules L_i over $A_n^!$. Then we construct chain maps between resolutions associated to the generators of the algebra A_n. The relations that these chain maps satisfy allow us to introduce differential graded algebras (E_n, d) which act as intermediaries between $(A_n^!, \partial)$ and A_n. More precisely,

(1) There are canonical inclusions of differential graded algebras: $E_n \hookrightarrow \mathrm{END}(\oplus_i \mathbb{P}(L_i))$ inducing isomorphisms $\mathrm{H}(E_n, d) \cong \mathrm{Ext}^*(\oplus_i L_i, \oplus_i L_i)$.
(2) There are canonical isomorphisms of algebras: $\mathrm{H}(E_n, d) \cong A_n$.

In this sense, the algebra (E_n, d) is a differential graded (DG) enhancement of the algebra A_n. A setting such as this is ripe for an application of the perturbation lemma [**Kel03, CS19**]: the algebra $A_n \cong \mathrm{H}(E_n, d)$ supports a minimal A_∞-structure (A_n, m) where $m = \{m_r : A_n^{\otimes r} \to A_n[2-r]\}_{r\geq 1}$ with $m_1 = 0$ which makes (A_n, m) equivalent to (E_n, d) as A_∞-algebras. The question posed above can be answered in equations as

$$(A_n^!, \partial)^! \cong (E_n, d) \cong (A_n, m).$$

Section 4 contains a computation of this minimal A_∞-algebra structure for the algebra (E_n, d). This stable minimal A_∞-algebra is non-trivial, see Theorem 4.4 for formulas, which shows that the A_∞-structures on unstable algebras A_n and C_{n-1} must be non-trivial as well. This result is a little bit surprising because Seidel and Thomas used formality, the vanishing of all A_∞-structures, on zigzag algebras to build their braid group representations see [**ST01**, Lemma 4.21]. Our A_∞-structure on zigzag algebras does not contradict Seidel and Thomas's theorem since a different grading is used, more precisely the operations $m_3 \in CC^3(C_n, C_n)^{-1}$ (see Section 4) live in a degree of Hochschild cohomology which is not covered by their vanishing result. Corresponding vanishing statements feature in related works [**LW23, BW22**].

In Section 5, we check that the derived category of the A_∞-algebra zigzag algebra supports a braid group action. This is done by transporting the braid group action defined in the context of $(A_n^!, \partial)$ using the derived Morita equivalence obtained from the Koszul duality construction. The fact that the A_∞-structure m on the zigzag algebra facilitates a braid group representation, is strong evidence to suggest its relevance and naturality within the context of the Crane and Frenkel conjectures.

Finally, zigzag algebras appear in a variety of contexts. It is natural to ask whether one can interpret this new A_∞-structure within one of those contexts. In Section 6, we recall how A_n or C_{n-1} can be found in the Fukaya category of a surface, why there is a braid group representation and see that the equations we discovered from the A_∞ perturbation lemma can be understood in terms of relationships among disks and curves on the surfaces.

Organization. We review some tools from homological algebra in Section 2. In Section 3, the dual zigzag algebra is introduced, resolutions $\mathbb{P}(L_i)$ of simples L_i are constructed as well as chain maps between these resolutions. In Section 4 the DG algebra (E_n, d) and its subalgebra (S_n, d) are introduced and the perturbation lemma is used to compute the A_∞-structure on the cohomology of S_n, which is the zigzag algebra C_{n-1}. In Section 5 the categorical braid group representation is constructed. In Section 6, the stable A_∞-structure is examined using curves on a surface.

Notations. The ground ring will be a field of characteristic 2. Below is a list of the algebras that will play a role in this paper:

(1) $A_n^!$ is the Koszul algebra (one of the simples is not spherical) which has the original DG structure.
(2) A_n is the Koszul dual of $A_n^!$ (one of the projectives is not spherical). This is almost the zigzag algebra.
(3) C_{n-1} is a subalgebra of A_n. This is the zigzag algebra and all of the $n-1$ projectives are spherical.
(4) E_n is the DG endomorphism algebra of simples of $A_n^!$, and its homology is A_n.
(5) E_n' is a certain subalgebra of E_n whose homology is A_n.
(6) S_{n-1} is a DG subalgebra of E_n. Its homology is C_{n-1} (the zigzag algebra).

Most of these algebras are defined at the beginning of Section 4. The diagram below may be useful.

$$\begin{array}{ccc} (S_{n-1}, d) & \xrightarrow{\subset} & (E_n, d) \\ \Big\downarrow \mathrm{H} & & \Big\downarrow \mathrm{H} \\ C_{n-1} & \xrightarrow{\subset} & A_n \end{array}$$

2. Preliminaries

The braid group. Let q be a formal variable, and $\mathbb{Z}[q, q^{-1}]$ be the ring of Laurent polynomials. The Temperley-Lieb algebra TL_n over $\mathbb{Z}[q, q^{-1}]$ is generated by elements u_i for $i = 1, \ldots, n-1$ subject to the relations that

(i) $u_i u_j = u_j u_i$ if $|i-j| > 1$,
(ii) $u_i^2 = (1+q)u_i$ for $i = 1, \ldots, n-1$,
(iii) $u_i u_{i+1} u_i = q u_i$ for $i = 1, \ldots, n-1$ and $u_i u_{i-1} u_i = q u_i$ for $i = 2, \ldots, n$.

Consider the $\mathbb{Z}[q, q^{-1}]$-lattice $V_{n-1} = \oplus_{i=1}^{n-1} \mathbb{Z}[q, q^{-1}] v_i$. Define

$$u_i(v_j) := \begin{cases} (1+q)v_i & i = j \\ v_i & j = i-1 \\ qv_i & j = i+1 \\ 0 & |i-j| > 1 \end{cases} \tag{2.1}$$

Base changing from $\mathbb{Z}[q, q^{-1}]$ to $\mathbb{C}$ by sending q to a chosen complex number, we call the resulting $(n-1)$-dimensional representation the (reduced) *Burau representation*, which is still denoted V_{n-1}.

The braid group Br_n is generated by elements t_i for $i = 1, \ldots, n-1$ subject to the relations that

(i) $t_i t_j = t_j t_i$ if $|i-j| > 1$,
(ii) $t_i t_{i+1} t_i = t_{i+1} t_i t_{i+1}$ for $i = 1, \ldots, n-2$.

For $i = 1, \ldots, n-1$, sending the braid group generators t_i to $1 - u_i$ defines a homomorphism from the group algebra of Br_n into TL_n. Via this homomorphism, we can also regard V_{n-1} as a representation of the braid group.

Homological algebra. We gather in this section some necessary homological algebra background that we shall use. We refer the reader to the references [**Kel94**, **Kel06b**] for more details.

Fix once and for all a base field $\Bbbk$ of characteristic 2. Unadorned tensor product $\otimes$ will stand for the tensor product over $\Bbbk$. Throughout we use the French convention $\mathbb{N} := \{0, 1, 2, \ldots\}$.

A *differential graded (DG) algebra* (A, d_A) over $\Bbbk$ consists of a $\mathbb{Z}$-graded algebra $A = \oplus_{i \in \mathbb{Z}} A^i$ and a degree one endomorphism d_A of A that is 2-nilpotent and satisfies the Leibniz rule:

$$d_A^2 = 0, \qquad d_A(ab) = d_A(a)b + a d_A(b),$$

for any $a, b \in A$. A left (resp. right) DG module (M, d_M) over (A, d_A) is a left (resp. right) A-module equipped with a degree one, 2-nilpotent endomorphism d_M,

$d_M^2 = 0$, which is compatible with the differential action on A:

$$d_M(ax) = d_A(a)x + ad_M(x) \qquad (\text{resp. } d_M(xa) = d_M(x)a + xd_A(a)).$$

For a DG module (M, d_M) we will also consider the shifted module $(M[i], d_{M[i]})$ where the degree j component $M[i]^j$ is equal to M^{i+j}. In what follows, we will be exclusively considering DG algebras of the following type.

DEFINITION 2.1. A DG algebra A is called *positive*, if it satisfies the following three conditions.

(i) The algebra A is non-negatively graded, $A \cong \oplus_{i\in\mathbb{N}} A^i$, and is finite-dimensional in each fixed degree.
(ii) The homogeneous degree-zero part A^0 is semi-simple.
(iii) The differential d_A acts trivially on A^0.

The subscripts in the differentials on the DG algebras and modules will be dropped when no confusion can be caused. We adapt some basic definitions from [**KQ15**, Section 2] in the situation when A is positive, and the reader is referred loc. cit. for a summary of the basic definitions and techniques of the homological theory of DG modules. Some further information on the subject in the context of *hopfological algebra* can be found in [**Qi14**]. Let $H = \Bbbk[d]/(d^2)$. We will use the notations A_d–mod (resp. $\mathcal{C}(A)$, $\mathcal{D}(A)$) to stand for the abelian (resp. homotopy, derived) category of DG modules over A. Here A_d denotes the *smash product algebra*,[1] which is isomorphic, as a vector space, to $A \otimes H$, subject to the relations that $A \cong A \otimes 1 \subset A \otimes H$, $H \cong 1 \otimes H \subset A \otimes H$ as subalgebras, and the rule for commuting elements

$$(1 \otimes d) \cdot (a \otimes 1) = (a \otimes 1) \cdot (1 \otimes d) + d(a) \otimes 1,$$

where a is an arbitrary element of A.

When A is a positive DG algebra, A_d is also a positively graded algebra with degree zero part isomorphic to A^0. Choose a complete set of pairwise non-isomorphic indecomposable idempotents $\{e_i | i \in I\}$ in A^0, and set $P_i := Ae_i$. By construction, P_i's are DG summands of the rank-one free module A.

As a matter of convention, given two graded modules M, N over a graded algebra A, we set $\mathrm{Hom}_A(M, N)$ to be the space of degree preserving A-module maps between M and N. We define

$$\mathrm{HOM}_A(M, N) := \oplus_{l\in\mathbb{Z}} \mathrm{Hom}_A(M, N[l]).$$

The categories $\mathcal{C}(A)$, $\mathcal{D}(A)$ are triangulated, equipped with the invertible *shift* endo-functors $[l]$ for any $l \in \mathbb{Z}$.

DEFINITION 2.2. Let A be a DG algebra, and M be a DG module over A.

(i) The module M is called *cofibrant* if for any surjective quasi-isomorphism of DG modules $N_1 \twoheadrightarrow N_2$, the induced map

$$\mathrm{HOM}_A(M, N_1) \longrightarrow \mathrm{HOM}_A(M, N_2)$$

of complexes is a homotopy equivalence.

[1] It is usually denoted by $A\#H$.

(ii) When A is positive, we say M is a *finite cell module* if it has a finite exhaustive increasing filtration $F^\bullet$ by DG submodules such that the subquotients
$$F^\bullet/F^{\bullet-1} \cong P_i[l_i]$$
for some projective A-modules P_i, $i = 1, \ldots, n$ and $l_i \in \mathbb{Z}$.

Cofibrant modules are "nice" in the sense that we have a good control of the morphism spaces between them in the derived category $\mathcal{D}(A)$. Namely, there is, for any cofibrant module M, a functorial-in-N isomorphism of $\Bbbk$-vector spaces
$$\mathrm{Hom}_{\mathcal{C}(A)}(M,N) \cong \mathrm{Hom}_{\mathcal{D}(A)}(M,N),$$
which is induced from the localization functor $\mathcal{C}(A) \longrightarrow \mathcal{D}(A)$. It is an easy exercise to show that finite cell modules are cofibrant. It is clear that if V is a finite-dimensional complex V, then $A \otimes V$ is a finite cell module. Finite cell modules in the DG setting are also known as "one-sided twisted complexes" of projectives [**BK90**].

EXAMPLE 2.3. Let A be a DG algebra. Assume that a_0 is an element of A with $d_A(a_0) = 0$. Here we construct an example of a finite cell module that is not of the form $A \otimes V$.

Consider a free A-module on two generators $M := Av_0 \oplus Av_1$, where $\deg(v_0) - \deg(v_1) = 1 - \deg(a_0)$. Define a differential d_M as follows. On generators, it acts by
$$d_M(v_0) = 0, \qquad d_M(v_1) = a_0 v_0.$$
It is extended to all of M by the Leibniz rule. We leave it as an exercise for the reader to check that $d_M^2 = 0$, and M is not isomorphic to the tensor product of A by a two-dimensional complex.

The morphism space $\mathrm{Hom}_{\mathcal{D}(A)}(M,N)$ between two DG modules $M, N \in \mathcal{D}(A)$ is canonically isomorphic to the degree zero cohomology of the complex $\mathbf{R}\mathrm{HOM}_A(M,N)$
$$\mathrm{Hom}_{\mathcal{D}(A)}(M,N) \cong \mathrm{H}^0(\mathbf{R}\mathrm{HOM}_A(M,N)) \tag{2.2}$$
where $\mathbf{R}\mathrm{HOM}_A(M,N) := \mathrm{HOM}_A(\mathbb{P}(M),N)$ is computed using a cofibrant replacement $\mathbb{P}(M) \to M$ of M. More generally, attached to any two objects in a triangulated category one has the "Ext-space" as follows.

DEFINITION 2.4. Given any DG modules $M, N \in \mathcal{D}(A)$ and $i \in \mathbb{Z}$, we set the *ith p-DG* Ext*-space*, or simply the Ext*-space* between them, to be
$$\mathrm{Ext}^i_{\mathcal{D}(A)}(M,N) := \mathrm{Hom}_{\mathcal{D}(A)}(M, N[i]),$$
and we define *the total* Ext*-space* as
$$\mathrm{Ext}^\bullet_{\mathcal{D}(A)}(M,N) := \bigoplus_{i\in\mathbb{Z}} \mathrm{Ext}^i_{\mathcal{D}(A)}(M,N).$$

One can also make these definitions in the homotopy category $\mathcal{C}(A)$ but we will not need them. It follows from (2.2) that there is a bi-functorial isomorphism
$$\mathrm{Ext}^\bullet_{\mathcal{D}(A)}(M,N) \cong \bigoplus_{i\in\mathbb{Z}} \mathrm{H}^i(\mathbf{R}\mathrm{HOM}_A(M,N)). \tag{2.3}$$

We next recall the notion of compact modules, which takes place in the derived category.

DEFINITION 2.5. Let A be a DG algebra. A DG module M over A is called *compact* (in the derived category $\mathcal{D}(A)$) if and only if, for any family of DG modules N_i where i takes value in some index set I, the natural map

$$\oplus_{i\in I} \mathrm{Hom}_{\mathcal{D}(A)}(M, N_i) \longrightarrow \mathrm{Hom}_{\mathcal{D}(A)}(M, \oplus_{i\in I} N_i)$$

is an isomorphism of $\Bbbk$-vector spaces.

The strictly full subcategory of $\mathcal{D}(A)$ consisting of compact modules will be denoted by $\mathcal{D}^c(A)$. It is triangulated and will be referred to as the *compact derived category*.

For positive DG algebras, we have a close relationship between finite cell modules and compact modules. Denote by $\mathcal{F}(A)$ the collection of all finite cell modules over A, and let $\overline{\mathcal{F}}(A)$ be the closure of $\mathcal{F}(A)$ inside $\mathcal{D}(A)$ under isomorphisms. Notice that the natural inclusion functor $\overline{\mathcal{F}}(A) \subset \mathcal{D}(A)$ factors through the compact derived category $\mathcal{D}^c(A)$.

PROPOSITION 2.6. *The natural inclusion functor*

$$\overline{\mathcal{F}}(A) \subset \mathcal{D}^c(A)$$

is an equivalence of triangulated categories. □

In other words, over a positive DG algebra, any compact module is quasi-isomorphic to a finite cell module. The proof of the Proposition can be found in [**Sch11**]. Applying the usual Grothendieck group construction to the triangulated category $\mathcal{D}^c(A)$, we obtain the DG Grothendieck group of a DG algebra A, which we denote by $K_0(A)$. The abelian group stucture on $K_0(A)$ arises from the module structure $K_0(\Bbbk) \otimes K_0(A) \to K_0(A)$ induced by the categorical action of $\mathcal{D}(\Bbbk)$ on $\mathcal{D}^c(A)$.

Resolutions. Let M be a DG module. By a *DG resolution* of M we mean a quasi-isomorphism of DG modules $P \longrightarrow M$ such that P is cofibrant. It is known that resolutions for DG modules always exist [**Kel94**, Theorem 3.1], although they are usually not unique. However, it is not true in general that any DG module has a resolution by a finite cell module. Proposition 2.6 says that, over a positive DG algebra, this happens precisely when the module is compact in the derived category.

NOTATION 2.7. In what follows, we introduce a convenient way to rewrite finite cell modules. Let M be a DG module with a finite increasing filtration $F^\bullet$ ($\bullet \in \mathbb{Z}$), with $F^i/F^{i-1} := P_i$. Then we will write this filtered module schematically as a direct sum of its associated graded pieces, and put an arrow from P_i to P_j decorated by a map ϕ if $d_M(p_i)$ projects onto $\phi(p_i) \in P_j$ (necessarily $j < i$), where p_i is a lift of an element of P_i inside M. For instance, the following diagram stands for a filtered DG module

$$P_n \xrightarrow{\phi_n} P_{n-1} \xrightarrow{\phi_{n-1}} \cdots \xrightarrow{\phi_{m+2}} P_{m+1} \xrightarrow{\phi_{m+1}} P_m$$

with the filtration implicitly understood as $F^i = \sum_{k=n}^{i} P_k$. On each F^i, the differential structure is defined inductively by the composition

$$d_{F^i} : P_j \longrightarrow P_j + P_{j-1} \subset F^i, \qquad p_j \mapsto (d_{P_j}(p_j), \phi_j(p_j)),$$

where $m \leq j \leq i$.

Alternatively, one can regard the original filtration on M as obtained from collapsing the bigrading of a bigraded A-module $M \cong \bigoplus_{i,j\in\mathbb{Z}} P^i_j$ into a single grading, which will always be referred to as the *q-grading*, and take $F^j := \sum_{k\geq j, i\in\mathbb{Z}} P^i_k$. Note that d_M does not necessarily preserve the bigrading.

For instance, in this notation, the filtered DG module in Example 2.3 would be represented in this notation by

$$0 \longrightarrow A \xrightarrow{\cdot a_0} A \longrightarrow 0.$$

LEMMA 2.8. *Let $\phi : M \longrightarrow N$ be a map of DG modules over A. The cone of the morphism ϕ is isomorphic to the filtered DG module*

$$M[1] \xrightarrow{\ \phi\ } N\ ,$$

where N sits in homological degree zero. The cocone of ϕ is isomorphic to the filtered module

$$M \xrightarrow{\ \phi\ } N[-1]\ ,$$

where M sits in homological degree zero.

For later convenience, we will simply write the (co)cone of ϕ as

$$M[1] \xrightarrow{\phi} N \qquad \left(\text{resp. } M \xrightarrow{\phi} N[-1]\right).$$

PROOF. This follows from the definition of the cone and cocone. See [**KQ15**, Definition 2.14]. □

LEMMA 2.9 ([**QS16**, Lemma 3.11]). *Let $0 \longrightarrow K \xrightarrow{\phi} L \xrightarrow{\psi} M \longrightarrow 0$ be a short exact sequence of graded modules over the smash product algebra A_d. Then the filtered DG module*

$$K[2] \xrightarrow{\phi} L[1] \xrightarrow{\psi} M \tag{2.4a}$$

is acyclic.

Smooth DG algebras. We now recall the notion of *smooth* DG algebras. Throughout, for simplicity, we will assume that A is a finite-dimensional DG algebra.

DEFINITION 2.10. The DG algebra A is called *left (resp. right) smooth* if all simple left (resp. right) DG modules are compact in the derived category of left (resp. right) DG modules.

Given a finite-dimensional DG algebra A, we have a strictly full subcategory $\mathcal{D}^f(A) \subset \mathcal{D}(A)$ which consists of DG left (resp. right) modules that are quasi-isomorphic to finite-dimensional left (resp. right) DG-modules. The natural inclusion of $\mathcal{D}^c(A)$ into $\mathcal{D}(A)$ factors through $\mathcal{D}^f(A)$, since A is finite-dimensional.

The following result is well-known. For completeness, we record a reference in which the result is proven in the context of hopfological algebra.

LEMMA 2.11 ([**QS16**, Lemma 3.15]). *If A is left (resp. right) smooth, then the natural inclusion functor $\mathcal{D}^c(A) \longrightarrow \mathcal{D}^f(A)$ is an equivalence of triangulated categories of left (resp. right) DG modules.*

We also point out the connection to the usual notion of smoothness for DG algebras, which is that A is required to be perfect in the derived category of (A, A)-bimodules. In other words, one can find a finite-length cofibrant replacement of A as a DG (A, A)-bimodule. Tensoring this replacement with any simple left (resp. right) DG A-module, one obtains a finite-length cofibrant replacement of the simple left (resp. right) module. In this sense, our requirement is weaker than the usual notion of smoothness, and is adapted only for this paper to discuss finite-length left (resp. right) DG modules categories.

Background on A_∞-algebras. We quickly review some basic material on A_∞-algebras over a field of characteristic two, following the references [**Kel01**] and [**Kel06a**].

An A_∞-algebra A over a field $\Bbbk$ is a $\mathbb{Z}$-graded vector space over $\Bbbk$ with operations $m_n \colon A^{\otimes n} \to A$ of degree $2 - n$ satisfying relations

$$\sum m_u(1^{\otimes r} \otimes m_s \otimes 1^{\otimes t}) = 0,$$

where the sum is over all decompositions $n = r + s + t$ and $u = r + 1 + t$.

A right A_∞-module over A is a $\mathbb{Z}$-graded vector space M with operations $m_n \colon M \otimes A^{\otimes n-1} \to M$ of degree $2 - n$ satisfying relations similar to those of an A_∞-algebra. For details see [**Kel01**, Section 4.2]. A left A_∞-module is defined in a similar way.

A morphism of right A_∞-modules over A is sequence of maps $f_n \colon M \otimes A^{\otimes n-1} \to N$ of degree $1 - n$, such that for each $n \geq 1$, there is a relation

$$\sum f_u \circ (1^{\otimes r} \otimes m_s \otimes 1^{\otimes t}) = \sum m_u(f_r \otimes 1^{\otimes s}),$$

where the left hand sum is over all decompositions $n = r + s + t, r, t \geq 0, s \geq 1$, $u = r+1+t$, and the right hand sum is over all decompositions $n = r+s, r \geq 1, s \geq 0$, and $u = 1 + s$. The composition of morphisms is defined in [**Kel01**, Section 4.2].

An A_∞-bimodule M is a $\mathbb{Z}$-graded vector space with operations $m_{i,j} \colon A^{\otimes i-1} \otimes M \otimes A^{\otimes j-1} \to M$ of degree $3 - i - j$ satisfying certain relations. For more details see [**Kel06a**, Section 3]. One can form a tensor product of a right A_∞-module M and a left A_∞-module N over an A_∞-algebra A, which we denote as $M \boxtimes_A N$. Again see [**Kel06a**, Section 3] for details.

3. The DG algebra $A_n^!$

The quiver presentation. Let $\Bbbk$ be a field of characteristic 2. Let n be a natural number greater than or equal to two, and Q_n be the following quiver:

(3.1) 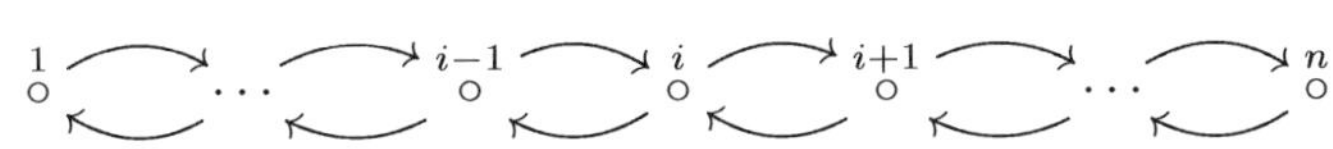

Let $\Bbbk Q_n$ be the path algebra associated to Q_n over the ground field $\Bbbk$. Following Khovanov-Seidel [**KS02**], we use, for instance, the symbol $(i|j|k)$, where i, j, k are vertices of the quiver Q_n, to denote the path which starts at a vertex i, then goes through j (necessarily $j = i \pm 1$) and ends at k. The composition of paths is given defined by

$$(i_i|i_2|\cdots|i_r) \cdot (j_1|j_2|\cdots|j_s) = \begin{cases} (i_i|i_2|\cdots|i_r|j_2|\cdots|j_s) & \text{if } i_r = j_1, \\ 0 & \text{otherwise,} \end{cases} \tag{3.2}$$

where $i_1, \ldots, i_r$ and $j_1, \ldots, j_s$ are sequences of neighboring vertices in Q_n.

DEFINITION 3.1. The algebra $A_n^!$ is the quotient of the path algebra $\Bbbk Q_n$ by the relations

$$(i|i-1|i) = (i|i+1|i)(i = 2, \ldots, n-1), (1|2|1) = 0.$$

This is a graded algebra where the degree of the generators are given as follows. The degree of $(i|i+1)$ is 1, the degree of $(i|i-1)$ is 0, and the degree of (i) is 0.

REMARK 3.2. In the literature, the grading on $A_n^!$ is usually taken to be the path length grading.

The algebra $A_n^!$ is Morita equivalent to a maximally singular block of category $\mathcal{O}(\mathfrak{gl}_n)$ and is isomorphic to a particular Webster algebra. For more details see [**QS16**].

The algebra $A_n^!$ is Koszul, whose quadratic dual is isomorphic to the algebra A_n considered in [**KS02**].

The quiver algebra $A_n^!$ has a differential d, given on the path generators by

$$d(i|i+1) = (i|i+1|i|i+1), \qquad d(i|i-1) = 0,$$

and extended to the whole algebra via the Leibniz rule. Sometimes for convenience, we will write the loop $(i|i+1|i) = (i|i-1|i)$ at the vertex i simply as c_i, so that $d(i|i+1) = c_i(i|i+1) = (i|i+1)c_{i+1}$.

In the rest of the paper, we will be exclusively focusing on a family of DG algebras that are positive (Definition 2.1). Now let A be a positive DG algebra, and $\{e_i | i = 1, \ldots, n\}$ be a complete list of pairwise non-isomorphic indecomposable idempotents in A. Define P_i (resp. ${}_iP$) to be the DG module $A \cdot e_i$ (resp. $e_i \cdot A$), and $L_i := P_i / A' \cdot P_i$ (resp. ${}_iL := {}_iP / {}_iP \cdot A'$), where $A' = \oplus_{i>0} A^i$ is the augmentation ideal of A. Then A is left (resp. right) smooth if and only if all L_i's (resp. ${}_iL$'s), $i = 1, \ldots, n$ are compact in the derived category. Furthermore, the natural categorical pairing, which exists for any finite-dimensional DG algebra,

$$\mathbf{R}\mathrm{HOM}_A(-,-) : \mathcal{D}^c(A) \times \mathcal{D}^f(A) \longrightarrow \mathcal{D}(\Bbbk) \tag{3.3}$$

gives rise to a pairing on $\mathcal{D}^c(A)$ via the equivalence $\mathcal{D}^c(A) \longrightarrow \mathcal{D}^f(A)$ of Lemma 2.11. It descends to a semi-linear non-degenerate pairing on the Grothendieck group

$$[\mathbf{R}\mathrm{HOM}_A(-,-)] : K_0(A) \times K_0(A) \longrightarrow \mathbb{Z}, \tag{3.4}$$

which is perfect if all L_i's ($i = 1, \ldots, n$) are absolutely irreducible.

LEMMA 3.3.

(i) *For $i = 1, \ldots, n-1$, the left DG module L_i is quasi-isomorphic to the finite cell module $\mathbb{P}(L_i)$*

(3.5) 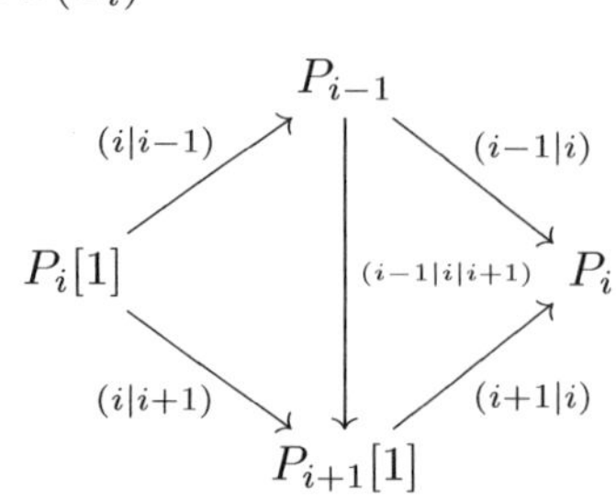

The right DG module ${}_iL$ is quasi-isomorphic to the finite cell module $\mathbb{P}({}_iL)$

(3.6)

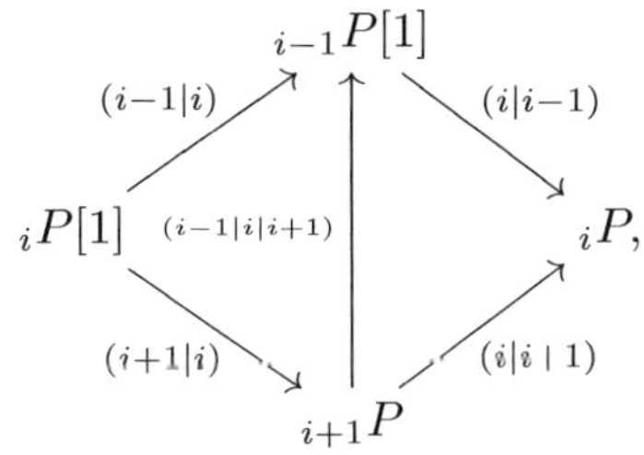

(ii) *The left DG module L_n is quasi-isomorphic to the finite cell module $\mathbb{P}(L_n)$*

(3.7)

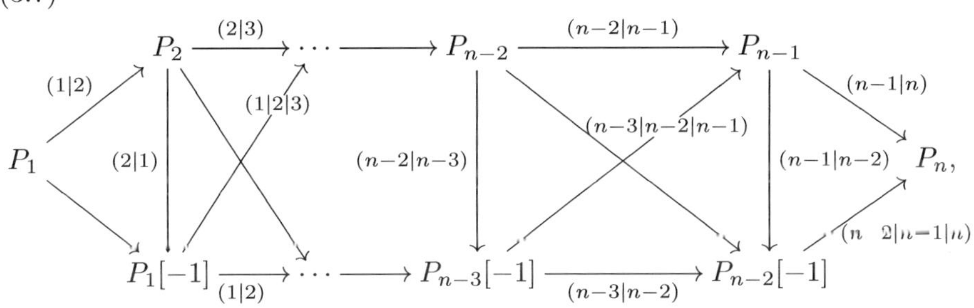

where all the undecorated arrows are identity maps. The right DG module ${}_nL$ is quasi-isomorphic to the finite cell module $\mathbb{P}({}_nL)$

$$ {}_{n-1}P[1] \xrightarrow{(n|n-1)} {}_nP\ . \tag{3.8}$$

PROOF. The first statement is the $p = 2$ case of [**QS16**].

It is straightforward to check that the module in (3.7) is in fact a DG module. This object $\mathbb{P}(L_n)$ could be filtered by starting from P_n and adding modules moving down and to the left and then up. More precisely:

$$F_1 = P_n, \quad F_2 = P_n \oplus P_{n-2}[-1], \ldots, F_{2j-1} = F_{2j-2} \oplus P_{n-j+1},$$
$$F_{2j} = F_{2j-1} \oplus P_{n-j-1}[-1], \ldots, F_{2n-2} = \mathbb{P}(L_n).$$
□

COROLLARY 3.4. *The* **R**HOM*-space between the simple left DG modules L_i ($i = 1, \ldots, n-1$) is given by*

$$\mathbf{R}\mathrm{HOM}_{A_n^!}(L_i, L_j) \cong \begin{cases} \Bbbk \oplus \Bbbk[1] & i = j, \\ \Bbbk & |i-j| = 1, \\ 0 & \textit{otherwise}. \end{cases}$$

The spaces $\mathbf{R}\mathrm{HOM}_{A_n^!}(L_n, L_{n-1})$, $\mathbf{R}\mathrm{HOM}_{A_n^!}(L_{n-1}, L_n)$ and $\mathbf{R}\mathrm{HOM}_{A_n^!}(L_n, L_n)$ are one-dimensional.

PROOF. By definition, one has

$$\mathbf{R}\mathrm{HOM}_{A_n^!}(L_i, L_j) \cong \mathrm{HOM}_{A_n^!}(\mathbb{P}(L_i), L_j).$$

The result follows by plugging in (3.5) into this equality. □

COROLLARY 3.5. *The DG algebra $(A_n^!, d)$ is smooth.*

PROOF. The lemma shows that simple left and right DG modules over $A_n^!$ are finite-cell, and thus are compact objects in the derived category. □

Temperley-Lieb action. Now we recall a categorical Temperley-Lieb algebra action on $\mathcal{D}^c(A_n^!)$ constructed in [**QS16**].

DEFINITION 3.6. Let i be an integer in $\{1, \dots, n-1\}$.

(i) The *ith cup functor* is the functor

$$\cup_i \colon \mathcal{D}(\Bbbk) \longrightarrow \mathcal{D}(A_n^!), \qquad \cup_i(V) := L_i \otimes V.$$

(ii) The *ith cap functor* is given by

$$\cap_i \colon \mathcal{D}(A_n^!) \longrightarrow \mathcal{D}(\Bbbk), \qquad \cap_i(M) := {}_iL \otimes^{\mathbf{L}}_{A_n^!} M.$$

Thanks to Lemma 3.3, the above functors interacting between $\mathcal{D}(A_n^!)$ and $\mathcal{D}(\Bbbk)$ send compact objects to compact objects. Furthermore, there is a functor isomorphism

$$\cap_i(M) = {}_iL \otimes^{\mathbf{L}}_{A_n^!} (M) \cong \mathbf{R}\mathrm{HOM}_{A_n^!}(L_i, M) \tag{3.9}$$

In turn, this allows one to define, for each $i \in \{1, \dots, n-1\}$, an endo-functor

$$\mathfrak{U}_i^! = \cup_i \circ \cap_i : \mathcal{D}^c(A_n^!) \longrightarrow \mathcal{D}^c(A_n^!), \tag{3.10}$$

which is usually diagrammatically presented as

$\mathfrak{U}_i^! =$ 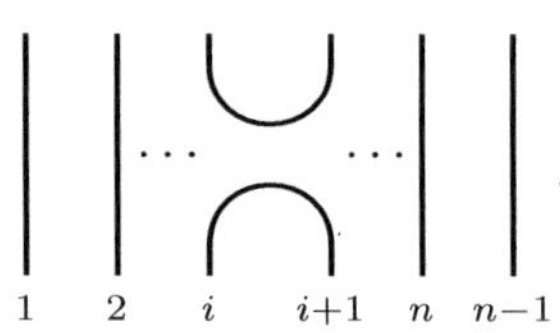

This diagrammatic notation explains the names "cups" and "caps" introduced earlier.

THEOREM 3.7 ([**QS16**, Theorem 5.10]). *For $i = 1, \dots, n-1$, the functors $\mathfrak{U}_i^!$ are self-biadjoint, and they enjoy the following functor-isomorphism relations.*

(i) $\mathfrak{U}_i^! \circ \mathfrak{U}_i^! \cong \mathfrak{U}_i^! \oplus \mathfrak{U}_i^![1]$,
(ii) $\mathfrak{U}_i^! \circ \mathfrak{U}_j^! \cong \mathfrak{U}_j^! \circ \mathfrak{U}_i^!$ *for* $|i-j| > 1$,
(iii) $\mathfrak{U}_i^! \circ \mathfrak{U}_j^! \circ \mathfrak{U}_i^! \cong \mathfrak{U}_i^!$ *for* $|i-j| = 1$.

Via the self-adjointness of the functors $\mathfrak{U}_i^!$, one can define, for each $i \in \{1, \dots, n-1\}$, the *braid functors* on $\mathcal{D}^c(A_n^!)$ as follows. For any $M \in \mathcal{D}(A_n^!)$, set $\mathfrak{T}_i^!(M)$ to be the cone of the following map

$$\mathfrak{T}_i^!(M) := L_i \otimes \mathbf{R}\mathrm{HOM}_{A_n^!}(L_i, M)[1] \xrightarrow{\lambda} M. \tag{3.11}$$

Here λ is the canonical pairing map. Alternatively, λ can be identified with the counit map of the biadjoint pair $\cup_i$ and $\cap_i$ (identifying $\cap_i(M)$ with $\mathbf{R}\mathrm{HOM}_{A_n^!}(L_i, M)$ via Equation (3.9)).

THEOREM 3.8 ([**QS16**, Theorem 5.14]). *The functors $\mathfrak{T}_i^!$, $i = 1, \dots, n-1$, are invertible on $\mathcal{D}^c(A_n^!)$. Furthermore, they satisfy the following categorical braid relations:*

(1) $\mathfrak{T}_i^!\mathfrak{T}_j^! \cong \mathfrak{T}_j^!\mathfrak{T}_i^!$ *if* $|i-j| > 1$,
(2) $\mathfrak{T}_i^!\mathfrak{T}_{i+1}^!\mathfrak{T}_i^! \cong \mathfrak{T}_{i+1}^!\mathfrak{T}_i^!\mathfrak{T}_{i+1}^!$.

Non-trivial maps between simples. Lemma 3.3 allows one to construct explicit DG module maps between the cofibrant replacements $\mathbb{P}(L_i)$. We record them here for later use.

By Corollary 3.4, there is a unique non-trivial map (up to homotopy) from L_{i+1} to L_i, which we denote by $\alpha_{i,i+1}$. It is explicitly given by, on the cofibrant replacemets of Lemma 3.3:

(3.12)

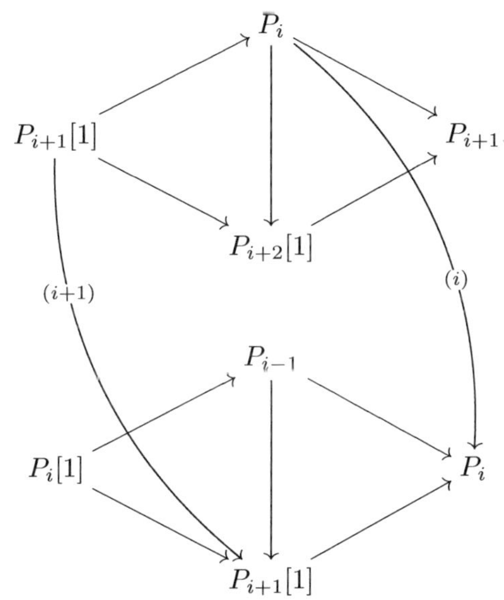

.

Similarly, there is a unique (up to homotopy) map $\alpha_{i+1,i}\colon L_i \to L_{i+1}$, which is given by:

(3.13)

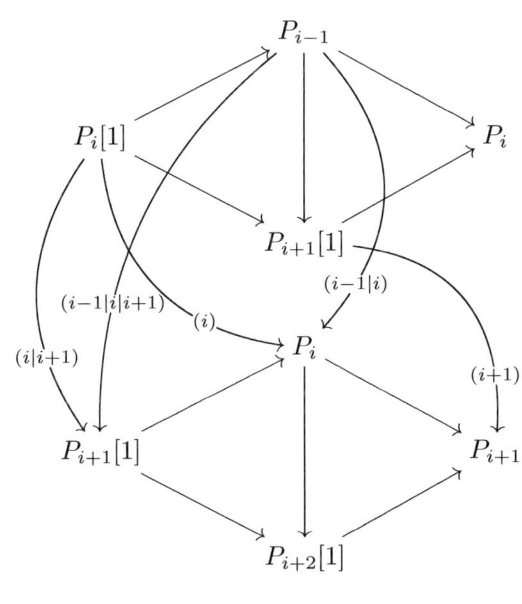

.

The unique (up to homotopy) non-trivial map $\alpha_{n-1,n}\colon L_n \to L_{n-1}$ is given as follows:

(3.14)

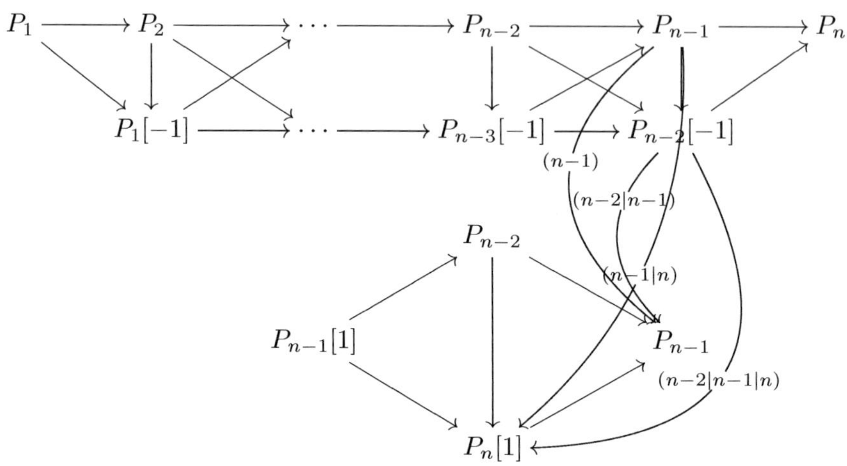

The unique (up to homotopy) non-trivial map $\alpha_{n,n-1}\colon L_{n-1} \to L_n$ is given by:

(3.15)

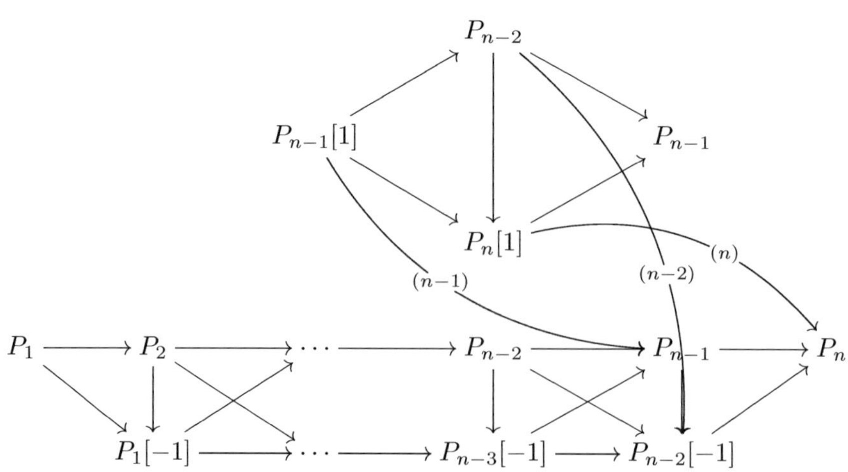

Some homotopies. Define the map $h_i\colon L_i \to L_i$ for $i = 1, \ldots, n-1$ by

(3.16)

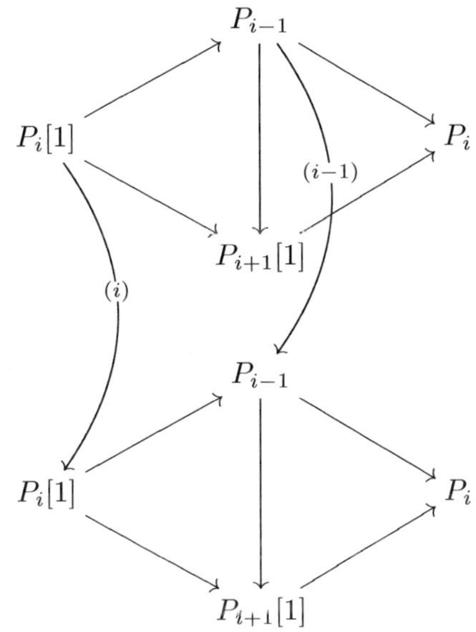

Let $h_n\colon L_n \to L_n$ be the map

(3.17)

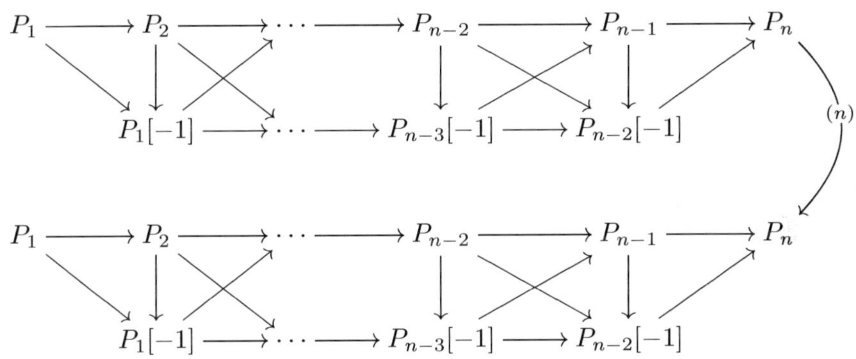

Finally, we define maps $h_{n,i}\colon L_i \to L_n$

(3.18)

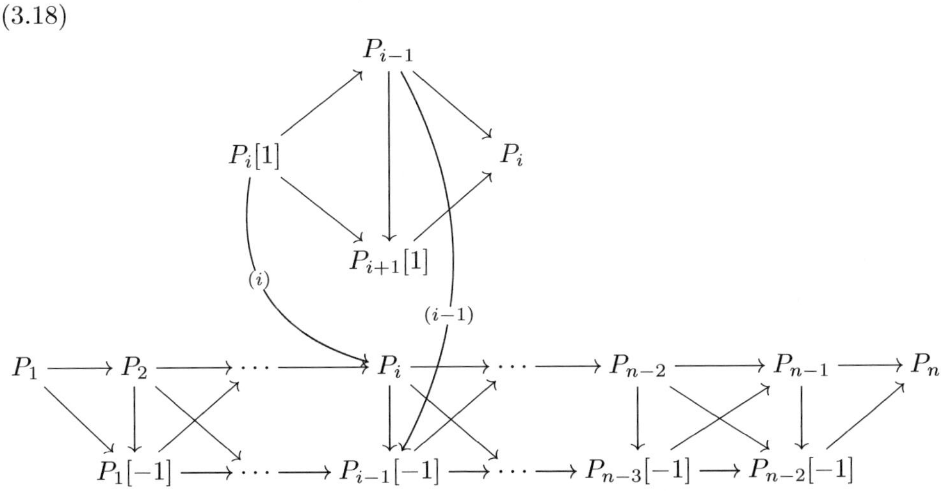

Auxiliary maps. It is notationally convenient to define

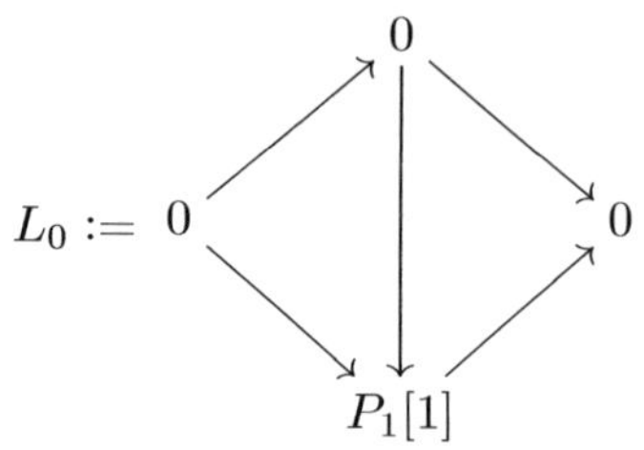

.

Now we define maps $\alpha_{1,0} \colon L_0 \to L_1$

(3.19)

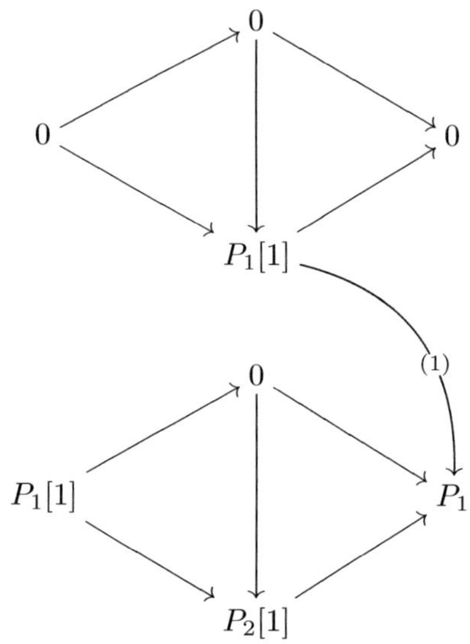

and $\alpha_{0,1} \colon L_1 \to L_0$

(3.20)

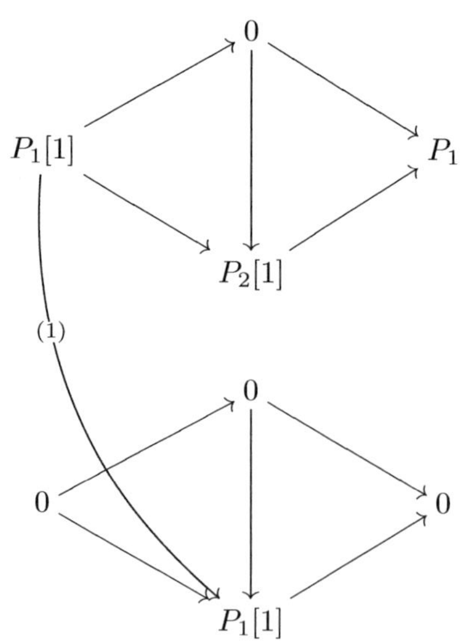

.

Note that $\alpha_{1,0}\alpha_{0,1}$ is null-homotopic.

4. The A_∞ zigzag algebra

The zigzag algebra. We now recall the definition of the algebra A_n that served a crucial role in [**KS02**]. When graded by path length, this algebra is Koszul dual to the algebra $A_n^!$ when it is also graded by path length.

DEFINITION 4.1. The ring A_n is the quotient of the path algebra $\Bbbk Q_n$ by the relations

$$(i|i\pm 1|i\pm 2)=0, (i|i+1|i)=(i|i-1|i), \text{ for } (i=2,\dots,n-1), (n|n-1|n)=0,$$
$$\deg((i))=0, \quad \deg((i|i+1))=0, \quad \deg((i+1|i))=1.$$

The zigzag algebra C_{n-1} is idempotented subalgebra given by

$$C_{n-1}=((1)+\cdots+(n-1))A_n((1)+\cdots+(n-1)).$$

For later use, we define the projective modules

$$Z_i:=A_n\cdot(i), \quad i=1,\dots,n, \tag{4.1}$$

which are the indecomposable direct summands of A_n. Then C_{n-1} can also be identified with

$$C_{n-1}\cong \mathrm{END}_{A_n}(\oplus_{i=1}^{n-1}Z_i). \tag{4.2}$$

The dual DG algebra E_n. We now define an intermediate DG algebra E_n as the graded endomorphism algebra

$$E_n:=\mathrm{END}_{A_n^!}(\oplus_{i=1}^n\mathbb{P}(L_i)). \tag{4.3a}$$

The differential on the endomorphism algebra, as usual, is given by

$$d(f)(x):=d(f(x))+f(d(x)) \tag{4.3b}$$

for any homogeneous $f\in E_n$ and $x\in\oplus_{i=1}^n\mathbb{P}(L_i)$. Here the usual sign that should appear in (4.3b) is dropped since we are working in characteristic two.

In E_n, we have the natural idempotents 1_i, $i=1,\dots,n$, which act as identity maps on $\mathbb{P}(L_i)$ and zero on $\mathbb{P}(L_j)$ for $j\neq i$. We will also use an idempotent truncation of E_n as follows. Consider $e=1_1+\cdots+1_{n-1}$. Then we have the idempotented subalgebra $S_{n-1}:=eE_ne$.

We will use the cofibrant DG modules

$$Q_i:=E_n1_i, \quad i=1,\dots,n, \tag{4.4}$$

which, by construction, is a direct DG summand of E_n.

For the computation of the A_∞-structure on the homology of E_n, we will consider a subalgebra of E_n defined by the explicit DG module maps recorded above. Let E_n' be the $\Bbbk$-algebra generated by

- 1_i for $i=1,\dots,n$, $\deg(1_i)=0$.
- $\alpha_{i+1,i},\alpha_{i,i+1}$ for $i=1,\dots,n-1$, $\deg(\alpha_{i+1,i})=1, \deg(\alpha_{i,i+1})=0$.
- h_i for $i=1,\dots,n$, $\deg(h_i)=0$.
- $h_{n,i}$ for $i=1,\dots,n-2$, $\deg(h_{n,i})=1$.

We set $h_{n,n-1}=\alpha_{n,n-1}$.

PROPOSITION 4.2. *These generators satisfy the following set of relations:*

- *The elements 1_i are orthogonal idempotents.*
- *The idempotents 1_i act by identity on the other generators.*
- $\alpha_{i,i+1}\alpha_{i+1,i+2}=0$, *for $i=1,\dots,n-2$.*

- $\alpha_{i+2,i+1}\alpha_{i+1,i} = 0$, *for* $i = 1, \ldots, n-3$.
- $h_i^2 = h_i$, *for* $i = 1, \ldots, n$.
- $h_i\alpha_{i,i+1} = 0$, *for* $i = 1, \ldots, n-1$.
- $\alpha_{i,i+1}h_{i+1} = \alpha_{i,i+1}$, *for* $i = 1, \ldots, n-2$.
- $\alpha_{i,i+1}\alpha_{i+1,i}\alpha_{i,i+1} = 0$.
- $\alpha_{n-1,n}h_n = 0$.
- $\alpha_{i+1,i}h_i = h_{i+1}\alpha_{i+1,i}$, *for* $i = 1, \ldots, n-2$.
- $\alpha_{n,n-1}h_{n-1} + h_n\alpha_{n,n-1} = \alpha_{n,n-1}$.
- $h_{n,i}h_i = h_{n,i}$, *for* $i = 1, \ldots, n-2$.
- $h_nh_{n,i} = 0$, *for* $i = 1, \ldots, n-2$.
- $h_{n,i}\alpha_{i,i+1} = 0$, *for* $i = 1, \ldots, n-2$.

The DG structure is defined on the generators by

- $d(1_i) = 0$,
- $d(\alpha_{i+1,i}) = d(\alpha_{i,i+1}) = 0$,
- $d(h_i) = \alpha_{i,i-1}\alpha_{i-1,i} + \alpha_{i,i+1}\alpha_{i+1,i}$, *for* $i = 1, \ldots, n-1$,
- $d(h_{n,i}) = h_{n,i+1}\alpha_{i+1,i}$.

PROOF. These are straightforward computations. □

One could easily compute bases of the idempotented subspaces $1_iE'_n1_j$ and their homologies $1_i\mathrm{H}(E'_n)1_j$ as follows.

First assume $i, j \neq n$.

$$1_iE'_n1_i = \Bbbk\langle 1_i, h_i, \alpha_{i,i+1}\alpha_{i+1,i}, \alpha_{i,i-1}\alpha_{i-1,i}\rangle \qquad d(h_i) = \alpha_{i,i+1}\alpha_{i+1,i} + \alpha_{i,i-1}\alpha_{i-1,i}$$
$$\mathrm{H}(1_iE'_n1_i) = \Bbbk\langle 1_i, c_i\rangle$$
$$1_iE'_n1_j = 0 \quad \text{for} \quad |i-j| > 1$$
$$1_iE'_n1_{i+1} = \Bbbk\langle \alpha_{i,i+1}\rangle$$
$$\mathrm{H}(1_iE'_n1_{i+1}) = \Bbbk\langle \alpha_{i,i+1}\rangle$$
$$1_{i+1}E'_n1_i = \Bbbk\langle \alpha_{i+1,i}, \alpha_{i+1,i}\alpha_{i,i+1}\alpha_{i+1,i}, \alpha_{i+1,i}h_i\rangle \qquad d(\alpha_{i+1,i}h_i) = \alpha_{i+1,i}\alpha_{i,i+1}\alpha_{i+1,i}$$
$$\mathrm{H}(1_{i+1}E'_n1_i) = \Bbbk\langle \alpha_{i+1,i}\rangle$$

Second, assume i or j is n.

$$1_nE'_n1_n = \Bbbk\langle 1_n, h_n, \alpha_{n,n-1}\alpha_{n-1,n}\rangle \qquad d(h_n) = \alpha_{n,n-1}\alpha_{n-1,n}$$
$$\mathrm{H}(1_nE'_n1_n) = \Bbbk\langle 1_n\rangle$$
$$1_nE'_n1_i = \Bbbk\langle h_{n,i}, h_{n,i+1}\alpha_{i+1,i}\rangle \qquad d(h_{n,i}) = h_{n,i+1}\alpha_{i+1,i} \qquad \text{for } i = 1, \ldots, n-2$$
$$\mathrm{H}(1_nE'_n1_i) = 0 \quad \text{for } i = 1, \ldots, n-2$$
$$1_iE'_n1_n = 0 \text{ for } i = 1, \ldots, n-2$$
$$\mathrm{H}(1_iE'_n1_n) = 0 \quad \text{for } i = 1, \ldots, n-2$$
$$1_nE'_n1_{n-1} = \Bbbk\langle \alpha_{n,n-1}, \alpha_{n,n-1}\alpha_{n-1,n}\alpha_{n,n-1}, h_n\alpha_{n,n-1}\rangle$$
$$d(h_n\alpha_{n,n-1}) = \alpha_{n,n-1}\alpha_{n-1,n}\alpha_{n,n-1}$$
$$\mathrm{H}(1_nE'_n1_{n-1}) = \Bbbk\langle \alpha_{n,n-1}\rangle$$
$$1_{n-1}E'_n1_n = \Bbbk\langle \alpha_{n-1,n}\rangle$$
$$\mathrm{H}(1_{n-1}E'_n1_n) = \Bbbk\langle \alpha_{n-1,n}\rangle$$

PROPOSITION 4.3. *There is an isomorphism of algebras* $A_n \cong \mathrm{H}(E'_n) \cong \mathrm{H}(E_n)$.

PROOF. This is clear from the relations in the algebras, the computations of the homology of $\mathrm{H}(E'_n)$ above, and the computation of the homology of $\mathrm{H}(E_n)$ using Corollary 3.4. □

A_∞-structure on C_{n-1}. Consider the idempotented subalgebra $C_{n-1} \subset \mathrm{H}(E_n)$. We compute the A_∞-structure on the idempotented algebra $C_{n-1} \cong \mathrm{H}(S_{n-1})$ using the perturbation lemma.

Let $\overrightarrow{\alpha}_i = \alpha_{i,i+1}\alpha_{i+1,i}$ and $\overleftarrow{\alpha}_i = \alpha_{i,i-1}\alpha_{i-1,i}$. Now we define linear maps p, j, and H as follows.

$$\begin{aligned}
p(h_i) &= 0\\
p(\alpha_{i,i+1}) &= (i|i+1)\\
p(\alpha_{i+1,i}) &= (i+1|i)\\
p(\overrightarrow{\alpha}_i) &= (i|i+1|i)\\
p(\overleftarrow{\alpha}_i) &= (i|i+1|i)\\
j((i|i+1)) &= \alpha_{i,i+1}\\
j((i+1|i)) &= \alpha_{i+1,i}\\
j((i|i+1|i)) &= \overrightarrow{\alpha}_i\\
H(\overrightarrow{\alpha}_i) &= 0\\
H(\overleftarrow{\alpha}_i) &= h_i\\
H(\alpha_{i+1,i}\overrightarrow{\alpha}_i) &= \alpha_{i+1,i}h_i\\
H(\overrightarrow{h}_i) &= 0\\
H(\alpha_{i+1,i}h_i) &= 0.
\end{aligned}$$

This data is summarized in (4.5).

(4.5)
$$H \circlearrowright S_{n-1} \underset{j}{\overset{p}{\rightleftarrows}} C_{n-1}$$

The perturbation lemma (see [**Kel06a**, Theorem 2.3]) determines an A_∞-structure on C_{n-1} is given by

(4.6)
$$m_k = \sum_T m_k^T$$

where T ranges over planar rooted binary trees with k leaves and m_k^T is given by composing the tree-shaped diagram obtained by labeling each leaf by j, each branch point by multiplication μ in S_{n-1}, each internal edge by H, and the root by p. Note that this A_∞-structure is unique up to A_∞-isomorphism.

THEOREM 4.4. *All higher operations m_k vanish for $k > 3$. There are three families of m_3 operations:*

$$\begin{aligned}
m_3((i|i+1),(i+1|i),(i|i+1|i)) &= (i|i+1|i)\\
m_3((i-1|i),(i|i-1),(i-1|i)) &= (i-1|i)\\
m_3((i|i+1|i),(i|i-1),(i-1|i)) &= (i|i+1|i).
\end{aligned}$$

The non-zero trees giving rise to these operations are:

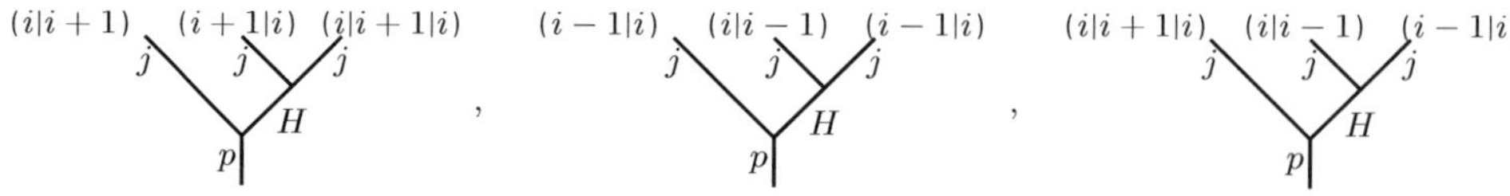

PROOF. It is clear that these trees give rise to non-trivial operations stated in the theorem and that the other terms m_k^T in (4.6) are zero in these cases.

In order to see that trees with more than 3 inputs are zero, first note that the possible outputs of H are h_i and $\alpha_{i+1,i}h_i$.

If there are only 3 leaves in a tree, then it is straightforward to see that the only non-trivial possibilities are the ones stated in the theorem.

If there are more than three leaves, then, at the root of the tree, two outputs of H must be multiplied in E_n' non-trivially and then mapped to something non-zero in C_n via p. Since the only non-zero outputs of $H(x)H(y)$ are h_i and $\alpha_{i+1,i}h_i$, and these are mapped to zero by p, we conclude that there could not be any non-trivial operations m_k with $k > 3$. □

Non-vanishing of Hochschild cohomology. In this section, we prove that the map $m_3 \in C^3(C_{n-1}, C_{n-1})$ is not a Hochschild coboundary when $n \geq 4$. In more detail, the perturbation lemma gave the assignments in Theorem 4.4. After extending by zero, the map $m_3 : C_{n-1}^{\otimes 3} \to C_{n-1}$ can be interpreted as a degree -1 element $m_3 \in C^3(C_{n-1}, C_{n-1})^{-1}$ of the Hochschild cochain complex,

$$\cdots \to C^2(C_{n-1}, C_{n-1})^{-1} \xrightarrow{\delta} C^3(C_{n-1}, C_{n-1})^{-1} \to \cdots,$$

computing the degree -1 part of Hochschild cohomology of the algebra C_{n-1}. We show here that there is no map $m \in C^2(C_{n-1}, C_{n-1})^{-1}$ which satisfies $\delta m = m_3$. Since the boundary map δ preserves degree and the perturbation lemma guarantees $\delta m_3 = 0$, the computation below suffices to show that m_3 determines a non-trivial class in $\mathrm{HH}^3(C_{n-1}, C_{n-1})^{-1}$.

Now, for any map

$$m \in C^2(C_{n-1}, C_{n-1})^{-1},$$

i.e., m is an element in $\mathrm{Hom}(C_{n-1}^{\otimes 2}, C_{n-1}[1])$, homogeneity implies that there are constants α_i, b_i, c_i, d_i and η_i indexed by the idempotents i of C_{n-1}, $1 \leq i \leq n-1$, such that

$$\begin{aligned} m((i|i+1), (i+1|i)) &= \alpha_i(i) \\ m((i|i+1|i), (i|i+1)) &= b_i(i|i+1) \\ m((i|i+1|i), (i|i-1)) &= c_i(i|i-1) \\ m((i|i+1|i), (i|i+1|i)) &= d_i(i|i+1|i) \\ m((i|i-1), (i-1|i|i-1)) &= \eta_i(i|i-1). \end{aligned}$$

By evaluating, the equation

$$(\delta m)(x, y, z) = xm(y, z) - m(xy, z) + m(x, yz) - m(x, y)z$$

on inputs $(x, y, z) \in C^{\otimes 3}$ of the form $((i|i+1), (i+1|i), (i|i-1))$, $((i|i-1), (i-1|i|i-1), (i-1|i))$, $((i|i+1), (i+1|i), (i|i+1|i))$ and $((i|i+1|i), (i|i+1), (i+1|i))$, we obtain the equations

$$\begin{aligned} c_i + \alpha_i &= 0 \\ b_{i-1} + \eta_i &= 0 \end{aligned}$$

$$\begin{aligned} \eta_{i+1} + d_i + \alpha_i &= 1 \\ \alpha_i + d_i + b_i &= 0 \end{aligned}$$

which are inconsistent. To produce these equations, $(i-1), (i), (i+1)$ must be in C_{n-1}, so $n \geq 4$.

5. Categorical symmetries

In this section, we construct a (weak) categorical Temperley-Lieb algebra and braid group action on the compact derived category of the A_∞-algebra C_{n-1}. For notational simplicity, we will often let $C = C_{n-1}$.

For $i = 1, \ldots, n-1$, let $C_i := C(i)$ and ${}_iC := (i)C$. Also set $B_i = C_i \otimes {}_iC$. This (C, C)-bimodule can be regarded as arising from the cohomology of the DG bimodule $S_{n-1}1_i \otimes 1_iS_{n-1}$ over $S_{n-1} \otimes S_{n-1}$. It follows that B_i has a unique minimal A_∞-structure over $C \otimes C$.

LEMMA 5.1. *The bimodule B_i has the minimal A_∞-structure*

$$m_{k,l}\colon C^{\otimes k-1} \otimes B_i \otimes C^{\otimes l-1} \to B_i$$

of degree $3-k-l$ by

$$m_{k,l} = 0 \text{ if } k \text{ and } l > 0 \qquad m_{k,1} = m_k \otimes \mathrm{Id} \qquad m_{1,l} = \mathrm{Id} \otimes m_l$$

PROOF. It is straightforward to check that this defines an A_∞-structure on the bimodule. □

LEMMA 5.2. *On the (C, C)-bimodule C, there is a minimal A_∞-structure*

$$m_{i,j}\colon C^{\otimes i-1} \otimes C \otimes C^{\otimes j-1} \to C$$

of degree $3-i-j$ by $m_{i,j} = m_{i+j-1}$.

PROOF. It is also straightforward to check that this is an A_∞-structure on the bimodule C. For more details see [**Tra08**, Lemma 4.1]. □

LEMMA 5.3. *There is an A_∞-bimodule map $f_{i,j}\colon B_k \to C$ defined by $f_{i,j} = 0$ unless $i = 2, j = 1$ or $i = 1, j = 2$. In these other cases*

$$\begin{aligned} f_{2,1}(\alpha \otimes [\beta(k) \otimes (k)\gamma)]) &= m_3(\alpha, \beta(k), (k)\gamma) \\ f_{1,2}([\alpha(k) \otimes (k)\beta)] \otimes \gamma) &= m_3(\alpha(k), (k)\beta, \gamma). \end{aligned}$$

PROOF. It is a straightforward, yet tedious, calculation that this defines an A_∞-map. □

DEFINITION 5.4. Consider the derived category $\mathcal{D}_\infty(C)$, and let

$$k \in \{1, \ldots, n-1\}.$$

(i) The functor $\mathcal{U}_k$ is the derived tensor functor

$$\mathfrak{U}_k : \mathcal{D}_\infty(C) \longrightarrow \mathcal{D}_\infty(C), \quad M \mapsto B_k \boxtimes_C M.$$

(ii) The functor $\mathcal{T}_k$ is the derived tensor functor with the one-sided twisted complex

$$\mathfrak{T}_k : \mathcal{D}_\infty(C) \longrightarrow \mathcal{D}_\infty(C), \quad M \mapsto (B_k \xrightarrow{f} C) \boxtimes_C M.$$

THEOREM 5.5. *Let $\mathcal{D}^c_\infty(C)$ be the compact derived category of the A_∞-algebra C.*

(1) *The functors* $\mathcal{U}_i$, $i \in \{1, \ldots, n-1\}$ *define a categorical Temperley-Lieb algebra action on* $\mathcal{D}^c_\infty(C)$.
(2) *The functors* $\mathcal{T}_i$, $i \in \{1, \ldots, n-1\}$ *define a categorical braid group action on* $\mathcal{D}^c_\infty(C)$.

PROOF. The proof essentially reduces to the categorical braid group action of Br_n on $\mathcal{D}^c(A_n^!)$ constructed in [**QS16**] specialized over a field of characteristic two.

By construction, there is an equivalence of DG derived categories

$$\mathcal{D}(A_n^!) \cong \mathcal{D}(E_n). \tag{5.1a}$$

On the other hand, the algebra A_n is the cohomology of E_n with its minimal A_∞-algebra structure. By standard arguments about A_∞-algebras (see, for instance, [**Kel02**, Section 3.3]), there is a quasi-isomorphism of A_∞-algebras between E_n and A_n, which induces a triangulated equivalence

$$\mathcal{D}(E_n) \cong \mathcal{D}_\infty(A_n). \tag{5.1b}$$

The derived equivalences (5.1a) and (5.1b) descend to equivalences of the corresponding compact derived categories.

Next, the A_∞-bimodule structures on the bimodules defining $\mathfrak{U}_i$ and $\mathfrak{T}_i$ can also be naturally identified with the A_∞-bimodule structure inherited from the functors $\mathcal{U}_i^!$ and $\mathfrak{T}_i^!$ on $\mathcal{D}(A_n^!)$.

In [**QS16**], the Temperley-Lieb algebra TL_n and the n-stranded braid group Br_n act on $\mathcal{D}(A_n^!)$, generated by $\mathfrak{U}_i^!$ and $\mathfrak{T}_i^!$, $i = 1, \ldots, n-1$. By Lemma 3.3, one can see that the categorical Temperley-Lieb generators preserve the smallest triangulated subcategory generated by the first $n-1$ simple DG modules $L_1, \ldots, L_{n-1}$:

$$\mathcal{U}_i^!(L_j) = \cup_i \circ \cap_i(L_j) = L_i \otimes (\mathbb{P}(L_i) \otimes_{A_n^!} L_j) = \begin{cases} L_i \oplus L_i[1] & i = j \\ L_i & j = i-1 \\ L_i[1] & j = i+1 \\ 0 & |i-j| > 1 \end{cases} \tag{5.2}$$

Denote this subcategory by $\langle L_1, \ldots, L_{n-1}\rangle$. Under the functor $\mathrm{HOM}_{A_n^!}(\oplus_{i=1}^n \mathbb{P}(L_i), \text{-})$, $\langle L_1, \ldots, L_{n-1}\rangle$ corresponds to the triangulated subcategory generated by $Q_1, \ldots, Q_{n-1}$ (equation (4.4)). This subcategory, by construction, is equivalent to the compact derived category over the idempotent truncated DG algebra S_{n-1}. Upon taking cohomology, Q_i has defines the A_∞-modules, Z_i, whose A_∞-endomorphism algebra is equal to C_{n-1}, which is the cohomology algebra of the truncated DG algebra S_{n-1}. We exhibit the containment relation of the categories involved as follows

$$\begin{array}{ccccc} \mathcal{D}(A_n^!) & \xrightarrow{\mathrm{HOM}(\oplus_{i=1}^n \mathbb{P}(L_i), \text{-})} & \mathcal{D}(E_n) & \xleftarrow{\cong} & \mathcal{D}_\infty(A_n) \\ \uparrow & & \uparrow & & \uparrow \\ \langle L_1, \ldots, L_{n-1}\rangle & & \langle Q_1, \ldots, Q_{n-1}\rangle & & \langle Z_1, \ldots, Z_{n-1}\rangle \\ & & \cong\uparrow & & \cong\uparrow \\ & & \mathcal{D}^c(S_{n-1}) & & \mathcal{D}^c_\infty(C_{n-1}) \end{array} . \tag{5.3}$$

Translating the corresponding action on $\mathcal{D}^c(A_n^!)$ to the A_∞-algebra C via the above diagram, the Temperley-Lieb algebra and braid group actions become actions on the derived category $\mathcal{D}_\infty(C_{n-1})$, preserving the full subcategory of compact objects. □

REMARK 5.6. The objects L_i $(i = 1, \ldots, n-1)$ are analogues of an A_{n-1} *chain of spherical objects* in the derived category $\mathcal{D}(A_n^!)$ in the sense of Seidel-Thomas [**ST01**]. Their images in $\mathcal{D}_\infty(A_n)$ under the chain of equivalences

$$\mathcal{D}(A_n^!) \simeq \mathcal{D}(E_n) \cong \mathcal{D}_\infty(A_n),$$

are given by the projective modules C_i. Thus the collection C_i, $i = 1, \ldots, n-1$ also constitues such an A_{n-1}-chain of spherical objects. The braid group action thus naturally arises from spherical twists along this chain. We will point out the symplecto-geometric meaning of this collection of spherical objects in the next section.

COROLLARY 5.7. *The categorical braid group action descends to the Burau representation at* $q = -1$ *in the Grothendieck group. More specifically, the following squares commute*

$$\begin{array}{ccc} K_0(\mathcal{D}^c_\infty(C)) & \xrightarrow{\cong} & V_{n-1} \\ \downarrow{\scriptstyle [\mathfrak{U}_i]} & & \downarrow{\scriptstyle u_i} \\ K_0(\mathcal{D}^c_\infty(C)) & \xrightarrow{\cong} & V_{n-1} \end{array} \qquad \begin{array}{ccc} K_0(\mathcal{D}^c_\infty(C)) & \xrightarrow{\cong} & V_{n-1} \\ \downarrow{\scriptstyle [\mathfrak{T}_i]} & & \downarrow{\scriptstyle t_i} \\ K_0(\mathcal{D}^c_\infty(C)) & \xrightarrow{\cong} & V_{n-1}. \end{array}$$

PROOF. This is a direct translation of [**QS16**, Proposition 5.16] via the equivalences in the diagram (5.3). □

6. Remarks on connections to symplectic topology

The zigzag algebras C_{n-1} appear geometrically as endomorphism algebras of Lagrangians $\oplus_i L_i$ in Fukaya categories $\mathcal{D}_\infty(\mathcal{F}(S))$ [**KS02**]. When the grading conventions are those of Section 4, the relevant symplectic manifold S is a surface and the identification

$$C_{n-1} \cong \mathrm{END}_{\mathcal{D}_\infty(\mathcal{F}(S))}(\oplus_i L_i) \tag{6.1}$$

endows the zigzag algebra with some auxiliary geometric content. The purpose of this section is to suggest how to locate the A_∞-structure of Theorem 4.4, obtained by algebraic means, within the geometric construction of the Fukaya category.

Slightly oversimplifying, the Fukaya category $\mathcal{F}(S)$ of a surface S has objects oriented embedded 1-manifolds $L \subset S$ with spin structure. When $L, L' \in Ob(\mathcal{F}(S))$ are transverse, morphisms $CF(L, L') := \mathbb{F}_2\langle L \pitchfork L' \rangle$ are spanned by points of intersection. The A_∞-structure is encoded by maps $m_r : \mathcal{F}(S)^{\otimes r} \to \mathcal{F}(S)$ for $r \geq 1$. If the collection $\{L_0, \ldots, L_r\}$ is pairwise transverse, then these maps

$$m_r : CF(L_r, L_{r-1}) \otimes \cdots \otimes CF(L_1, L_0) \to CF(L_r, L_0) \qquad \text{are given by}$$

$$m_r(x_r, \ldots, x_1) := \sum_{x_0 \in L_n \pitchfork L_1} \#\mathcal{M}(x_r, \ldots, x_0)x_0 \tag{6.2}$$

a (signed) count of immersed disks representing points in moduli spaces $\mathcal{M}(x_r, \ldots, x_0)$ of virtual dimension zero [**Sei08**, II (13b)]. Figure 1 depicts such a disk among Lagrangians which have been perturbed to be transverse. Each map

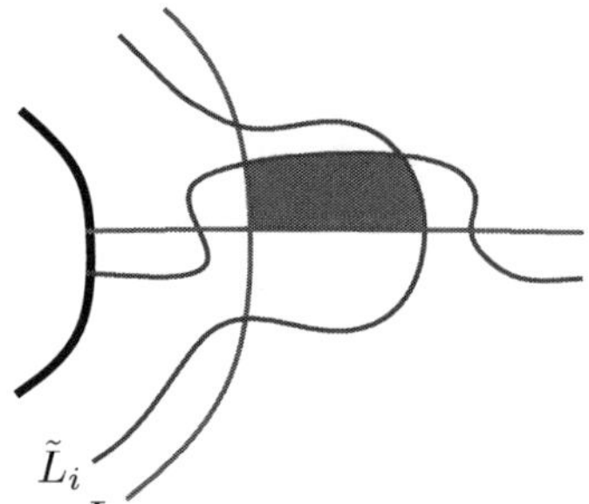

FIGURE 1. A disk among perturbations

m_r can be extended to a map $m_r : \oplus_n \mathcal{F}(S)^{\otimes n} \to \mathcal{F}(S)$ so that the A_∞-relations become $d^2 = 0$ after setting $d := \sum_n m_n$. Since $\mathcal{F}(S)$ is too small to be triangulated, focus is placed on the triangulated hull $\mathcal{D}_\infty(\mathcal{F}(S))$ [**Sei08**, I (3h)].

Within this setting, consider simple closed curves arranged in the pattern depicted by the figure below.

Each curve being a circle determines a spherical object in the Fukaya category. The braid group representation is generated by the twist functors

$$T_{L_i} : \mathcal{D}_\infty(\mathcal{F}(S)) \to \mathcal{D}_\infty(\mathcal{F}(S)) \qquad T_{L_i}(X) := \mathrm{Cone}(\mathrm{Hom}(L_i, X) \otimes Y \xrightarrow{ev} X).$$

When L_i has non-trivial spin structure, the action of T_{L_i} agrees with the Dehn twist on $\mathcal{F}(S)$ [**Sei08**, Corollary 17.18]. For two curves, L_i and L_j above, the associated Dehn twists either commute or satisfy the braid relation depending upon whether the geometric intersection number is 0 or 1 respectively [**FM12**, page 84]. These objects determine the zigzag algebra C_{n-1} in Equation (6.1) above. The algebra A_n can obtained by adjoining an embedded interval to $\oplus_i L_i$ [**KS02**, Figure 3].

When the curves $\{L_0, \ldots, L_r\}$ are not pairwise transverse, some care is needed to define the A_∞-structure. One approach is to limit over all chain complexes associated to (isomorphic) Hamiltonian perturbations of Lagrangians, in this manner the map m_r can be computed for any non-transverse collection $\{L_0, \ldots, L_r\}$ by any transverse family obtained by Hamiltonian isotopies. With this in mind, to compute $m_3((i|i+1), (i+1|i), (i|i+1))$ in Theorem 4.4 we perturb both L_i and L_{i+1} to obtain isomorphic objects $\tilde{L}_i$ and $\tilde{L}_{i+1}$. Under the identifications made in the figure above, maps, such as $(i|i+1)$ correspond to distinct points of intersection, this results in the figure below. Here we suggest that, since we are working with coefficients in characteristic 2, we can define a new map $\tilde{m}_3$ which counts the entire moduli space associated to the green disk. For a related comment see [**Sei08**, Remark 13.4]. Setting $\tilde{m}_r = 0$ for $r \neq 3$ and $\delta = \sum_r \tilde{m}_r$ the algebraic portion of this paper shows that $(d+\delta)^2 = 0$ and so an A_∞-structure is obtained. We expect that

δ admits a geometric definition as a sum over moduli spaces parallel to Equation (6.2).

Acknowledgments

The authors would like to thank Igor Frenkel for always sharing with us his panoramic view of mathematics and physics, as well as his constant encouragement.

The third author would like to thank Ben Elias and Anthony Licata for discussions related to the DG enhanced zigzag algebra.

The first author thanks K. Kawamuro and M. Tehrani for their cordial collegiality.

References

[Bir74] Joan S. Birman, *Braids, links, and mapping class groups*, Annals of Mathematics Studies, No. 82, Princeton University Press, Princeton, NJ; University of Tokyo Press, Tokyo, 1974. MR375281

[BK90] A. I. Bondal and M. M. Kapranov, *Framed triangulated categories* (Russian), Mat. Sb. **181** (1990), no. 5, 669–683, DOI 10.1070/SM1991v070n01ABEH001253; English transl., Math. USSR-Sb. **70** (1991), no. 1, 93–107. MR1055981

[BW22] S. Barmeier and Z. Wang, *A_∞ deformations of extended Khovanov arc algebras and Stroppel's conjecture*, 2022. arXiv:2211.03354.

[CF94] Louis Crane and Igor B. Frenkel, *Four-dimensional topological quantum field theory, Hopf categories, and the canonical bases*, J. Math. Phys. **35** (1994), no. 10, 5136–5154, DOI 10.1063/1.530746. Topology and physics. MR1295461

[CK14] Yanfeng Chen and Mikhail Khovanov, *An invariant of tangle cobordisms via subquotients of arc rings*, Fund. Math. **225** (2014), no. 1, 23–44, DOI 10.4064/fm225-1-2. MR3205563

[CS19] Benjamin Cooper and Joshua Sussan, *Preprojective analogue of the cone construction*, Homology Homotopy Appl. **21** (2019), no. 2, 171–198, DOI 10.4310/HHA.2019.v21.n2.a10. MR3921615

[EQ16] Ben Elias and You Qi, *An approach to categorification of some small quantum groups II*, Adv. Math. **288** (2016), 81–151, DOI 10.1016/j.aim.2015.10.009. MR3436383

[FKS05] Igor Frenkel, Mikhail Khovanov, and Olivier Schiffmann, *Homological realization of Nakajima varieties and Weyl group actions*, Compos. Math. **141** (2005), no. 6, 1479–1503, DOI 10.1112/S0010437X05001727. MR2188446

[FM12] Benson Farb and Dan Margalit, *A primer on mapping class groups*, Princeton Mathematical Series, vol. 49, Princeton University Press, Princeton, NJ, 2012. MR2850125

[Kel94] Bernhard Keller, *Deriving DG categories*, Ann. Sci. École Norm. Sup. (4) **27** (1994), no. 1, 63–102. MR1258406

[Kel01] Bernhard Keller, *Introduction to A-infinity algebras and modules*, Homology Homotopy Appl. **3** (2001), no. 1, 1–35, DOI 10.4310/hha.2001.v3.n1.a1. MR1854636

[Kel02] Bernhard Keller, *A-infinity algebras in representation theory*, Representations of algebra. Vol. I, II, Beijing Norm. Univ. Press, Beijing, 2002, pp. 74–86. MR2067371

[Kel03] B. Keller, *Koszul duality and coderived categories (after K. Lefèvre)*, 2003. Available at https://webusers.imj-prg.fr/~bernhard.keller/publ/kdc.pdf.

[Kel06a] Bernhard Keller, *A-infinity algebras, modules and functor categories*, Trends in representation theory of algebras and related topics, Contemp. Math., vol. 406, Amer. Math. Soc., Providence, RI, 2006, pp. 67–93, DOI 10.1090/conm/406/07654. MR2258042

[Kel06b] Bernhard Keller, *On differential graded categories*, International Congress of Mathematicians. Vol. II, Eur. Math. Soc., Zürich, 2006, pp. 151–190. MR2275593

[Kho16] Mikhail Khovanov, *Hopfological algebra and categorification at a root of unity: the first steps*, J. Knot Theory Ramifications **25** (2016), no. 3, 1640006, 26, DOI 10.1142/S021821651640006X. MR3475073

[KQ15] Mikhail Khovanov and You Qi, *An approach to categorification of some small quantum groups*, Quantum Topol. **6** (2015), no. 2, 185–311, DOI 10.4171/QT/63. MR3354331

[KQS17] Mikhail Khovanov, You Qi, and Joshua Sussan, *p-DG cyclotomic nilHecke algebras*, Mem. Amer. Math. Soc. **293** (2024), no. 1462, v+94, DOI 10.1090/memo/1462. MR4684340

[KS02] Mikhail Khovanov and Paul Seidel, *Quivers, Floer cohomology, and braid group actions*, J. Amer. Math. Soc. **15** (2002), no. 1, 203–271, DOI 10.1090/S0894-0347-01-00374-5. MR1862802

[LW23] J. Liu and Z. Wang, *A_∞-deformations of zigzag algebras via Ginzburg algebras*, arXiv:2301.06757, 2023.

[Qi14] You Qi, *Hopfological algebra*, Compos. Math. **150** (2014), no. 1, 1–45, DOI 10.1112/S0010437X13007380. MR3164358

[QS16] You Qi and Joshua Sussan, *A categorification of the Burau representation at prime roots of unity*, Selecta Math. (N.S.) **22** (2016), no. 3, 1157–1193, DOI 10.1007/s00029-015-0216-8. MR3518548

[QS17] You Qi and Joshua Sussan, *Categorification at prime roots of unity and hopfological finiteness*, Categorification and higher representation theory, Contemp. Math., vol. 683, Amer. Math. Soc., Providence, RI, 2017, pp. 261–286, DOI 10.1090/conm/683. MR3611717

[QS18] You Qi and Joshua Sussan, *p-DG cyclotomic nilHecke algebras II*, Mem. Amer. Math. Soc. **293** (2024), no. 1463, v+106, DOI 10.1090/memo/1463. MR4684341

[RZ03] Raphaël Rouquier and Alexander Zimmermann, *Picard groups for derived module categories*, Proc. London Math. Soc. (3) **87** (2003), no. 1, 197–225, DOI 10.1112/S0024611503014059. MR1978574

[Sch11] Olaf M. Schnürer, *Perfect derived categories of positively graded DG algebras*, Appl. Categ. Structures **19** (2011), no. 5, 757–782, DOI 10.1007/s10485-010-9221-4. MR2836543

[Sei08] Paul Seidel, *Fukaya categories and Picard-Lefschetz theory*, Zurich Lectures in Advanced Mathematics, European Mathematical Society (EMS), Zürich, 2008, DOI 10.4171/063. MR2441780

[ST01] Paul Seidel and Richard Thomas, *Braid group actions on derived categories of coherent sheaves*, Duke Math. J. **108** (2001), no. 1, 37–108, DOI 10.1215/S0012-7094-01-10812-0. MR1831820

[Str09] Catharina Stroppel, *Parabolic category $\mathcal{O}$, perverse sheaves on Grassmannians, Springer fibres and Khovanov homology*, Compos. Math. **145** (2009), no. 4, 954–992, DOI 10.1112/S0010437X09004035. MR2521250

[Tra08] Thomas Tradler, *Infinity-inner-products on A-infinity-algebras*, J. Homotopy Relat. Struct. **3** (2008), no. 1, 245–271. MR2426181

Department of Mathematics, University of Iowa, Iowa City, Iowa, 52242
Email address: ben-cooper@uiowa.edu

Department of Mathematics, University of Virginia, Charlottesville, Virginia, 22904
Email address: yq2dw@virginia.edu

Department of Mathematics, CUNY Medgar Evers, Brooklyn, New York, 11225; and Mathematics Program, The Graduate Center, CUNY, New York, New York 10016
Email address: jsussan@mec.cuny.edu

Contemporary Mathematics
Volume **817**, 2025
https://doi.org/10.1090/conm/817/16325

Looking for a refined Monster

Nora Ganter

ABSTRACT. We discuss some categorical aspects of the objects that appear in the construction of the Monster and other sporadic simple groups. We define the basic representation of the categorical torus $\mathcal{T}$ classified by an even symmetric bilinear form I and of the semi-direct product of $\mathcal{T}$ with its canonical involution. We compute the centraliser of the basic representation of $\mathcal{T}\rtimes\{\pm 1\}$ and find it to be a categorical extension of the extraspecial 2-group with commutator $I \mod 2$. We study the inertia groupoid of a categorical torus and find that it is given by the torsor of the topological Looijenga line bundle, so that 2-class functions on $\mathcal{T}$ are canonically theta-functions. We discuss how discontinuity of the categorical character in our formalism means that the character of the basic representation fails to be a categorical class function. We compute the automorphisms of $\mathcal{T}$ and of $\mathcal{T}\rtimes\{\pm 1\}$ and relate these to the Conway groups.

1. Introduction

The largest of the sporadic finite simple groups in the Atlas classification [**CCN$^+$85**] is known by the name *the Monster*. Its irreducible representations have dimensions

$$1, \quad 196883, \quad 21296876, \quad 842609326, \quad \ldots.$$

These characters were found before the Monster was first constructed, and it was then that John McKay observed that the coefficients of the j-function

$$j(q) - 744 = q^{-1} + 196884q + 21493760q^2 + 864229970q^3 + \ldots$$

appeared to be counting the graded dimension of an infinite series of Monster representations. The conjectural picture describing this phenomenon, which would come to be known as Monstrous Moonshine, was further developed by Conway and Norton [**CN79**] and generalised to involve pairs of commuting elements of the Monster by Norton [**Mas87**]. The conjecture was proved in the classical setting by Borcherds [**Bor92**] and in the generalised setting by Carnahan [**Car07**], [**Car10**], [**Car12**]. The Monster itself was constructed by Griess [**Gri82**], see also [**Tit85**]. Its most conceptual construction to date is due to Frenkel, Lepowsky and Meurman, who brought together the young disciplines of infinite dimensional Lie theory

2020 *Mathematics Subject Classification.* Primary 18G45, 20D08, 20J05, 11H56; Secondary 22E67, 18N25, 17B69, 20C99.

Key words and phrases. Sporadic groups, categorification, moonshine.

The author was supported by ARC grants DP160104912 and DP210103081.

and string theory, in order to conceptualise the appearance of theta functions and integrate them into the construction of the Monster. This lead them to develop the notion of a of vertex operator algebra, realising the Monster as the symmetry group of the Moonshine Module [**FLM88**]. For a detailed account of the early history of Monstrous Moonshine, we refer the reader to the introduction of [**FLM88**].

From the first encounters with the Monster, the subtle role of the number 24 was apparent. In the context of classical Moonshine, this manifests itself in the level of the modular functions involved; in Generalised Moonshine, it manifests itself in the occurrence of 24th roots of unity and in the role of certain central extensions of the centraliser subgroups of the Monster. It was conjectured by Mason [**Mas02**] that these were governed by a 3-cocycle of order 24 in the group cohomology of the Monster; this three-cocycle is called the moonshine anomaly. We will write M for the Monster group. In [**JF19**], Johnson-Freyd studies a cocycle

$$\alpha_{\mathbb{C}} \in H^3_{gp}(M; U(1)),$$

conjecturally that of Mason,[1] and shows it to be of order 24.

In the days of the ATLAS project, Schur's work on the role of cocycles of degree 2 [**Sch04**] [**Sch07**] as defining central extensions was already part of the standard repertoire of the group theorist. The analogous classification of categorical central extensions by degree 3 cocycles, on the other hand, was only just emerging, having been the topic of the unpublished PhD thesis of Sính [**Sí75**]. It would be some time before her *gr-catégories*, nowadays better known under the name categorical groups or, among non-group theorists, as 2-groups,[2] became more mainstream. In any case, the ATLAS project was concerned with the classification of finite simple groups; the community had no interest in categorical groups. It is therefore not surprising that some of the intrinsic categorical features of key objects in the construction of the sporadic groups went unnoticed and, despite their elementary nature, were not studied very deeply at the time. In recent years, the question has been posed whether the Monster might, in fact, occur in nature as a categorical group. Put differently, we know $\alpha_{\mathbb{C}}$ to classify a categorical group $\mathcal{M}$ sitting inside an exact sequence

$$1 \longrightarrow \mathbb{B}C_{24} \longrightarrow \mathcal{M} \longrightarrow M \longrightarrow 1,$$

where C_{24} denotes the cyclic group of order 24, and $\mathbb{B}C_{24}$ is the one object category with automorphism group C_{24}. We can see what appears to be a shadow of $\alpha_{\mathbb{C}}$ in the Monster's natural habitat: Moonshine and the Moonshine module. Defining a categorical group from a 3-cocycle, however, is the most tedious and unintuitive construction one could imagine, involving triples of elements of the Monster group in the case of $\mathcal{M}$. One would instead hope to realise $\mathcal{M}$ as the automorphisms of an object of a bicategory. To be interesting, this object should be natural and easily constructed and have the potential to simplify our understanding of the Monster. The paper at hand will not provide an answer to this problem. Instead, I will try to give an idea of where to look and how to think about these structures.

[1] Johnson-Freyd's moonshine anomaly is obtained using conformal nets, while Mason's conjecture lives in the world of vertex operator algebras. The referee pointed out that, while these two formalisms are conjecturally equivalent to each other, the full VOA-conformal nets dictonary has yet to be worked out.

[2] In this paper, the term 2-group will be reserved for a finite group whose order is a power of two.

Let (Λ, I) be an even unimodular lattice, and let $O(\Lambda, I)$ be its group of isometries. These data are already of a categorical nature: let $\mathfrak{t} = \Lambda \otimes_{\mathbb{Z}} \mathbb{R}$ and consider the torus

$$T = \Lambda \otimes_{\mathbb{Z}} U(1) \cong \mathfrak{t} \,/\, \Lambda,$$

where on the right Λ is identified with $\Lambda \otimes 1$. Then the Lie group cohomology [**WW15**] of T is computed by the isomorphism

$$H^3_{gp}(T; U(1)) \cong H^4(BT; \mathbb{Z})$$

[**SP11**, Corollary 97]. The elements of $H^4(BT;\mathbb{Z})^{O(\Lambda,I)}$ are exactly the even symmetric bilinear forms on Λ invariant under the action of $O(\Lambda, I)$, see for instance [**Gan18**]. In particular, I itself defines such an element. As in the finite case, elements of the third Lie group cohomology classify categorical central extensions. The categorical torus

$$1 \longrightarrow \mathbb{B}U(1) \longrightarrow \mathcal{T} \longrightarrow T \longrightarrow 1$$

corresponding to I under this classification is easily constructed. We review two equivalent constructions in Sections 3.1 and 3.3 below. The motivating example is the categorical Leech torus $\mathcal{T}_{Leech}$, where (Λ, I) is taken to be the Leech lattice with its usual biliear form. In this setting, the isometry group is Conway's group $O(\Lambda, I) = Co_0$.

In the paper at hand, we shall begin to explore the representation and character theory of categorical tori. While far from presenting a mature theory, we will see glimpses of how the categorical picture captures important features, such as a theta function formalism, from a geometric point of view. At the same time, it is analytically much less intricate than loop group representations, which have long been studied as geometric counterpart to representations of affine Lie algebras, such as, for instance, the (untwisted) affine Heisenberg algebra $\widehat{\mathfrak{h}}$ in [**FLM88**]. The emerging philosophy views categorical Lie groups as a third pillar of the theory. The relationship between loop groups and categorical groups has been studied, for instance, in [**BSCS07**], [**Wal12**], [**BM94**], [**LW23**], [**Gan18**], and there are already instances, where the finite dimensional categorical picture sheds light on the choices one ought to make in the loop group setting.

We will be interested in symmetries. The invariance of the cohomology class I under the action of its isometry group $O(\Lambda, I)$ realises the centre preserving automorphisms of the categorical torus $\mathcal{T}$ classified by I as a categorical central extension of $O(\Lambda, I)$. This computation can also be found in [**Wal22**].We will also be interested in the categorical group $\mathcal{T} \rtimes \{\pm 1\}$, which can be thought of as a semidirect product and plays the role of 'orbifoldisation'. The outer automorphisms of $\mathcal{T} \rtimes \{\pm 1\}$ form a categorical extension of

$$PO(\Lambda, I) = O(\Lambda, I)/\{\pm \mathrm{id}\}.$$

In the case of the Leech lattice, this group is the Conway group Co_1. A categorical central extension of the extraspecial 2-group 2^{1+24} in ATLAS notation turns up as the centraliser of the basic representation of $\mathcal{T}_{\mathbb{C}} \rtimes \{\pm 1\}$. It is therefore expected that the normaliser of this basic representation is closely related to the subgroup $2^{1+24}.Co_1$ of the Monster.

2. Background on crossed modules and strict categorical groups

2.1. Crossed modules and strict categorical groups. By a *categorical group* we will mean a small monoidal groupoid with weakly invertible objects. A categorical group is called *strict* if its objects are strictly invertible and all structure morphisms (units and associators) are identity maps. Strict categorical groups are most easily described using crossed modules. We will largely follow the conventions in [**Noo07**].

DEFINITION 2.1. A crossed module Ψ consists of two groups G_0 and G_1, equipped with an action of G_0 on G_1 from the right, denoted

$$\begin{aligned} G_1 \times G_0 &\longrightarrow G_1 \\ (\alpha, x) &\longmapsto \alpha^x \end{aligned}$$

and a group homomorphism $\psi\colon G_1 \to G_0$, written $\alpha \mapsto \underline{\alpha}$, such that the following two conditions are satisfied:

(CM1) $\underline{\alpha^x} = x^{-1}\underline{\alpha}x$ (equivariance)
(CM2) $\beta^{\underline{\alpha}} = \alpha^{-1}\beta\alpha$ (Pfeiffer identity).

Associated to such a crossed module Ψ is the four term exact sequence

$$0 \longrightarrow \ker(\psi) \longrightarrow G_1 \xrightarrow{\ \psi\ } G_0 \longrightarrow \operatorname{coker}(\psi) \longrightarrow 0,$$

called the *crossed module extension* corresponding to Ψ.

DEFINITION 2.2. A homomorphism from a crossed module Ψ to another crossed module Φ is a commuting square

$$\begin{array}{ccc} G_1 & \xrightarrow{f_1} & H_1 \\ {\scriptstyle\psi}\downarrow & & \downarrow{\scriptstyle\phi} \\ G_0 & \xrightarrow{f_0} & H_0, \end{array}$$

where the vertical maps are the structure maps of the two crossed modules Ψ and Φ, and the horizontal maps f_0 and f_1 are group homomorphisms that are compatible with the actions in the sense that

$$f_1(\alpha^x) = f_1(\alpha)^{f_0(x)}.$$

DEFINITION 2.3. In the situation of the Definition 2.1, the strict categorical group associated to Ψ is the groupoid

$$\mathcal{S}(\Psi) \simeq \begin{array}{c} G_0 \ltimes G_1 \\ \downarrow\ \downarrow \\ G_0 \end{array}$$

with source pr_1 and target $pr_1 \cdot \psi$, equipped with the strict monoidal structure given by the group multiplications.

The construction of Definition 2.3 defines a functor $\mathcal{S}$ from the category of crossed modules and crossed module homomorphisms to the category of strict categorical groups and strict monoidal functors. If $\mathcal{S}$ is a categorical group, and 1 is its unit object, then we will write $\pi_0(\mathcal{S})$ for the group of isomorphism classes of $\mathcal{S}$ and $\pi_1(\mathcal{S}) = Aut_{\mathcal{S}}(1)$ for the Bernstein centre of $\mathcal{S}$. One has $\pi_1(\mathcal{S}(\Psi)) = \ker(\psi)$

and $\pi_0(\mathcal{S}(\Psi)) = \mathrm{coker}(\psi)$. When we speak of a categorical group extension of the form

$$1 \longrightarrow \mathbb{B}A \longrightarrow \mathcal{G} \longrightarrow G \longrightarrow 1,$$

where G is a group and A is an abelian group, we mean a categorical group $\mathcal{G}$ with $\pi_0(\mathcal{G}) \cong G$ and $\pi_1(\mathcal{G}) \cong A$.

2.2. Strict automorphisms and centre of a crossed module. Let Ψ be a crossed module as in Definition 2.1. Equip the set of crossed homomorphisms

$$Cross(G_0, G_1) = \{\chi\colon G_0 \to G_1 \mid \chi(xy) = \chi(x)^y \chi(y)\}$$

with the composition rule

$$(\chi_1 \circ \chi_2)(x) = \chi_2(x) \cdot \chi_1(x \cdot \underline{\chi_2(x)}), \tag{2.1}$$

and write $Cross^{\times}(G_0.G_1)$ for the group of invertible elements in this semi-group. In the Lie setting, we will require crossed homomorphisms to be analytic.

DEFINITION 2.4 ([**Nor90**] [**Bre92**]). The *actor crossed module* $Act(\Psi)$ consists of the groups $Act_0(\Psi) = Aut(\Psi)$ of crossed module automorphisms and $Act_1(\Psi) = Cross^{\times}(G_0, G_1)$, along with the group homomorphism $\delta\colon Act_1 \to Act_0$ sending the crossed homomorphism χ to the pair of group automorphisms $x \mapsto x\underline{\chi(x)}$ and $\alpha \mapsto \alpha\chi(\underline{\alpha})$. The action of Act_0 on Act_1 is given by

$$\chi^{(f_0, f_1)} = f_1^{-1}\chi f_0.$$

By the adjoint representation of Ψ we will mean the crossed square

$$\begin{array}{cccc}
 & \alpha & \longmapsto & \chi_\alpha \\
 & G_1 & \longrightarrow & Cross^{\times}(G_0, G_1) \\
\boldsymbol{Ad}: & \downarrow{\scriptstyle \psi} & & \downarrow{\scriptstyle \delta} \\
 & G_0 & \longrightarrow & Aut(\Psi) \\
 & x & \longmapsto & \left(x(-)x^{-1}, (-)^{x^{-1}}\right),
\end{array}$$

where $\chi_\alpha(x) = \alpha^x\alpha^{-1}$. We will also write $c_x = Ad_0(x)$ for conjugation with x.

We will mostly be working with a variation of Definition 2.4 that considers only the centre-preserving crossed module homomorphisms; i.e., we will work with the crossed module $Act^{+}(\Psi)$ and the crossed square Ad^{+} obtained by replacing $Aut(\Psi)$ with the automorphisms of Ψ that act as the identity on $\pi_1(\Psi)$.

To define the Drinfeld centre of a crossed module, we follow Pirashvili,[3] see [**Pir23**]. Given Ψ as in Definition 2.1, we form the semi-direct product

$$G_0 \ltimes Cross(G_0, G_1),$$

where the action of G_0 is via Ad_0 and the action in $Act(\Psi)$, so,

$$\chi^x(z) = \chi(xzx^{-1})^x.$$

[3]Note that our crossed module conventions differ from those in the work of Norrie and Pirashvili referenced here, who both use left actions. A variation of the definition of centraliser also appears in [**CG04**, Page 41].

Consider the subgroup

$$Z_0(\Psi) = \left\{(x,\xi) \in G_0 \ltimes Cross(G_0, G_1) \mid Ad_0(x^{-1}) = \delta(\xi)\right\}.$$

In $Z_0(\Psi)$, the multiplication in the semi-direct product simplifies to

$$(x,\xi)(y,\upsilon) = (xy, z \mapsto \upsilon(z)\xi(z)^y),$$

and in particular, $Z_0(\Psi)$ is a group.

DEFINITION 2.5 ([**Pir23**]). The centre of Ψ is the crossed module $Z(\Psi)$ given by the map

$$\begin{aligned} \zeta\colon G_1 &\longrightarrow Z_0(\Psi) \\ \alpha &\longmapsto (\underline{\alpha}, Ad_1(\alpha^{-1})), \end{aligned}$$

together with the action $\alpha^{(x,\xi)} = \alpha^x$.

The crossed module $Z(\Psi)$ encodes the Drinfeld centre of $\mathcal{S}(\Psi)$,

$$\mathcal{S}(Z(\Psi)) \cong \mathcal{Z}(\mathcal{S}(\Psi)).$$

2.3. Weak maps between crossed modules. In [**Noo07**], Noohi worked out how (not necessarily strict) monoidal functors and monoidal natural transformations between strict categorical groups are encoded in terms of the relevant crossed modules. All of Noohi's crossed modules are discrete. In the Lie setting, the butterfly formalism of [**AN09**] is the more suitable notion of weak morphism. Since all the relevant groups in our examples are homotopically discrete, the two formalisms turn out to be equivalent; we do not concern ourselves with the distinctions here and refer the reader to [**Wal22**] for a detailed comparison of different notions in the Lie case.

DEFINITION 2.6 (compare [**Noo07**, Definition 8.4]). Let Φ and Ψ be crossed modules with structure maps $\phi\colon H_1 \longrightarrow H_0$ and $\psi\colon G_1 \longrightarrow G_0$. A weak morphism from Φ to Ψ consist of a pair of pointed set maps $p_i\colon H_i \longrightarrow G_i$, for $i \in \{0,1\}$, together with a map

$$\kappa\colon H_0 \times H_0 \longrightarrow G_1$$

satisfying

$$p_0(\underline{\alpha}) = \underline{p_1(\alpha)} \tag{W1}$$

$$p_0(x)p_0(y)\underline{\kappa_{x,y}} = p_0(xy) \quad \text{(Künneth 1)} \tag{W2}$$

$$p_1(\alpha)p_1(\beta)\kappa_{\underline{\alpha},\underline{\beta}} = p_1(\alpha\beta) \quad \text{(Künneth 2)} \tag{W3}$$

$$\kappa_{x,y}^{p_0(z)}\kappa_{xy,z} = \kappa_{y,z}\kappa_{x,yz} \quad \text{(cocycle), and} \tag{W4}$$

$$(p_1(\alpha))^{p_0(x)}\kappa_{\underline{\alpha},x} = p_1(\alpha^x)\kappa_{x,\underline{\alpha^x}} \quad \text{(equivariance).} \tag{W5}$$

We will use the notation $\kappa_{x,y,z}$ to refer to either side of (W4).

Combined with the other four, our condition (W5) is equivalent to that in [**Noo07**]. Note that (W3) implies $\kappa_{1,1} = 1$, and hence (W4) implies $\kappa_{x,1} = 1 = \kappa_{1,y}$. It follows that the restriction of p_1 to $\pi_1(\mathcal{S}(\Psi))$ is a group homomorphism, and so is $\pi_0(p_0)$. Horizontal composition of weak morphisms is given by

$$(f,\kappa)(g,\gamma) = (f \circ g, \beta) \tag{2.2}$$

with

$$\beta_{x,y} = f_1(\gamma_{x,y})\kappa_{g_0(x),g_0(y),\underline{\gamma_{x,y}}}.$$

It is a tedious yet straightforward computation to check that β satisfies the cocycle condition (W4), and we have

$$\beta_{x,y,z} = f_1(\gamma_{x,y,z})\kappa_{g_0(x),g_0(y),g_0(z),\underline{\gamma_{x,y,z}}}.$$

Associativity and well-definedness of (2.2) follow. The monoidal functor corresponding to $(p,\kappa) \in wHom(\Phi,\Psi)$ is the functor

$$P\colon \mathcal{S}(\Phi) \longrightarrow \mathcal{S}(\Psi)$$

which is given by p_0 on objects, by

$$(x,\alpha) \longmapsto (p_0(x), p_1(\alpha)\kappa_{x,\underline{\alpha}})$$

on arrows and has Künneth isomorphisms

$$K_{x,y} = (p_0(x)p_0(y), \kappa_{x,y}).$$

DEFINITION 2.7. Given a crossed module Ψ as in Definition 2.1, we define the weak actor crossed module $wAct(\Psi)$ of Ψ as follows:

$$wAct(\Psi)_0 = \{(f,\kappa) \in wHom(\Psi,\Psi) \mid f_0 \text{ and } f_1 \text{ are bijections}\}$$

is the group of invertible weak endomorphisms of Ψ with the multiplication (2.2), while

$$wAct(\Psi)_1 = Maps_*(G_0, G_1)^\times$$

is the group invertible elements in the semi-group of pointed maps from G_0 to G_1 with multiplication given by (2.1). The structure map $\delta\colon wAct(\Psi)_1 \longrightarrow wAct(\Psi)_0$ sends η to (f,κ) with

$$\begin{aligned} f_0(x) &= x\underline{\eta(x)} \\ f_1(\alpha) &= \alpha\eta(\underline{\alpha}) \\ \kappa_{x,y} &= \eta(y)^{-1}\left(\eta(x)^y\right)^{-1}\eta(xy). \end{aligned}$$

The action of $wAct(\Psi)_0$ on $wAct(\Psi)_1$ is given by

$$\begin{aligned} \eta^{(f,\kappa)}(x) &= f_1^{-1}\left(\eta\left(f_0(x)\right)\right)\gamma_{f_0(x),\underline{\eta(f_0(x))}} \\ &= f_1^{-1}\left(\eta\left(f_0(x)\right)\kappa^{-1}_{x,\underline{\eta^{(f,\kappa)}(x)}}\right), \end{aligned}$$

where in the first row, $(f^{-1},\gamma) = (f,\kappa)^{-1}$, and in the second row,

$$\underline{\eta^{(f,\kappa)}(x)} = x^{-1}f_0^{-1}\left(f_0(x)\underline{\eta(f_0(x))}\right).$$

We will call a weak endomorphism of Ψ *centre-preserving* if its restriction to $\pi_1(\Psi)$ is the identity. Since all the weak automorphisms in the image of δ are centre preserving, the centre-preserving weak automorphisms of Ψ form a sub-crossed module of $wAct(\Psi)$, which we will denote $wAct^+(\Psi)$. The strict categorical group $\mathcal{S}(wAct(\Psi))$ encodes the strictly invertible weakly monoidal endofunctors of $\mathcal{S}(\Psi)$. We will will use the notation[4] $\mathcal{A}ut(\mathcal{S}(\Psi))$ for this categorigal group and $\mathcal{A}ut^+(\mathcal{S}(\Psi))$ for $\mathcal{S}(wAct^+(\Psi))$.

[4]Note the notation clash with [**CG04**], where $\mathcal{A}ut$ is reserved for strict automorphisms, while our $\mathcal{S}(wAct(\Psi))$ is contained in their $\mathcal{E}q(\mathcal{G})$.

2.4. Automorphisms. Let G be a group. Then the crossed module extension

$$\begin{array}{c} 1 \longrightarrow Z(G) \longrightarrow G \xrightarrow{Ad} Aut(G) \longrightarrow Out(G) \longrightarrow 1 \\ g \longmapsto c_g = g(-)g^{-1}, \end{array} \tag{2.3}$$

encodes, via the construction of Definition 2.3, the categorical group

$$\mathcal{A}ut(\mathbb{B}G) \simeq \mathcal{S}(Ad).$$

EXAMPLE 2.8. For $G = T \rtimes \{\pm 1\}$ the extension (2.3) takes the form

$$\begin{array}{ccccccccc} 1 \longrightarrow & Z(T\rtimes\{\pm1\}) & \longrightarrow & T\rtimes\{\pm1\} & \xrightarrow{Ad} & Aut(T\rtimes\{\pm1\}) & \longrightarrow & Out(T\rtimes\{\pm1\}) & \longrightarrow 1 \\ & \| & & \| & & \| & & \| & \\ 1 \longrightarrow & T[2] & \longrightarrow & T\rtimes\{\pm1\} & \xrightarrow{(-)^2} & T\rtimes GL(\Lambda^\vee) & \longrightarrow & PGL(\Lambda^\vee) & \longrightarrow 1, \end{array}$$

where $PGL(\Lambda^\vee) = GL(\Lambda^\vee)/\{\pm id\}$.

Given a strict categorical group $\mathcal{G} = \mathcal{S}(\Psi)$, we consider the composite

$$\Psi \xrightarrow{\mathbf{Ad}^+} Act^+(\Psi) \longrightarrow wAct^+(\Psi),$$

where $\boldsymbol{Ad}$ is the crossed square of Definition 2.4. Applying $\mathcal{S}$, we obtain the categorical crossed module [**Bre92**] [**CGV06**]

$$\mathcal{A}d^+ \colon \mathcal{G} \longrightarrow \mathcal{A}ut^+(\mathcal{G}),$$

encoding the automorphism 3-group

$$\mathcal{A}ut^+(\mathbb{B}\mathcal{G}) = \mathcal{A}ut^+(\mathcal{G})/\!\!/\mathcal{G}$$

[**Rou03**, 4] with 2-isomorphism classes

$$\mathcal{C}oker(\mathcal{A}d^+) \simeq \mathcal{O}ut^+(\mathcal{G})$$

(see [**CG04**, p. 42]), and Bernstein centre equal to the Drinfeld centre of $\mathcal{G}$,

$$\mathcal{K}er(\mathcal{A}d^+) \simeq \mathcal{Z}(\mathcal{G}).$$

Its homotopy groups fit into the exact sequence

$$\begin{array}{l} 1 \longrightarrow (\pi_1\mathcal{G})^{\pi_0\mathcal{G}} \longrightarrow \pi_1\mathcal{G} \longrightarrow Cross(\pi_0\mathcal{G}, \pi_1\mathcal{G}) \\ \longrightarrow \pi_0\mathcal{Z}\mathcal{G} \longrightarrow \pi_0\mathcal{G} \longrightarrow \pi_0\mathcal{A}ut^+(\mathcal{G}) \longrightarrow \pi_0\mathcal{O}ut^+(\mathcal{G}) \longrightarrow 1 \end{array} \tag{2.4}$$

[**CG04**, Corollary 3.3]. It is sometimes helpful to restrict our attention to the automorphisms over $G = \pi_0(\mathcal{G})$, i.e., the automorphisms of the extension. Given a central extension of finite dimensional Lie groups

$$1 \longrightarrow A \longrightarrow \widetilde{G} \xrightarrow{\pi} G \longrightarrow 1$$

defined by a locally continuous A-valued 2-cocycle β on G, the automorphisms of π fixing the central subgroup are identified with the first cohomology group

$$H^1_{gp}(G; A) = Hom(G, A).$$

We therefore have a commuting diagram with exact rows

$$\begin{array}{ccccccccc} 1 & \longrightarrow & Z(\widetilde{G}) & \longrightarrow & \widetilde{G} & \longrightarrow & Inn(\widetilde{G}) & \longrightarrow & 1 \\ & & \downarrow & & \downarrow{\scriptstyle Ad} & & \downarrow & & \\ 1 & \longrightarrow & H^1_{gp}(G;A) & \longrightarrow & Aut^+(\widetilde{G}) & \longrightarrow & Stab([\beta]) & \longrightarrow & 1, \end{array} \tag{2.5}$$

whose last term refers to the stabilizer of the class $[\beta]$ in $Aut(G)$. Let now $\mathcal{Q}$ be the multiplicative A-bundle gerbe over G, corresponding to the categorical central extension

$$\mathbb{B}A \longrightarrow \mathcal{G} \longrightarrow G,$$

where $\mathcal{G}$ is a strict categorical group. Then the auto-equivalences of $\mathcal{Q}$ are identified with the auto-equivalences of $\mathcal{G}$ over G that fix the centre. In other words, the categorical group $\mathcal{A}ut(\mathcal{Q})$ is the kernel of the homomorphism

$$\begin{aligned} Cl\colon \mathcal{A}ut^+(\mathcal{G}) &\longrightarrow Aut(G) \\ f &\longmapsto \pi_0(f) \end{aligned}$$

sending an auto-equivalence F to its effect on isomorphism classes. If $\mathcal{Q}$ is classified by the Čech-simplicial 3-class $[\alpha]$ then the image of Cl consists of the stabilizer of $[\alpha]$ inside $Aut(G)$,

$$im(Cl) = Stab([\alpha]).$$

Further, $\mathcal{A}ut(\mathcal{Q})$ is identified with the strict categorical group associated to

$$\check{C}^1(G;A) \xrightarrow{d} \check{Z}^2(G;A).$$

The universal property of kernel induces the dashed arrow in the diagram

$$\begin{array}{ccccc} \mathcal{G}|_{Z(G)} & \longrightarrow & \mathcal{G} & \xrightarrow{Cl\circ Ad} & Inn(G) \\ \downarrow & & \downarrow{\scriptstyle Ad^+} & & \downarrow \\ \mathcal{A}ut(\mathcal{Q}) & \longrightarrow & \mathcal{A}ut^+(\mathcal{G}) & \xrightarrow{Cl} & Aut(G), \end{array}$$

and we obtain isomorphisms

$$\pi_1(\mathcal{G}|_{Z(G)}) \cong \pi_1(\mathcal{G})$$

and

$$H^1_{gp}(G;A)) \cong \pi_1(\mathcal{A}ut^+(\mathcal{G}))$$

and a map of short exact sequences

$$\begin{array}{ccccccccc} 1 & \longrightarrow & Z(G) & \longrightarrow & G & \xrightarrow{Ad} & Inn(G) & \longrightarrow & 1 \\ & & \downarrow & & \downarrow{\scriptstyle \pi_0(Ad^+)} & & \downarrow & & \\ 0 & \longrightarrow & H^2_{gp}(G,A) & \longrightarrow & \pi_0(\mathcal{A}ut^+(\mathcal{G})) & \longrightarrow & Stab([\alpha]) & \longrightarrow & 1. \end{array} \tag{2.6}$$

3. The basic representation of a categorical torus

3.1. Categorical tori. Let I be an even symmetric bilinear form on a lattice $\Lambda^\vee$, not necessarily unimodular. We write $\Lambda = \mathrm{Hom}(\Lambda^\vee, \mathbb{Z})$ for the dual lattice of $\Lambda^\vee$, and view Λ as the weight lattice of the torus $T = \Lambda^\vee \otimes_{\mathbb{Z}} U(1)$ with Lie algebra $\mathfrak{t} = \Lambda^\vee \otimes_{\mathbb{Z}} \mathbb{R}$. Let J be any choice of integral bilinear form on Λ satisfying

$$I(m,n) = J(m,n) + J(n,m).$$

By mild abuse of notation, we write J also for the bilinear form $J_{\mathbb{R}} = J \otimes \mathbb{R}$ on $\mathfrak{t}$. Then we have the crossed module Θ given by

$$\begin{aligned} \theta\colon \Lambda^\vee \times \mathbb{R}/\mathbb{Z} &\longrightarrow \mathfrak{t} \\ (m,z) &\longmapsto m, \end{aligned}$$

where $x \in \mathfrak{t}$ acts on $\Lambda^\vee \times \mathbb{R}/\mathbb{Z}$ via

$$(m,[a])^x = (m,[a+J(m,x)])$$

(c.f. [**Gan18**]). The categorical torus classified by I is the strict categorical group $\mathcal{T} = \mathcal{S}(\Theta)$. So, $\mathcal{T}$ has objects $\mathfrak{t}$ and arrows

$$x \xrightarrow{[a]} x+m,$$

with source $x \in \mathfrak{t}$, target $x+m$ with $m \in \Lambda^\vee$, and label $[a] \in \mathbb{R}/\mathbb{Z}$. Composition is given by addition of labels

$$x \xrightarrow{[a]} x+m, \xrightarrow{[b]} x+m+n, \qquad x \xrightarrow{[a+b]} x+m+n$$

and the monoidal structure is given by

$$(x \xrightarrow{[a]} x+m) \bullet (y \xrightarrow{[b]} y+n) = (x+y \xrightarrow{[a+b+J(m,y)]} x+y+m+n)$$

with trivial associators and unit maps. We will sometimes find it convenient to identify the circle group $\mathbb{R}/\mathbb{Z}$ with $U(1)$ via the exponential map $a \mapsto e^{2\pi i a}$ and to write the labels multiplicatively.

We similarly construct $\mathcal{T} \rtimes \{\pm 1\}$ as the strict categorical group associated to the crossed module Θ' with structure map

$$\theta'\colon \Lambda^\vee \times \mathbb{R}/\mathbb{Z} \longrightarrow \mathfrak{t} \rtimes \{\pm 1\},$$

which is obtained by replacing $\mathfrak{t}$ with $\mathfrak{t} \rtimes \{\pm 1\}$ in the construction of Θ. In this setting the action of $\mathfrak{t}$ on arrows is as before and -1 acts on everything by the involution sending $x \in \mathfrak{t}$ to $-x$ and sending the arrow

$$x \xrightarrow{[a]} x+m \qquad \text{to} \qquad -x \xrightarrow{[a]} -x-m. \tag{3.1}$$

Finally, we define the complexifications $\Theta_{\mathbb{C}}$ of Θ and $\Theta'_{\mathbb{C}}$ of Θ' by replacing $\mathfrak{t}$ with the complex Lie algebra $\mathfrak{h} = \Lambda^\vee \otimes_{\mathbb{Z}} \mathbb{C}$ and replacing the circle group $U(1)$ with the group of complex units $\mathbb{C}^\times$. These crossed modules define the complex categorical torus

$$\mathcal{T}_{\mathbb{C}} = \mathcal{S}(\Theta_{\mathbb{C}})$$

and the semi-direct product

$$\mathcal{T}_{\mathbb{C}} \rtimes \{\pm 1\} = \mathcal{S}(\Theta'_{\mathbb{C}}).$$

3.2. The basic representation of a categorical torus. We write

$$\widehat{T} = \mathrm{Hom}(T, U(1)) \quad \text{and} \quad \check{T} = \mathrm{Hom}(U(1), T)$$

for the character and cocharacter lattice of T. These are isomorphic to Λ and $\Lambda^\vee$, and their elements are written multiplicatively. We write $e^{2\pi i\lambda} \in \widehat{T}$ for the character corresponding to the weight λ. The basic representations of $\mathcal{T}$ is most easily defined as a strict map of crossed modules. Let

$$\begin{aligned} \mathbb{C}[\widehat{T}] &= \mathbb{C}\{e^{2\pi i\lambda}\}_{\lambda\in\Lambda} \\ e^{2\pi i\lambda}\cdot e^{2\pi i\mu} &= e^{2\pi i(\lambda+\mu)} \end{aligned}$$

be the group algebra of $\widehat{T}$, and similarly for $\check{T}$. Let $\mathrm{Aut}_{var}(T_\mathbb{C})$ be the group of automorphisms of the complex variety

$$T_\mathbb{C} = \mathrm{spec}(\mathbb{C}[\widehat{T}]).$$

Viewing $\mathbb{C}[\widehat{T}]$ as the algebra of polynomial functions on $T_\mathbb{C}$, the group $\mathrm{Aut}_{var}(T_\mathbb{C})$ acts on $\mathbb{C}[\widehat{T}]$ by precomposition. Before we state the definition, we introduce some further notation: we will write

$$\begin{aligned} J^\sharp\colon \Lambda^\vee &\longrightarrow \Lambda \\ m &\longmapsto J(m,-) \end{aligned}$$

and

$$\begin{aligned} J^\flat\colon \Lambda^\vee &\longrightarrow \Lambda \\ m &\longmapsto J(-,m) \end{aligned}$$

for the adjoints of J, and

$$\exp\colon \mathfrak{h} \longrightarrow T_\mathbb{C}$$

for the exponential map

$$\exp = \Lambda^\vee \otimes_\mathbb{Z} e^{2\pi i(-)}.$$

DEFINITION 3.1. The basic representation of $\Theta'_\mathbb{C}$ is the strict crossed module homomorphism

$$r_{bas}\colon \quad \begin{array}{ccc} (m,z) & \longmapsto & z e^{2\pi i J^\sharp(m)} \\ & & \\ \Lambda^\vee \times \mathbb{C}^\times & \longrightarrow & \mathbb{C}[\widehat{T}]^\times \\ \Big\downarrow \theta'_\mathbb{C} & & \Big\downarrow 1 \\ \mathfrak{h} \rtimes \{\pm 1\} & \longrightarrow & Aut_{var}(T_\mathbb{C}) \\ & & \\ (x,1) & \longmapsto & \exp(x) \\ & & \\ (x,-1) & \longmapsto & \exp(x)\circ inv. \end{array}$$

Here we view the units in $\mathbb{C}[\widehat{T}]$ as a crossed module over $\mathrm{Aut}_{var}(T_{\mathbb{C}})$ via the trivial map, and we have identified $T_{\mathbb{C}}$ with the subgroup of $Aut_{var}(T_{\mathbb{C}})$ given by translations. We use the notation t for multiplication with the element t and write inv for the automorphism $s \mapsto s^{-1}$.

The exponential map exp used in the definition of r_{bas} is an analytic map that does not have an algebraic counterpart. It is therefore natural to consider the analytic version of the basic representation. For this, we view $T_{\mathbb{C}}$ as a complex analytic variety and replace the algebraic sections $\mathbb{C}[\widehat{T}] = \Gamma\mathcal{O}_{T_{\mathbb{C}}}$ with the holomorphic sections $\Gamma\mathcal{O}^{an}_{T_{\mathbb{C}}}$ and $Aut_{var}(T_{\mathbb{C}})$ with the holomorphic automorphisms $Aut^{an}_{var}(T_{\mathbb{C}})$. This results in the definition of r^{an}_{bas}, the analytic basic representation of $\Theta'_{\mathbb{C}}$.

Given integers k and n, we define the representation of level k and weight n,

$$
\begin{array}{rccc}
 & (m,z) & \longmapsto & z^{kn}e^{2\pi i k J^{\sharp}(m)} \\
 & \Lambda^{\vee} \times \mathbb{C}^{\times} & \longrightarrow & \mathbb{C}[\widehat{T}]^{\times} \\
r_{k,n}\colon & \Big\downarrow{\scriptstyle \theta'_{\mathbb{C}}} & & \Big\downarrow \\
 & \mathfrak{h} \rtimes \{\pm 1\} & \longrightarrow & Aut_{var}(T_{\mathbb{C}}) \\
 & (x,1) & \longmapsto & \exp(nx) \\
 & (x,-1) & \longmapsto & \exp(nx) \circ inv.
\end{array}
$$

It, too, has an analytic counterpart $r^{an}_{k,n}$. The basic representation is the special case $r_{bas} = r_{1,1}$. The strict categorical group $\mathcal{A}$ associated to the crossed module

$$\mathbb{C}[\widehat{T}] \xrightarrow{\ 1\ } \mathrm{Aut}_{Var}(T_{\mathbb{C}})$$

acts, in a natural manner, on the $\mathbb{C}$-linear category $Coh(T_{\mathbb{C}})$, and similarly in the analytic setting. Thus, we may interpret $r_{k,n}$ as mapping to a categorical group of $\mathbb{C}$-linear functors and linear natural transformations,

$$\varrho_{k,n} : \mathcal{T}_{\mathbb{C}} \rtimes \{\pm 1\} \longrightarrow \mathcal{GL}(\mathcal{V})$$

with $\mathcal{V} = Coh(T_{\mathbb{C}})$. Restricted to $\mathcal{T}_{\mathbb{C}}$, the basic representation $\varrho_{bas} = \varrho_{1,1}$ sends the object x to the functor

$$\exp(x)^{*} \colon Coh(T_{\mathbb{C}}) \longrightarrow Coh(T_{\mathbb{C}}),$$

and it sends the arrow $x \xrightarrow{\ z\ } x+m$ to the natural transformation

$$\varrho_{bas}\left(x \xrightarrow{\ z\ } x+m\right) = z\exp(x)^{*}e^{2\pi i J^{\sharp}(m)}.$$

We do have some freedom in the choice of interpretation of $r_{k,n}$ as action on a category.

3.3. Multiplicative bundle gerbe point of view. We will write

$$T^* = Hom(\Lambda^\vee, U(1)) \cong \Lambda \otimes U(1) \cong \mathfrak{t}^*/\Lambda$$

for the Pontryagin dual of $\Lambda^\vee$ and refer to T^* as the dual torus. The complexification

$$T^*_{\mathbb{C}} = Hom(\Lambda^\vee, \mathbb{C}^\times) = \Lambda \otimes \mathbb{C}^\times$$

may then be viewed as the (analytification of the) complex algebraic variety

$$T^*_{\mathbb{C}} = spec\left(\mathbb{C}[\check{T}]\right).$$

We have the Poincaré line bundle

$$\mathcal{P} = \mathfrak{h} \times T^*_{\mathbb{C}} \times \mathbb{C}/\sim$$

over $T_{\mathbb{C}} \times T^*_{\mathbb{C}}$, with $\sim$ defined by

$$(x, \eta, z) \sim (x + m, \eta, \eta(m)z).$$

Writing

$$J^\flat_{U(1)} = J^\flat \otimes U(1),$$

we let

$$\mathbf{L} = \mathbf{L}^J = \left(\mathrm{id}_T \times J^\flat_{U(1)}\right)^* (\mathcal{P}|_{T\times T^*})$$

be the pull-back of $\mathcal{P}$ to $T \times T$. For $t \in T$, we let $\mathbb{L}_t$ be the line bundle

$$\mathbb{L}_t = \mathbf{L}^J_{t\times T}.$$

This bundle comes equipped with canonical isomorphisms

$$\varphi_x \colon T \times \mathbb{C} \cong \mathbb{L}_t,$$

for any choice of x with $\exp(x) = t$, satisfying

$$\varphi_{x+m} \cdot e^{2\pi i J^\sharp(m)} = \varphi_x.$$

The bundle $\mathbf{L}$ was used in [**Gan18**] to equip the trivial bundle gerbe over T with a multiplicative structure. The categorical group corresponding to this multiplicative bundle gerbe is equivalent to the categorical torus classified by the even bilinear form I. We may think of this as follows. Let $Lines_{\mathbb{C}}$ be the category of 1-dimensional $\mathbb{C}$-vector spaces, and equip

$$\mathcal{I} = T \times Lines_{\mathbb{C}},$$

with the multiplication

$$(s, L) \bullet (t, L') = (st, L \otimes L' \otimes \mathbf{L}_{s,t})$$

(canonical associators and unit $(1, \mathbb{C})$).

In this setting the basic representation takes the following shape. Let $\mathcal{V}$ be the category of complex vector bundles on T. The monoidal functor

$$R_{bas} \colon (\mathcal{I}, \bullet) \longrightarrow (\mathcal{GL}(\mathcal{V}), \circ)$$

sends the object (t, L) to the endofunctor

$$t_* \circ (\mathbb{L}_t \otimes L \otimes -)$$

and the pair $((s.L), (t, L'))$ to the natural isomorphism induced by

$$\begin{aligned} s_* \left(\mathbb{L}_s \otimes t_* \left(\mathbb{L}_t \otimes -\right)\right) &= s_* t_* \left(t^* \mathbb{L}_s \otimes \left(\mathbb{L}_t \otimes -\right)\right) \\ &= (st)_* \left(\mathbb{L}_{st} \otimes \left(\mathbf{L}^J_{s,t} \otimes -\right)\right). \end{aligned}$$

This is related to the basic representation r_{bas} of $\mathcal{T}$ by the diagram

$$\begin{array}{ccc} \mathcal{T} & \xrightarrow{\;r_{bas}\;} & \mathcal{A} \\ {\scriptstyle E}\downarrow & \overset{\phi}{\swarrow\!\!\!\!=} & \downarrow \\ \mathcal{I} & \xrightarrow[\;R_{bas}\;]{} & \mathcal{A}ut(\mathcal{V}), \end{array}$$

where E takes the object x to $(\exp(x), \mathbb{C})$ and ϕ is the (monoidal) natural transformation

$$\phi\colon \varrho_{bas} \Longrightarrow R_{bas} \circ E$$

given by

$$\phi_x = \exp(x)_* \varphi_x.$$

Finally, the bundle $\boldsymbol{L}$ inherits from $\mathcal{P}|_{T\times T^*}$ a connection with holonomy

$$Hol_{(\mathbf{L},\nabla)}(\bar{f},\bar{g}) = e^{2\pi i\left(\int_0^1 J(\dot{f}(t),g(t))dt\right) - J(f(0),\Delta_g)},$$

where $\Delta_g = g(1) - g(0)$. Compare this to [**Gan18**], but with slightly different conventions. This offers various ways one could hope to reinvent our representations, using complex vector bundles with connection or flat connection in the place of coherent sheaves.

3.4. The centre of a categorical torus.

LEMMA 3.2. *The centre of the categorical torus $\mathcal{T}$ is the strict categorical group associated to the crossed module*

$$\begin{aligned} \zeta\colon \Lambda^\vee \times U(1) &\longrightarrow \mathfrak{t} \times \Lambda \\ (m,z) &\longrightarrow (m, -I^\sharp(m)), \end{aligned}$$

with the action given by

$$(m,z)^{(x,\lambda)} = \left(m, ze^{2\pi i J(m,x)}\right).$$

PROOF. Recall that the Pontryagin dual of $\mathfrak{t}$ is identified with the dual vector space via composition with the exponential map,

$$\begin{aligned} \mathfrak{t}^* &\cong Hom(\mathfrak{t}, U(1)) \\ \eta &\mapsto e^{2\pi i \eta(-)}, \end{aligned}$$

see for instance [**Con90**, VII 9.11]. It follows that we have

$$Z_0(\Theta) \cong \mathfrak{t} \times_{T^*} \mathfrak{t}^*,$$

where we have made the identification $T^* \cong Hom(\Lambda^\vee, U(1))$, and the map from $\mathfrak{t}$ to T^* sends x to $e^{2\pi i J^b(x)}$. The structure map

$$\begin{aligned} \Lambda^\vee \times U(1) &\longrightarrow \mathfrak{t} \times_{T^*} \mathfrak{t}^* \\ (m,z) &\longmapsto (m, -J^\sharp(m)), \end{aligned}$$

is identified with the map ζ in the lemma using the isomorphism

$$\mathfrak{t} \times \Lambda \xrightarrow{\cong} \mathfrak{t} \times_{T^*} \mathfrak{t}^*$$
$$(x, \lambda) \longmapsto (x, J^b(x) + \lambda). \qquad \square$$

COROLLARY 3.3. *In the case where* $\mathcal{G} = \mathcal{T}$ *is the categorical torus, the long exact sequence* (2.4) *takes the shape*

$$1 \longrightarrow U(1) \xrightarrow{\cong} U(1) \longrightarrow 1$$
$$1 \longrightarrow \widehat{T} \longrightarrow \pi_0(\mathcal{Z}(\Theta)) \longrightarrow T \longrightarrow 1$$
$$1 \longrightarrow \pi_0 \mathcal{A}ut^+(\mathcal{T}) \xrightarrow{\cong} \pi_0 \mathcal{O}ut^+(\mathcal{T}) \longrightarrow 1.$$

In the self-dual case, i.e., when $I^\sharp$ is an isomorphism, we have a canonical isomorphism $\mathfrak{t} \cong \pi_0(\mathcal{Z}(\Theta))$, and the short exact sequence in the middle row is identified with the usual short exact sequence

$$0 \longrightarrow \Lambda^\vee \longrightarrow \mathfrak{t} \xrightarrow{\exp} T \longrightarrow 1.$$

4. Centralisers and normalisers

Let k be a field, and let

$$\varrho \colon G \longrightarrow GL(V)$$

be a k-linear representation of a finite group. Then we have the centraliser and the normaliser of ϱ, defined as

$$C(\varrho) = \{h \in GL(V) \mid \varrho = c_h \circ \varrho\}$$

and

$$N(\varrho) = N(\varrho) = \{(f, h) \in Aut(G) \times GL(V) \mid \varrho \circ f = h\varrho(-)h^{-1}\}.$$

These sit inside a commuting diagram with exact rows

$$\begin{array}{ccccccccc} 1 & \longrightarrow & Z(G) & \longrightarrow & G & \longrightarrow & Inn(G) & \longrightarrow & 1 \\ & & \downarrow & & \downarrow & & \downarrow & & \\ 1 & \longrightarrow & C(\varrho) & \longrightarrow & N(\varrho) & \longrightarrow & Aut(G), & & \end{array}$$

where $Z(G)$ denotes the centre of G. If ϱ is irreducible, then Schur's lemma implies

$$C(\varrho) = k^\times.$$

Let (Λ, I) be the Leech lattice equipped with its even symmetric bilinear form, and consider the extraspecial 2-group $\widetilde{\Lambda/2\Lambda}$, with commutator given by

$$e_\alpha e_\beta = (-1)^{I(\alpha,\beta)} e_\beta e_\alpha, \qquad \alpha, \beta \in \Lambda/2\Lambda.$$

Let ϕ be the quadratic form

$$\phi \colon \Lambda \otimes_{\mathbb{Z}} \mathbb{F}_2 \longrightarrow \mathbb{F}_2$$
$$\alpha \longmapsto \frac{1}{2} I(\alpha, \alpha),$$

and $O(\phi)$ its isometry group. The construction of the monster begins by considering the action of $\widetilde{\Lambda/2\Lambda}$ on its unique centre-faithful irreducible representation ϱ.

This representation is a 2^{12}-dimensional real representation, whose centraliser and normaliser fit into the diagram

$$\begin{array}{ccccccccc}
 & & & & 1 & & 1 & & \\
 & & & & \downarrow & & \downarrow & & \\
1 & \longrightarrow & \{\pm 1\} & \longrightarrow & \widetilde{\Lambda/2\Lambda} & \longrightarrow & \Lambda/2\Lambda & \longrightarrow & 1 \\
 & & \| & & \downarrow & & \downarrow & & \\
1 & \longrightarrow & C(\varrho) & \longrightarrow & N(\varrho) & \longrightarrow & Aut(\widetilde{\Lambda/2\Lambda}) & \longrightarrow & 1 \\
 & & & & \downarrow & & \downarrow & & \\
 & & & & O(\phi) & = & Out(\widetilde{\Lambda/2\Lambda}) & & \\
 & & & & \downarrow & & \downarrow & & \\
 & & & & 1 & & 1 & &
\end{array}$$

[**FLM88**, 10.4.15]. The exactness on the right in the second row follows from the uniqueness of ϱ. The group $O(\phi)$ receives a map from Co_1 and hence Co_0, pulling back the middle column to an extension

$$1 \longrightarrow \widetilde{\Lambda/2\Lambda} \longrightarrow \widetilde{C} \longrightarrow Co_0 \longrightarrow 1.$$

One then identifies the central elements $-1 \in \widetilde{\Lambda/2\Lambda}$ and $-\operatorname{id} \in Co_0$ to obtain the group

$$C = \widetilde{C} \,/\, (-1) \sim (-\operatorname{id}).$$

This is an extension of Co_1 by the extraspecial 2-group $\widetilde{\Lambda/2\Lambda}$. The Monster can be generated by C and one more element, the triality symmetry.

4.1. The centraliser of a crossed module homomorphism. Given a strict homomorphism of crossed modules

$$\begin{array}{ccc}
G_1 & \xrightarrow{f_1} & H_1 \\
{\scriptstyle\psi}\downarrow & & \downarrow{\scriptstyle\phi} \\
G_0 & \xrightarrow{f_0} & H_0,
\end{array}$$

as in Definition 2.2, we define the group $C_0(f)$ as follows. Elements of $C_0(f)$ are pairs (h, χ), with $h \in H_0$ and $\chi\colon G_0 \longrightarrow H_1$, such that

$$\begin{aligned}
\underline{\chi(s)} &= f_0(s^{-1})h^{-1}f_0(s)h, \\
\chi(\underline{\alpha}) &= f_1(\alpha^{-1})f_1(\alpha)^h, \\
\chi(st) &= \chi(s)^{f_0(t)}\chi(t).
\end{aligned}$$

In the Lie setting, we require the map χ to be analytic. The multiplication in $C_0(f)$ is given by

$$(h, \chi)(k, \sigma) = (hk, s \mapsto \sigma(s)\chi(s)^k).$$

DEFINITION 4.1. In the above situation, the centraliser of f is the crossed module $C(f)$ given by

$$\begin{aligned} \gamma\colon H_1 &\longrightarrow C_0(f) \\ \zeta &\longmapsto \left(\underline{\zeta}, s \mapsto \left(\zeta^{f_0(s)}\right)^{-1}\zeta\right) \end{aligned}$$

where $C_0(f)$ acts on H_1 via $\zeta^{(h,\chi)} = \zeta^h$.

A computation closely following [**Pir23**] shows that $C(f)$ is indeed a crossed module. The centre of a crossed module is a special case of this construction,

$$Z(\Psi) = C\left(\mathrm{id}_\Psi\right),$$

and there is a crossed module homomorphism

$$\begin{aligned} Z(\Psi) &\longrightarrow C(f) \\ (s,\xi) &\longmapsto (f_0(x), f_1\circ\xi) \\ \alpha &\longmapsto f_1(\alpha). \end{aligned}$$

4.2. The centraliser of the basic representation of a categorical torus. We are interested in the centraliser $C(r^{an}_{bas})$, where r^{an}_{bas} is the crossed module homomorphism of Definition 3.1, encoding the analytic basic representation of $\mathcal{T}_\mathbb{C} \rtimes \{\pm1\}$.

DEFINITION 4.2. We will write E for the crossed module

$$\begin{aligned} \eta\colon \Lambda^\vee \times \mathbb{C}^\times &\longrightarrow \widetilde{\Lambda^\vee} \\ (m,z) &\longmapsto 2m, \end{aligned}$$

where the central extension is defined by the 2-cocycle J^t modulo 2,

$$\begin{aligned} \Lambda^\vee \times \Lambda^\vee &\longrightarrow \mathbb{Z}/2\mathbb{Z} \\ (m,n) &\longmapsto [J(n,m)], \end{aligned}$$

and acts on $\Lambda^\vee \times \mathbb{C}^\times$ via $(m,z)^{(n,\iota)} = (m, (-1)^{J(n,m)}z)$.

The strict categorical group associated to E is the extraspecial categorical 2-group

$$\mathcal{S}(E) = \mathcal{T}_\mathbb{C}^{\{\pm1\}}$$

with

$$\pi_0(E) \cong \widetilde{T[2]} \cong \widetilde{\Lambda^\vee/2\Lambda^\vee}$$

and $\pi_1(E) = \mathbb{C}^\times$, which turned up in [**Gan18**, Theorem 6.3] as the categorical fixed points of the action of $\{\pm1\}$ on the complexification of $\mathcal{T} \simeq \mathcal{T}_{J^t}$. This is the same action as the one in (3.1) above. We find it convenient to use additive notation for the centre of $\widetilde{\Lambda^\vee}$.

THEOREM 4.3. *We have an equivalence of crossed modules* $F\colon E \longrightarrow C_0\left(r^{an}_{bas}\right)$.

PROOF. Define F to be the strict crossed-module homomorphism

$$\begin{array}{ccccc}
(m,z) & \longmapsto & & & ze^{-2\pi i J^b(m)} \\
\Big\downarrow & \Lambda^\vee \times \mathbb{C}^\times & \xrightarrow{F_1} & \left(\Gamma\mathcal{O}_{T_\mathbb{C}}^{an}\right)^\times & \zeta \\
 & \Big\downarrow \eta & & \Big\downarrow \gamma & \Big\downarrow \\
(2m,0) & \widetilde{\Lambda^\vee} & \xrightarrow{F_0} & C_0\left(r_{bas}^{an}\right) & (\mathrm{id},\chi_\zeta) \\
 & (n,\iota) & \longmapsto & \left(\exp\left(\frac{n}{2}\right), \chi_{(n,\iota)}\right) &
\end{array}$$

where

$$\chi_{(n,\iota)}(x,\epsilon) = \epsilon^\iota e^{\pi i\left(J(x,n)+(\epsilon-1)J^b(n)\right)}$$

and

$$\chi_\zeta(x,\epsilon)(t) = \frac{\zeta(t)}{\zeta(\exp(x)\cdot t^\epsilon)}.$$

If $\zeta = ze^{2\pi i\lambda}$ then

$$\chi_\zeta(x,\epsilon) = e^{-2\pi i(\lambda(x)+(\epsilon-1)\lambda)},$$

so the diagram commutes. The section $\zeta\colon T_\mathbb{C} \longrightarrow \mathbb{C}^\times$ is in $\pi_1\left(C\left(r_{bas}^{an}\right)\right)$ if and only if ζ satisfies

$$\zeta(\exp(x)t^\epsilon) = \zeta(t)$$

for all $(x,\epsilon) \in \mathfrak{h} \rtimes \{\pm 1\}$ and $t \in T_\mathbb{C}$. Taking $t = 1$, we see that if ζ is an element of $\pi_1\left(C\left(r_{bas}^{an}\right)\right)$, then ζ is constant on $T_\mathbb{C}$. Hence $\pi_1(F)$ is an isomorphism. To see that $\pi_0(F)$ is injective, assume we are given $(n,\iota) \in \widetilde{\Lambda^\vee}$ and $\zeta \in \left(\Gamma\mathcal{O}_{T_\mathbb{C}}^{an}\right)^\times$ such that $\exp\left(\frac{n}{2}\right) = 1$ and $\chi_{(n,\iota)} = \chi_\zeta$. Then $n \in 2\Lambda^\vee$, and we have

$$\zeta(\exp(x)\cdot t^\epsilon) = \zeta(t)\cdot \epsilon^\iota e^{-\pi i\left(J(x,n)+(\epsilon-1)J^b(n)\right)}(t)$$

for all $(x,\epsilon) \in \mathfrak{h} \rtimes \{\pm 1\}$ and $t \in T_\mathbb{C}$. Setting $t = 1$ and $z = \zeta(1)$, we obtain

$$\zeta(\exp(x)) = z\cdot \epsilon^\iota e^{-2\pi i J\left(x,\frac{n}{2}\right)}.$$

The left-hand side is independent of ϵ, and we conclude that $\iota = 0$ and $\zeta = ze^{-2\pi i J^b\left(\frac{n}{2}\right)}$. It remains to show surjectivity of $\pi_0(F)$. Assume that we are given an automorphism f of $T_\mathbb{C}$ commuting with multiplication by $\exp(x)$ for all $x \in \mathfrak{h}$. Then

$$f(\exp(x)) = \exp(x)\cdot f(1),$$

so f is given by multiplication with $f(1)$. If, in addition, we assume that $f(t)^{-1} = f(t^{-1})$ holds for all $t \in T_\mathbb{C}$, it follows that $f(1) \in T[2]$ is an element of order 2. Choose $n \in \Lambda^\vee$ such that $f(1) = \exp\left(\frac{n}{2}\right)$, and let

$$\chi\colon \mathfrak{h} \rtimes \{\pm 1\} \longrightarrow \left(\Gamma\mathcal{O}_{T_\mathbb{C}}^{an}\right)^\times$$

be such that $\left(\exp\left(\frac{n}{2}\right), \chi\right)$ is an element of $C_0\left(r^{an}_{bas}\right)$. If $m \in \Lambda^\vee$, then χ sends the element $(m,1)$ to the constant function $(-1)^{J(m,n)}$. Hence

$$\begin{aligned}\zeta\colon T_{\mathbb{C}} &\longrightarrow \mathbb{C}^\times\\ t &\longmapsto \frac{e^{\pi i J(y,n)}}{\chi(y,1)(1)},\end{aligned}$$

where $t = \exp(y)$, is a well-defined element of $\left(\Gamma\mathcal{O}^{an}_{T_{\mathbb{C}}}\right)^\times$. Further, we have

$$\chi(0,-1)\left(t^{-1}\right)\cdot\chi(0,-1)(t) = \chi(0,1)(t) = 1,$$

and hence $\chi(0,\epsilon)(1) = \epsilon^\iota$ for some $\iota \in \mathbb{Z}/2\mathbb{Z}$, and

$$\chi(x,\epsilon)(1) = \chi(x,1)\,(1)\cdot\epsilon^\iota.$$

Letting $t = \exp(y)$, we compute

$$\begin{aligned}\frac{\zeta(t)}{\zeta(\exp(x)\cdot t^\epsilon)} &= \frac{\chi(x+\epsilon y,1)}{\chi(y,1)}(1)\cdot e^{\pi i J(-x-\epsilon y+y,n)}\\ &= \frac{\chi(x+\epsilon y,\epsilon)}{\chi(y,1)}(1)\cdot\epsilon^\iota e^{-\pi i J(x+\epsilon y-y,n)}\\ &= \frac{\chi(x,\epsilon)}{\chi_{(n,\iota)}}(t).\end{aligned}$$

It follows that the classes of $\left(\exp\left(\frac{n}{2}\right), \chi\right)$ and $\left(\exp\left(\frac{n}{2}\right), \chi_{n,\iota}\right)$ in $\pi_0\left(C\left(r^{an}_{bas}\right)\right)$ agree, and we have shown that $\pi_0(F)$ is surjective. □

5. Character theory

The character theory of categorical representations was developed in [**Bar11**] and [**GK08**] and has recently seen a revival in a slightly diffferent setting: the theory relies on a notion of categorical trace, and there are two such notions in use. These are different, but related; see [**Wil08b**] for a comparison and [**GV20**] for a recent application of categorical traces. We will be working with the definition

$$\mathbb{T}r(F) = 2Hom(\mathrm{id}, F)$$

for the categorical trace of a 1-endomorphism F. By the inertia groupoid

$$G/\!\!/G^{conj} \cong Fun(\mathbb{BZ}, \mathbb{B}G)$$

we will mean the action groupoid of G acting on itself by conjugation. Characters of classical representations are invariant functions on $G/\!\!/G^{conj}$. If a finite group G acts either by functors on a k-linear category, or more generally, by 1-morphisms in a k-linear bicategory, we speak of a k-linear categorical representation of G. In this context, there are two levels of character theory. Applying the categorical trace, one obtains the categorical character X_ϱ. This is a k-vector bundle over $G/\!\!/G^{conj}$, meaning that one has k-vector spaces

$$X_\varrho(g) = \mathbb{T}r(\varrho(g)),$$

and compatible isomorphisms

$$\psi_{g,h}\colon X_\varrho(g) \cong X_\varrho(hgh^{-1}).$$

If g and h commute, then $\psi_{g,h}$ is an endomorphism of $\mathbb{T}r(g)$, and in many interesting settings one can take the trace again. This results in the definition of 2-character

$$\chi(g,h) = tr(\psi_{g,h}),$$

which is a function on pairs of commuting elements of G, invariant under simultaneous conjugation. In other words, the 2-character is an invariant function on the interated inertia groupoid $Fun(\mathbb{B}\mathbb{Z}^2, \mathbb{B}G)$.

Consider next an action of a finite categorical group $\mathcal{G}$. This may be viewed as a projective categorical representation of the group $G = \pi_0(\mathcal{G})$. By a categorical class function on $\mathcal{G}$ we shall mean a bundle on the inertia groupoid of $\mathcal{G}$,

$$\mathcal{G}/\!\!/\mathcal{G}^{conj} := Bifun(\mathbb{B}\mathbb{Z}, \mathbb{B}\mathcal{G})/2\text{-isos}.$$

It was shown in [**GU16**] that the categorical character of a k-linear representation ϱ of a finite categorical group $\mathcal{G}$ is a categorical class function on $\mathcal{G}$. If $\mathcal{G}$ is classified by $[\alpha] \in H^3(G; k^\times)$, then $\mathcal{G}/\!\!/\mathcal{G}^{conj}$ is equivalent to the central extension of

$$G/\!\!/G^{conj} \simeq \coprod_{[g]} 1/\!\!/\widetilde{C}_g$$

classified by the transgression of α. This consists of central extensions $\widetilde{C}_g$ of the centralisers by $k^\times$, one for each conjugacy class $[g]$ in G. The categorical character of ϱ is thus a module over the α-twisted Drinfeld double of G in the sense of [**Wil08a**]. When $G = M$ is the Monster and $\alpha = \alpha_{\mathbb{C}}$ is the Moonshine anomaly, this resembles the formalism governing generalised Moonshine.

Turning our attention to the torus, the inertia groupoid of the commutative group T is easily described: it has objects T and arrows $T \times T$ with source and target both given by the projection to the first factor, the fibre $\{t\} \times T$ being the centraliser of t in T. There are no other arrows, since everything commutes. In light of the picture for finite categorical groups, we expect $\mathcal{T}/\!\!/\mathcal{T}^{conj}$ to be a central extension of $T/\!\!/T^{conj}$ by $U(1)$.

This means that we expect to find each centraliser $T = C_t$ replaced by a central extension by $U(1)$. As individual groups, these are trivial central extensions, so we have non-canonical isomorphisms $\widetilde{C}_t \cong T \times U(1)$. We will, however, find that with varying t the global topology of the inertia groupoid of $\mathcal{T}$ is non-trivial.

5.1. Inertia 2-groupoids. For a brief introduction to the language of bicategories we refer the reader to [**Lei98**], noting that we will follow the opposite convention for pseudo-natural transformations, found, for instance, in [**GPS95**]. Let $\mathcal{G}$ be a categorical group. By the inertia 2-groupoid of $\mathcal{G}$ we will mean the bifunctor 2-groupoid $\mathbf{Bicat}(\mathbb{B}\mathbb{Z}, \mathbb{B}\mathcal{G})$. Its objects may be thought of as strong monoidal functors from $\mathbb{Z}$ to $\mathcal{G}$. The small inertia 2-groupoid $\mathbf{Bicat}^{strict}(\mathbb{B}\mathbb{Z}, \mathbb{B}\mathcal{G})$ is defined as the full sub-2-groupoid whose objects are strict monoidal functors. We will see that the inclusion of the small inertia 2-groupoid inside the full inertia 2-groupoid for $\mathcal{T}$ is an equivalence of Lie 2-groupoids. The inertia groupoid $\mathcal{G}/\!\!/\mathcal{G}^{conj}$ is obtained from the inertia 2-groupoid by taking 2-isomorphism classes of 1-arrows. The goal of this section is to describe $\mathcal{T}/\!\!/\mathcal{T}^{conj}$.

LEMMA 5.1. *Let M be a trivial $\mathbb{Z}$-module, and let N be a submodule of M.*

(i) *Any 1-cochain $\mu\colon \mathbb{Z} \longrightarrow M$ is uniquely determined by $d\mu$ and $\mu(1)$.*
(ii) *The 1-cochain μ as above takes values in N if and only if $d\mu$ does and $\mu(1) \in N$.*
(iii) *For any 2-cocycle $\gamma\colon \mathbb{Z} \times \mathbb{Z} \longrightarrow M$ and integer $a \in \mathbb{Z}$, we have*

$$\gamma(0, a) = \gamma(0, 0) = \gamma(a, 0).$$

PROOF.

(i) If μ_1 and μ_2 are 1-cochains with $d\mu_1 = d\mu_2$, then their difference is a group homomorphism $\mu_1 - \mu_2 \colon \mathbb{Z} \longrightarrow M$. If in addition $\mu_1(1) = \mu_2(1)$, it follows that $\mu_1 - \mu_2 = 0$.

(ii) Assume that $d\mu$ takes values in N. Since $H^2(\mathbb{Z}, N) = 0$, there exists a 1-cochain $\nu \colon \mathbb{Z} \longrightarrow N$ with $d\nu = d\mu$. Using (i), we deduce

$$\mu(a) = \nu(a) + a(\mu(1) - \nu(1)).$$

So, if $\mu(1) \in N$, it follows that μ takes values in N.

(iii) As in (ii), choose any 1-cochain $\alpha \colon \mathbb{Z} \longrightarrow M$ with $d\alpha = \gamma$. Let a be an integer. Then

$$\gamma(0, a) = \alpha(0) = \gamma(a, 0).$$

□

THEOREM 5.2. *The inertia 2-groupoid* **Bicat**$(\mathbb{BZ}, \mathbb{B}\mathcal{T})$ *is isomorphic to the Lie 2-groupoid with*

objects: *triples* (x, γ, c) *with* $x \in \mathfrak{t}$, *and*

$$\gamma \colon \mathbb{Z} \times \mathbb{Z} \longrightarrow \Lambda^\vee$$
$$c \colon \mathbb{Z} \times \mathbb{Z} \longrightarrow U(1)$$

2-cocycles;

1-arrows: *six-tuples* $(x, \gamma, c, y, \mu, w)$, *with source and target as follows*

$$(x, \gamma, c) \xrightarrow{(y,\mu,w)} (x + \mu(1), \gamma + d\mu, c')$$

Here we have (x, γ, c) *as above,* $y \in \mathfrak{t}$,

$$\mu \colon \mathbb{Z} \longrightarrow \Lambda^\vee$$
$$w \colon \mathbb{Z} \longrightarrow U(1)$$

1-cochains, and

$$c'(a, b) = c(a, b) \cdot dw(a, b) \cdot \left(e^{2\pi i J(\mu(1), x)}\right)^{ab};$$

2-arrows: *eight-tuples* $(x, \gamma, c, y, \mu, w, n, u)$, *viewd as 2-arrow*

$$(x, \gamma, c) \underset{(y+n,\mu,w')}{\overset{(y,\mu,w)}{\Downarrow u}} (x + \mu(1), \gamma + d\mu, c')$$

with $n \in \Lambda^\vee$, $u \in U(1)$, *and*

$$w'(a) = w(a) \cdot \left(e^{2\pi i I(x,n)}\right)^a;$$

Horizontal composition: *the following diagram commutes strictly*

$$\begin{array}{ccc} & (x+\mu(1), \gamma + d\mu, c') & \\ {\scriptstyle (y,\mu,w)}\nearrow & & \searrow{\scriptstyle (y',\mu',w')} \\ (x,\gamma,c) & \xrightarrow[(y+y',\mu+\mu',w\cdot w')]{} & (x+(\mu+\mu')(1), \gamma + d(\mu+\mu'), c'') \end{array}$$

where

$$c''(a,b) = c(a,b) \cdot dw''(a,b) \cdot \left(e^{2\pi i J\left((\mu+\mu')(1),x\right)}\right)^{ab}.$$

PROOF. **The objects:** a strong monoidal functor $\mathbb{Z} \longrightarrow \mathcal{T}$ consists of a sequence $a \longmapsto x_a$, indexed by $\mathbb{Z}$, together with arrows

$$\varphi_{a,b}\colon x_a + x_b \xrightarrow{\cong} x_{a+b}$$

and

$$\phi_0\colon 0 \xrightarrow{\cong} x_0,$$

satisfying

$$\phi_{a+b,c} \circ (\phi_{a,b} \bullet \mathrm{id}_{x_c}) = \phi_{a,b+c} \circ (\mathrm{id}_{x_a} \bullet \phi_{b,c}) \tag{5.1}$$

and

$$\phi_{0,a} \circ (\phi_0 \bullet \mathrm{id}_{x_a}) = \mathrm{id}_{x_a} = \phi_{a,0} \circ (\mathrm{id}_{x_a} \bullet \phi_0). \tag{5.2}$$

Given such data, we set $x := x_1$ and $\lambda(a) := x_a - ax$ and $\gamma = d\lambda$. This allows us to write $\phi_{a,b}$ and ϕ_0 in the following form

$$\phi_{a,b}: \quad x_a + x_b \xrightarrow{c(a,b)\cdot e^{J(\lambda(a),bx)}} x_a + x_b + \gamma(a,b).$$

and

$$\phi_0: \quad 0 \xrightarrow{c(0,0)^{-1}} \lambda(0).$$

The condition (5.1) is satisfied if and only if

$$c\colon \mathbb{Z} \times \mathbb{Z} \longrightarrow U(1)$$

is a 2-cocycle. By Lemma 5.1, the 1-cochain λ takes values in $\Lambda^\vee$ and is uniquely determined by γ and by $\lambda(1) = 0$. To summarize, the data of a strong monoidal functor $\mathbb{Z} \longrightarrow \mathcal{T}$ are determined by the triple (x, γ, c), where $x \in \mathfrak{t}$, and $\gamma\colon \mathbb{Z}\times\mathbb{Z} \longrightarrow \Lambda^\vee$ and $c\colon \mathbb{Z} \times \mathbb{Z} \longrightarrow U(1)$ are 2-cocycles.

The 1-arrows: a pseudo-natural transformation from $((x_a)_a, \phi_{a,b}, \phi_0)$ to $((x'_a)_a, \phi'_{a,b}, \phi'_0)$, viewed as bifunctors $\mathbb{BZ} \longrightarrow \mathbb{B}G$, consists of an element $y \in \mathfrak{t}$, together with a sequence of arrows

$$\beta_a\colon y + x_a \xrightarrow{\cong} x'_a + y$$

satisfying

$$\beta_{a+b} \circ (\mathrm{id}_y \bullet \phi_{a,b}) = \left(\phi'_{a,b} \bullet \mathrm{id}_y\right) \circ \left(\mathrm{id}_{x'_a} \bullet \beta_b\right) \circ (\beta_a \bullet \mathrm{id}_{x_b}) \tag{5.3}$$

and

$$\beta_0 \circ (\mathrm{id}_y \bullet \phi_0) = \phi'_0 \bullet \mathrm{id}_y\,. \tag{5.4}$$

Writing β_a in the form

$$\beta_a: \quad y + x_a \xrightarrow{w(a)\cdot e^{-J(x_a+\mu(a),y)}} x_a + \mu(a) + y,$$

condition (5.3) spells out to

$$d\mu = d\lambda' - d\lambda, \quad \text{and}$$
$$dw(a,b) = \frac{c'(a,b)}{c(a,b)} \cdot \left(e^{-J(\mu(1),x)}\right)^{ab}.$$

Let $m = \mu(1)$ and $z = w(1)$. Then μ and w are uniquely determined by their boundaries together with m and z. To summarize, a 1-arrow in $\mathbf{Bicat}(\mathbb{BZ}, \mathbb{B}\mathcal{T})$ between the monoidal functors determined by (x, γ, c) and $(x+m, \gamma', c')$ is determined by its source and target together with a pair $(y, z) \in \mathfrak{t} \times U(1)$. Any choice of (y, z) is valid.

The 2-arrows These are the modifications. □

LEMMA 5.3. *The inclusion of the strict inertia 2-groupoid inside the full inertia 2-groupoid of $\mathcal{T}$ is an equivalence.*

PROOF. The set of 1-arrows from $(x, 0, 1)$ to (x, γ, c) is non-empty. Given a choice of (y, z), the arrows β_a look as follows:

$$\beta_a : \quad y + ax \xrightarrow{w(a) \cdot e^{-J(ax+\lambda(a),y)}} ax + \lambda(a) + y.$$

Here $\lambda \colon \mathbb{Z} \longrightarrow \Lambda^\vee$ is the 1-cochain with $d\lambda = \gamma$ and $\lambda(1) = 0$, and $w \colon \mathbb{Z} \longrightarrow U(1)$ is the 1-cochain with $dw = c$ and $w(1) = z$, as in Lemma 5.1 (i). □

COROLLARY 5.4. *The small inertia 2-groupoid of $\mathcal{T}$ is isomorphic to the Lie 2-groupoid with*

objects: *elements $x \in \mathfrak{t}$;*

1-arrows: *quadruples*

$$(x, y, m, w) \in \mathfrak{t} \times \mathfrak{t} \times \Lambda^\vee \times U(1)$$

written

$$x \xrightarrow{(y,w)} x + m.$$

2-arrows: *as follows*

$$x \underset{(y+n,\, w \cdot e^{I(x,n)})}{\overset{(y,w)}{\Downarrow u}} x + m$$

with $n \in \Lambda^\vee$ and $u \in U(1)$.

Horizontal composition: *the following diagram commutes strictly*

$$\begin{array}{ccccc} & & x+m, & & \\ & \overset{(y,w)}{\nearrow} & & \overset{(y',w')}{\searrow} & \\ x & & \xrightarrow{\quad (y+y',\, m+m',\, w \cdot w') \quad} & & x+m+m'. \end{array}$$

PROOF. The full inertia 2-groupoid is described in Theorem 5.2. In the notation of Theorem 5.2, set

$$m := \mu(1) \quad \text{and} \quad w := w(1).$$

Then the 1-cocycle μ and the 1-cochain w are uniquely determined by m and w. $\square$

THEOREM 5.5. *The inertia groupoid* $\mathcal{T}/\!\!/\mathcal{T}^{conj}$ *is equivalent to*

(i) *the Lie groupoid with objects* T *and arrows*

$$T \times \mathfrak{t} \times U(1)/\sim$$

with

$$(t, y+n, w) \sim \left(t, y, w \cdot e^{-I(x,n)}\right), \qquad n \in \Lambda^\vee,$$

where $x \in \mathfrak{t}$ *is any element with* $\exp(x) = t$*, source and target of* (t, y, w) *equal* t*, and composition of arrows is*

$$[t, y, w] \circ [t, y', w'] = [t, y + y', w \cdot w'];$$

(ii) *the Lie groupoid with objects* T *and arrows*

$$\mathfrak{t} \times T \times U(1)/\sim$$

with

$$(x+m, s, z) \sim \left(x, s, z \cdot e^{I(m,y))}\right), \qquad m \in \Lambda^\vee,$$

where $y \in \mathfrak{t}$ *is any element with* $\exp(y) = s$*, source and target of* (x, s, z) *equal* $\exp(x)$*, and composition of arrows is*

$$[x, s, z] \circ [x, s', z'] = [x, s \cdot s', z \cdot z'].$$

PROOF. Part (i) is immediate from Corollary 5.4. An isomorphism between the groupoids described in the two parts is given by

$$\begin{aligned} \mathfrak{t} \times \mathfrak{t} \times U(1) &\longrightarrow \mathfrak{t} \times \mathfrak{t} \times U(1) \\ (x, y, w) &\longmapsto (x, y, z) \end{aligned}$$

with $z = w \cdot e^{-I(x,y)}$. This isomorphism is $\Lambda^\vee \times \Lambda^\vee$ equivariant with respect to the actions described in the Theorem. $\square$

If we view $\mathcal{T}/\!\!/\mathcal{T}^{conj}$ as family of $U(1)$-central extensions of centralisers, $\widetilde{C}_t$, parametrized by $t \in T$, then Theorem 5.5(ii) shows that each $\widetilde{C}_t$ is isomorphic to the trivial central extension

$$\widetilde{C}_t \cong T \times U(1),$$

but there is no global trivialization over T. Instead, the trivialization for $\widetilde{C}_t$ depends on a choice of $x \in \mathfrak{t}$ with $\exp(x) = t$. Different choices yield trivializations that differ by an automorphisms of $T \times U(1)$. More precisely, we have a commuting diagram

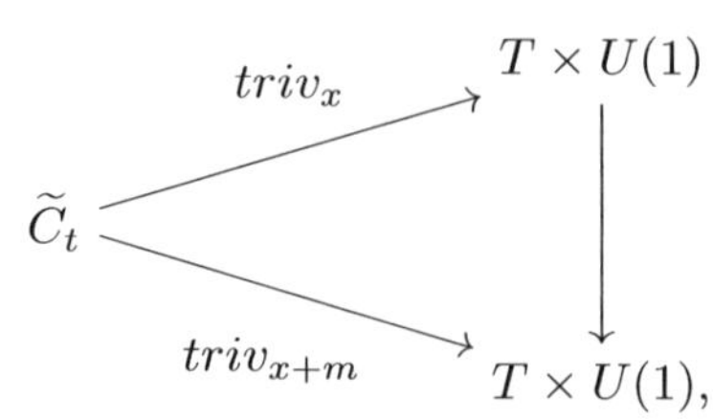

where the vertical arrow is obtained by exponentiating the map

$$\begin{pmatrix} \mathbb{1} & 0 \\ -I^\sharp(m) & 1 \end{pmatrix} : \mathfrak{t} \times \mathbb{R} \longrightarrow \mathfrak{t} \times \mathbb{R}.$$

5.2. Class functions on categorical tori. Consider the semi-direct product

$$H := (T \times U(1)) \rtimes \Lambda^\vee$$

with multiplication

$$(s, z, m)(s', z', m') = \left(s \cdot s', z \cdot z' \cdot e^{-2\pi i I(m)}(s'), m + m'\right).$$

COROLLARY 5.6. *We have an equivalence of groupoids*

$$\mathcal{T} /\!\!/ \mathcal{T}^{conj} \simeq \mathfrak{t} /\!\!/ H$$

where $T \times U(1)$ acts trivially, and $\Lambda^\vee$ acts by addition.

So, vector bundles on the inertia groupoid of $\mathcal{T}$ may be thought of as H-equivariant bundles on $\mathfrak{t}$. A source of such bundles are representations of H. Writing $\mathbb{C}_{\lambda,k}$ for the irreducible representation of the torus $T \times U(1)$ with weight (λ, k), any representation of H decomposes as

$$V \cong \bigoplus_{k \in \mathbb{Z}} V_k \otimes \mathbb{C}_k,$$

such that $U(1)$ acts on the kth summand with winding number k, and further

$$V_k \cong \bigoplus_{\lambda \in \Lambda} V_{\lambda,k} \otimes \mathbb{C}_{\lambda,k},$$

where T acts on the λ summand with weight λ. The action of $\Lambda^\vee$ on V defines isomorphisms

$$\psi_m \colon V_{\lambda,k} \xrightarrow{\cong} V_{\lambda - kI^\sharp(m),k}.$$

It follows that the character of V_k is a linear combination of the expressions

$$\sum_{m \in \Lambda^\vee} e^{2\pi i \lambda - kI^\sharp(m)}.$$

5.3. The Looijenga line bundle. Fix a complex number τ with imaginary part $im(\tau) > 0$, and set $q = e^{2\pi i \tau}$. Over

$$T_{\mathbb{C}} \,/\, q^{\Lambda^\vee} \cong \mathfrak{t}_{\mathbb{C}} \,/\, (\tau\Lambda^\vee + \Lambda^\vee),$$

we have the *Looijenga line bundle for I*

$$\mathcal{L}_{Lo}(I) := (T_{\mathbb{C}} \times \mathbb{C}) / \sim,$$

with

$$(5.5) \qquad (h, c) \sim \left(hq^m, ce^{-2\pi i I^\sharp(m)}(h) \cdot q^{-\frac{1}{2}I(m,m)}\right), \qquad m \in \Lambda^\vee.$$

LEMMA 5.7. *Let $\mathbf{L}^I$ be the (topological) complex line bundle on $T \times T$ associated to the principal $U(1)$-bundle described in Theorem 5.5 (ii). Then we have an isomorphism of line bundles*

$$\mathbf{L}^I \longrightarrow \mathcal{L}_{Lo}(I).$$

PROOF. We use the map

$$\begin{aligned} \mathfrak{t} \times \mathfrak{t} &\longrightarrow \mathfrak{t}_{\mathbb{C}} \\ (x, y) &\longrightarrow \tau x + y \end{aligned}$$

to identify the base spaces. To identify the line bundles, we note that the map

$$\begin{aligned} \mathfrak{t} \times T \times \mathbb{C} &\longrightarrow T_{\mathbb{C}} \times \mathbb{C} \\ (x, t, z) &\longmapsto \left(tq^x, zq^{-\frac{1}{2}I(x,x)}\right) \end{aligned}$$

covers this identification and sends the equivalence relation of Theorem 5.5 (ii) to (5.5). □

5.4. Discontinuity of the categorical character. One could hope that the categorical character of a representation of $\mathcal{T}$ was a vector bundle over $\mathcal{T}/\!\!/\mathcal{T}^{conj}$. In the case of the basic representation ϱ_{bas} as in Section 3, however, it turns out that the categorical character does not have the desired format: the vector space $X_{\varrho_{bas}}(t)$ does not vary continuously with $t \in T$. For t^* acting on $\mathcal{C}oh(T_{\mathbb{C}})$, one has

$$\mathbb{T}r(t^*) = Nat(\mathrm{id}, t^*) = \begin{cases} \mathbb{C}[\widehat{T}] & t = 1 \\ 0 & \text{else.} \end{cases}$$

Other variations of the theory exhibit the same discontinuity behaviour: if $f\colon X \longrightarrow X$ is an automorphism of an algebraic variety over k, then besides the action of f^* on $\mathcal{C}oh(X)$ one can consider the action on the bounded derived category $\mathcal{D}^b_{coh}(X)$ or the action by convolution with the Fourier-Mukai kernel $\mathcal{O}_{\Gamma^t_f}$ associated to the (transpose) graph of f. It is difficult to get a handle on traces of the derived functors Lf^*, but the categorical trace $\mathbb{T}r(\mathcal{O}_{\Gamma^t_f})$ was computed in [**Gan15**] and found to depend on the fixed point locus X^f. This fixed point locus is empty when f is given by multiplication with an element $t \in T_{\mathbb{C}} \setminus \{1\}$. On the other hand, for $t = 1$, we have the trace of the identity kernel. This is given by the Hochschild cohomology group

$$\mathbb{T}r^\bullet(\mathcal{O}_\Delta) = HH^\bullet(T_{\mathbb{C}}).$$

Changing the definition of categorical trace to the second notion (*"round trace"* in [**Wil08b**]), one arrives at a similar picture with Hochschild homology in place of Hochschild cohomology. In the setting of Section 3.3, where t^* acts on vector bundles with flat connection on the analytic variety $T_{\mathbb{C}}$, parallel transport along any piecewise smooth path from 1 to t defines an element of $\mathbb{T}r(t^*)$. Homotopic paths yield the same element. Writing $\mathcal{L}_t T_{\mathbb{C}}$ for the space of piecewise smooth paths from 1 to t inside $T_{\mathbb{C}}$, we therefore have an inclusion

$$\mathbb{C}\pi_0\left(\mathcal{L}_t T_{\mathbb{C}}\right) \subseteq \mathbb{T}r(t^*),$$

and further a non-canonical bijection

$$\Lambda^\vee \cong \pi_0\left(\mathcal{L}_t T\right).$$

In this setting, the discontinuity of R_{bas} expresses itself in the non-trivial monodromy of the connection on the bundle $\mathbb{L}_t$.

On the "twisted sector", i.e., when $\epsilon = -1$, the fixed points under $s \mapsto ts^{-1}$ are exactly the square roots of t in $T_{\mathbb{C}}$.

It would be interesting to have an interpretation of our basic representation that does not suffer from such discontinuity issues. In this context it should be noted that the homotopy theoretic counterpart to the derived fixed points $\Gamma^t_g \cap^R \Delta$ in the computation of categorical traces in the Fourier-Mukai setting are the twisted sectors of the loop space, $\mathcal{L}_g X$.

6. Automorphisms of categorical tori

6.1. Automorphisms of categorical tori. We will write $\Gamma_2\Lambda = (\Lambda \otimes \Lambda)^{S_2}$ for the second divided power of the weight lattice Λ. This is the dual of the second symmetric power $Sym^2(\Lambda^\vee) = (\Lambda^\vee \otimes \Lambda^\vee)/S_2$ of the coweight lattice; so elements of $\Gamma_2\Lambda$ may be thought of as symmetric bilinear forms on $\Lambda^\vee$. Associated to such a symmetric bilinear form B is the quadratic form $\phi_B(m) = B(m,m)$. There is an exact sequence

$$0 \longrightarrow (\Lambda \otimes \Lambda)^{sgn} \longrightarrow \Lambda \otimes \Lambda \longrightarrow \Gamma_2\Lambda \longrightarrow \Lambda/2\Lambda \longrightarrow 0,$$

where the second map is the symmetrisation map sending a bilinear form J to $J + J^t$ and the third map sends a symmetric bilinear form B to the class $[\phi_B]$ modulo 2. Its kernel (the image of the symmetrisation map) consists of the even symmetric bilinear forms on $\Lambda^\vee$. We will denote it by $(\Gamma_2\Lambda)_{ev}$. The elements of $(\Lambda \otimes \Lambda)^{sgn}$ are the skew symmetric bilinear forms on $\Lambda^\vee$. Fix a bilinear form J on $\Lambda^\vee$ and let $I = J + J^t$ be its symmetrisation. Then an automorphism f of $\Lambda^\vee$ is in the isometry group $O(\Lambda^\vee, I)$ if an only if $J - f^*J$ is skew symmetric. Inside the semi-direct product $O(\Lambda^\vee, I) \ltimes (\Lambda \otimes \Lambda)$, consider the subgroup with elements

$$\widetilde{O(\Lambda^\vee, I)} = \{(f, B) \mid B - B^t = J - f^*J\}.$$

This is a non-split extension of the form

$$0 \longrightarrow \Gamma_2\Lambda \longrightarrow \widetilde{O(\Lambda^\vee, I)} \longrightarrow O(\Lambda^\vee, I) \longrightarrow 1.$$

DEFINITION 6.1. We will write E for the quotient group

$$E := \widetilde{O(\Lambda^\vee, I)} \,/\, (\Gamma_2\Lambda)_{ev}$$

and Ξ for the crossed module

$$\begin{aligned} \xi\colon \Lambda &\longrightarrow E \\ \lambda &\longmapsto (\mathrm{id}, [\lambda \otimes \lambda]), \end{aligned}$$

where (f, B) acts on Λ via f^*.

The sits inside a crossed module extension

$$0 \longrightarrow 2\Lambda \longrightarrow \Lambda \xrightarrow{\xi} E \longrightarrow O(\Lambda^\vee, I) \longrightarrow 1.$$

Let $\mathcal{T}$ be the categorical torus associated to J, as in Section 3.1.

THEOREM 6.2. *The categorical group of centre preserving weak automorphism of $\mathcal{T}$ is equivalent to the strict categorical group $\mathcal{S}(\Xi)$.*

PROOF. We have an equivalence of crossed module extensions

$$\begin{array}{ccccccccccc} 0 & \longrightarrow & 2\Lambda & \longrightarrow & \Lambda \times_{\Lambda/2\Lambda} \Gamma_2\Lambda & \xrightarrow{\tilde{\xi}} & \widetilde{O(\Lambda^\vee, I)} & \longrightarrow & O(\Lambda^\vee, I) & \longrightarrow & 1 \\ & & \| & & \downarrow & & \downarrow & & \| & & \\ 0 & \longrightarrow & 2\Lambda & \longrightarrow & \Lambda & \xrightarrow{\xi} & E & \longrightarrow & O(\Lambda^\vee, I) & \longrightarrow & 1, \end{array}$$

where $\widetilde{\xi}(\lambda, B) = (\mathrm{id}, B)$, and the action on the top is via the action of $O(\Lambda^\vee, I)$ on Λ. Write $\widetilde{\Xi}$ for the crossed module on the top. The promised equivalence is given by the strict crossed module homomorphism

$$A\colon \widetilde{\Xi} \longrightarrow wAct^+(\Theta),$$

where A_0 sends the element (f, B) of $\widetilde{O(\Lambda^\vee, I)}$ to the weak automorphism

$$\begin{aligned} f_0(x) &= f(x) \\ f_1(m, [a]) &= (f(m), [a]) \\ \beta_{x,y} &= (0, [B(x, y)]), \end{aligned}$$

and A_1 sends the element (λ, B) of $\Lambda \times_{\Lambda/2\Lambda} \Gamma_2\Lambda$ to

$$\eta(x) = \frac{1}{2}\left(B(x, x) - \lambda(x)\right).$$

To see that A is an isomorphism on π_1 and hence fully faithful, note that

$$\pi_1 wAct(\Theta) = \{\eta\colon \mathfrak{t} \to \mathbb{R}/\mathbb{Z} \mid \eta|_{\Lambda^\vee} = 0 \text{ and } \eta(x)\eta(y) = \eta(xy)\}$$

is canonically identified with $\widehat{T}$. Under this identification, the restriction of A to $\pi_1(\widetilde{\Xi})$ becomes the standard isomorphism $\Lambda \cong \widehat{T}$. Since

$$H^2_{gp}(T; U(1)) \cong H^3(BT; \mathbb{Z}) = 0,$$

we obtain from (2.6) an isomorphism

$$\pi_0(wAct^+(\Theta)) \cong O(\Lambda^\vee, I).$$

More precisely, the class of $A_0(f, B)$ is mapped to f under the isomorphism (2.6). Hence A is an equivalence of crossed modules. □

6.2. The unimodular case. If $I^\sharp\colon \Lambda^\vee \longrightarrow \Lambda$ is an isomorphism, then the centre of $\widetilde{T[2]}$ consists of $\{\pm 1\}$, and we have an isomorphism of extensions

$$\begin{array}{ccccccccc} 1 & \longrightarrow & Inn(\widetilde{T[2]}) & \longrightarrow & Aut(\widetilde{T[2]}) & \longrightarrow & Out(\widetilde{T[2]}) & \longrightarrow & 1 \\ & & {\scriptstyle I^\sharp_{\mathbb{F}_2}}\downarrow{\scriptstyle\cong} & & \downarrow{\scriptstyle\cong} & & \downarrow{\scriptstyle\cong} & & \\ 1 & \longrightarrow & T^*[2] & \longrightarrow & E' & \longrightarrow & O(T[2], [\phi_I]) & \longrightarrow & 1. \end{array}$$

where

$$E' \subset O(T[2], [\phi_I]) \ltimes C^1(T[2], \mathbb{F}_2)$$

is the subgroup

$$E' = \{(f, c) \mid dc = J + f^*J\}.$$

In this situation, E is isomorphic to the pull-back of the extension E' along the composite

$$O(\Lambda^\vee, I) \longrightarrow PO(\Lambda^\vee, I) \longrightarrow O(T[2], [\phi_I]).$$

In the case of the Leech lattice, it follows that E is equal to the extension $\widetilde{C}$ defined on Page 160.

6.3. Automorphisms of $\mathcal{T}\rtimes\{\pm 1\}$. Write $\widetilde{\Lambda^\vee}$ for the central extension of $\Lambda^\vee$ by $\mathbb{Z}/2\mathbb{Z}$ with cocycle J^t (modulo 2). It will be convenient to use additive notation for the elements of its centre. Throughout this section, $I^\sharp$ is assumned to be an isomorphism.

DEFINITION 6.3. We will write Ξ' for the crossed module

$$\begin{aligned} \xi'\colon \widetilde{\Lambda^\vee}/2\Lambda^\vee &\longrightarrow T\rtimes E \\ ([n],\iota) &\longmapsto \left(\exp\left(\frac{n}{2}\right), \mathrm{id}, B\right), \end{aligned}$$

where $I^\sharp(n)\equiv \phi_B \mod 2$. The action of $T\rtimes E$ on $\widetilde{\Lambda^\vee}/2\Lambda^\vee$ is given by

$$([n],\iota)^{(t,f,B)} = ([f^{-1}n], B(f^{-1}n, f^{-1}n)) .$$

THEOREM 6.4. *The categorical group of centre preserving weak automorphism of $\mathcal{T}\rtimes\{\pm 1\}$ is equivalent to the strict categorical group $\mathcal{S}(\Xi')$.*

PROOF. We will write $\widetilde{\Xi}'$ for the crossed module

$$\begin{aligned} \widetilde{\xi}'\colon \widetilde{\Lambda^\vee}\times_{\Lambda/2\Lambda}\Gamma_2\Lambda &\longrightarrow \mathfrak{t}\rtimes \widetilde{O(\Lambda^\vee, I)} \\ (n,\iota,B) &\longmapsto \left(\frac{n}{2}, \mathrm{id}, B\right), \end{aligned}$$

where the fibred product in the source is defined using the map $I^\sharp\colon \Lambda^\vee \longrightarrow \Lambda$ to map from $\widetilde{\Lambda^\vee}$ to $\Lambda/2\Lambda$ and the map $\Gamma_2\Lambda\longrightarrow\Lambda/2\Lambda$ sending B to ϕ_B modulo 2. The action of $\mathfrak{t}\rtimes\widetilde{O(\Lambda^\vee, I)}$ on $\widetilde{\Lambda^\vee}\times_{\Lambda/2\Lambda}\Gamma_2\Lambda$ is given by

$$(n,\iota,S)^{(a,f,B)} = (f^{-1}n, B(f^{-1}n, f^{-1}n), f^*S) .$$

The strict homomorphism

$$\begin{array}{ccc} \widetilde{\Lambda^\vee}\times_{\Lambda/2\Lambda}\Gamma_2\Lambda & \longrightarrow & \widetilde{\Lambda^\vee}/2\Lambda^\vee \\ {\scriptstyle\widetilde{\xi}'}\downarrow & & \downarrow{\scriptstyle\xi'} \\ \mathfrak{t}\rtimes\widetilde{O(\Lambda^\vee, I)} & \longrightarrow & T\rtimes E \end{array}$$

obtained by modding out $2\Lambda^\vee\times(\Gamma_2\Lambda)_{ev}$ in degree 1 and $\Lambda^\vee\times(\Gamma_2\Lambda)_{ev}$ in degree 0 induces isomorphisms on π_0 and π_1. We will now define a weak homomorphism

$$(A',\kappa)\colon \widetilde{\Xi}' \longrightarrow wAct^+(\Theta').$$

Given $a\in\mathfrak{t}$, we will write c_a for the inner automorphism associated to $(a,1)$ as in Definition 2.4. Explicitly, c_a is the strict automorphism of $\mathcal{T}\rtimes\{\pm 1\}$ given by

$$\begin{aligned} (c_a)_0\colon (x,\epsilon) &\mapsto (x+(1-\epsilon)a,\epsilon) \\ (c_a)_1\colon (m,[r]) &\mapsto (m,[r-J(m,a)]). \end{aligned}$$

Given $(f,B)\in\widetilde{O(\Lambda^\vee, I)}$, we define the weak automorphism (f,β) of $\mathcal{T}\rtimes\{\pm 1\}$, given by

$$\begin{aligned} f_0 : (x,\epsilon) &\mapsto (f(x),\epsilon), \\ f_1\colon (m,[r]) &\mapsto (f(m),[r]) \\ \beta_{(x,\epsilon),(y,\delta)} &= (0,[\epsilon B(x,y)]). \end{aligned}$$

This extends the weak automorphism of $\mathcal{T}$ in the proof of Theorem 6.2. We define (A', κ) to be given by the pointed maps

$$\begin{array}{ccc} (n,\iota,B) & \longmapsto & \eta = \eta_{(n,\iota,B)} \\ \widetilde{\Lambda^\vee} \times_{\Lambda/2\Lambda} \Gamma_2\Lambda & \xrightarrow{A_1'} & Maps_*(\mathfrak{t} \rtimes \{\pm 1\}, \Lambda^\vee \times \mathbb{R}/\mathbb{Z})^\times \\ \Big\downarrow \tilde{\xi}' & & \Big\downarrow \delta \\ \mathfrak{t} \rtimes \widetilde{O(\Lambda^\vee, I)} & \xrightarrow{A_0'} & wAct^+(\Theta')_0 \\ (a,f,B) & \longmapsto & c_a \circ (f,\beta), \end{array}$$

with

$$\eta(x,\epsilon) = \left(\frac{\epsilon - 1}{2} n, \left[\frac{1}{2}\left(B(x,x) + J(n,x) + \frac{\epsilon-1}{2}\iota\right)\right]\right),$$

along with the Künneth

$$\kappa_{(a,f,B),(b,g,B')}(x,\epsilon) = (0, [\epsilon B(g(x),b) - B(b,g(x)) + (\epsilon - 1)B(b,b)]).$$

By a lengthy yet straight-forward computation, one checks that the pair (A', κ) satisfies Axioms (W1)–(W5). The Bernstein centre of $\widetilde{\Xi}'$ is given by the central subgroup of $\widetilde{\Lambda^\vee}$, so $\pi_1(\widetilde{\Xi}') \cong \mathbb{Z}/2\mathbb{Z}$, while

$$\pi_1(wAct^+(\Theta') = \mathrm{Hom}(\mathfrak{t} \rtimes \{\pm 1\}, \Lambda^\vee \times \mathbb{R}/\mathbb{Z}) = Hom(\{\pm 1\}, \mathbb{R}/\mathbb{Z}).$$

Explicitly, the map between these two groups equals

$$A_1'(0,\iota,0)(x,\epsilon) = \left(0, \left[\frac{(\epsilon-1)\iota}{4}\right]\right)$$

(or in multiplicative notation, $(x,\epsilon) \mapsto \epsilon^\iota$). So, (A', κ) induces an isomorphism on π_1. To determine the effect of (A', κ) on π_0, we use the short exact sequence

$$0 \longrightarrow H^2_{gp}(T \rtimes \{\pm 1\}, U(1)) \longrightarrow \pi_0(\mathcal{A}ut^+(\mathcal{T} \rtimes \{\pm 1\})) \longrightarrow Stab([\alpha]) \longrightarrow 1$$

from (2.6). The Lyndon-Hochschild-Serre spectral sequence

$$H^p_{gp}\left(\{\pm 1\}; H^q_{gp}(T; U(1))\right) \Longrightarrow H^{p+q}_{gp}(T \rtimes \{\pm 1\}; U(1))$$

yields the isomorphism

$$H^2_{gp}(T \rtimes \{\pm 1\}; U(1)) \cong H^1_{gp}(\{\pm 1\}; \Lambda_-) \cong \Lambda/2\Lambda.$$

These elements are the centre preserving automorphisms covering the identity of $T \rtimes \{\pm 1\}$, and they are represented by the strict automorphisms

$$\begin{aligned} h_\lambda \colon (x,\epsilon) &\mapsto (x,\epsilon) \\ (m,[r]) &\mapsto \left(m, \left[r + \frac{1}{2}\lambda(m)\right]\right). \end{aligned}$$

Given $\lambda \in \Lambda$, let $B \in \Gamma_2\Lambda$ be such that $\phi_B \equiv \lambda \mod 2$, and set $\eta(x,\epsilon) = (0, [\frac{1}{2}B(x,x)])$. Then $\delta(\eta) = (h_\lambda, \beta)$, with

$$\beta_{(x,\epsilon),(y,\delta)} = (0, [\epsilon B(x,y)]).$$

It follows that $A_0(0, \mathrm{id}, B)$ represents the same class as h_λ in $\pi_0(wAct^+(\Theta'))$. Hence $\pi_0(A)$ maps $\Gamma_2\Lambda/(\Gamma_2\Lambda)_{ev}$ isomorphically onto $H^2_{gp}(T\rtimes\{\pm1\}; U(1))$, and we have a map of short exact sequences

$$\begin{array}{ccccccccc} 0 & \longrightarrow & \Gamma_2\Lambda/(\Gamma_2\Lambda)_{ev} & \longrightarrow & \pi_0(\widetilde{\Xi}') & \longrightarrow & T \rtimes O(I,\Lambda^\vee) & \longrightarrow & 1 \\ & & \downarrow\cong & & \downarrow\pi_0(A') & & \downarrow & & \\ 0 & \longrightarrow & \Lambda/2\Lambda & \longrightarrow & \pi_0(\mathcal{A}ut^+(T\rtimes\{\pm1\})) & \longrightarrow & Stab([\alpha]) & \longrightarrow & 0. \end{array}$$

Here

$$Stab([\alpha]) = T \rtimes O(\Lambda^\vee, I) \subseteq T \rtimes GL(\Lambda^\vee) \cong Aut(T\rtimes\{\pm1\}).$$

The left-most vertical arrow in (2.6) is the restriction of the map $\pi_0(\mathcal{A}d)$ to the centre, and it is given by

$$\begin{aligned} T[2] &\longrightarrow \Lambda/2\Lambda \\ \exp(\frac{n}{2}) &\longmapsto [I^\sharp(n)]. \end{aligned}$$

In other words, if n is such that $I^\sharp(n) = \lambda$, then $c_{\frac{n}{2}}$ also represents the same class as h_λ, and the subgroup $T \rtimes \{\pm1\}$ of $T \rtimes O(\Lambda^\vee, I)$ is identified, via (2.6) with $Inn(T \rtimes \{\pm1\})$. The class of $[(a, f, B)] \in \pi_0(\widetilde{\Xi}')$ maps to the element $(\exp(2a), f)$ of $T \rtimes O(\Lambda^\vee, I)$. It follows that the induced map on cokernels is an isomorphism, and hence, by the Five Lemma, so is $\pi_0(A')$. □

COROLLARY 6.5. *We have*

$$\pi_0(\mathcal{O}ut(\mathcal{T}\rtimes\{\pm1\})) \cong PO(\Lambda^\vee, I).$$

PROOF. Since $I^\sharp$ is an isomorphism, the proof of the Theorem implies that the leftmost vertical map of (2.6) is also an isomorphism, so we have an isomorphism between the cokernels of the middle and the right vertical arrows. □

COROLLARY 6.6. *The exact sequence* (2.4) *for* $\mathcal{T}\rtimes\{\pm1\}$ *takes the form*

$$1 \longrightarrow U(1) \xrightarrow{\cong} U(1) \longrightarrow 1$$

$$1 \longrightarrow \{\pm1\} \xrightarrow{\cong} \pi_0\mathcal{Z}(\mathcal{T}\rtimes\{\pm1\}) \longrightarrow 1$$

$$1 \longrightarrow T\rtimes\{\pm1\} \longrightarrow \pi_0\mathcal{A}ut^+(\mathcal{T}\rtimes\{\pm1\}) \longrightarrow PO(\Lambda^\vee, I) \longrightarrow 1.$$

It would be interesting to see whether the cokernel of a canonical map

$$\mathcal{T}\rtimes\{\pm1\} \longrightarrow \mathcal{N}(\varrho_{bas})$$

(with a suitable definition of categorical normaliser $\mathcal{N}(\varrho_{bas})$) can be related to the group

$$C = 2^{1+24}.Co_1$$

in the construction of the Monster.

Acknowledgments

It is a pleasure to thank Matthew Ando and Gerd Laures, John McKay and Igor Frenkel, Geoffrey Mason, Konrad Waldorf and Thomas Nikolaus, Theo Johnson-Freyd, and David Treumann for helpful conversations. Special thanks go to Igor Frenkel for convincing me to bring this circle of ideas to paper. Many thanks go to the referee and the referee's PhD student for extensive feedback and help in reorganising the exposition.

References

[AN09] Ettore Aldrovandi and Behrang Noohi, *Butterflies. I. Morphisms of 2-group stacks*, Adv. Math. **221** (2009), no. 3, 687–773, DOI 10.1016/j.aim.2008.12.014. MR2511036

[Bar11] Bruce Bartlett, *The geometry of unitary 2-representations of finite groups and their 2-characters*, Appl. Categ. Structures **19** (2011), no. 1, 175–232, DOI 10.1007/s10485-009-9189-0. MR2772605

[BM94] J.-L. Brylinski and D. A. McLaughlin, *The geometry of degree-four characteristic classes and of line bundles on loop spaces. I*, Duke Math. J. **75** (1994), no. 3, 603–638, DOI 10.1215/S0012-7094-94-07518-2. MR1291698

[Bor92] Richard E. Borcherds, *Monstrous moonshine and monstrous Lie superalgebras*, Invent. Math. **109** (1992), no. 2, 405–444, DOI 10.1007/BF01232032. MR1172696

[Bre92] Lawrence Breen, *Théorie de Schreier supérieure* (French), Ann. Sci. École Norm. Sup. (4) **25** (1992), no. 5, 465–514. MR1191733

[BSCS07] John C. Baez, Danny Stevenson, Alissa S. Crans, and Urs Schreiber, *From loop groups to 2-groups*, Homology Homotopy Appl. **9** (2007), no. 2, 101–135. MR2366945

[Car07] Scott Huai-Lei Carnahan, *Monstrous Lie algebras and generalized moonshine*, ProQuest LLC, Ann Arbor, MI, 2007. Thesis (Ph.D.)–University of California, Berkeley. MR2717439

[Car10] Scott Carnahan, *Generalized moonshine I: genus-zero functions*, Algebra Number Theory **4** (2010), no. 6, 649–679, DOI 10.2140/ant.2010.4.649. MR2728485

[Car12] Scott Carnahan, *Generalized moonshine, II: Borcherds products*, Duke Math. J. **161** (2012), no. 5, 893–950, DOI 10.1215/00127094-1548416. MR2904095

[CCN+85] J. H. Conway, R. T. Curtis, S. P. Norton, R. A. Parker, and R. A. Wilson, *ATLAS of finite groups*, Oxford University Press, Eynsham, 1985. Maximal subgroups and ordinary characters for simple groups; With computational assistance from J. G. Thackray. MR827219

[CG04] P. Carrasco and A. R. Garzón, *Obstruction theory for extensions of categorical groups*, Appl. Categ. Structures **12** (2004), no. 1, 35–61, DOI 10.1023/B:APCS.0000013810.93405.c8. Homotopy theory. MR2057410

[CGV06] P. Carrasco, A. R. Garzón, and E. M. Vitale, *On categorical crossed modules*, Theory Appl. Categ. **16** (2006), No. 22, 585–618. MR2259264

[CN79] J. H. Conway and S. P. Norton, *Monstrous moonshine*, Bull. London Math. Soc. **11** (1979), no. 3, 308–339, DOI 10.1112/blms/11.3.308. MR554399

[Con90] John B. Conway, *A course in functional analysis*, 2nd ed., Graduate Texts in Mathematics, vol. 96, Springer-Verlag, New York, 1990. MR1070713

[FLM88] Igor Frenkel, James Lepowsky, and Arne Meurman, *Vertex operator algebras and the Monster*, Pure and Applied Mathematics, vol. 134, Academic Press, Inc., Boston, MA, 1988. MR996026

[Gan15] Nora Ganter, *Inner products of 2-representations*, Adv. Math. **285** (2015), 301–351, DOI 10.1016/j.aim.2015.05.015. MR3406502

[Gan18] Nora Ganter, *Categorical tori*, SIGMA Symmetry Integrability Geom. Methods Appl. **14** (2018), Paper No. 014, 18, DOI 10.3842/SIGMA.2018.014. MR3764535

[GK08] Nora Ganter and Mikhail Kapranov, *Representation and character theory in 2-categories*, Adv. Math. **217** (2008), no. 5, 2268–2300, DOI 10.1016/j.aim.2007.10.004. MR2388094

[GPS95] R. Gordon, A. J. Power, and Ross Street, *Coherence for tricategories*, Mem. Amer. Math. Soc. **117** (1995), no. 558, vi+81, DOI 10.1090/memo/0558. MR1261589

[Gri82] Robert L. Griess Jr., *The friendly giant*, Invent. Math. **69** (1982), no. 1, 1–102, DOI 10.1007/BF01389186. MR671653

[GU16] Nora Ganter and Robert Usher, *Representation and character theory of finite categorical groups*, Theory Appl. Categ. **31** (2016), Paper No. 21, 542–570. MR3518979

[GV20] Dennis Gaitsgory and Yakov Varshavsky, *Local terms for the categorical trace of Frobenius*, 2020.

[JF19] Theo Johnson-Freyd, *The Moonshine anomaly*, Comm. Math. Phys. **365** (2019), no. 3, 943–970, DOI 10.1007/s00220-019-03300-2. MR3916985

[Lei98] Tom Leinster, *Basic bicategories*, 1998.

[LW23] Matthias Ludewig and Konrad Waldorf, *Lie 2-groups from loop group extensions*, 2023.

[Mas87] Geoffrey Mason, *Finite groups and modular functions*, The Arcata Conference on Representations of Finite Groups (Arcata, Calif., 1986), Proc. Sympos. Pure Math., vol. 47, Amer. Math. Soc., Providence, RI, 1987, pp. 181–210, DOI 10.1090/pspum/047.1/933359. With an appendix by S. P. Norton. MR933359

[Mas02] Geoffrey Mason, *Orbifold conformal field theory and cohomology of the monster*, Unpublished manuscript, 2002.

[Noo07] Behrang Noohi, *Notes on 2-groupoids, 2-groups and crossed modules*, Homology Homotopy Appl. **9** (2007), no. 1, 75–106. MR2280287

[Nor90] Katherine Norrie, *Actions and automorphisms of crossed modules* (English, with French summary), Bull. Soc. Math. France **118** (1990), no. 2, 129–146. MR1087375

[Pir23] Mariam Pirashvili, *On the centre of crossed modules of groups*, 2023.

[Rou03] Alain Rousseau, *Bicatégories monoïdales et extensions de gr-catégories* (French, with English summary), Homology Homotopy Appl. **5** (2003), no. 1, 437–547. MR2072344

[Sch04] J. Schur, *Über die Darstellung der endlichen Gruppen durch gebrochen lineare Substitutionen* (German), J. Reine Angew. Math. **127** (1904), 20–50, DOI 10.1515/crll.1904.127.20. MR1580631

[Sch07] J. Schur, *Untersuchungen über die Darstellung der endlichen Gruppen durch gebrochene lineare Substitutionen* (German), J. Reine Angew. Math. **132** (1907), 85–137, DOI 10.1515/crll.1907.132.85. MR1580715

[SP11] Christopher J. Schommer-Pries, *Central extensions of smooth 2-groups and a finite-dimensional string 2-group*, Geom. Topol. **15** (2011), no. 2, 609–676, DOI 10.2140/gt.2011.15.609. MR2800361

[Sí75] Hoàng Xuân Sính, *Gr-catégories*, PhD thesis, Université Paris Diderot, 1975.

[Tit85] Jacques Tits, *Le Monstre (d'après R. Griess, B. Fischer et al.)* (French), Astérisque **121-122** (1985), 105–122. Seminar Bourbaki, Vol. 1983/84. MR768956

[Wal12] Konrad Waldorf, *A construction of string 2-group models using a transgression-regression technique*, Analysis, geometry and quantum field theory, Contemp. Math., vol. 584, Amer. Math. Soc., Providence, RI, 2012, pp. 99–115, DOI 10.1090/conm/584/11588. MR3013040

[Wal22] Konrad Waldorf, *Categorical symmetries of t-duality*, 2022.

[Wil08a] Simon Willerton, *The twisted Drinfeld double of a finite group via gerbes and finite groupoids*, Algebr. Geom. Topol. **8** (2008), no. 3, 1419–1457, DOI 10.2140/agt.2008.8.1419. MR2443249

[Wil08b] Simon Willerton, *Two 2-traces*, Talk, 2008.

[WW15] Friedrich Wagemann and Christoph Wockel, *A cocycle model for topological and Lie group cohomology*, Trans. Amer. Math. Soc. **367** (2015), no. 3, 1871–1909, DOI 10.1090/S0002-9947-2014-06107-2. MR3286502

School of Mathematics and Statistics, The University of Melbourne, Parkville Victoria 3010, Wurundjeri Country, Australia

Email address: `nganter@unimelb.edu.au`

Contemporary Mathematics
Volume **817**, 2025
https://doi.org/10.1090/conm/817/16331

From finite state automata to tangle cobordisms: a TQFT journey from one to four dimensions

Mee Seong Im and Mikhail Khovanov

To Igor B. Frenkel, on his anniversary, with admiration.

ABSTRACT. This is a brief introduction to link homology theories that categorify Reshetikhin–Turaev $\mathsf{SL}(N)$-quantum link invariants. A recently discovered surprising connection between finite state automata and Boolean TQFTs in dimension one is explained as a warm-up.

CONTENTS

1. Introduction

This note is based on the talk that one of us (M.K.) gave at the First International Congress of Basic Science which was held at the Yanqi Lake Beijing Institute of Mathematical Sciences and Applications (BIMSA) in July 2023.

We briefly discuss topological quantum field theories (TQFTs) and explain a recent surprising observation [**GIK⁺23**] that Boolean-valued one-dimensional TQFTs with defects correspond to nondeterministic finite state automata. The precise statement is given in Theorem 2.2 below and in [**GIK⁺23**].

We then review various approaches to categorification of Reshetikhin–Turaev link invariants for the fundamental representations of quantum $\mathsf{SL}(N)$ emphasizing

2020 *Mathematics Subject Classification.* Primary 57K16, 57K18, 57K45, 68Q45, 18M30; Secondary 68Q70, 18B20.

Key words and phrases. Topological quantum field theory, TQFT, finite-state automata, Reshetikhin–Turaev invariants, categorification, link homology, Khovanov–Rozansky homology, foam evaluation.

The second author was partially supported by NSF grant DMS-2204033 and Simons Collaboration Award 994328 while working on this paper. The second author was supported by the Beijing Institute of Mathematical Sciences and Applications (BIMSA).

the original approach via matrix factorizations [**KR08a, KR08b, Yon11, Wu12a**] and Robert–Wagner foam evaluation approach [**RW20a**].

Reshetikhin–Turaev link invariants [**RT90, Res91**] are part of the Chern–Simons three-dimensional TQFT discovered by Witten and Reshetikhin–Turaev [**Wit89, RT91**]. Categorification of Reshetikhin–Turaev link invariants can be viewed as a 4-dimensional TQFT restricted to links in $\mathbb{R}^3$ and to link cobordisms. Boolean one-dimensional TQFTs and finite state automata connection is explained first, as a warm-up to discussing these sophisticated TQFTs in dimensions three and four.

This is a relatively short paper which reviews Khovanov–Rozansky link homology in its second part. We provide key references but can not give a fully comprehensive coverage of the substantial amount of literature on this and closely related subjects due to the immense and rapid development over the last twenty-five years.

2. TQFTs, Boolean TQFTs and automata

In this section, we survey the correspondence between automata and one-dimensional Boolean TQFTs explained in [**GIK⁺23**].

2.1. TQFTs. An n-dimensional TQFT (topological quantum field theory) over a field $\mathbf{k}$ is a tensor functor from the category of n-dimensional cobordisms to the category of vector spaces

$$\mathcal{F} : \mathsf{Cob}_n \longrightarrow \mathbf{k}\text{–}\mathsf{vect}.$$

The category Cob_n has oriented $(n-1)$-manifolds as objects and n-dimensional cobordisms between them as morphisms. This category is symmetric tensor, and the functor $\mathcal{F}$ must respect this structure so that, in particular,

$$\mathcal{F}(K_1 \sqcup K_2) \cong \mathcal{F}(K_1) \otimes \mathcal{F}(K_2)$$

for $(n-1)$-manifolds K_1 and K_2.

2.2. One-dimensional TQFTs. A TQFT $\mathcal{F}$ in one dimension assigns a vector space V to a point with positive orientation, a space W to a point with negative orientation, which we can write as

$$\mathcal{F}(+) = V, \qquad \mathcal{F}(-) = W,$$

and maps $\mathbf{k} \longrightarrow V \otimes W$ and $W \otimes V \longrightarrow \mathbf{k}$ to the *cup*, respectively, *cap*, cobordisms shown in the top row of Figure 2.2.1. Transposition cobordisms, see Figure 2.2.1 top right, are mapped to transposition maps between products of V and W, such as $V \otimes W \longrightarrow W \otimes V$, $v \otimes w \mapsto w \otimes v$.

Isotopy relations shown in the second row of that figure imply that vector space V is finite-dimensional and vector space $W \cong V^*$ can be taken to be the dual of V, so that cup and cap cobordisms are sent by $\mathcal{F}$ to the *evaluation* and *coevaluation* maps:

$$\mathsf{coev} : \mathbf{k} \longrightarrow V \otimes V^*, \qquad \mathsf{ev} : V^* \otimes V \longrightarrow \mathbf{k}$$

given in a basis $\{v_1, \dots, v_n\}$ of V and the dual basis $\{v^1, \dots, v^n\}$ of V^* by

$$\mathsf{coev}(1) = \sum_{i=1}^{n} v_i \otimes v^i,$$

$$\mathsf{ev}(v^i \otimes v_j) = \delta_{i,j}, \quad \mathsf{ev}(v^* \otimes v) = v^*(v), \quad v^* \in V^*, \quad v \in V.$$

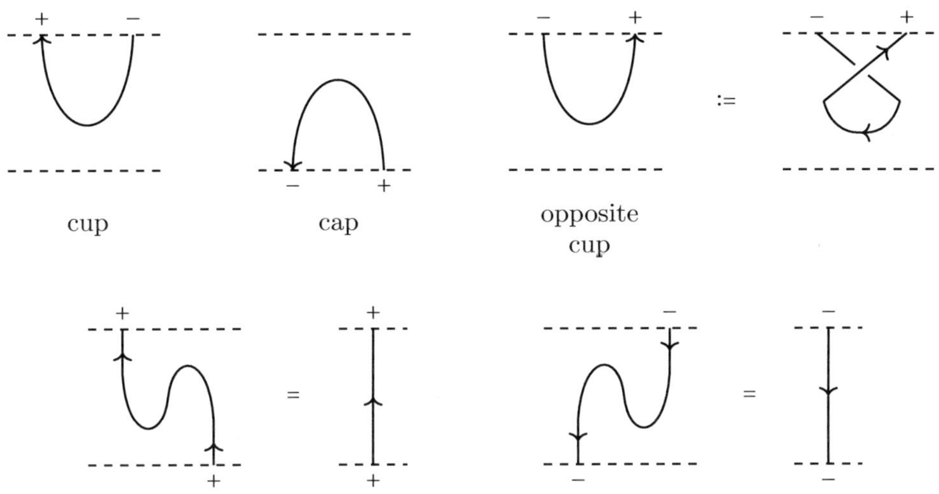

FIGURE 2.2.1. Top left: the cup and cap cobordisms. The cup and cap cobordisms for the opposite orientation are obtained by composing these two with the permutation cobordisms, see top right for the cup cobordism with the opposite orientation. Bottom row shows isotopy relations on the cup and cap cobordisms.

Composing with the transposition morphism in Figure 2.2.1 in top right gives the cup and cap morphisms for the opposite orientation of the arc.

The TQFT $\mathcal{F}$ evaluated on a circle gives a linear map $\mathbf{k} \longrightarrow \mathbf{k}$ which is multiplication by $\dim V$, and that is the only invariant of a one-dimensional TQFT. Two such TQFTs

$$\mathcal{F} : \mathsf{Cob}_1 \longrightarrow \mathbf{k}\text{–}\mathsf{vect}, \qquad \mathcal{F}(+) = V, \qquad \mathcal{F}(-) = V^*$$

are isomorphic if and only if the corresponding vector spaces are isomorphic, that is, if they have the same dimension (evaluate the same on the circle).

To spice up this simple classification, let us add 0-dimensional defects, which are points on one-manifolds, labelled by elements of a finite set Σ, and also allow one-manifolds to end in the middle of a cobordism, see Figure 2.2.2. In this way, the boundary of a one-manifold cobordism splits into the *outer boundary* and *inner boundary*. Connected components of a morphism in $\mathsf{Cob}_{\Sigma,1}$ can be classified into four types:

- arcs with two outer endpoints,
- *half-intervals*, which are arcs with one outer and one inner endpoint,
- *floating intervals*, which are arcs with two inner endpoints,
- circles.

The morphism in Figure 2.2.2 consists of two arcs, three half-intervals, one floating interval and one circle. Each connected component of a morphism may be decorated by Σ-labelled dots.

This bigger category $\mathsf{Cob}_{\Sigma,1}$ is also a symmetric tensor category, containing Cob_1 as a subcategory. How can one classify TQFTs (tensor functors)

$$\mathcal{F} : \mathsf{Cob}_{\Sigma,1} \longrightarrow \mathbf{k}\text{–}\mathsf{vect}?$$

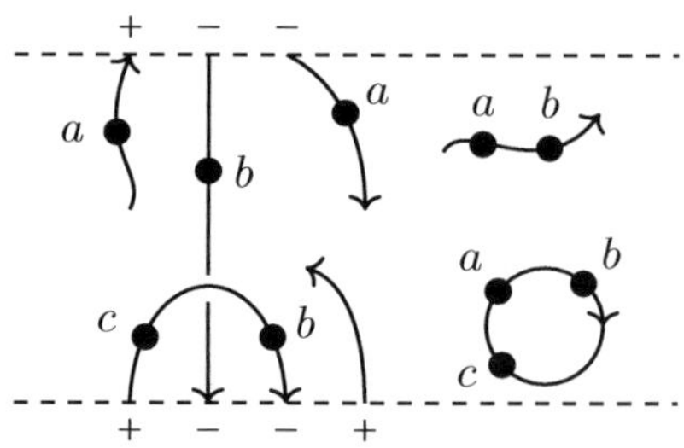

FIGURE 2.2.2. A morphism from oriented 1-manifold $(+--+)$ to $(+--)$ in category $\mathsf{Cob}_{\Sigma,1}$. This morphism has seven outer and five inner boundary points. It has two arcs, three half-intervals, one floating interval and one circle.

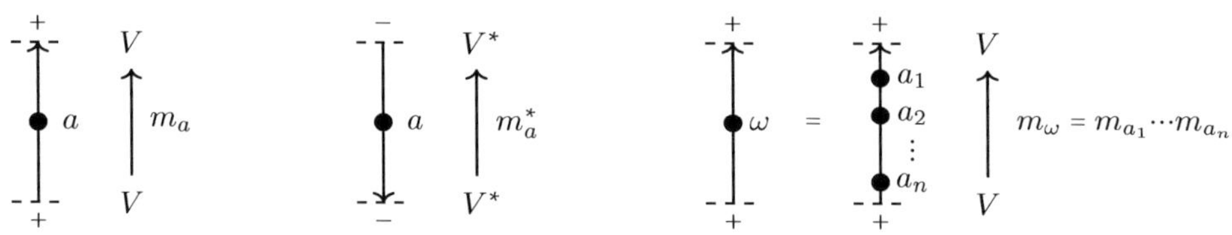

FIGURE 2.2.3. Left: to a labelled dot on an upward-oriented interval, functor $\mathcal{F}$ associates an endomorphism m_a of V. The dual endomorphism of V^* (given by the transposed matrix of that of m_a) is associated by $\mathcal{F}$ to the downward-oriented interval with an a-dot. Right: a sequence of labelled dots on an upward-oriented interval defines a word $\omega = a_1 \cdots a_n$ and the induced endomorphism $m_\omega = m_{a_1} \cdots m_{a_n}$ of V.

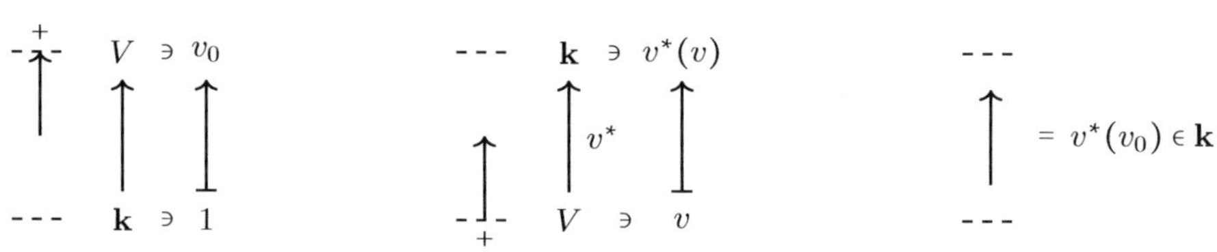

FIGURE 2.2.4. Left: two half-intervals and associated maps. Right: evaluation of a floating interval (without dot decorations).

Necessarily $\mathcal{F}(+) \cong V$ and $\mathcal{F}(-) \cong V^*$ for some finite-dimensional vector space V, with cup and cap maps given by evaluation and coevaluation.

Applying $\mathcal{F}$ to a dot labelled a on an upward-oriented strand gives a morphism $\mathcal{F}(+) \longrightarrow \mathcal{F}(+)$, that is, a linear map $m_a : V \longrightarrow V$, see Figure 2.2.3. Applying $\mathcal{F}$ to a dot labelled a on a downward-oriented strand then necessarily gives the dual linear map $m_a^* : V^* \longrightarrow V^*$ (to check the duality property, move a dot across a local maximum or minimum of a cobordism shown in Figure 2.2.1 top left). A collection of labelled dots on a strand can be encoded by a word $\omega = a_1 \cdots a_n$, with the corresponding map on V given by the product, see Figure 2.2.3.

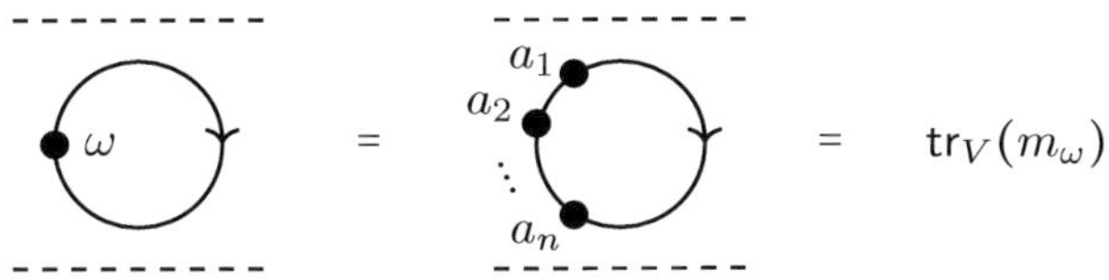

FIGURE 2.2.5. An ω-decorated circle evaluates to the trace of m_ω on V.

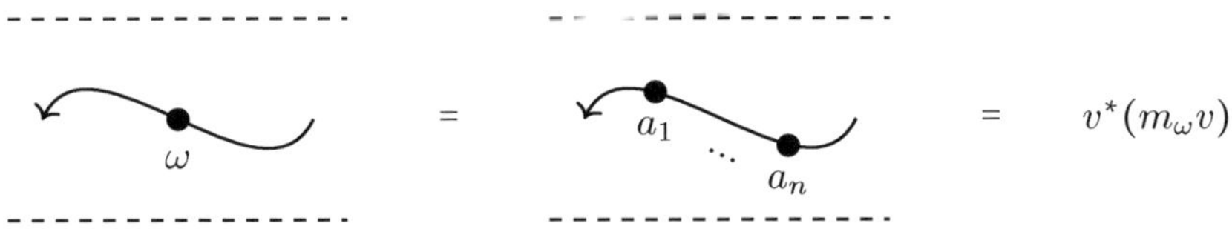

FIGURE 2.2.6. Evaluation of a floating ω-decorated interval (both endpoints are inner).

A *half-interval* (an interval with one outer and one inner endpoint) oriented upward with the outer endpoint at the top, upon applying $\mathcal{F}$, gives a linear map $\mathbf{k} \cong \mathcal{F}(\varnothing) \longrightarrow \mathcal{F}(+) \cong V$, that is, a vector $v_0 \in V$, see Figure 2.2.4. For the other half-interval with a + endpoint, also see Figure 2.2.4, applying $\mathcal{F}$ results in a morphism $V \longrightarrow \mathbf{k}$, described by a covector $v^* \in V^*$.

No relations on m_a's, v_0 and v^* are imposed. Thus, a TQFT for the category $\mathsf{Cob}_{\Sigma,1}$ is given by a finite-dimensional vector space V, a collection of linear maps $m_a : V \longrightarrow V$, for $a \in \Sigma$, a vector v_0 and a covector v^*. Two such theories are isomorphic if there is an isomorphism between the corresponding vector spaces that intertwines maps m_a, for all a, vectors v_0 and covectors v^* for the two theories.

Topological quantum field theory $\mathcal{F}$ has a large number of invariants, which are values of $\mathcal{F}$ on:

- A circle carrying a word $\omega = a_1 \cdots a_n$, up to cyclic permutation. The functor $\mathcal{F}$ evaluates this decorated circle to the trace $\mathsf{tr}_V(m_{a_n} \cdots m_{a_1}) \in \mathbf{k}$.
- A floating interval carrying a word $\omega = a_1 \cdots a_n$. The functor $\mathcal{F}$ evaluates this interval to $v^*(m_{a_1} \cdots m_{a_n} v_0) \in \mathbf{k}$.

These evaluations are depicted in Figures 2.2.5 and 2.2.6. For instance, for an interval, start with $v_0 \in V$ at the endpoint where the orientation looks into the interval, then apply the map m_{a_n} to v_0 and so on until the opposite end of the interval is reached. Then we apply the covector v^* to the product.

2.3. Automata and regular languages. By an *alphabet* or *set of letters* we mean a finite set Σ. A *language* L is any subset of the free monoid Σ^* on Σ.

A (nondeterministic) finite state automaton (Q) over alphabet Σ consists of a finite set of states Q, a *transition function*

$$\delta : \Sigma \times Q \longrightarrow \mathcal{P}(Q), \tag{2.1}$$

where $\mathcal{P}(Q)$ is the powerset of Q, a subset $Q_{\mathsf{in}} \subset Q$ of *initial* states and a subset $Q_{\mathsf{t}} \subset Q$ of *accepting* or *terminal* states.

To an automaton there is associated a graph $\Gamma_{(Q)}$ with Q as the set of vertices, an oriented edge $q \longrightarrow q'$ labelled $a \in \Sigma$ if and only if $q' \in \delta(a,q) \subset Q$ and two subsets

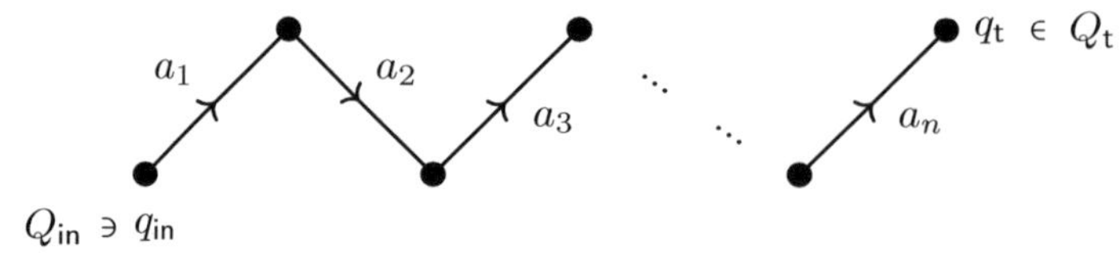

FIGURE 2.3.1. Oriented path $a_1a_2\cdots a_n$ in the graph of an automaton, starting at an initial vertex and terminating at an accepting vertex.

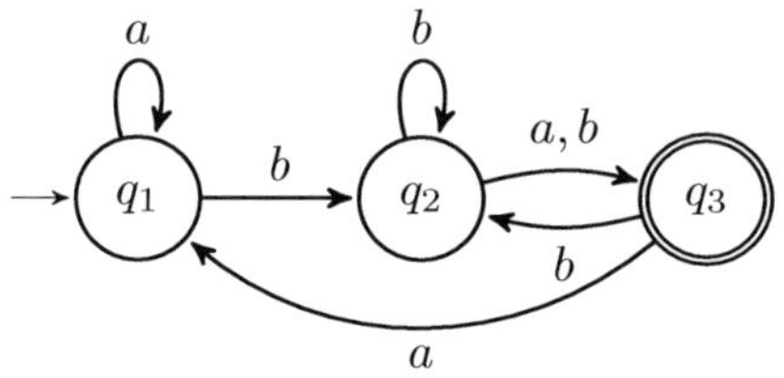

FIGURE 2.3.2. A nondeterministic automaton on 3 states q_1, q_2, and q_3 that accepts the language $L = (a+b)^*b(a+b)$ in Example 2.1. It has a single initial state q_1, indicated by a short arrow into the state, and a single accepting state q_3, indicated by the double border.

$Q_{\mathsf{in}}, Q_{\mathsf{t}}$ of distinguished vertices. Vice versa, such a data of a decorated oriented graph determines an automaton.

A word $\omega = a_1\cdots a_n \in \Sigma^*$ is *accepted* by the automaton (Q) if there exists an initial state $q_{\mathsf{in}} \in Q_{\mathsf{in}}$, an accepting state $q_{\mathsf{t}} \in Q_{\mathsf{t}}$ and an oriented path from q_{in} to q_{t} where consecutive labels of oriented edges are $a_1, \dots, a_n$, see Figure 2.3.1.

The set of words $\omega \in \Sigma^*$ accepted by (Q) is called *the language of automaton* (Q). We denote this language by $L_{(Q)}$. A language $L \subset \Sigma^*$ is called *regular* if it is the language of some automaton. One can check that a language is regular if and only if it can be described by a regular expression.

EXAMPLE 2.1. Consider the language L for the alphabet $\Sigma = \{a, b\}$ given by the regular expression $L = (a+b)^*b(a+b)$. This expression describes all words which have b as the second from the last letter. An example of a nondeterministic automaton for L is shown in Figure 2.3.2.

2.4. Boolean TQFT from an automaton. In the definition of a TQFT, we can replace field $\mathbf{k}$ by any commutative semiring. Commutativity is needed for the following reason: floating components of a cobordism are evaluated to elements of a ground ring or semiring. These components can change relative position, that is, float up or down past each other, which correspond to requiring that the ground (semi)ring is *commutative*.

By analogy with the category $\mathsf{Cob}_{\Sigma,1}$ and a TQFT $\mathcal{F}$ on it we can imagine encoding words ω in some language L by placing their labels next to dots along a one-manifold. A floating interval that carries a word ω can evaluate to two values, depending on whether or not ω is in the language L. We make these values $0, 1$, with ω evaluating to 1 if it belongs to a language L and to 0 otherwise. It is then natural to replace a field $\mathbf{k}$ by the Boolean semiring $\mathbb{B} = \{0,1 : 1+1=1\}$ which consists of these two elements, with the additional rule that $1+1=1$. In particular, $\mathbb{B}$ is not a ring, but a commutative semiring.

We specialize to the commutative semiring $\mathbb{B}$ and fix an automaton (Q) as above. Replace a $\mathbf{k}$-vector space $V \cong \mathbf{k}^n$ by a free module $\mathbb{B}Q$ over $\mathbb{B}$ with a set of generators Q. Elements of $\mathbb{B}Q$ are formal finite sums of distinct elements of Q or, equivalently, subsets of Q, so we can identify $\mathbb{B}Q \cong \mathcal{P}(Q)$ with the set of subsets of Q. The zero element of $\mathbb{B}Q$ corresponds to the empty subset of Q. Note that $q+q=q$ for $q \subset Q$ since $1+1=1$. We view $\mathbb{B}Q$ as a free $\mathbb{B}$-module with a basis Q. It has $2^{|Q|}$ elements. Unlike the case of a field, where the group $\mathsf{GL}(V)$ acts freely and transitively on the set of bases of V, the only basis of $\mathbb{B}Q$, up to changing the order of elements, is Q.

To the automaton (Q) we associate a Boolean-valued TQFT

$$\mathcal{F} : \mathsf{Cob}_{\Sigma,1} \longrightarrow \mathbb{B}\text{–}\mathsf{fmod} \tag{2.2}$$

taking values in the category of free $\mathbb{B}$-modules by assigning $\mathbb{B}Q$ to the + point and its dual $\mathbb{B}Q^*$ to the − point:

$$\mathcal{F}(+) = \mathbb{B}Q, \qquad \mathcal{F}(-) = \mathbb{B}Q^*,$$

$$\mathsf{ev} : \mathbb{B}Q^* \otimes \mathbb{B}Q \longrightarrow \mathbb{B}, \qquad \mathsf{ev}(q^* \otimes q') = \delta_{q,q'}$$

$$\mathsf{coev} : \mathbb{B} \longrightarrow \mathbb{B}Q \otimes \mathbb{B}Q^*, \qquad \mathsf{coev}(1) = \sum_{q\in Q} q \otimes q^*.$$

Here Q^* is a copy of the set Q with elements labelled q^*, over $q \in Q$, with the evaluation and coevaluation maps ev and coev given by the above formulas.

For each $a \in \Sigma$ the transition map $\delta_a : Q \longrightarrow \mathcal{P}(Q)$ extends to a $\mathbb{B}$-linear endomorphism δ_a of the free module $\mathbb{B}Q$ that takes q to $\delta(a,q)$, using the identification $\mathcal{P}(Q) \cong \mathbb{B}Q$, see above. Note that δ_a denotes the transition map while $\delta_{q,q'}$ stands for the delta function on a pair of states.

For example, in the basis $\{q_1,q_2,q_3\}$ of $\mathbb{B}Q$ in Example 2.1 the two maps for $a,b \in \Sigma$ are given by $\mathbb{B}$-valued matrices

$$\delta_a = \begin{pmatrix} 1 & 0 & 1 \\ 0 & 0 & 0 \\ 0 & 1 & 0 \end{pmatrix}, \qquad \delta_b = \begin{pmatrix} 0 & 0 & 0 \\ 1 & 1 & 1 \\ 0 & 1 & 0 \end{pmatrix},$$

see Figure 2.3.2.

To an upward-oriented arc cobordism from + to + with an a-dot, $a \in \Sigma$, we associate the endomorphism δ_a of $\mathbb{B}Q$, see Figure 2.4.1.

To a downward-oriented arc, which is a cobordism from − to −, we associate the dual map $\delta_a^* : \mathbb{B}Q^* \longrightarrow \mathbb{B}Q^*$ given by the transposed matrix to that of δ_a, see Figure 2.4.1.

To upward-oriented half-intervals (cobordisms from the empty 0-manifold $\varnothing_0$ to + and from + to $\varnothing_0$) associate maps between free $\mathbb{B}$-modules $\mathbb{B}$ and $\mathbb{B}Q$ given by

$$\mathbb{B} \longrightarrow \mathbb{B}Q, \quad 1 \longmapsto \sum_{q\in Q_{\mathsf{in}}} q,$$

$$\mathbb{B}Q \longrightarrow \mathbb{B}, \quad q \longmapsto \begin{cases} 1 & \text{if } q \in Q_{\mathsf{t}}, \\ 0 & \text{if } q \notin Q_{\mathsf{t}}, \end{cases}$$

see Figure 2.4.2, and likewise for downward-oriented half-intervals. These maps are determined by subsets Q_{in} of initial and Q_{t} of accepting states. Alternatively the

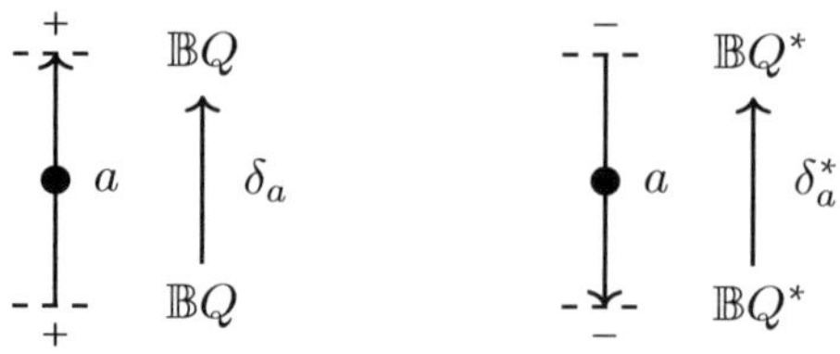

FIGURE 2.4.1. To a labelled dot on an upward-oriented interval functor $\mathcal{F}$ associates endomorphism δ_a of $\mathbb{B}Q$. To a dot on a downward-oriented interval functor $\mathcal{F}$ associates the dual operator δ_a^* on $\mathbb{B}Q^*$.

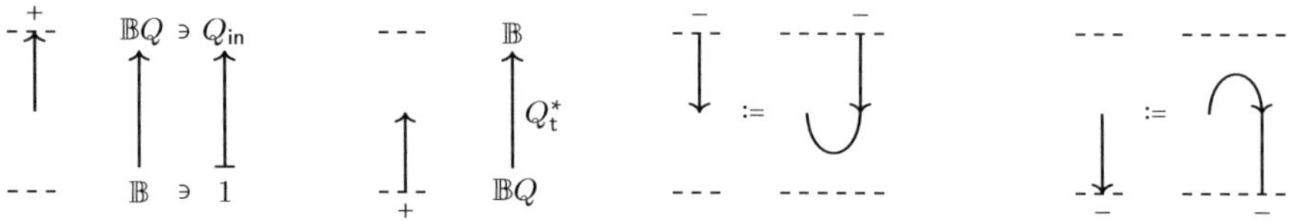

FIGURE 2.4.2. Left: maps assigned to the half-intervals with a + boundary points. Right: defining maps for half-intervals with a – boundary point.

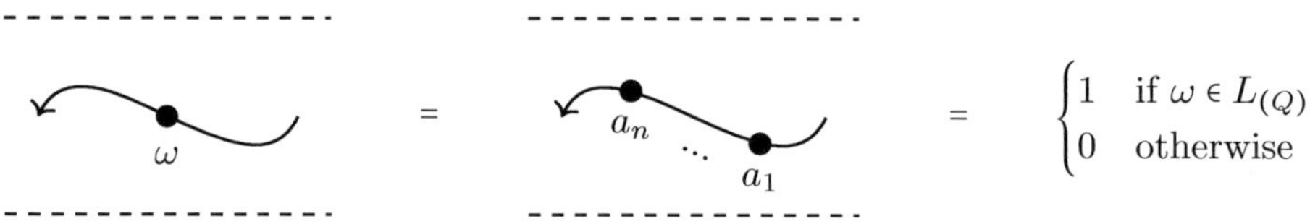

FIGURE 2.4.3. Evaluation of a floating ω-decorated interval (both endpoints are inner). Note that word ω is written in the opposite direction from that in the linear case (over a field) in Figure 2.2.6. This is due to the opposite conventions, where in linear algebra the actions are usually on the left, while in automata theory the actions are on the right.

image of 1 under the first map can be written as Q_{in}, using identification $\mathbb{B}Q \cong \mathcal{P}(Q)$, and the second map can be denoted Q_{t}^*.

Given a floating interval with the word $\omega = a_1 \cdots a_n$ written on it so that a_1 is at the tail and a_n at the head, the interval evaluates to $1 \in \mathbb{B}$ if and only if ω is in the language $L_{(Q)}$ defined by the automaton (Q), see Figure 2.4.3. If ω is not in $L_{(Q)}$ the interval evaluates to 0. The evaluation can also be written as $Q_{\mathsf{t}}^*(\delta_{a_n} \cdots \delta_{a_1} Q_{\mathsf{in}})$, that is, evaluating the product $\delta_{a_n} \cdots \delta_{a_1} Q_{\mathsf{in}} \in \mathbb{B}Q$ on the functional Q_{t}^*.

A circle with a circular word $\omega = a_1 \cdots a_n$ placed on it evaluates to the trace of operator $\delta_\omega := \delta_{a_n} \cdots \delta_{a_1}$ on $\mathbb{B}Q$, see Figure 2.4.4. Equivalently, it evaluates to 1 if and only if there is a state $q \in Q$ and a sequence of arrows labelled $a_1, a_2, \ldots, a_n$ terminating back in q (*i.e.*, and oriented cycle with word ω in the graph of the automaton).

We state a main result in [**GIK**$^+$**23**]:

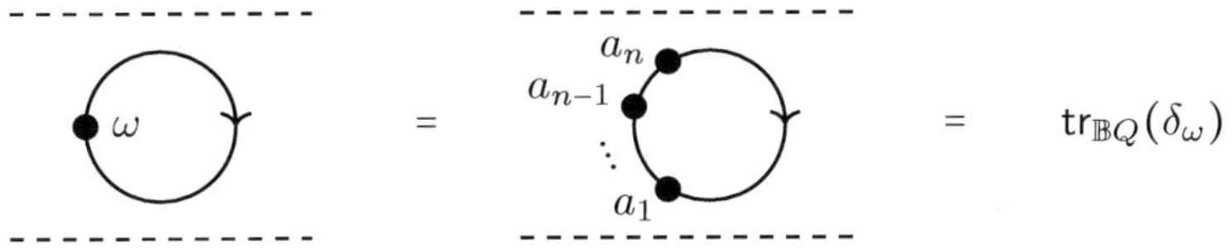

FIGURE 2.4.4. Evaluation of an ω-decorated circle is 1 if and only if there is a closed path ω in the graph of (Q).

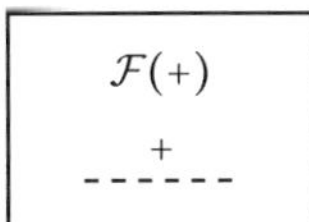	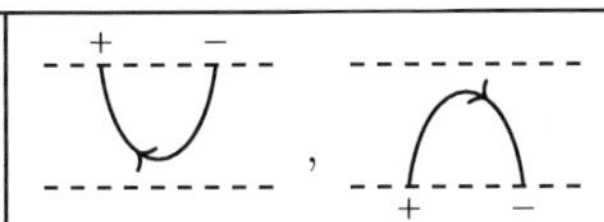	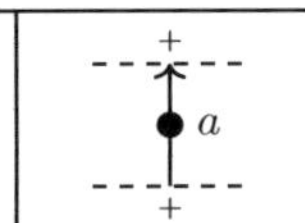	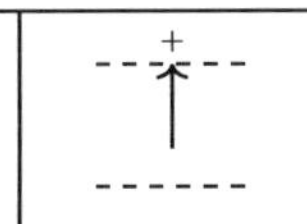	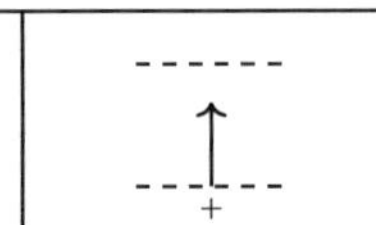
free $\mathbb{B}$-module $\mathbb{B}Q$	coevaluation, evaluation maps $\mathbb{B} \rightleftarrows \mathbb{B}Q \otimes \mathbb{B}Q^*$	transition map δ_a	set of initial states Q_{in}	set of accepting states Q_{t}

FIGURE 2.4.5. Summary table for the automata $\longleftrightarrow$ Boolean TQFT correspondence.

THEOREM 2.2 (Gustafson–Im–Kaldawy–Khovanov–Lihn). *A nondeterministic finite state automaton (Q) on alphabet Σ defines a one-dimensional Boolean-valued TQFT*

$$\mathcal{F} : \mathsf{Cob}_{\Sigma,1} \longrightarrow \mathbb{B}\text{–}\mathsf{fmod} \tag{2.3}$$

with $\mathcal{F}(+) = \mathbb{B}Q$, transition function of (Q) encoding TQFT maps for Σ-labelled defects on strands, and sets of initial and accepting states encoding the maps for undecorated half-intervals. This correspondence gives a bijection between isomorphism classes of Boolean-valued TQFTs for $\mathsf{Cob}_{\Sigma,1}$ with $\mathcal{F}(+)$ a free $\mathbb{B}$-module and isomorphism classes of nondeterministic finite state automata on alphabet Σ.

$\mathbb{B}\text{–}\mathsf{fmod}$ in (2.3) denotes the category of free $\mathbb{B}$-semimodules and semimodule maps.

Regular language $L_{(Q)}$ of the automaton (Q) describes evaluation of floating intervals decorated by words in Σ^* (with ω-decorated interval evaluating to 1 if and only if $\omega \in L_{(Q)}$), while evaluation of circles with defects in the TQFT $\mathcal{F}$ is determined by oriented cycles in the graph of (Q).

The table in Figure 2.4.5 summarizes the correspondence between generating morphisms in $\mathsf{Cob}_{\Sigma,1}$ and structural parts of an automaton (the set of states, transition maps, and sets of initial and accepting states).

Consider a one-dimensional TQFT

$$\mathcal{F} : \mathsf{Cob}_{\Sigma,1} \longrightarrow \mathbb{B}\text{–}\mathsf{mod} \tag{2.4}$$

valued, more generally, in the category of all $\mathbb{B}$-modules rather than free ones. Then necessarily $P = \mathcal{F}(+)$ is a finitely-generated semimodule which is *projective* in the sense of being a retract of a free module [**GIK**[+]**23, IK22**]. Namely, there are semimodule maps

$$P \xrightarrow{\iota} \mathbb{B}^n \xrightarrow{p} P, \qquad p \circ \iota = \mathsf{id}_P \tag{2.5}$$

for some n. Note that P is a direct summand of $\mathbb{B}^n$ only if P is free, otherwise it is just a retract. Then $P^* \cong \mathcal{F}(-)$ is finitely-generated projective as well, with the retract maps p^*, ι^* given by dualizing those for P.

Finitely-generated projective $\mathbb{B}$-modules P are described by finite distributive lattices, *i.e.*, see **[IK22]**: it is a theorem that goes back at least to Birkhoff that any such P is isomorphic to the $\mathbb{B}$-semimodule $\mathcal{U}(X)$ of open sets in a finite topological space X, with the empty set $\varnothing$ the 0 element of the semimodule and addition $U_1 + U_2 := U_1 \cup U_2$ given by the union of sets.

The structure of a TQFT $\mathcal{F}$ in this case is given by

- a collection of endomorphisms $\delta_a : \mathcal{U}(X) \longrightarrow \mathcal{U}(X)$ for $a \in \Sigma$, taking open sets to open sets, $\varnothing$ to $\varnothing$, and preserving unions of sets,
- initial element $Q_{\mathsf{in}} \in \mathcal{U}(X)$ and a terminal map $Q_{\mathsf{t}} : \mathcal{U}(X) \longrightarrow \mathbb{B}$ taking $\varnothing$ to 0 and intertwining union of sets with addition in $\mathbb{B}$.

Such structures are called *quasi-automata* in **[GIK⁺23]**. It is an interesting question whether they can be of use in computer science.

2.5. Extending to arbitrary commutative semirings. It is straightforward to extend the above TQFT construction from vector spaces over a field $\mathbf{k}$ and $\mathbb{B}$-semimodules to semimodules over a commutative semiring R. A commutative semiring R is an abelian group under addition, a commutative monoid under multiplication, and distributivity property holds, $a(b+c) = ab + ac$. Semiring R has the zero element 0 and the unit element 1. Subtraction operation $a - b$ is usually not available in semirings. It is straightforward to define the notion of a module M over R (alternatively called a *semimodule*) and introduce the category $R\text{–}\mathsf{dmod}$ of R-modules.

In the definition of a TQFT over R the tensor product of vector spaces is replaced by the tensor product of R-modules, and vector space $V = \mathcal{F}(+)$ is replaced by an R-module $P = \mathcal{F}(+)$. To have cup and cap morphisms subject to the isotopy relations above requires P to be a projective R-module of finite rank, see **[GIK⁺23]**, for instance.

This observation quickly leads to the following in **[GIK⁺23]**:

THEOREM 2.3 (Gustafson–Im–Kaldawy–Khovanov–Lihn). *One-dimensional TQFTs*

$$\mathcal{F} : \mathsf{Cob}_1 \longrightarrow R\text{–}\mathsf{mod}$$

over a commutative semiring R correspond to finitely-generated projective R-modules. One-dimensional TQFTs with defects, that is, tensor functors

$$\mathcal{F} : \mathsf{Cob}_{\Sigma,1} \longrightarrow R\text{–}\mathsf{mod} \tag{2.6}$$

are in a correspondence with finitely-generated projective R-modules P equipped with endomorphisms $m_a : P \longrightarrow P$ for $a \in \Sigma$, an element $v_0 \in P$ and a covector $v^* \in \mathsf{Hom}_R(P, R)$. This correspondence is a bijection between isomorphism classes of TQFTs and isomorphism classes of data $(P, \{m_a\}_{a\in\Sigma}, v_0, v^*)$.

Here, an R-semimodule P is defined to be finitely-generated projective if it is a retract of free semimodule R^n for some n:

$$P \xrightarrow{\iota} R^n \xrightarrow{p} P, \qquad p \circ \iota = \mathsf{id}_P . \tag{2.7}$$

It may be interesting to look at examples of such TQFTs when R is, for instance, the tropical semiring, see [**JK14, IJK16**] where projective modules over the tropical semiring are studied.

Another question is whether the notion of a *finite state machine*, which extends the notion of a finite state automaton, has the TQFT counterpart.

The authors are not aware of any studies or results on Boolean TQFTs (and, generally, TQFTs over commutative semirings that are not rings) in dimension two and higher. The above correspondence between finite state automata and one-dimensional Boolean TQFTs with defects, observed in [**GIK$^+$23**], see also [**IK22**], remains a curiosity, for now.

Two-dimensional TQFTs for oriented surfaces, without defects and over a field $\mathbf{k}$, are classified by commutative Frobenius $\mathbf{k}$-algebras, and one open problem is to find a supply of commutative Frobenius $\mathbf{k}$-semialgebras when $\mathbf{k}$ is the Boolean semiring $\mathbb{B}$ or the tropical semiring.

3. Reshetikhin–Turaev $\mathsf{SL}(N)$-invariants and their categorification

Quantum link polynomials were discovered by Vaughan Jones [**Jon85, Jon97**] (the Jones polynomial), Louis Kauffman [**Kau90**] (the Kauffman polynomial and bracket), J. Hoste, A. Ocneanu, K. Millet, P. Freyd, W.B.R. Lickorish, D. Yetter [**FYH$^+$85**], J. Przytycki and P. Traczyk [**PT88**] (the HOMFLYPT polynomial) and others (Alexander polynomial, discovered several decades prior, was an outlier). N. Reshetikhin and V. Turaev [**RT90**] put these polynomials into the framework of quantum groups and their representations (V. Drinfeld [**Dri86, Dri87**], M. Jimbo [**Jim85**]), see also earlier preprints of N. Reshetikhin [**Res87a, Res87b**] and many other references.

Furthermore, while at generic values of the parameter q of the quantum group $U_q(\mathfrak{g})$ of a simple Lie algebra $\mathfrak{g}$, its representation theory produces link invariants, root of unity values give rise to Witten–Reshetikhin–Turaev [**Wit89, RT91, Res91**] and related invariants of 3-manifolds, see [**TV92, Kup96a, KL01**], which can also be thought of as three-dimensional TQFTs.

Quantum group $U_q(\mathfrak{g})$ is a Hopf algebra deformation of the universal enveloping algebra $U(\mathfrak{g})$ of a simple Lie algebra $\mathfrak{g}$. For generic q, representation theory of $U_q(\mathfrak{g})$ gives rise to Reshetikhin–Turaev invariants $P_{\mathfrak{g}}(L) \in q^{\ell}\mathbb{Z}[q,q^{-1}]$ of knots and links L in $\mathbb{R}^3$, where ℓ is a rational number that depends on $\mathfrak{g}$ and the linking number of L, see [**Le00**]. A key property of $U_q(\mathfrak{g})$ is that it is *quasitriangular*.

To define the Reshetikhin–Turaev invariant of a link, its connected components need to be labelled by irreducible representations of $U_q(\mathfrak{g})$. The latter are parametrized by positive integral weights $\lambda \in \Lambda^+$ of $\mathfrak{g}$. Quantum invariants of 3-manifolds (Witten–Reshetikhin–Turaev invariants) are given by an appropriate sum of these invariants when q is a root of unity [**RT91, KM91**]. A 3-manifold is given by surgery on a link.

3.1. Tangles and Reshetikhin–Turaev invariants. Reshetikhin–Turaev invariants [**RT90**] are defined for tangles, which are links with boundary (more carefully, the invariant is usually defined for framed links and tangles). Tangles constitute a braided monoidal category Tan. Composition of tangles is given by concatenation, while the tensor product on morphisms is given by placing tangles in parallel. Objects in the category of tangles are finite sequences of signs, which are orientations of a tangle at its endpoints.

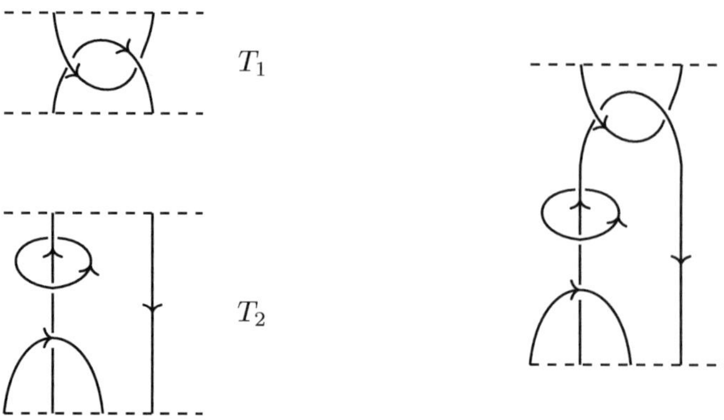

FIGURE 3.1.1. Two tangles and their composition.

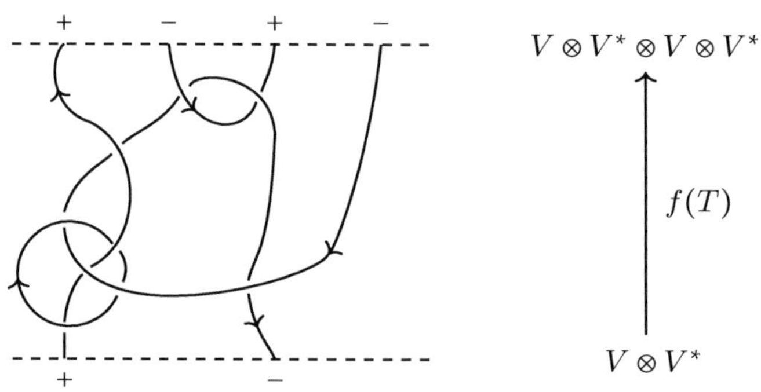

FIGURE 3.1.2. This tangle T is a morphism from $(+-)$ to $(+-+-)$. Tensor products of representations for the interwiner $f(T)$ are shown on the right.

The Reshetikhin–Turaev invariant $f(T)$ of a tangle T is an intertwiner (homomorphism of representations) between tensor products of representations, read off from the endpoints of a tangle. If one picks a representation V of the quantum group $U_q(\mathfrak{g})$, the invariant is an intertwiner between tensor products of V and its dual V^*, according to the orientations of the endpoints, see an example in Figure 3.1.2.

To define the Reshetikhin–Turaev invariant in full generality, one first modifies the category Tan by labeling the components of a tangle by positive integral weights $\lambda \in \Lambda^+$. The Reshetikhin–Turaev invariant for a labelled tangle is built from intertwiners between tensor products of irreducible representations V_λ of $U_q(\mathfrak{g})$.

A link L is a tangle with the empty boundary. Reshetikhin–Turaev invariant $f(L) : \mathbb{C} \longrightarrow \mathbb{C}$ is then a scalar, depending on q, see Figure 3.1.3, and $f(L) \in \mathbb{Z}[q^{1/D}, q^{-1/D}]$, where D divides the determinant of the Cartan matrix of $\mathfrak{g}$, see [**Le00**]. This integrality is a special property of Reshetikhin–Turaev invariants.

3.2. Crane–Frenkel conjecture. Around 1994, Igor B. Frenkel (the graduate advisor of the second author) and Louis Crane proposed in [**CF94**]:

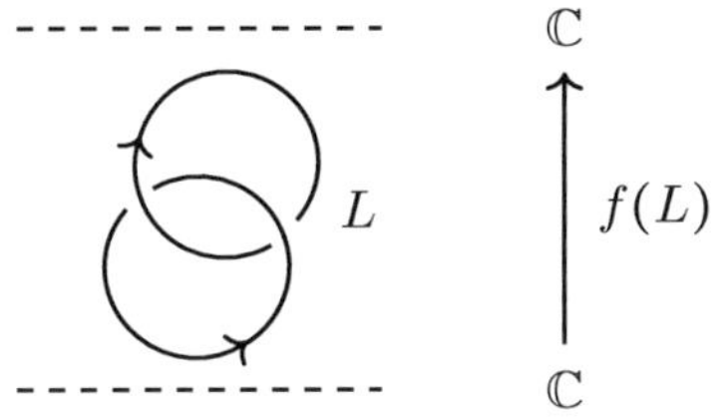

FIGURE 3.1.3. The invariant of a link L is a scalar $f(L) \in \mathbb{Z}[q^{1/D}, q^{-1/D}]$.

CONJECTURE 3.1 (Crane–Frenkel). *There exists a categorification of the quantum group $U_q(\mathfrak{sl}_2)$ at roots of unity giving rise to a 4D TQFT.*

One motivation for the conjecture was that Floer homology was already known at the time. It has the Casson invariant as its Euler characteristic. Floer homology can be viewed as a 4-dimensional TQFT, defined for at least some 3-manifolds and 4-cobordisms. It was natural to wonder whether other quantum invariants of 3-manifolds can be realized as Euler characteristics.

Another motivation came from geometric representation theory, with the discovery of the Kazhdan–Lusztig basis in the Hecke algebra in [**KL79**] and its geometric interpretation via sheaves on flag varieties. This was followed by the Beilinson–Lusztig–MacPherson's [**BLM90**] geometric interpretation of $V^{\otimes k}$, for V a fundamental representation of $U_q(\mathfrak{sl}_n)$, via sheaves on partial flag varieties, with generators E_i, F_i of the quantum group acting by correspondences. Lusztig's geometric realization [**Lus91**] of Kashiwara–Lusztig bases [**Kas94**] of $U_q(\mathfrak{g})$ and bases of its irreducible representations and Lusztig's discovery of his bases of idempotented quantum groups [**Lus10**].

I. Frenkel's insight, beyond the above conjecture, was that positivity and integrality structure of these bases should be used to systematically lift Hecke algebra and quantum group elements to functors acting on categories that replace representations of quantum groups. On the TQFT level this should correspond to lifting quantum invariants one dimension up, from 3D to 4D (categorification).

While Conjecture 3.1 is still open, significant progress in the past thirty years has been made on:

- Link homology theories, which are four-dimensional counterparts of Reshetikhin–Turaev quantum link invariants,
- Categorification of quantum groups at generic q (A. Lauda, A. Lauda–Khovanov, R. Rouquier and further foundational work by many researchers),
- Categorification at prime roots of unity (Y. Qi, J. Sussan, B. Elias, Khovanov).

3.3. Semisimple versus triangulated. Reshetikhin–Turaev link invariants are governed by *semisimple* categories of representations of quantum groups $U_q(\mathfrak{g})$. In one dimension up (4D), categories cannot be semisimple. They are also unlikely to be abelian.

In four dimensions, an *extended* TQFT for link cobordisms requires one to assign:

- a category $\mathcal{C}_n$ to a plane with n points (ignoring orientations, for simplicity),

- a functor $F(T) : \mathcal{C}_n \longrightarrow \mathcal{C}_m$ between these categories to a tangle T with n bottom and m top boundary points, and
- a natural transformation $F(T_1) \Rightarrow F(T_2)$ between these functors to a tangle cobordism between T_1 and T_2.

This assignment should assemble into a 2-functor

$$\mathcal{F} : \mathsf{Tan}_2 \longrightarrow \mathcal{NT}$$

from the 2-category of tangle cobordisms to the 2-category $\mathcal{NT}$ of natural transformations (between functors between appropriate categories). Usually one wants to convert topological structures into something algebraic, so functors and natural transformations must be additive and defined over a field or a commutative ring.

In particular, the braid group B_n on n strands (the mapping class group of a plane with n points) needs to act on $\mathcal{C}_n$, the category assigned to the plane with n points. An action of a group G on a semisimple category essentially just permutes its simple objects, thus reduces to a homomorphism $G \longrightarrow S_n$ into the symmetric group. Homomorphisms from braid groups B_n to symmetric groups S_n are unlikely to be part of a sophisticated structure that carries interesting information about four-dimensional topology. The same argument applies to full four-dimensional TQFTs, replacing braid groups by mapping class groups of closed surfaces. In some cases one can expect that endomorphism rings $\mathsf{End}(L_i)$ of simple objects are not the ground field $\mathbf{k}$ but a field or a division ring D over $\mathbf{k}$, in which case homomorphisms of B_n into $\mathsf{Aut}_{\mathbf{k}}(D)$ or into cross-products of its direct products with the symmetric group may be available, but the crux of this informal argument (even just a gut feeling) that such homomorphisms cannot be upgraded to a sophisticated 4D TQFT remains.

This informal argument about unsuitability of semisimple categories and the need for triangulated categories in four-dimensional TQFTs can be found in the old paper of one of us [**Kho02**, Section 6.5] which further argues that interesting four-dimensional TQFTs are unlikely to assign abelian categories to surfaces. For a much more recent and precise work we refer to Reutter [**Reu20**] and [**RSP22**].

Replacing semisimple or abelian categories by triangulated categories removes the obstacle which is the lack of interesting categorical actions. Consider a ring A, take its category of modules $A\text{–}\mathsf{mod}$ and form the category $\mathcal{H}(A\text{–}\mathsf{mod})$ of finite length complexes of A-modules up to chain homotopies:

$$\cdots \longrightarrow M^i \xrightarrow{d^i} M^{i+1} \xrightarrow{d^{i+1}} M^{i+2} \xrightarrow{d^{i+2}} \cdots \qquad \text{where } M^i \in A\text{–}\mathsf{mod}, \quad d \circ d = 0.$$

One can come up with a specific finite-dimensional ring A (interestingly, graded A-modules are essentially the same as double complexes) and a *faithful* action of the braid group on the category $\mathcal{H}(A\text{–}\mathsf{mod})$, see [**KS02**], also [**ST01**] in the context of algebraic geometry. Passing to the Grothendieck group of $\mathcal{H}(A\text{–}\mathsf{mod})$ recovers the Burau representation of the braid group (or the permutation representation, ignoring the extra grading. An additional grading on the category of modules turns the Grothendieck group into a $\mathbb{Z}[q, q^{-1}]$-module). Faithfulness holds in either case.

There are many ways to construct algebras with a faithful braid group action on their homotopy categories but the above example seems minimal, in a sense. One can, for instance, ask for a finite-dimensional algebra A_n over a field $\mathbf{k}$ with a faithful action of the braid group B_n on n strands on its homotopy category. In the example in [**KS02**] $\dim_{\mathbf{k}}(A_n) = 4n-6$, and we do not know examples of algebras of dimension less than $4n-6$ for $n \geq 3$ with a faithful action of B_n on their homotopy categories of complexes.

Graded A_n and A-modules correspond to bicomplexes, and one is working in the homotopy category of complexes over them. The latter category can be replaced by the stable category of tricomplexes, with a faithful braid group action on it [**KQ20**]. It is an open question whether this step into stable categories and tricomplexes can be extended beyond categorified Burau representation, to categorification of other braid group and quantum group representations.

3.4. The HOMFLYPT polynomial. The Reshetikhin–Turaev invariant for the quantum group $U_q(\mathfrak{sl}(N))$ of $\mathfrak{sl}(N)$ and its fundamental N-dimensional representation V is determined by the skein relation

(III) $q^N P_N($  $) - q^{-N} P_N($  $) = (q-q^{-1}) P_N($  $).$

and normalization on the unknot:

$P_N($  $) = [N] = \dfrac{q^N - q^{-N}}{q - q^{-1}} = q^{N-1} + q^{N-3} + \ldots + q^{1-N}.$

Sometimes the normalization on the unknot is taken to be 1, since otherwise the invariant of any nonempty link is divisible by $[N]$. The disjoint union with the unknot multiplies either invariant by $[N]$ and the above normalization is natural from the categorification viewpoint.

The skein relation above is due to the space of intertwiners (homomorphisms of quantum group representations) $V^{\otimes 2} \longrightarrow V^{\otimes 2}$ being two-dimensional, so that any three maps are related by a linear equation.

The HOMFLYPT polynomial [**FYH⁺85,PT88**] is a 2-variable invariant $P(L) \in \mathbb{Z}[a^{\pm 1}, b^{\pm 1}]$ given by replacing the coefficients in the above skein relation by $a = q^N, b = q - q^{-1}$. One-variable specializations $P_N(L)$ of $P(L)$ have representation-theoretic interpretation via the quantum group $U_q(\mathfrak{sl}(N))$, as briefly explained earlier. Replacing N by $-N$ and q by q^{-1} preserves the invariant, so one can restrict to $N \geq 0$. We record special cases:

- $N = 0$: $P_0(L)$ is the Alexander polynomial.
- $N = 1$: $P_1(L)$ is a trivial invariant.
- $N = 2$: $P_2(L)$ is the Jones polynomial.
- $N = 3$: $P_3(L)$ is the Kuperberg $\mathfrak{sl}(3)$ quantum invariant [**Kup96b**].

3.5. MOY graphs and their invariants. The above specializations $P_N(L)$ take values in the ring $\mathbb{Z}[q, q^{-1}]$ of Laurent polynomials with *integer* coefficients. It is possible to reduce links to linear combinations of planar objects (webs or graphs) on which the invariant takes values in $\mathbb{Z}_+[q, q^{-1}]$, the semiring of Laurent polynomials with *non-negative integer* coefficients. Decompose $V^{\otimes 2} \cong \Lambda_q^2(V) \oplus S_q^2(V)$ into the sum of two irreducible representations—the second quantum exterior and symmetric powers of V—and consider projection and inclusion operators

$$V^{\otimes 2} \xrightarrow{\ p\ } \Lambda_q^2(V) \xrightarrow{\ \iota\ } V^{\otimes 2}$$

scaled so that $p \circ \iota = (q+q^{-1})\,\mathsf{id}$ is the identity map of $\Lambda^2_q(V)$ times $q+q^{-1}$, see Figure 3.5.1.

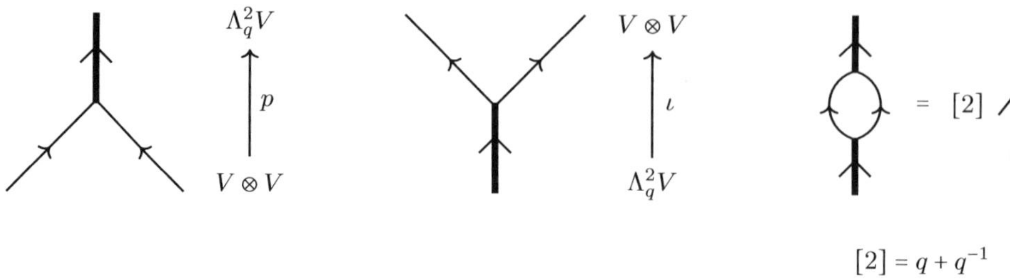

FIGURE 3.5.1. Left: diagram of projection p onto $\Lambda^2_q V$. Middle: diagram of inclusion ι back into $V \otimes V$. Right: these maps are scaled so that the composition $p \circ \iota$ is the identity times $q + q^{-1}$.

From these two basic pieces one can assemble oriented planar graphs with thin and thin edges and all vertices of valency three, as in Figure 3.5.1 left and center, with one thick and two thin edges at each vertex. Vertices correspond to the intertwiners p and ι. These are the simplest instances of Murakami–Ohtsuki–Yamada (MOY) graphs. We think of thin edges as carrying label one (for V) and thick edges as carrying label two (for $\Lambda^2_q V$). Such a graph with boundary defines an intertwiner between tensor products of V, $\Lambda^2_q V$ and their duals. A closed MOY graph Γ defines an endomorphism of the trivial representation of $U_q(\mathfrak{sl}(N))$, thus a function $P_N(\Gamma) \in \mathbb{C}(q)$.

The invariant $P_N(\Gamma)$ is both integral and positive, see [**MOY98**]:

THEOREM 3.2 (Murakami–Ohtsuki–Yamada). *For any planar MOY graph* Γ *as above, its invariant* $P_N(\Gamma) \in \mathbb{Z}_+[q,q^{-1}]$.

Integrality and positivity properties of $P_N(\Gamma)$ are key to its categorification and categorification of the corresponding link invariants $P_N(L)$.

Let us first observe that the link invariant $P_N(L)$ reduces to the invariants $P_N(\Gamma)$ of MOY planar graphs via skein relations:

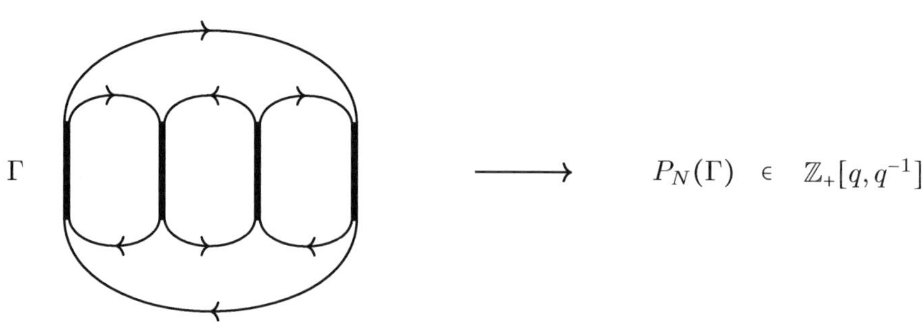

FIGURE 3.5.2. A MOY graph Γ.

(I) 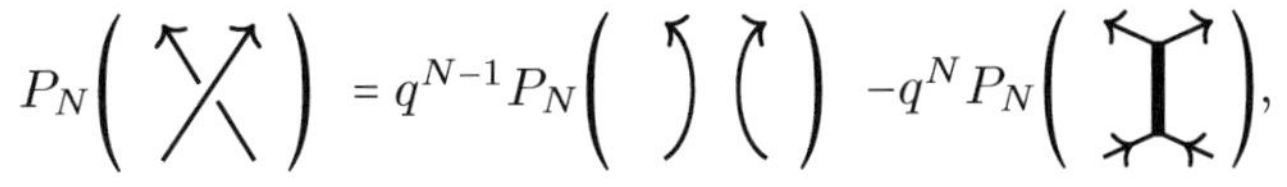

(II) 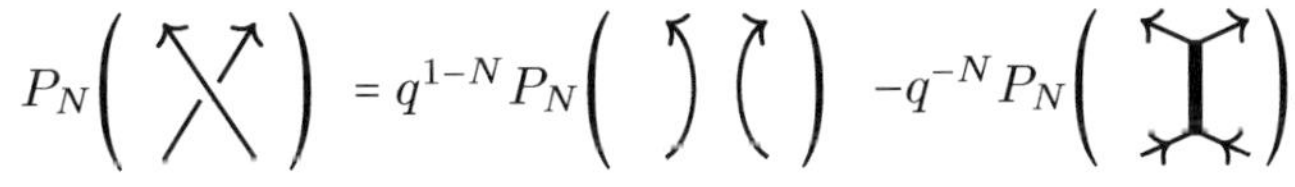

implying the HOMFLYPT relation

(III) 

More generally, one can allow lines of thickness a, $1 \leq a \leq N-1$, which correspond to fundamental representations $\Lambda_q^a V$ of the quantum group $U_q(\mathfrak{sl}(N))$ and $(a, b, a+b)$-vertices, see the top row of Figure 3.5.3, which denote the scaled projection and inclusion between the corresponding representations

$$\Lambda_q^a V \otimes \Lambda_q^b V \xrightarrow{p_{a,b}} \Lambda_q^{a+b} V, \qquad \Lambda_q^{a+b} V \xrightarrow{\iota_{a,b}} \Lambda_q^a V \otimes \Lambda_q^b V.$$

normalized so that the composition

$$p_{a,b} \circ \iota_{a,b} = \begin{bmatrix} a+b \\ b \end{bmatrix} \mathsf{id} = \frac{[a+b]!}{[a]![b]!} \cdot \mathsf{id}, \qquad [a]! := \prod_{c=1}^{a} [c], \qquad [c] = \frac{q^c - q^{-c}}{q - q^{-1}}$$

is the identity endomorphism of $\Lambda_q^{a+b} V$ scaled by the q-binomial coefficient. It is also natural to allow thickness N as well and replace $\mathfrak{sl}(N)$ by the Lie algebra $\mathfrak{gl}(N)$. The exterior power $\Lambda^N V$ is a trivial representation of $\mathfrak{sl}(N)$ but not of $\mathfrak{gl}(N)$, and likewise for their quantum deformations, and vertices $(a, N-a)$ are now allowed, with lines of thickness a and $N-a$ merging into a line of thickness N.

Murakami–Ohtsuki–Yamada planar graphs or webs Γ with edges of arbitrary thickness from 1 to N have vertices (a, b) as above ($a + b \leq N$). The quantum invariant $P_N(\Gamma)$ of an MOY graph with edges of arbitrary thickness takes values in $\mathbb{Z}_+[q, q^{-1}]$, and Theorem 3.2 holds for such graphs as well.

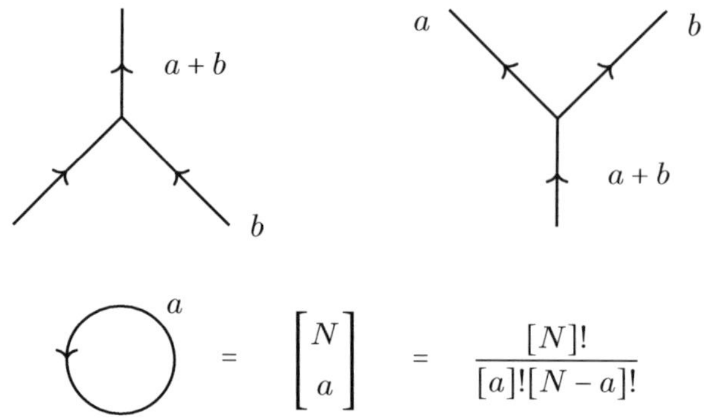

FIGURE 3.5.3. Top: (a, b)-vertices of general MOY graphs. Bottom: Circle of thickess a evaluates to a quantum binomial coefficient.

Consider a link L with components labelled by numbers in $\{1, 2, \ldots, N\}$ and pick a planar projection. The analogue of the skein relations above (I) and (II) are the relations in Figures 3.5.4, 3.5.5, see for instance [**Wu12b**]. These skein relations allow to define $P_N(L) \in \mathbb{Z}[q, q^{-1}]$ for links with components labelled by numbers from 1 to N.

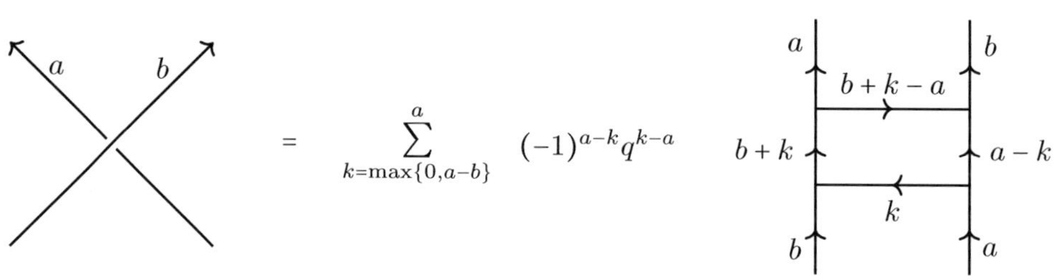

FIGURE 3.5.4. Decomposing a negative crossing for two strands of arbitrary thickness $a, b \le N$.

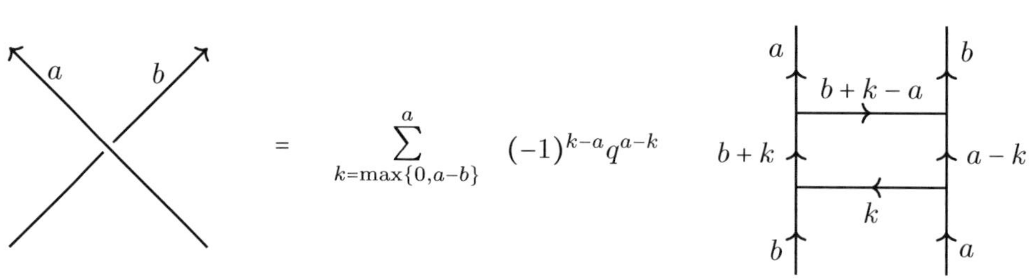

FIGURE 3.5.5. Decomposition of a positive crossing with thicknesses a, b.

3.6. Matrix factorizations and link homology.

Idea: To realize $P_N(L)$ as the Euler characteristic of a bigraded homology theory

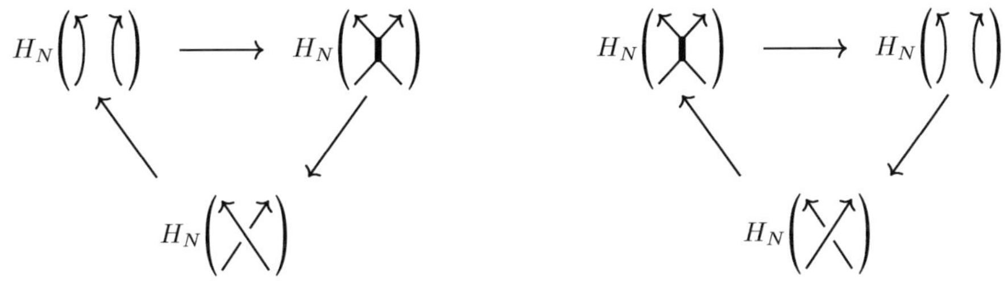

FIGURE 3.6.1. Long exact sequences to compute homology of a crossing given those of its planar resolutions.

$H_N(L)$ of links, first build homology $H_N(\Gamma)$ for planar graphs Γ as *singly-graded* vector spaces. There must be an equality of Laurent polynomials

$$P_N(\Gamma) = \mathsf{gdim}\, H_N(\Gamma),$$

where for a $\mathbb{Z}$-graded vector space $V = \oplus_i V_i$, the graded dimension $\mathsf{gdim}\, V := \sum_i \dim(V_i) q^i$. Then lift skein relations above to the long exact sequence in Figure 3.6.1.

Commutative ground ring R of the theory may be different from a field, in which case one expects $H_N(\Gamma)$ to be a free graded R-module of graded rank $P_N(\Gamma)$.

This idea was successfully realized by L. Rozansky and one of us back in 2004 in [**KR08a**]. To define homology groups (or state spaces) $H_N(\Gamma)$ of planar graphs Γ *matrix factorizations* were used. Start with a polynomial ring $S = \mathbf{k}[x_1, \ldots, x_k]$ in several variables over a field $\mathbf{k}$, where $\mathsf{char}(\mathbf{k}) = 0$. A polynomial $\omega \in S$ is called a *potential* if the ideal $(\partial\omega/\partial x_1, \ldots, \partial\omega/\partial x_k)$ has finite codimension in S. Informally, this means that ω is sufficiently generic. In this case the quotient algebra $J_\omega := S/(\partial\omega/\partial x_1, \ldots, \partial\omega/\partial x_k)$ is known to be Frobenius, via the Grothendieck residue construction. The quotient algebra is called *the Milnor algebra* of ω.

For a potential $\omega \in S$, consider 2-periodic generalized complexes M of free S-modules

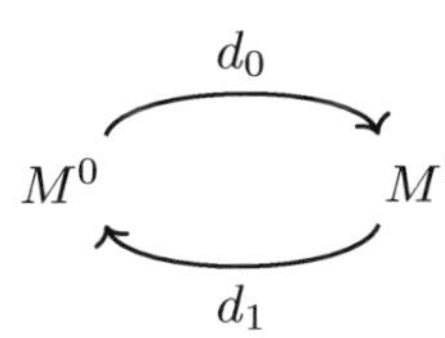

and maps between them such that $d^2(m) = \omega m$. Modulo homotopies, these constitute a triangulated category of matrix factorizations MF_ω. When matrix factorizations M, M' have finite ranks (meaning M^0, M'^0 are finite rank free S-modules) hom spaces $\mathsf{Hom}(M, M')$ in the category MF_ω are finite-dimensional $\mathbf{k}$-vector spaces, due to multiplications by $\partial\omega/\partial x_i$ being homotopic to 0. The Milnor algebra J_ω then acts on $\mathsf{Hom}(M, M')$, implying finite-dimensionality of hom spaces.

To a potential ω there is assigned a two-dimensional TQFT with corners, built out of categories of matrix factorizations for potentials that are signed sums of copies of ω over several sets of variables. This construction goes back to Kapustin–Li [**KL03**], see [**CM16**] for a more recent treatment, while the one-variable case is thoroughly explained in [**KR08a**]. Without going into full details and specializing to $\mathbf{k} = \mathbb{Q}$ and $\omega = x^{N+1}$, to an arc (viewed as a one-manifold with boundary) one assigns a factorization L for $\omega = x_1^{N+1} - x_2^{N+1}$, with $S = \mathbb{Q}[x_1, x_2]$:

$$S \xrightarrow{\pi_{12}} S \xrightarrow{x_1 - x_2} S \qquad \text{where} \qquad \pi_{12} = \frac{x_1^{N+1} - x_2^{N+1}}{x_1 - x_2} = \sum_{i=0}^{N} x_1^i x_2^{N-i}.$$

Term $x_1 - x_2$ ensures that x_1, x_2 act the same up to homotopy, when viewed as endomorphisms of factorization L in the homotopy category. One thinks of L as the identity factorization, implementing the identity functor on the category of matrix factorizations. Namely, tensoring L, say over the subring $\mathbb{Q}[x_1]$ with a factorization M, in variables x_1 and some other variables (not x_2, say x_3, x_4) for the potential $-x_1^{N+1} + \omega'(x_3, x_4)$ results in the factorization $M' := L \otimes_{\mathbb{Q}[x_1]} M$ over $\mathbb{Q}[x_2, x_3, x_4]$, and one can check that upon substitution $x_2 \mapsto x_1$ factorizations M, M' are naturally isomorphic in the homotopy category. Factorization M' is of infinite rank over $\mathbb{Q}[x_2, x_3, x_4]$, but upon removing contractible summands it becomes finite rank and isomorphic to M.

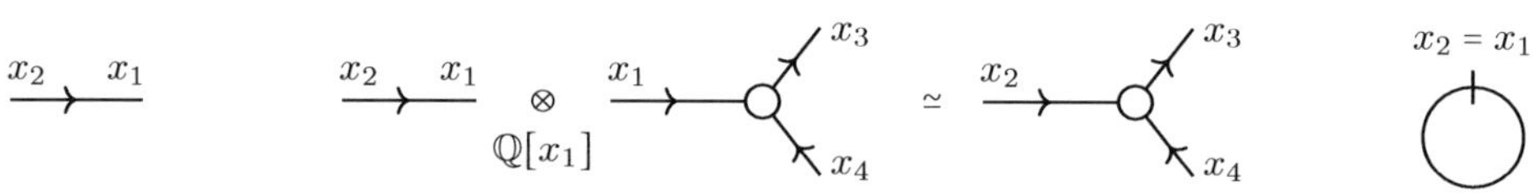

FIGURE 3.6.2. Left: an arc with variables as labels at the end point. Middle: tensoring with a factorization M, forgetting variable x_1, and relabeling x_2 back to x_1 gives a functor isomorphic to the identity functor in the homotopy category MF_ω. Right: closing up an arc into a circle and equating $x_1 = x_2$.

Closing up the arc into a circle and equating $x_1 = x_2$, see Figure 3.6.2 on the right, gives a complex

$$\mathbb{Q}[x_1] \xrightarrow{(N+1)x^N} \mathbb{Q}[x_1] \xrightarrow{\quad 0 \quad} \mathbb{Q}[x_1],$$

where the ring is reduced to $\mathbb{Q}[x_1, x_2]/(x_1 - x_2) \cong \mathbb{Q}[x_1]$. This a general feature of building topological theories from matrix factorizations: closing up a factorization with boundary points and equating the variables, as long as the potentials at the endpoints match, results in a two-periodic complex, with $d^2 = 0$, rather than just a factorization, due to the cancellation of terms in ω upon identifying the variables.

For the circle as above, the homology group of the complex is $\mathbb{Q}[x]/(x^N) \cong H^*(\mathbb{CP}^{N-1})$ (the other homology group is 0). Algebra $\mathbb{Q}[x]/(x^N)$ is commutative Frobenius and gives rise to a 2D TQFT once a nondegenerate trace on it is chosen. Building an extended TQFT from matrix factorizations is discussed in [**KR08a**], also see [**CM16**] and references there.

In the above example one can work with graded polynomial rings with $\deg(x_j) = 2$ and factorizations $\sum_j \pm x_j^{N+1}$. Then the graded degree of the vector space of the circle is

$$\mathsf{gdeg}(H_N(\mathsf{circle})) = 1 + q^2 + \ldots + q^{2(N-1)} = q^{N-1}[N] = q^{N-1} P_N(\mathsf{unknot}).$$

This equality points to a possible match between the 2D TQFT for the one-variable potential x^{N+1} and homology of links and planar graphs, for the simplest possible link and MOY graph, which is the circle in the plane (of thickness 1). Up to a power of q, the quantum invariant of the unknot equals the graded dimension of the commutative Frobenius algebra that the matrix factorization TQFT for the potential $\omega = x^{N+1}$ assigns to the circle. That algebra is also isomorphic to the cohomology ring of $\mathbb{CP}^{N-1}$.

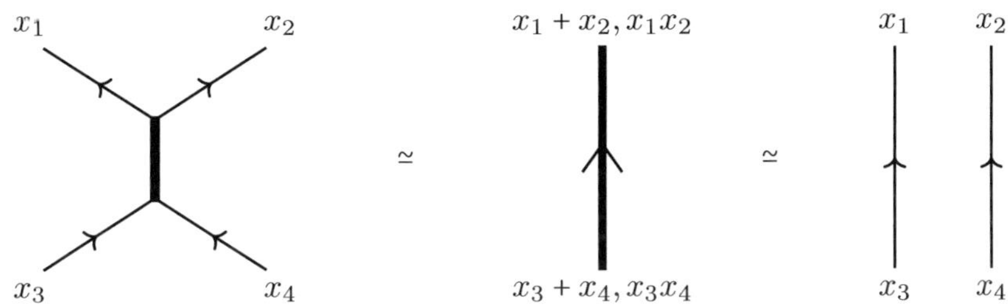

FIGURE 3.6.3. Double edge factorization can be thought of as the identity factorization for the subring of symmetric functions in two variables.

To move beyond circles, recall that, for link diagrams with components colored by V, the corresponding MOY graphs have edges of thickness one and two only, with any thickness two edge having two thin "in" edges and two thin "out" edges, see Figure 3.6.3 on the left.

Matrix factorization associated to a neighbourhood of a double edge should have the potential

$$\omega = x_1^{N+1} + x_2^{N+1} - x_3^{N+1} - x_4^{N+1}, \tag{3.1}$$

see Figure 3.6.3 on the left. The four variables x_1, x_2, x_3, x_4 are assigned to the endpoints of the diagram. The term x_i^{N+1} enters ω with the plus sign, respectively the minus sign, if the orientation at that point is out, respectively into, the diagram. Potential ω is the difference of two terms, and the first term $x_1^{N+1} + x_2^{N+1}$ can be written as a polynomial in the elementary symmetric functions: $x_1^{N+1} + x_2^{N+1} = g(x_1 + x_2, x_1x_2)$.

To write the identity factorization for the 2-variable polynomial g we decompose the difference of two g's for two different sets of variables:

$$\begin{aligned} g(y_1,y_2) - g(z_1,z_2) &= (g(y_1,y_2) - g(z_1,y_2)) + (g(z_1,y_2) - g(z_1,z_2)) \\ &= (y_1 - z_1)u_1(\underline{y},\underline{z}) + (y_2 - z_2)u_2(\underline{y},\underline{z}) \end{aligned}$$

for some polynomials u_1, u_2 in the four variables y_1, y_2, z_1, z_2. Tensor the two factorizations

$$\left(S \xrightarrow{u_1} S \xrightarrow{y_1 - z_1} S\right) \quad \text{and} \quad \left(S \xrightarrow{u_2} S \xrightarrow{y_2 - z_2} S\right)$$

to get a factorization associated to a double edge j of Γ:

$$M_j := \left(S \xrightarrow{u_1} S \xrightarrow{y_1 - z_1} S\right) \otimes_S \left(S \xrightarrow{u_2} S \xrightarrow{y_2 - z_2} S\right).$$

Notice that M_j has the potential given by (3.1).

For a general planar MOY graph Γ with edges of thickness one and two, place marks on thin edges, with at least one mark on each edge. Denote by I the set of marks and consider the ring $\mathbb{Q}[x_i]_{i\in I}$. To a thick edge there is assigned a factorization with the potential $x_{i_1}^{N+1} + x_{i_2}^{N+1} - x_{i_3}^{N+1} - x_{i_4}^{N+1}$, to a thin arc – an identity factorization as described earlier, see Figure 3.6.4 on the left. In the Figure 3.6.4 example, the factorization for the thin arc shown has potential $x_{i_5}^{N+1} - x_{i_1}^{N+1}$.

Given a graph Γ and a set of markings I of thin edges as described, tensor together factorizations M_j over all double edges j and factorizations for thin arcs (if some thin edge carries more than one mark) to get the product factorization M_Γ. Then the differential $D = \sum_j d_j$ in M_Γ satisfies $D^2 = \sum_j d_j^2 = \sum_i (x_i^{N+1} - x_i^{N+1}) = 0$.

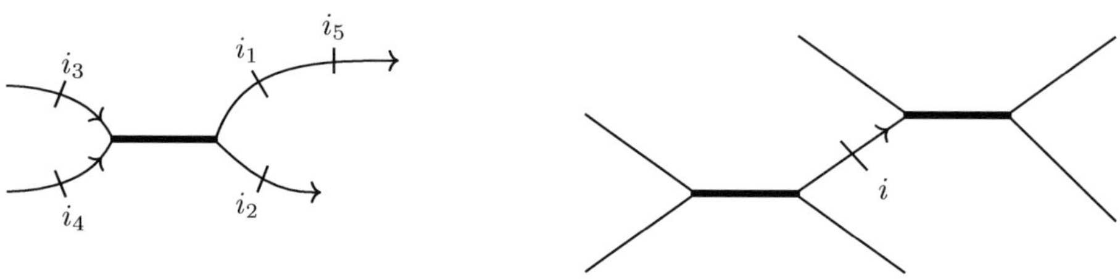

FIGURE 3.6.4. Left: markings around a thick edge and an additional mark on an adjacent thin edge. Right: potentials ω add up to 0 for a closed diagram, due to cancellation of terms. For a mark labelled i, the contribution of potentials is $x_i^{N+1} - x_i^{N+1} = 0$, with the signs opposite due to opposite orientations.

Define $H_N(\Gamma) := H_N(M_\Gamma, D)$. This is our homology (or state space) of the graph Γ. Placing extra marks on any thin edges of Γ does not change $H_N(\Gamma)$.

From [**KR08a**], we cite the following:

THEOREM 3.3 (Khovanov–Rozansky). *$H_N(\Gamma)$ lives is a single $\mathbb{Z}/2$ degree and* $\mathsf{gdim}(H_N(\Gamma)) = P_N(\Gamma)$.

To extend the homology to links, consider resolutions Γ_0, Γ_1 of a crossing in Figure 3.6.5. There are homomorphisms of factorizations

$$\chi_0 : M_{\Gamma_0} \longrightarrow M_{\Gamma_1}, \qquad \chi_1 : M_{\Gamma_1} \longrightarrow M_{\Gamma_0},$$

and to positive and negative crossings one assigns two-step complexes of factorizations with the differential given by χ_0, χ_1, respectively, see Figure 3.6.5 on the right.

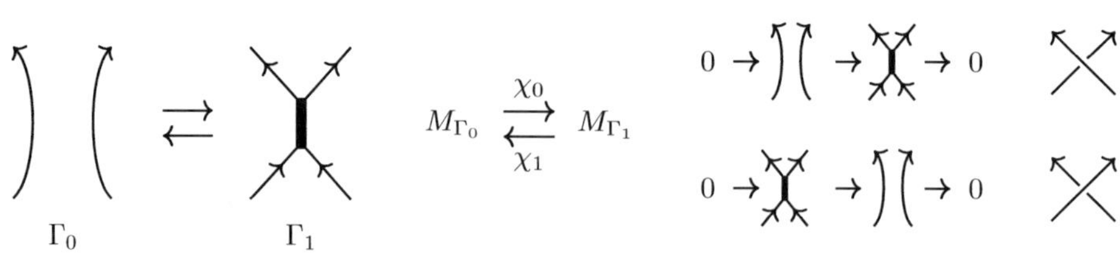

FIGURE 3.6.5. Left: MOY graphs Γ_0, Γ_1 and corresponding factorizations. Right: forming complexes for positive and negative crossings out of factorizations M_{Γ_0} and M_{Γ_1}.

Homology of a link diagram D is given by tensoring these two-step complexes over all crossings of D, computing the homology of all terms (for the inner differential in factorizations), and then taking the homology again for the differential build from maps χ_0, χ_1 over all crossings.

From [**KR08a**], we also recall:

THEOREM 3.4 (Khovanov–Rozansky). *The resulting homology does not depend on the choice of a link diagram D of an oriented link L and can be denoted $H_N(L)$.*

(1) *$H_N(L)$ is bigraded and its Euler characteristic*

$$\chi_N(L) = \sum_{i,j}(-1)^i q^j \dim H_N^{i,j}(L) = P_N(L)$$

is the Reshetikhin–Turaev link invariant for the fundamental $\mathsf{SL}(N)$ representation.

(2) *$H_N(L)$ is functorial under link cobordisms.*

Homology theory H_N of links and link cobordisms can be viewed as a categorification of quantum invariant P_N, the latter a one-variable specialization of the HOMFLYPT polynomial. The functoriality under link cobordisms, shown in [**KR08a**], is up to overall scaling by elements of $\mathbb{Q}^*$. The theory extends to tangles and tangle cobordisms as well.

Y. Yonezawa [**Yon11**] and H. Wu [**Wu14**] extended the homology from *thin* MOY graphs to arbitrary MOY graphs, with general $(a, b, a+b)$ trivalent vertices, and from a categorification of the link invariant P_N to categorification of Reshetikhin–Turaev invariants where components of a link are colored by arbitrary quantum exterior powers $\Lambda_q^a V$ of the fundamental representation $V \cong \mathbb{C}^N$, $1 \le a \le N$.

In their construction the homology of the unknot colored a is the cohomology of the Grassmannian of a-planes in $\mathbb{C}^N$:

$$H_N(\bigcirc^a) = \mathsf{H}^*(\mathsf{Gr}(a, N), \mathbb{Q}).$$

This cohomology ring is a commutative Frobenius algebra and gives rise to a 2D TQFT. Standard merge, split, birth and death cobordisms between unnested circles in the plane and relations on compositions of these cobordisms show that in a functorial link homology theory homology of the unknot is a commutative Frobenius algebra, and for the theory above that algebra is the cohomology ring of the complex Grassmannian.

This theory admits a generalization where the potential x^{N+1} is replaced by the potential $\omega = x^{N+1} + a_1 X^N + \ldots + a_N x$, where a_i is a formal variable of degree $2i$. The potential is then homogeneous and defined over the ground ring $R_N = \mathbb{Q}[a_1, \ldots, a_N]$. The 2D TQFT for matrix factorizations and link homology construction can be generalized to the ground ring R_N. In the resulting homology theory $\widetilde{H}_N$ the ring R_N, which is the homology of the empty link, can be interpreted as the $U(N)$-equivariant cohomology of a point p:

$$\widetilde{H}_N(\varnothing) = R_N \cong \mathsf{H}^*_{U(N)}(p, \mathbb{Q}).$$

Homology of an a-labelled unknot is then the $U(N)$-equivariant cohomology of the complex Grassmannian:

$$\widetilde{H}_N\Big($$  $$\Big) = \mathsf{H}^*_{U(N)}(\mathsf{Gr}(a,N),\mathbb{Q}).$$

We refer to papers [**Gor04, Kra10, Wu12a, Wu15**] for the construction of equivariant link homology, see also [**LL16, Ras15, MY19**] for some applications. Deforming the potential proved to be a prolific idea, also leading to a categorification of the bigraded HOMFLYPT polynomial [**KR08b**] rather than of its singly-graded specializations P_N. This triply-graded link homology theory is closely related to the category of Soergel bimodules [**Kho07**], which is a fundamental object at the center of modern geometric representation theory [**EMTW20**]. We refer to [**GSV05, Wit18, Aga22**] and many other papers for physical interpretations of $\mathsf{SL}(N)$ link homology.

Bigraded and triply-graded categorifications of the HOMFLYPT polynomial and its specializations, including for torus and algebraic links, relate to deep structures in representation theory and geometry [**GORS14, GHM21**].

A refinement of bigraded and triply-graded homology groups, now known as *y-ified homology*, was discovered by J. Batson and C. Seed [**BS15**] in the $N = 2$ case, S. Cautis and J. Kamnitzer [**CK17**] in the geometric language for any N and E. Gorsky and M. Hogancamp for the HOMFLYPT homology [**GH22**], also see [**CLS20**] for a generalization and an approach via the KLR categories and a related paper [**HRW21**]. This homology exhibits additional symmetries and relates link homology to Hilbert schemes. Sophisticated relations between flavors of link homology and equivariant matrix factorizations have been investigated by A. Oblomkov and L. Rozansky in [**OR10**] and follow-up papers, see [**Obl19**] for a review.

3.7. Model scenario for categorification of Reshetikhin–Turaev link invariants. Reshetikhin–Turaev link invariant functor [**RT90**] assigns the tensor product $V_{\underline{\lambda}} = V_{\lambda_1} \otimes \cdots \otimes V_{\lambda_n}$ of irreducible $U_q(\mathfrak{g})$ representations to a plane with n points marked by positive integral weights $\lambda_1, \dots, \lambda_n$, where $\underline{\lambda} := (\lambda_1, \dots, \lambda_n)$, and taking orientations into account. The invariant of a decorated tangle T with $\underline{\lambda}$ and $\underline{\mu}$ endpoint label sequences at the bottom and top planes is an intertwiner

$$f(T) : V_{\underline{\lambda}} \longrightarrow V_{\underline{\mu}}. \tag{3.2}$$

Let us explain the best possible scenario for categorification of Reshetikhin–Turaev link invariants. Upon a categorification, the tensor product $V_{\underline{\lambda}}$ of representations should be replaced by a triangulated category $\mathcal{C}_{\underline{\lambda}}$. The Grothendieck group $K_0(\mathcal{C}_{\underline{\lambda}})$ of that category should be related to $V_{\underline{\lambda}}$, for instance, be a $\mathbb{Z}[q,q^{-1}]$-lattice in $V_{\underline{\lambda}}$, so that

$$K_0(\mathcal{C}_{\underline{\lambda}}) \otimes_{\mathbb{Z}[q,q^{-1}]} \mathbb{C}(q) \cong V_{\underline{\lambda}} \tag{3.3}$$

(assuming $U_q(\mathfrak{g})$ is defined over the field of rational functions $\mathbb{Q}(q)$).

To a collection of n points on the plane labelled by $\underline{\lambda}$ assign the category $\mathcal{C}_{\underline{\lambda}}$ as above. To a tangle T there should be assigned an exact functor $\mathcal{F}(T) : \mathcal{C}_{\underline{\lambda}} \longrightarrow \mathcal{C}_{\underline{\mu}}$, see Figure 3.7.1 on the left. On the Grothendieck group level the functor $\mathcal{F}(T)$ should give the Reshetikhin–Turaev tangle invariant, $[\mathcal{F}(T)] = f(T)$. This equality can be written as a commutative diagram

$$\begin{array}{ccc} K_0(\mathcal{C}_{\underline{\mu}}) & \hookrightarrow & V_{\underline{\mu}} \\ \uparrow {\scriptstyle [\mathcal{F}(T)]} & & \uparrow {\scriptstyle f(T).} \\ K_0(\mathcal{C}_{\underline{\lambda}}) & \hookrightarrow & V_{\underline{\lambda}} \end{array}$$

Horizontal inclusions are those of $\mathbb{Z}[q,q^{-1}]$-modules and become isomorphisms upon tensoring with $\mathbb{Q}(q)$. Alternatively, $K_0(\mathcal{C}_{\underline{\lambda}}) \otimes_{\mathbb{Z}[q,q^{-1}]} \mathbb{C}(q)$ could be a subspace of $V_{\underline{\lambda}}$, such as the subspace $\mathsf{Inv}_{U_q(\mathfrak{g})}(V_{\underline{\lambda}})$ of quantum group invariants in the tensor product.

For composable tangles T', T there should be fixed an isomorphism of functors $F(T' \circ T) \cong F(T') \circ F(T)$.

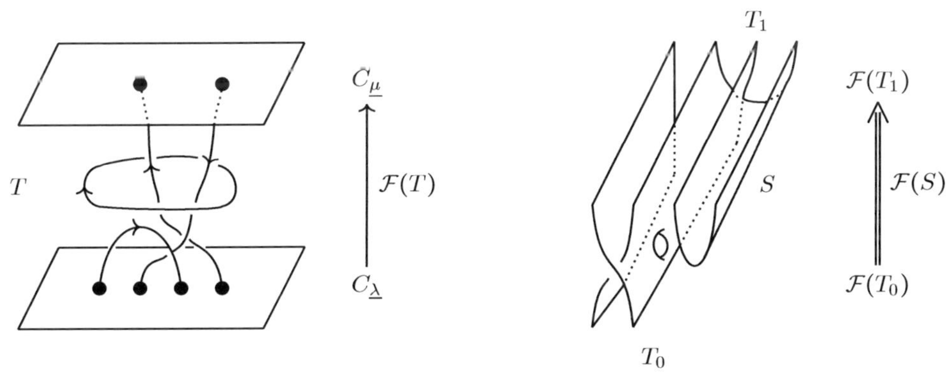

FIGURE 3.7.1. Left: To a tangle T there is assigned an exact functor $\mathcal{F}(T)$. Middle: A tangle cobordism S between tangles T_0 and T_1. Right: A natural transformation $\mathcal{F}(S)$ is associated to S.

A tangle is an oriented decorated one-manifold properly embedded in $\mathbb{R}^2 \times [0,1]$. A tangle cobordism between oriented tangles T_0, T_1 is a surface S with boundary and corners properly embedded in $\mathbb{R}^2 \times [0,1]^2$ such that its boundary is the union of T_0, T_1 on the intersections $S \cap (\mathbb{R}^2 \times [0,1] \times \{i\})$, $i = 0,1$, and the union of product tangles (finite sets of points times $[0,1]$) on the intersections $\mathbb{R}^2 \times \{i\} \times [0,1]$, $i = 0,1$. A tangle cobordism is depicted rather schematically in Figure 3.7.1, in the middle. Since tangles are additionally decorated by positive integral weights λ, connected components of S must be decorated by integral weights as well, with boundary decorations induced by those of S.

To a tangle cobordism S between tangles T_0, T_1 there should be assigned a natural transformation of functors $F(S) : \mathcal{F}(T_0) \longrightarrow \mathcal{F}(T_1)$.

As one varies over all collections of labelled points in the plane, functors $\mathcal{F}(T)$ and natural transformations $\mathcal{F}(S)$, these should fit into a *2-functor* $\mathcal{F}$ from the *2-category of tangle cobordisms* to the 2-category of natural transformations between exact functors. (Tangle cobordisms are oriented and decorated by positive integral weights of $\mathfrak{g}$.) Having such a 2-functor is the most functorial scenario for a link homology theory.

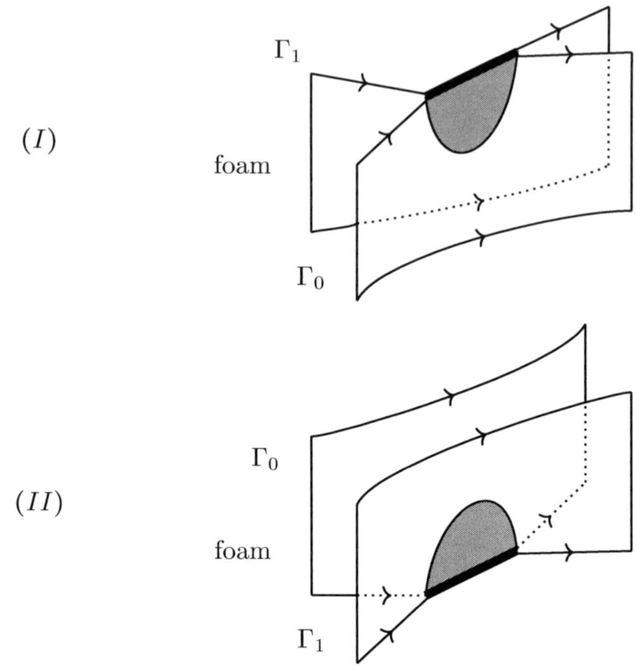

FIGURE 3.7.2. Foam cobordisms between two resolutions Γ_0, Γ_1 of a crossing, also see Figure 3.6.5.

In the matrix factorization approach to link homology, described earlier, the link homology does extend to a 2-functor from the category of tangle cobordisms to the category of bigraded vector spaces (or bigraded R_N-modules, for the equivariant theory).

The Lie algebra $\mathfrak{g} = \mathfrak{sl}(N)$ and all components are colored by V. For n points on the plane one forms the category MF_ω with the potential $\omega = \sum_{i=1}^{n} \pm x_i^{N+1}$, where the signs depend on orientations of points. Then one forgets that MF_ω is a triangulated category, views it only as an additive category, and forms the homotopy category of complexes $\mathcal{C}_n = \mathcal{H}(MF_\omega)$ over it. To a tangle T there is assigned a complex of matrix factorizations and to a tangle cobordism – a homomorphism of complexes. The entire construction results in a 2-functor as above:

$$\begin{array}{ccc} \text{2-category } 2\,\mathsf{Tan} \text{ of} & \xrightarrow{\ \mathcal{F}\ } & \text{2-category } \mathsf{CMF} \text{ of (complexes of)} \\ \text{Tangle cobordisms} & \Longrightarrow & \text{matrix factorizations.} \end{array}$$

Objects of the category CMF are potentials ω in finite sets of variables $\underline{x}$, morphisms from $\omega_1(\underline{x})$ to $\omega_2(\underline{y})$ are complexes of matrix factorizations with potential $\omega_2(\underline{y}) - \omega_1(\underline{x})$, and two-morphisms are homomorphisms of complexes of factorizations modulo chain homotopies.

The 2-functor $\mathcal{F}$ above is $\mathbb{Q}^*$-projective, with the map associated to a link cobordism well-defined up to scaling by elements of $\mathbb{Q}^*$. Extending T. Sano's approach to strict cobordism invariance [**San08**] from $N = 2$ to any N should be one way to resolve the $\mathbb{Q}^*$-indeterminacy.

To get a homomorphism from the Grothendieck group of the category $\mathcal{C}_n$ to the tensor product $V_{\underline{\lambda}}$, more precisely, to its subspace of invariants $\mathsf{Inv}(V_{\underline{\lambda}})$, one should restrict to the subcategory generated by matrix factorizations associated to planar graphs with a given boundary.

Versions of $\mathsf{SL}(N)$ homology and the 2-functor $\mathcal{F}$ can also be recovered from

- Parabolic-singular blocks of highest weight categories for $\mathfrak{sl}(k)$, where k depends on $\underline{\lambda}$, see [**Sus07, MS09**], and, in the $\mathfrak{sl}(2)$ case, [**BFK99, Str05, FKS06, Str09, BS11, CK14**].
- Fukaya–Floer categories of quiver varieties for $\mathsf{SL}(k)$, k a function of weights $\underline{\lambda}$ ($N = 2$ case [**SS06, Rez09, AS19**]) and arbitrary N case restricted to braid and braid closures [**Man07**], see also [**WW15**].
- Derived categories of coherent sheaves on the convolution varieties of affine Grassmannians; $N = 2$ case: [**CK08a**] and arbitrary N: [**CK08b**], and also on quiver varieties [**AN23**]. For a recent physics-motivated approach, see M. Aganagic [**Aga22, Aga21**].
- Various categories of importance in representation theory, see [**Web17**] and related papers.

The principle here is that in each of these cases there is (or expected to exist) a categorical action of the 2-category of tangle cobordisms Tan_2 on the corresponding derived or homotopy categories, giving a 2-functor similar to (I) and (II) in Figure 3.7.2. Tangles act by functors on the categories and tangle cobordisms act by natural transformations. Often, the corresponding categories for various examples are equivalent [**MW18**], sometimes with minor modifications. One expects the equivalences to respect the 2-functor actions.

Motivation to relate quiver varieties (Kronheimer–Nakajima varieties) and link homology came from observing that (a) homology groups of $\mathsf{SL}(k)$ quiver varieties carry an action of that Lie algebra and can be identified with weight spaces of $\mathsf{SL}(k)$ representations [**Nak94**], (b) due to the level-rank duality, those weight spaces can be identified with the invariants of tensor products of exterior powers $\Lambda^a V$ of the fundamental $\mathfrak{sl}(N)$ representation V, where k depends on the sequence $(a_1, \dots, a_n)$ of these parameters in the tensor product. One then expects that replacing homology by the derived category of coherent sheaves will result in a categorical action of the Lie algebra and a commuting categorical action of tangles and tangle cobordisms. This program was realized in [**CK08b**] restricting to tensor products of V and V^* and using related varieties which are suitable convolution varieties of the affine Grassmannian and should contain quiver varieties as open subvarities.

Extended TQFTs and biadjoint functors: When one has an extended TQFT in dimension n, to an $(n-2)$-manifold M assign a category $\mathcal{F}(M)$, to an $(n-1)$-cobordism K a functor $\mathcal{F}(K) : \mathcal{F}(\partial_0 K) \longrightarrow \mathcal{F}(\partial_1 K)$, to an n-cobordism L with corners a natural transformation $\mathcal{F}(\partial_0 L) \longrightarrow \mathcal{F}(\partial_1 L)$.

For each $(n-1)$-cobordism K there is a dual cobordism K^* from $\partial_1 K$ to $\partial_0 K$ (reflect K). There are four canonical n-cobordisms between compositions KK^*,

K^*K and identity cobordisms $\mathsf{id}_{\partial_1 K}$ and $\mathsf{id}_{\partial_0 K}$. Applying $\mathcal{F}$ to them tells us that functors $\mathcal{F}(K)$ and $\mathcal{F}(K^*)$ are biadjoint (both left and right adjoint).

Examples of biadjoint functors appear in algebraic geometry (Fourier–Mukai kernels between Calabi–Yau manifolds) and in symplectic topology (convolutions with Lagrangians in Fukaya–Floer categories). In particular, quiver varieties are Calabi–Yau (around their compact part) and their derived categories of coherent sheaves admit plenty of biadjoint functors. In three dimensions, the model example of a TQFT is the Witten–Reshetikhin–Turaev theory, for which the categories assigned to $3-2=1$-dimensional manifolds are semisimple. Any linear functor between these categories has a biadjoint functor. In higher dimensions, for an extended TQFT, one expects non-semisimple, likely triangulated, categories associated to $n-2$-dimensional manifolds. For an exact functor between triangulated categories having a biadjoint is a strong condition. It is then natural to pay special attention to Calabi–Yau varieties (and their derived categories of coherent sheaves), Fukaya–Floer categories, and suitable categories of representations built out of symmetric Frobenius algebras, for these types of categories provide large supply of biadjoint functors that can be constructed naturally [**Kho02**].

Reshetikhin–Turaev $\mathsf{SL}(N)$ (or $\mathsf{GL}(N)$) invariants for fundamental representations $\Lambda_q^a V$ are distinguished by the existence of *positive integral* diagrammatical calculus of intertwiners (*MOY calculus*). It guides the categorification of Reshetikhin–Turaev invariants. Positivity and integrality property is missing already for other representations of quantum $\mathsf{SL}(2)$, since q-spin networks are not positive and have denominators [**KL94**]. This creates serious problems trying to extend the above approaches beyond $(\mathsf{SL}(N), \Lambda_q^* V)$ case.

In the remarkable work [**Web17**], Webster categorified Reshetikhin–Turaev invariants for any simple Lie algebra $\mathfrak{g}$ and any labelling of a link's components by irreducible $\mathfrak{g}$-modules. Categorification of quantum groups [**KL01, Rou08**] was one of the motivations for Webster's construction. It is likely that for $\mathfrak{sl}(N)$ and coloring by minuscule representations his construction gives homology isomorphic to those coming from matrix factorizations (and those from Robert–Wagner foam evaluation).

Beyond the minuscule representation case, Webster homology and a number of other known link homology theories do not fully extend or not known to extend to link cobordisms.

For a subset of these theories [**FSS12, Roz14, CK12, Web17, GH22, SS22, OR19**], the reason is the following. Take a functorial link homology theory (and subject to TQFT assumption) and consider standardly embedded cobordisms in $\mathbb{R}^3 \times [0,1]$ between unlinks (disjoint unions of unknots). Restricting a link homology theory to these cobordisms results in a 2D TQFT. Consequently, homology A of the unknot must be a commutative Frobenius algebra over the homology R of the empty link. In particular, if R is a field, A must be finite dimensional over R, and this property fails in the above examples.

Other theories [**KR08b, Cau17, QRS18, RW20b, OR22**] have finite-dimensional homology groups over R (upon normalization, if needed), but their construction is only available for braid closures, making it hard to extend them to link cobordism. For an early categorification of the colored Jones polynomial [**Kho05**] functoriality is not known either, see also [**BW08**].

It is an important open problem to (in the first case) modify these link homology theories, including the Webster homology, to make them functorial under link cobordisms and (in the second) find whether they can redefined in a more functorial way, for all link diagrams and extending to link cobordisms or even to tangle cobordisms.

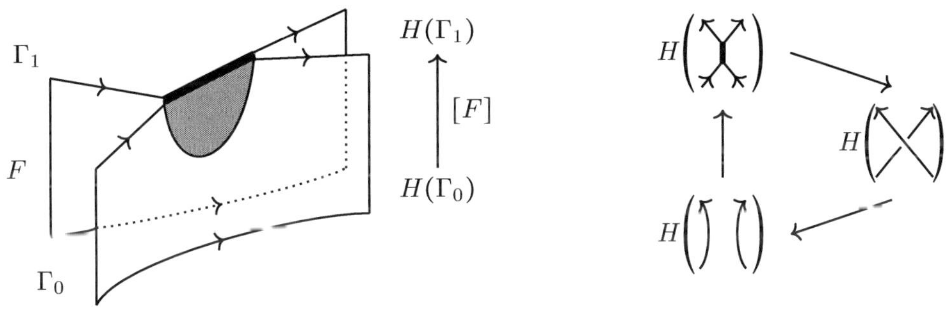

FIGURE 3.7.3. Foam F, a sort of "singular saddle" cobordism between webs Γ_0 and Γ_1, induces a map of state spaces for these webs which enters the inductive construction of link homology groups via cones or SES shown on the right.

3.8. Foam evaluation and link homology. L.-H. Robert and E. Wagner [**RW20a**] found a purely combinatorial approach to $\mathsf{SL}(N)$ link homology (again, for minuscule representations $\Lambda_q^a V$). Their approach is based on foam evaluations. Foams are cobordisms between planar graphs Γ as above and implicitly appear in most approaches to $\mathsf{SL}(N)$ link homology. Maps in the long exact sequences in Figure 3.6.1 between the two planar graphs should be induced by foam cobordisms shown in Figure 3.7.2 between graphs Γ_0, Γ_1, also see Figure 3.7.3. To build a combinatorial theory, Robert and Wagner construct a subtle evaluation of closed foams to symmetric functions $R = \mathbb{Z}[x_1, \dots, x_N]^{S_N}$ in N variables.

Let us briefly review foams and Robert–Wagner foam evaluation. $\mathsf{GL}(N)$ foams can have facets of thickness $a \in \{1, \dots, N\}$, seams where facets of thickness $a, b, a+b$ come together and vertices with 6 adjoint facets of thickness $a, b, c, a+b, b+c, a+b+c$, see Figure 3.8.1. At first, consider *closed foams* embedded in $\mathbb{R}^3$, that is, foams without boundary, while later we will need foam with boundary, viewed as cobordisms in $\mathbb{R}^2 \times [0,1]$ between MOY graphs.

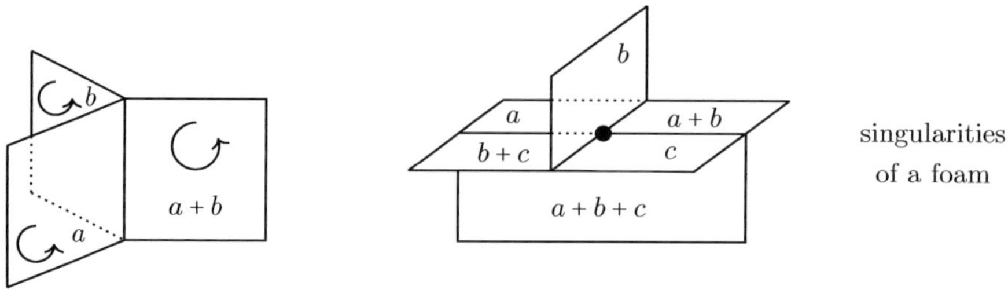

FIGURE 3.8.1. Left: an (a, b) seam of a GL_N foam and adjacent three facets. Facets are oriented in a compatible way, with orientation preserved when moving from an a or b facet to the adjacent $a+b$ facet. Orientation is reversed between adjacent a and b facets. Right: a singular vertex where four seams meet.

A *coloring* of foam F is a map $c : \mathsf{facets} \longrightarrow \mathcal{P}(N)$, where $|c(f)|$ is the thickness of a facet f, with the flow condition at each seam, $c(f_3) = c(f_1) \sqcup c(f_2)$, see Figures 3.8.2, 3.8.3. In particular, $c(f_1), c(f_2)$ are disjoint sets. Here $\mathcal{P}(N)$ is the set of subsets of $\{1, \ldots, N\}$, so that a coloring maps facets to subsets of the set of colors from 1 to N. A facet of thickness a is mapped to a subset of cardinality a; subsets for a, b thickness facets that meet along a seam and become an $a+b$ facet are disjoint and their union is the subset for the $a+b$ facet.

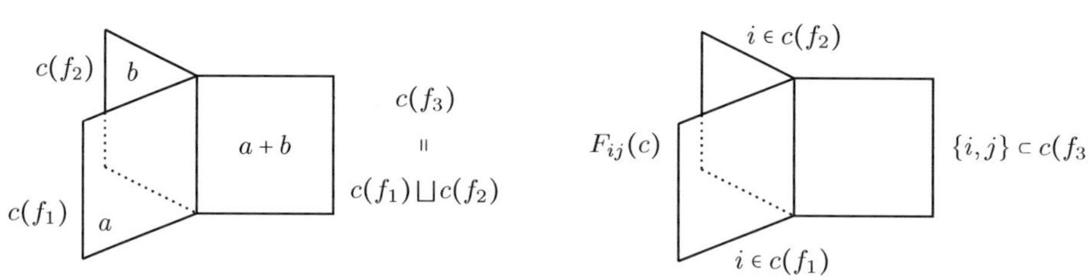

FIGURE 3.8.2. Left: Facets f_1, f_2 of thickness a and b, respectively, merging along a seam are colored by disjoint subsets $c(f_1)$ and $c(f_2)$ of $\{1, \ldots, N\}$ of cardinality a, respectively b. Right: if $i \in c(f_1), j \in c(f_2)$, the two thin facets belong to $F_{ij}(c)$ but the thick facet does not.

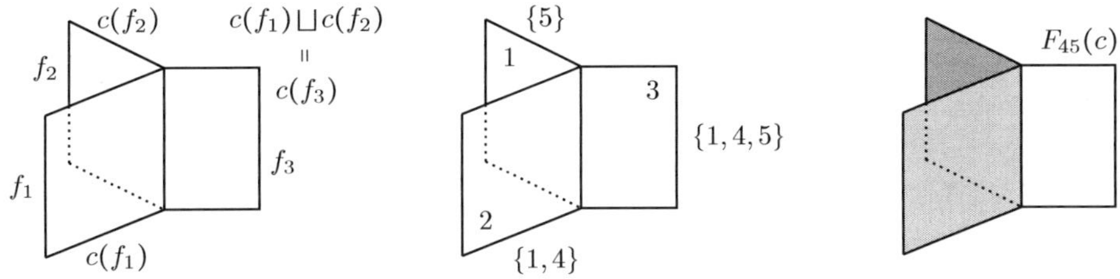

FIGURE 3.8.3. Left: Thin facets f_1, f_2 are colored by disjoint subsets $c(f_1), c(f_2)$. Thick facet f_3 is colored by the union of these subsets. Middle: an example of a coloring when f_1, f_2 have thickness 2, 1 and are colored by disjoint subsets $\{1,4\}$ and $\{5\}$, respectively. Facet f_3 is then colored by their union $\{1,4,5\}$. Right: part of the bicolored surface $F_{45}(c)$ for this coloring, containing both thin but not the thick facet along that seam.

Define the *bicolored surface* $F_{ij}(c)$, $1 \le i < j \le N$ as the union of facets that contain exactly one color from $\{i,j\}$. For a closed foam F, its bicolored surface is a closed compact surface without boundary embedded in $\mathbb{R}^3$. Figures 3.8.2, 3.8.3 on the right explain why $F_{ij}(c)$ has no singularities along seams of the foam, always containing either none or two facets along a seam. A similar computation shows that $F_{ij}(c)$ has no singularities at vertices of the foam.

Thus, $F_{ij}(c)$ is an orientable surface and the union of its connected components, each of Euler characteristic $2-2g$, for a component of genus $g \ge 0$. Robert–Wagner evaluation $\langle F, c\rangle$ of a closed foam F and its coloring c has the form

$$\langle F, c\rangle = \pm \prod_{i<j}(x_i - x_j)^{-\chi(F_{ij}(c))/2} \prod(\text{facet decoration contributions}).$$

We refer to [**RW20a**] and a review in [**KK24**] for the subtle formula for the minus sign, facet decorations and their contributions. Facets are decorated by dots (observables) labelled by homogeneous symmetric functions f in a variables, for a facet of thickness a. A coloring tells one which a variables out of N to select from $x_1, \ldots, x_N$ to turn f into a function in x's.

In general $\langle F, c\rangle$ has denominators $x_i - x_j$ but the sum

$$\langle F\rangle = \sum_c \langle F, c\rangle \in R$$

is a symmetric polynomial in $x_1, \ldots, x_N$.

With that sophisticated yet beautiful foam evaluation $\langle F\rangle$ at hand, Robert and Wagner build the state space (or homology) for a planar MOY graph Γ using the universal construction.

Universal construction of topological theories [**BHMV95, Kho20, KQR21, IK24, IK22**] begins with an evaluation of closed objects (closed foams, in our case) and builds state spaces for generic cross-sections of these objects. A generic cross-section Γ of a foam in $\mathbb{R}^3$ by a plane $\mathbb{R}^2$ is a planar MOY graph Γ. Fix Γ and consider the free R-module $\mathsf{Fr}(\Gamma)$ with a basis $[F]$ of all foams F with $\partial F = \Gamma$, see Figure 3.8.4 on the left. Notice that we no longer restrict to closed foams, instead looking at foams F in $\mathbb{R}^2 \times (-\infty, 0]$ with Γ as the boundary.

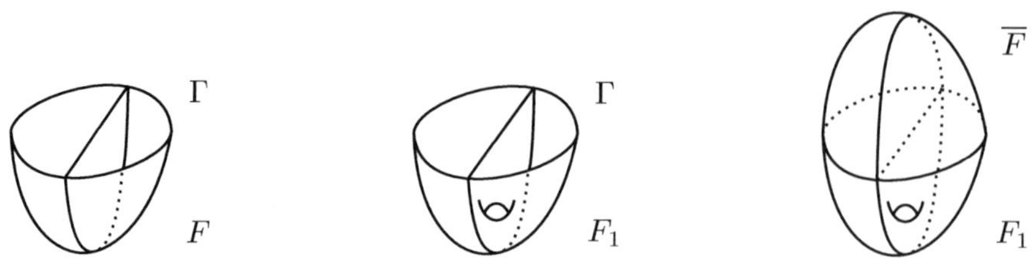

FIGURE 3.8.4. Left: A foam F with boundary Γ. Middle: another foam F_1 with $\partial F_1 = \Gamma$. Right: gluing F and F_1 along Γ into a closed foam. Here $\overline{F}$ is F reflected in the z-plane.

Define an R-bilinear form

$$(\ \ ,\ \ ) : \mathsf{Fr}(\Gamma) \times \mathsf{Fr}(\Gamma) \longrightarrow R$$

by $([F],[F_1]) = \langle \overline{F} F_1 \rangle$ and extending via R-bilinearity. Here we glue F and F_1 along the common boundary Γ into a closed foam $\overline{F}F_1$, where $\overline{F}$ denotes the reflection of F in a horizontal plane. Define the state space

$$H(\Gamma) := \mathsf{Fr}(\Gamma)/\ker((\ \ ,\ \ )_\Gamma)$$

of an MOY graph Γ as the quotient of the (large) free module $\mathsf{Fr}(\Gamma)$ by the kernel of the bilinear form $(\ \ ,\ \ )_\Gamma$. This means that an R-linear combination of foams $\sum_i \lambda_i F_i$, with $\lambda_i \in R$ and $\partial F_i = \Gamma$ is 0 in $H(\Gamma)$ if and only if for any foam F with $\partial F = \Gamma$, we have

$$\sum_i \lambda_i \langle \overline{F} F_i \rangle = 0 \in R.$$

One thinks of $\sum_i \lambda_i F_i = 0$ as a linear skein relation on foams with boundary Γ.

It is easy to see that state spaces $H(\Gamma)$, over all Γ, form a functorial topological theory. Namely, given a foam $F \in \mathbb{R}^2 \times [0,1]$ with top and bottom boundary, so that $\partial F = \partial_1 F \sqcup (-\partial_0 F)$, composition with F induces a map from $\mathsf{Fr}(F_0)$ to $\mathsf{Fr}(F_1)$. The map takes the kernel of the bilinear form $(\ \ ,\ \ )_{\Gamma_0}$ to the kernel of the bilinear form $(\ \ ,\ \ )_{\Gamma_1}$. Consequently, there is an induced map on state spaces

$$H(F) : H(\partial_0 F) \longrightarrow H(\partial_1 F).$$

Varying over all foams F with boundary, this results in a functor H from the category of foams to the category of R-modules.

For a general foam evaluation $\langle F \rangle$ such a functor is not interesting. In particular, one wants state spaces $H(\Gamma)$ to be sufficiently small, for instance, be finitely-generated R-modules. Even under such an assumption, functor H will not be a TQFT in general, with the natural map $H(\Gamma_1) \otimes_R H(\Gamma_2) \longrightarrow H(\Gamma_1 \sqcup \Gamma_2)$ not an isomorphism of R-modules.

For their evaluation sketched above, Robert and Wagner in [**RW20a**] proved:

THEOREM 3.5 (Robert–Wagner). *H is a functorial TQFT from* Foams *to free graded R-modules, with graded ranks the Murakami–Ohtsuki–Yamada planar graph invariants:*

$$\mathsf{grank}_R(\Gamma) = P_N(\Gamma).$$

Methods of Yonezawa–Wu [**Yon11, Wu12b**] for constructing homology from matrix factorizations for MOY graphs with edges of arbitrary thickness apply to Robert–Wagner state spaces. Robert–Wagner foam TQFT gives rise to a link homology theory which categorifies Reshetikhin–Turaev invariant for $\mathsf{SL}(N)$ and link components labelled by exterior powers of the fundamental representation [**ETW18**].

Furthermore, in [**ETW18**] this result is extended to:

THEOREM 3.6 (M. Ehrig–D. Tubbenhauer–P. Wedrich). *Robert–Wagner link homology theory is functorial for link cobordisms.*

Robert–Wagner's approach [**RW20a**, **ETW18**] to categorification of Reshetikhin–Turaev $\mathsf{SL}(N)$ invariants is *complementary* to all the others, which require *specific categories* (highest weight categories, coherent sheaves or Fukaya–Floer categories on quiver varieties, particular representation theory categories). In the Robert–Wagner construction, categories appear at the last step only (when extending to graphs with boundary, tangles, and their cobordisms). It is a *categorification of the state sum approach to quantum invariants* and should also be relevant to some models of 3-dimensional statistical mechanics.

Prior to Robert–Wagner's work, foams were heavily used in link homology and categorified quantum groups, see for example [**MSV09, Mac09, QR16, Cap13, RW16, Wed19**]. Foam evaluation should help to streamline and clarify a significant amount of prior work on the subject, including replacing the ground ring $\mathbb{Q}$ by $\mathbb{Z}$ or by the ring of symmetric functions $\mathbb{Z}[x_1, \ldots, x_N]^{S_N}$.

4. Interactions and applications

4.1. Some interactions.

Soergel category. A version of Robert–Wagner evaluation [**RW20b**] allows one to describe the category of Soergel bimodules [**EMTW20**] for the symmetric group S_N, with homs given by foams in $\mathbb{R}^3$ between braid-like graphs in the plane modulo universal construction relations [**RW20b, KRW**]. The Soergel category is central in geometric representation theory and categorifies the Hecke algebra of the symmetric group. Originally, the relation between foams and the Soergel category was established by D. Rose and P. Wedrich [**RW16**] and P. Wedrich [**Wed19**]. An earlier diagrammatic approach to the Soergel category by Elias–Khovanov [**EK10**] is via a two-dimensional graphical calculus rather than foams in $\mathbb{R}^3$, with the missing dimension encoded by labels in the ordered set $\{1, \ldots, N-1\}$.

Kronheimer–Mrowka theory. Inspired by the Robert–Wagner foam evaluation, Robert and one of us [**KR21**] related *unoriented* $\mathsf{SL}(3)$ foams and Kronheimer–Mrowka gauge $SO(3)$ theory for 3-orbifolds [**KM19**], proving one of Kronheimer–Mrowka conjectures. Kronheimer–Mrowka theory, in a rather special case, assigns homology groups to planar trivalent graphs Γ, and similar groups can be recovered from an unoriented version of Robert–Wagner foam evaluation. (Kronheimer–Mrowka theory is much more general, and, in particular, assigns homology groups to spacial trivalent graphs, for which a combinatorial counterpart is unknown.) Kronheimer–Mrowka theory and its combinatorial counterpart in [**KR21**] for planar trivalent graphs relate to the 4-Color Theorem, aiming to prove and rethink the 4-Color Theorem in a more conceptual way and via its relations to TQFTs, gauge theory, and low-dimensional topology. These two theories have been investigated by D. Boozer [**Boo19, Boo23**], who have shown, in particular, that they give rise to two non-isomorphic functors on foams in $\mathbb{R}^3$.

APS homology. Foam evaluation allows for a natural extension of the Asaeda–Przytycki–Sikora [**APS04**] annular SL(2) homology to the equivariant setting of a larger ground ring as well as its extension to annular SL(N) homology [**AK21**, **Akh23**].

4.2. SL(2) and SL(3) homology theories. SL(2) homology (aka Khovanov homology) is noticeably simpler than SL(N) homology and was discovered first [**Kho00**]. Foams are replaced by surfaces, and homology of links is built from a 2-dimensional TQFT where homology of the unknot has rank two over homology of the empty link. Khovanov homology categorifies the Jones polynomial. Odd Khovanov homology, discovered by P. Ozsvath, J. Rasmussen and Z. Szabo [**ORS13**], is another bigraded categorification of the Jones polynomial.

SL(3) homology, which categorifies the Kuperberg quantum $\mathfrak{sl}(3)$ invariant [**Kup96b**], is in-between $N = 2$ and $N \geq 4$ cases complexity-wise [**Kho04, MV07, MN08, Cla09**]. SL(3) foams do not have vertices, and closed SL(3) foams can be evaluated via "localization along singular circles" and manipulation of foams with a single such circle. Single-circle foam evaluation is encoded in the cohomology groups of the flag variety of $\mathbb{C}^3$ (commutative algebra structure plus the trace map).

4.3. Two applications to 4D topology.

I. *Rasmussen invariant.* The Rasmussen invariant $s(K)$ comes from an equivariant version of Khovanov homology ($N = 2$) given by replacing the ground ring $\mathbb{Z}$ by $\mathbb{Q}[t]$ and the Frobenius algebra $A = \mathbb{Z}[X]/(X^2)$ of homology of the unknot by $A_t = \mathbb{Q}[t, X]/(X^2 - t)$ over $\mathbb{Q}[t]$. Then SL(2)-homology $H_t(K)$ of a knot K becomes a $\mathbb{Q}[t]$-module, and it can be written as a sum of its t-torsion and a free summand, where the free summand is a free rank one A_t-module in homological degree 0 and q-degree $s(K) - 1$:

$$H_t(K) \cong \mathsf{Tor}(H_t(K)) \oplus A_t\{s(K) - 1\}.$$

From [**Ras10**], one obtains the following:

THEOREM 4.1 (J. Rasmussen). *The value $s(K)$ is an invariant of knot concordance and gives a lower bound on the slice genus of K. This bound is explicitly computable and sharp on positive knots.*

The Rasmussen invariant provides a combinatorial proof of Kronheimer–Mrowka–Milnor theorem (formerly Milnor's conjecture) that the slice genus of (p, q)-torus knot is $\frac{(p-1)(q-1)}{2}$.

Replacing $\mathbb{Q}$ by $\mathbb{F}_p$ or extending to SL(N) homology, $N \geq 3$ and its deformations leads to families of Rasmussen-like concordance invariants [**MTV07, MTV13, Lob12, Lew14**]. Some linear independence results on these invariants are known and there is even a postcard on the topic [**LZ21**], but the general theory of such concordance invariants remains a mystery.

II. *Exotic surfaces in $\mathbb{R}^4$.* K. Haiden and I. Sundberg in [**HS21**] give examples of surfaces $S_1, S_2 \subset \mathbb{D}^4$ that bound a knot $K \subset \mathbb{S}^3$ such that S_1, S_2 are homeomorphic but not diffeomorphic rel boundary. Non-diffeomorphic property follows by checking that maps induced by S_1, S_2 on Khovanov homology

$$H(S_i) : H(K) \longrightarrow H(\text{empty link}) \cong \mathbb{Z}$$

are not equal.

S. Akbulut showed in [**Akb91**] that S_1, S_2 have this property by using Donaldson theory. Thirty years later, there is a combinatorial proof of his result.

Link homology relates smooth 4D topology and geometric representation theory in a highly intricate way. Braid groups and tangle categories act by exact functors on key categories in geometric representation theory. Tangle cobordisms act by natural transformations between these functors. These tangle cobordism invariants are subtle enough to distinguish between distinct smooth structures on properly embedded surfaces in the four-ball $\mathbb{D}^4$ with the same underlying topological structure.

Going beyond mere homology groups, we should mention the spectrification of Khovanov homology by R. Lipshitz and S. Sarkar [**LS14, LLS20**] and [**HKK16**]; see N. Kitchloo [**Kit19**] for spectrifications of $\mathsf{SL}(N)$ homology.

Heegaard–Floer theory [**OS06b, OS06a**] and link Floer homology of Ozsvath–Szabo and Rasmussen [**Man16, OS18**] play a fundamental role in modern low-dimensional topology and should relate to yet-unknown categorified quantum group $\mathsf{GL}(1|1)$ and its categorified representations, as studied by Y. Tian, A. Ellis–I. Petkova–V. Vertesi, A. Manion–R. Rouquier and others [**Tia14, EPV19, MR20**].

This informal write-up discusses only a fraction of the rich developments in the past twenty-five years that were inspired by the work [**CF94**] of Igor Frenkel and Louis Crane and other related ideas of Igor Frenkel. We are delighted to celebrate Igor Frenkel's anniversary and wish him many more years of inspiring and exciting mathematical discoveries.

Acknowledgments

The authors are grateful to Tsinghua University and Yau Mathematical Sciences Center for their hospitality. The second author would like to acknowledge support from Beijing Institute of Mathematical Sciences and Applications (BIMSA) and the opportunity to give a plenary lecture at the first International Congress of Basic Science (ICBS). The authors would like to thank Joshua Sussan for valuable feedback on the paper.

References

[Aga21] Mina Aganagic, *Knot Categorification from Mirror Symmetry, Part II: Lagrangians*, arXiv preprint `arXiv:2105.06039` (2021), 1–211.

[Aga22] Mina Aganagic, *Homological Knot Invariants from Mirror Symmetry*, arXiv preprint `arXiv:2207.14104` (2022), 1–37.

[AK21] Rostislav Akhmechet and Mikhail Khovanov, *Anchored foams and annular homology*, Algebr. Geom. Topol. **23** (2023), no. 7, 3129–3204, DOI 10.2140/agt.2023.23.3129. MR4647674

[Akb91] Selman Akbulut, *A fake compact contractible 4-manifold*, J. Differential Geom. **33** (1991), no. 2, 335–356. MR1094459

[Akh23] Rostislav Akhmechet, *Anchored $\mathfrak{gl}_n$ foams and annular Khovanov–Rozansky homology*, Preprint, `arXiv:2305.08944`, 1–23, 2023.

[AN23] Rina Anno and Vinoth Nandakumar, *Exotic t-structures for two-block Springer fibres*, Trans. Amer. Math. Soc. **376** (2023), no. 3, 1523–1552, DOI 10.1090/tran/8765. MR4549684

[APS04] Marta M. Asaeda, Józef H. Przytycki, and Adam S. Sikora, *Categorification of the Kauffman bracket skein module of I-bundles over surfaces*, Algebr. Geom. Topol. **4** (2004), 1177–1210, DOI 10.2140/agt.2004.4.1177. MR2113902

[AS19] Mohammed Abouzaid and Ivan Smith, *Khovanov homology from Floer cohomology*, J. Amer. Math. Soc. **32** (2019), no. 1, 1–79, DOI 10.1090/jams/902. MR3867999

[BFK99] Joseph Bernstein, Igor Frenkel, and Mikhail Khovanov, *A categorification of the Temperley-Lieb algebra and Schur quotients of $U(\mathfrak{sl}_2)$ via projective and Zuckerman functors*, Selecta Math. (N.S.) **5** (1999), no. 2, 199–241, DOI 10.1007/s000290050047. MR1714141

[BHMV95] C. Blanchet, N. Habegger, G. Masbaum, and P. Vogel, *Topological quantum field theories derived from the Kauffman bracket*, Topology **34** (1995), no. 4, 883–927, DOI 10.1016/0040-9383(94)00051-4. MR1362791

[BLM90] A. A. Beilinson, G. Lusztig, and R. MacPherson, *A geometric setting for the quantum deformation of* GL_n, Duke Math. J. **61** (1990), no. 2, 655–677, DOI 10.1215/S0012-7094-90-06124-1. MR1074310

[Boo19] David Boozer, *Computer bounds for Kronheimer-Mrowka foam evaluation*, Exp. Math. **32** (2023), no. 4, 615–630, DOI 10.1080/10586458.2021.1982078. MR4669283

[Boo23] David Boozer, *The combinatorial and gauge-theoretic foam evaluation functors are not the same*, arXiv preprint `arXiv:2304.07659` (2023), 1–6.

[BS11] Jonathan Brundan and Catharina Stroppel, *Highest weight categories arising from Khovanov's diagram algebra III: category $\mathcal{O}$*, Represent. Theory **15** (2011), 170–243, DOI 10.1090/S1088-4165-2011-00389-7. MR2781018

[BS15] Joshua Batson and Cotton Seed, *A link-splitting spectral sequence in Khovanov homology*, Duke Math. J. **164** (2015), no. 5, 801–841, DOI 10.1215/00127094-2881374. MR3332892

[BW08] Anna Beliakova and Stephan Wehrli, *Categorification of the colored Jones polynomial and Rasmussen invariant of links*, Canad. J. Math. **60** (2008), no. 6, 1240–1266, DOI 10.4153/CJM-2008-053-1. MR2462446

[Cap13] Carmen Caprau, *The $sl(2)$ foam cohomology via a TQFT*, New York J. Math. **19** (2013), 61–90. MR3028136

[Cau17] Sabin Cautis, *Remarks on coloured triply graded link invariants*, Algebr. Geom. Topol. **17** (2017), no. 6, 3811–3836, DOI 10.2140/agt.2017.17.3811. MR3709661

[CF94] Louis Crane and Igor B. Frenkel, *Four-dimensional topological quantum field theory, Hopf categories, and the canonical bases*, J. Math. Phys. **35** (1994), no. 10, 5136–5154, DOI 10.1063/1.530746. Topology and physics. MR1295461

[CK08a] Sabin Cautis and Joel Kamnitzer, *Knot homology via derived categories of coherent sheaves. I. The $\mathfrak{sl}(2)$-case*, Duke Math. J. **142** (2008), no. 3, 511–588, DOI 10.1215/00127094-2008-012. MR2411561

[CK08b] Sabin Cautis and Joel Kamnitzer, *Knot homology via derived categories of coherent sheaves. II. $\mathfrak{sl}_m$ case*, Invent. Math. **174** (2008), no. 1, 165–232, DOI 10.1007/s00222-008-0138-6. MR2430980

[CK12] Benjamin Cooper and Vyacheslav Krushkal, *Categorification of the Jones-Wenzl projectors*, Quantum Topol. **3** (2012), no. 2, 139–180, DOI 10.4171/QT/27. MR2901969

[CK14] Yanfeng Chen and Mikhail Khovanov, *An invariant of tangle cobordisms via subquotients of arc rings*, Fund. Math. **225** (2014), no. 1, 23–44, DOI 10.4064/fm225-1-2. MR3205563

[CK17] Sabin Cautis and Joel Kamnitzer, *Knot homology via derived categories of coherent sheaves IV, coloured links*, Quantum Topol. **8** (2017), no. 2, 381–411, DOI 10.4171/QT/93. MR3659494

[Cla09] David Clark, *Functoriality for the $\mathfrak{su}_3$ Khovanov homology*, Algebr. Geom. Topol. **9** (2009), no. 2, 625–690, DOI 10.2140/agt.2009.9.625. MR2482322

[CLS20] Sabin Cautis, Aaron D. Lauda, and Joshua Sussan, *Curved Rickard complexes and link homologies*, J. Reine Angew. Math. **769** (2020), 87–119, DOI 10.1515/crelle-2019-0044. MR4178751

[CM16] Nils Carqueville and Daniel Murfet, *Adjunctions and defects in Landau-Ginzburg models*, Adv. Math. **289** (2016), 480–566, DOI 10.1016/j.aim.2015.03.033. MR3439694

[Dri86] V. G. Drinfel′d, *Quantum groups* (Russian, with English summary), Zap. Nauchn. Sem. Leningrad. Otdel. Mat. Inst. Steklov. (LOMI) **155** (1986), no. Differentsial′naya Geometriya, Gruppy Li i Mekh. VIII, 18–49, 193, DOI 10.1007/BF01247086; English transl., J. Soviet Math. **41** (1988), no. 2, 898–915. MR869575

[Dri87] Vladimir G. Drinfel′d, *Quantum groups*, Proceedings of the International Congress of Mathematicians, Vol. 1, 2 (Berkeley, Calif., 1986), Amer. Math. Soc., Providence, RI, 1987, pp. 798–820.

[EK10] Ben Elias and Mikhail Khovanov, *Diagrammatics for Soergel categories*, Int. J. Math. Math. Sci., posted on 2010, Art. ID 978635, 58, DOI 10.1155/2010/978635. MR3095655

[EMTW20] Ben Elias, Shotaro Makisumi, Ulrich Thiel, and Geordie Williamson, *Introduction to Soergel bimodules*, RSME Springer Series, vol. 5, Springer, Cham, [2020] ©2020, DOI 10.1007/978-3-030-48826-0. MR4220642

[EPV19] Alexander P. Ellis, Ina Petkova, and Vera Vértesi, *Quantum $\mathfrak{gl}_{1|1}$ and tangle Floer homology*, Adv. Math. **350** (2019), 130–189, DOI 10.1016/j.aim.2019.04.023. MR3946307

[ETW18] Michael Ehrig, Daniel Tubbenhauer, and Paul Wedrich, *Functoriality of colored link homologies*, Proc. Lond. Math. Soc. (3) **117** (2018), no. 5, 996–1040, DOI 10.1112/plms.12154. MR3877770

[FKS06] Igor Frenkel, Mikhail Khovanov, and Catharina Stroppel, *A categorification of finite-dimensional irreducible representations of quantum $\mathfrak{sl}_2$ and their tensor products*, Selecta Math. (N.S.) **12** (2006), no. 3-4, 379–431, DOI 10.1007/s00029-007-0031-y. MR2305608

[FSS12] Igor Frenkel, Catharina Stroppel, and Joshua Sussan, *Categorifying fractional Euler characteristics, Jones-Wenzl projectors and 3j-symbols*, Quantum Topol. **3** (2012), no. 2, 181–253, DOI 10.4171/QT/28. MR2901970

[FYH+85] P. Freyd, D. Yetter, J. Hoste, W. B. R. Lickorish, K. Millett, and A. Ocneanu, *A new polynomial invariant of knots and links*, Bull. Amer. Math. Soc. (N.S.) **12** (1985), no. 2, 239–246, DOI 10.1090/S0273-0979-1985-15361-3. MR776477

[GH22] Eugene Gorsky and Matthew Hogancamp, *Hilbert schemes and y-ification of Khovanov-Rozansky homology*, Geom. Topol. **26** (2022), no. 2, 587–678, DOI 10.2140/gt.2022.26.587. MR4444266

[GHM21] Eugene Gorsky, Matthew Hogancamp, and Anton Mellit, *Tautological classes and symmetry in Khovanov-Rozansky homology*, arXiv preprint `arXiv:2103.01212` (2021), 1–50.

[GIK+23] Paul Gustafson, Mee Seong Im, Remy Kaldawy, Mikhail Khovanov, and Zachary Lihn, *Automata and one-dimensional TQFTs with defects*, Lett. Math. Phys. **113** (2023), no. 5, Paper No. 93, 38, DOI 10.1007/s11005-023-01701-y. MR4636793

[Gor04] Bojan Gornik, *Note on Khovanov link cohomology*, arXiv preprint `arXiv:math/0402266` (2004), 1–12.

[GORS14] Eugene Gorsky, Alexei Oblomkov, Jacob Rasmussen, and Vivek Shende, *Torus knots and the rational DAHA*, Duke Math. J. **163** (2014), no. 14, 2709–2794, DOI 10.1215/00127094-2827126. MR3273582

[GSV05] Sergei Gukov, Albert Schwarz, and Cumrun Vafa, *Khovanov-Rozansky homology and topological strings*, Lett. Math. Phys. **74** (2005), no. 1, 53–74, DOI 10.1007/s11005-005-0008-8. MR2193547

[HKK16] Po Hu, Daniel Kriz, and Igor Kriz, *Field theories, stable homotopy theory, and Khovanov homology*, Topology Proc. **48** (2016), 327–360. MR3465966

[HRW21] Matthew Hogancamp, David E. V. Rose, and Paul Wedrich, *Link splitting deformation of colored Khovanov–Rozansky homology*, arXiv preprint `arXiv:2107.09590` (2021), 1–110.

[HS21] Kyle Hayden and Isaac Sundberg, *Khovanov homology and exotic surfaces in the 4-ball*, J. Reine Angew. Math. **809** (2024), 217–246, DOI 10.1515/crelle-2024-0001. MR4726569

[IJK16] Zur Izhakian, Marianne Johnson, and Mark Kambites, *Pure dimension and projectivity of tropical polytopes*, Adv. Math. **303** (2016), 1236–1263, DOI 10.1016/j.aim.2016.08.033. MR3552549

[IK22] Mee Seong Im and Mikhail Khovanov, *Topological theories and automata*, arXiv preprint `arXiv:2202.13398` (2022), 1–70.

[IK24] Mee Seong Im and Mikhail Khovanov, *One-dimensional topological theories with defects: the linear case*, Algebraic and topological aspects of representation theory, Contemp. Math., vol. 791, Amer. Math. Soc., [Providence], RI, [2024] ©2024, pp. 105–146, DOI 10.1090/conm/791/15874. MR4694170

[Jim85] Michio Jimbo, *A q-difference analogue of $U(\mathfrak{g})$ and the Yang-Baxter equation*, Lett. Math. Phys. **10** (1985), no. 1, 63–69, DOI 10.1007/BF00704588. MR797001

[JK14] Marianne Johnson and Mark Kambites, *Idempotent tropical matrices and finite metric spaces*, Adv. Geom. **14** (2014), no. 2, 253–276, DOI 10.1515/advgeom-2013-0034. MR3263426

[Jon85] Vaughan F. R. Jones, *A polynomial invariant for knots via von Neumann algebras*, Bull. Amer. Math. Soc. (N.S.) **12** (1985), no. 1, 103–111, DOI 10.1090/S0273-0979-1985-15304-2. MR766964

[Jon97] Vaughan F. R. Jones, *A polynomial invariant for knots via von Neumann algebras [MR0766964 (86e:57006)]*, Fields Medallists' lectures, World Sci. Ser. 20th Century Math., vol. 5, World Sci. Publ., River Edge, NJ, 1997, pp. 448–458, DOI 10.1142/9789812385215_0048. MR1622916

[Kas94] Masaki Kashiwara, *Crystal bases of modified quantized enveloping algebra*, Duke Math. J. **73** (1994), no. 2, 383–413, DOI 10.1215/S0012-7094-94-07317-1. MR1262212

[Kau90] Louis H. Kauffman, *An invariant of regular isotopy*, Trans. Amer. Math. Soc. **318** (1990), no. 2, 417–471, DOI 10.2307/2001315. MR958895

[Kho00] Mikhail Khovanov, *A categorification of the Jones polynomial*, Duke Math. J. **101** (2000), no. 3, 359–426, DOI 10.1215/S0012-7094-00-10131-7. MR1740682

[Kho02] Mikhail Khovanov, *A functor-valued invariant of tangles*, Algebr. Geom. Topol. **2** (2002), 665–741, DOI 10.2140/agt.2002.2.665. MR1928174

[Kho04] Mikhail Khovanov, *sl(3) link homology*, Algebr. Geom. Topol. **4** (2004), 1045–1081, DOI 10.2140/agt.2004.4.1045. MR2100691

[Kho05] Mikhail Khovanov, *Categorifications of the colored Jones polynomial*, J. Knot Theory Ramifications **14** (2005), no. 1, 111–130, DOI 10.1142/S0218216505003750. MR2124557

[Kho07] Mikhail Khovanov, *Triply-graded link homology and Hochschild homology of Soergel bimodules*, Internat. J. Math. **18** (2007), no. 8, 869–885, DOI 10.1142/S0129167X07004400. MR2339573

[Kho20] Mikhail Khovanov, *Universal construction of topological theories in two dimensions*, arXiv preprint `arXiv:2007.03361` (2020), 1–56.

[Kit19] Nitu Kitchloo, *Symmetry Breaking and Link Homologies I, II, III*, arXiv preprint `arXiv:1910.07443`, `arXiv:1910.07444`, `arXiv:1910.07516` (2019), 1–29, 1–34, 1–42.

[KK24] Mikhail Khovanov and Nitu Kitchloo, *A deformation of Robert-Wagner foam evaluation and link homology*, Algebraic and topological aspects of representation theory, Contemp. Math., vol. 791, Amer. Math. Soc., [Providence], RI, [2024] ©2024, pp. 147–204, DOI 10.1090/conm/791/15875. MR4694171

[KL79] David Kazhdan and George Lusztig, *Representations of Coxeter groups and Hecke algebras*, Invent. Math. **53** (1979), no. 2, 165–184, DOI 10.1007/BF01390031. MR560412

[KL94] Louis H. Kauffman and Sóstenes L. Lins, *Temperley-Lieb recoupling theory and invariants of* 3*-manifolds*, Annals of Mathematics Studies, vol. 134, Princeton University Press, Princeton, NJ, 1994, DOI 10.1515/9781400882533. MR1280463

[KL01] Thomas Kerler and Volodymyr V. Lyubashenko, *Non-semisimple topological quantum field theories for 3-manifolds with corners*, Lecture Notes in Mathematics, vol. 1765, Springer-Verlag, Berlin, 2001. MR1862634

[KL03] Anton Kapustin and Yi Li, *D-branes in Landau-Ginzburg models and algebraic geometry*, J. High Energy Phys. **12** (2003), 005, 44, DOI 10.1088/1126-6708/2003/12/005. MR2041170

[KM91] Robion Kirby and Paul Melvin, *The* 3*-manifold invariants of Witten and Reshetikhin-Turaev for* $\mathrm{sl}(2, \mathbf{C})$, Invent. Math. **105** (1991), no. 3, 473–545, DOI 10.1007/BF01232277. MR1117149

[KM19] P. B. Kronheimer and T. S. Mrowka, *Tait colorings, and an instanton homology for webs and foams*, J. Eur. Math. Soc. (JEMS) **21** (2019), no. 1, 55–119, DOI 10.4171/JEMS/831. MR3880205

[KQ20] Mikhail Khovanov and You Qi, *A faithful braid group action on the stable category of tricomplexes*, SIGMA Symmetry Integrability Geom. Methods Appl. **16** (2020), Paper No. 019, 32, DOI 10.3842/SIGMA.2020.019. MR4080798

[KQR21] Mikhail Khovanov, You Qi, and Lev Rozansky, *Evaluating thin flat surfaces*, Comm. Math. Phys. **385** (2021), no. 3, 1835–1870, DOI 10.1007/s00220-021-04011-3. MR4284003

[KR08a] Mikhail Khovanov and Lev Rozansky, *Matrix factorizations and link homology*, Fund. Math. **199** (2008), no. 1, 1–91, DOI 10.4064/fm199-1-1. MR2391017

[KR08b] Mikhail Khovanov and Lev Rozansky, *Matrix factorizations and link homology. II*, Geom. Topol. **12** (2008), no. 3, 1387–1425, DOI 10.2140/gt.2008.12.1387. MR2421131

[KR21] Mikhail Khovanov and Louis-Hadrien Robert, *Foam evaluation and Kronheimer-Mrowka theories*, Adv. Math. **376** (2021), Paper No. 107433, 59, DOI 10.1016/j.aim.2020.107433. MR4178907

[Kra10] Daniel Krasner, *Equivariant* sl(n)*-link homology*, Algebr. Geom. Topol. **10** (2010), no. 1, 1–32, DOI 10.2140/agt.2010.10.1. MR2580427

[KRW] Mikhail Khovanov, Louis-Hadrien Robert, and Emmanuel Wagner, *In-progress*, On foams in $\mathbb{R}^3$ between braid-like graphs in the plane.

[KS02] Mikhail Khovanov and Paul Seidel, *Quivers, Floer cohomology, and braid group actions*, J. Amer. Math. Soc. **15** (2002), no. 1, 203–271, DOI 10.1090/S0894-0347-01-00374-5. MR1862802

[Kup96a] Greg Kuperberg, *Noninvolutory Hopf algebras and* 3*-manifold invariants*, Duke Math. J. **84** (1996), no. 1, 83–129, DOI 10.1215/S0012-7094-96-08403-3. MR1394749

[Kup96b] Greg Kuperberg, *Spiders for rank* 2 *Lie algebras*, Comm. Math. Phys. **180** (1996), no. 1, 109–151. MR1403861

[Le00] Thang T. Q. Le, *Integrality and symmetry of quantum link invariants*, Duke Math. J. **102** (2000), no. 2, 273–306, DOI 10.1215/S0012-7094-00-10224-4. MR1749439

[Lew14] Lukas Lewark, *Rasmussen's spectral sequences and the* $\mathfrak{sl}_N$*-concordance invariants*, Adv. Math. **260** (2014), 59–83, DOI 10.1016/j.aim.2014.04.003. MR3209349

[LL16] Lukas Lewark and Andrew Lobb, *New quantum obstructions to sliceness*, Proc. Lond. Math. Soc. (3) **112** (2016), no. 1, 81–114, DOI 10.1112/plms/pdv068. MR3458146

[LLS20] Tyler Lawson, Robert Lipshitz, and Sucharit Sarkar, *Khovanov homotopy type, Burnside category and products*, Geom. Topol. **24** (2020), no. 2, 623–745, DOI 10.2140/gt.2020.24.623. MR4153651

[Lob12] Andrew Lobb, *A note on Gornik's perturbation of Khovanov-Rozansky homology*, Algebr. Geom. Topol. **12** (2012), no. 1, 293–305, DOI 10.2140/agt.2012.12.293. MR2916277

[LS14] Robert Lipshitz and Sucharit Sarkar, *A Khovanov stable homotopy type*, J. Amer. Math. Soc. **27** (2014), no. 4, 983–1042, DOI 10.1090/S0894-0347-2014-00785-2. MR3230817

[Lus91] G. Lusztig, *Quivers, perverse sheaves, and quantized enveloping algebras*, J. Amer. Math. Soc. **4** (1991), no. 2, 365–421, DOI 10.2307/2939279. MR1088333

[Lus10] George Lusztig, *Introduction to quantum groups*, Modern Birkhäuser Classics, Birkhäuser/Springer, New York, 2010. Reprint of the 1994 edition, DOI 10.1007/978-0-8176-4717-9. MR2759715

[LZ21] Lukas Lewark and Claudius Zibrowius, *Rasmussen invariants*, Mathematical Research Postcards `https://secure.math.ubc.ca/Links/mrp/cards/mrp2.pdf` (2021), 1–2.

[Mac09] Marco Mackaay, $\mathfrak{sl}(3)$*-foams and the Khovanov-Lauda categorification of quantum* $\mathfrak{sl}(k)$, arXiv preprint `arXiv:0905.2059` (2009), 1–20.

[Man07] Ciprian Manolescu, *Link homology theories from symplectic geometry*, Adv. Math. **211** (2007), no. 1, 363–416, DOI 10.1016/j.aim.2006.09.007. MR2313538

[Man16] Ciprian Manolescu, *An introduction to knot Floer homology*, Physics and mathematics of link homology, Contemp. Math., vol. 680, Amer. Math. Soc., Providence, RI, 2016, pp. 99–135, DOI 10.1090/conm/680. MR3591644

[MN08] Scott Morrison and Ari Nieh, *On Khovanov's cobordism theory for* $\mathfrak{su}_3$ *knot homology*, J. Knot Theory Ramifications **17** (2008), no. 9, 1121–1173, DOI 10.1142/S0218216508006555. MR2457839

[MOY98] Hitoshi Murakami, Tomotada Ohtsuki, and Shuji Yamada, *Homfly polynomial via an invariant of colored plane graphs*, Enseign. Math. (2) **44** (1998), no. 3-4, 325–360. MR1659228

[MR20] Andrew Manion and Raphael Rouquier, *Higher representations and cornered Heegaard Floer homology*, Preprint, `arXiv:2009.09627`, 1–147, 2020.

[MS09] Volodymyr Mazorchuk and Catharina Stroppel, *A combinatorial approach to functorial quantum $\mathfrak{sl}_k$ knot invariants*, Amer. J. Math. **131** (2009), no. 6, 1679–1713, DOI 10.1353/ajm.0.0082. MR2567504

[MSV09] Marco Mackaay, Marko Stošić, and Pedro Vaz, *$\mathfrak{sl}(N)$-link homology ($N \geq 4$) using foams and the Kapustin-Li formula*, Geom. Topol. **13** (2009), no. 2, 1075–1128, DOI 10.2140/gt.2009.13.1075. MR2491657

[MTV07] Marco Mackaay, Paul Turner, and Pedro Vaz, *A remark on Rasmussen's invariant of knots*, J. Knot Theory Ramifications **16** (2007), no. 3, 333–344, DOI 10.1142/S0218216507005312. MR2320159

[MTV13] Marco Mackaay, Paul Turner, and Pedro Vaz, *Erratum: A remark on Rasmussen's invariant of knots*, J. Knot Theory Ramifications **22** (2013), no. 1, 1392001, 1.

[MV07] Marco Mackaay and Pedro Vaz, *The universal sl_3-link homology*, Algebr. Geom. Topol. **7** (2007), 1135–1169, DOI 10.2140/agt.2007.7.1135. MR2336253

[MW18] Marco Mackaay and Ben Webster, *Categorified skew Howe duality and comparison of knot homologies*, Adv. Math. **330** (2018), 876–945, DOI 10.1016/j.aim.2018.03.034. MR3787560

[MY19] Marco Mackaay and Yasuyoshi Yonezawa, *$\mathfrak{sl}_N$-web categories and categorified skew Howe duality*, J. Pure Appl. Algebra **223** (2019), no. 5, 2173–2229, DOI 10.1016/j.jpaa.2018.07.013. MR3906545

[Nak94] Hiraku Nakajima, *Instantons on ALE spaces, quiver varieties, and Kac-Moody algebras*, Duke Math. J. **76** (1994), no. 2, 365–416, DOI 10.1215/S0012-7094-94-07613-8. MR1302318

[Obl19] Alexei Oblomkov, *Notes on matrix factorizations and knot homology*, Geometric representation theory and gauge theory, Lecture Notes in Math., vol. 2248, Springer, Cham, [2019] ©2019, pp. 83–127, DOI 10.1007/978-3-030-26856-5_3. MR4286062

[OR10] Alexei Oblomkov and Lev Rozansky, *Soergel bimodules and matrix factorizations*, arXiv preprint `arXiv:2010.14546` (2010), 1–51.

[OR19] Alexei Oblomkov and Lev Rozansky, *Dualizable link homology*, arXiv preprint `arXiv:1905.06511` (2019), 1–30.

[OR22] Alexei Oblomkov and Lev Rozansky, *Matrix factorizations and $\mathfrak{gl}(m|k)$-quantum invariants*, arXiv preprint `arXiv:2212.02665` (2022), 1–17.

[ORS13] Peter S. Ozsváth, Jacob Rasmussen, and Zoltán Szabó, *Odd Khovanov homology*, Algebr. Geom. Topol. **13** (2013), no. 3, 1465–1488, DOI 10.2140/agt.2013.13.1465. MR3071132

[OS06a] Peter Ozsváth and Zoltán Szabó, *An introduction to Heegaard Floer homology*, Floer homology, gauge theory, and low-dimensional topology, Clay Math. Proc., vol. 5, Amer. Math. Soc., Providence, RI, 2006, pp. 3–27. MR2249247

[OS06b] Peter Ozsváth and Zoltán Szabó, *Lectures on Heegaard Floer homology*, Floer homology, gauge theory, and low-dimensional topology, Clay Math. Proc., vol. 5, Amer. Math. Soc., Providence, RI, 2006, pp. 29–70. MR2249248

[OS18] Peter Ozsváth and Zoltán Szabó, *An overview of knot Floer homology*, Modern geometry: a celebration of the work of Simon Donaldson, Proc. Sympos. Pure Math., vol. 99, Amer. Math. Soc., Providence, RI, 2018, pp. 213–249, DOI 10.1090/pspum/099/08. MR3838884

[PT88] Józef H. Przytycki and Paweł Traczyk, *Invariants of links of Conway type*, Kobe J. Math. **4** (1988), no. 2, 115–139. MR945888

[QR16] Hoel Queffelec and David E. V. Rose, *The $\mathfrak{sl}_n$ foam 2-category: a combinatorial formulation of Khovanov-Rozansky homology via categorical skew Howe duality*, Adv. Math. **302** (2016), 1251–1339, DOI 10.1016/j.aim.2016.07.027. MR3545951

[QRS18] Hoel Queffelec, David E. V. Rose, and Antonio Sartori, *Annular Evaluation and Link Homology*, arXiv preprint `arXiv:1802.04131` (2018), 1–47.

[Ras10] Jacob Rasmussen, *Khovanov homology and the slice genus*, Invent. Math. **182** (2010), no. 2, 419–447, DOI 10.1007/s00222-010-0275-6. MR2729272

[Ras15] Jacob Rasmussen, *Some differentials on Khovanov-Rozansky homology*, Geom. Topol. **19** (2015), no. 6, 3031–3104, DOI 10.2140/gt.2015.19.3031. MR3447099

[Res87a] Nicolai Reshetikhin, *Quantized universal enveloping algebras, the Yang-Baxter equation and invariants of links, I*, USSR Academy of Sciences, Steklov Mathematical Institute, Leningrad Department `https://math.berkeley.edu/~reshetik/Publications/QGInv-1-1987.pdf` (1987), 1–68.

[Res87b] Nicolai Reshetikhin, *Quantized universal enveloping algebras, the Yang-Baxter equation and invariants of links, II*, USSR Academy of Sciences, Steklov Mathematical Institute, Leningrad Department `https://math.berkeley.edu/~reshetik/Publications/QGInv-2-1987.pdf` (1987), 1–24.

[Res91] Nicolai Reshetikhin, *Invariants of links and 3-manifolds related to quantum groups*, Proceedings of the International Congress of Mathematicians, Vol. I, II (Kyoto, 1990), Math. Soc. Japan, Tokyo, 1991, pp. 1373–1375.

[Reu20] David Reutter, *Semisimple four-dimensional topological field theories cannot detect exotic smooth structure*, J. Topol. **16** (2023), no. 2, 542–566. MR4577071

[Rez09] Reza Rezazadegan, *Seidel-Smith cohomology for tangles*, Selecta Math. (N.S.) **15** (2009), no. 3, 487–518, DOI 10.1007/s00029-009-0002-6. MR2551191

[Rou08] Raphael Rouquier, *2-Kac-Moody algebras*, arXiv preprint `arXiv:0812.5023` (2008), 1–66.

[Roz14] Lev Rozansky, *An infinite torus braid yields a categorified Jones-Wenzl projector*, Fund. Math. **225** (2014), no. 1, 305–326, DOI 10.4064/fm225-1-14. MR3205575

[RSP22] David Reutter and Christopher Schommer-Pries, *Semisimple field theories detect stable diffeomorphism*, arXiv preprint `arXiv:2206.10031` (2022), 1–63.

[RT90] N. Yu. Reshetikhin and V. G. Turaev, *Ribbon graphs and their invariants derived from quantum groups*, Comm. Math. Phys. **127** (1990), no. 1, 1–26. MR1036112

[RT91] N. Reshetikhin and V. G. Turaev, *Invariants of 3-manifolds via link polynomials and quantum groups*, Invent. Math. **103** (1991), no. 3, 547–597, DOI 10.1007/BF01239527. MR1091619

[RW16] David E. V. Rose and Paul Wedrich, *Deformations of colored $\mathfrak{sl}_N$ link homologies via foams*, Geom. Topol. **20** (2016), no. 6, 3431–3517, DOI 10.2140/gt.2016.20.3431. MR3590355

[RW20a] Louis-Hadrien Robert and Emmanuel Wagner, *A closed formula for the evaluation of foams*, Quantum Topol. **11** (2020), no. 3, 411–487, DOI 10.4171/qt/139. MR4164001

[RW20b] Louis-Hadrien Robert and Emmanuel Wagner, *Symmetric Khovanov-Rozansky link homologies* (English, with English and French summaries), J. Éc. polytech. Math. **7** (2020), 573–651, DOI 10.5802/jep.124. MR4086581

[San08] Taketo Sano, *Fixing the functoriality of Khovanov homology: a simple approach*, J. Knot Theory Ramifications **30** (2021), no. 11, Paper No. 2150074, 12, DOI 10.1142/S0218216521500747. MR4376719

[SS06] Paul Seidel and Ivan Smith, *A link invariant from the symplectic geometry of nilpotent slices*, Duke Math. J. **134** (2006), no. 3, 453–514, DOI 10.1215/S0012-7094-06-13432-4. MR2254624

[SS22] Catharina Stroppel and Joshua Sussan, *A Lie theoretic categorification of the coloured Jones polynomial*, J. Pure Appl. Algebra **226** (2022), no. 10, Paper No. 107043, 43, DOI 10.1016/j.jpaa.2022.107043. MR4385197

[ST01] Paul Seidel and Richard Thomas, *Braid group actions on derived categories of coherent sheaves*, Duke Math. J. **108** (2001), no. 1, 37–108, DOI 10.1215/S0012-7094-01-10812-0. MR1831820

[Str05] Catharina Stroppel, *Categorification of the Temperley-Lieb category, tangles, and cobordisms via projective functors*, Duke Math. J. **126** (2005), no. 3, 547–596, DOI 10.1215/S0012-7094-04-12634-X. MR2120117

[Str09] Catharina Stroppel, *Parabolic category $\mathcal{O}$, perverse sheaves on Grassmannians, Springer fibres and Khovanov homology*, Compos. Math. **145** (2009), no. 4, 954–992, DOI 10.1112/S0010437X09004035. MR2521250

[Sus07] Joshua Sussan, *Category O and sl(k) link invariants*, ProQuest LLC, Ann Arbor, MI, 2007. Thesis (Ph.D.)–Yale University. MR2710319

[Tia14] Yin Tian, *A categorification of $U_T(\mathfrak{sl}(1|1))$ and its tensor product representations*, Geom. Topol. **18** (2014), no. 3, 1635–1717, DOI 10.2140/gt.2014.18.1635. MR3228460

[TV92] V. G. Turaev and O. Ya. Viro, *State sum invariants of 3-manifolds and quantum 6j-symbols*, Topology **31** (1992), no. 4, 865–902, DOI 10.1016/0040-9383(92)90015-A. MR1191386

[Web17] Ben Webster, *Knot invariants and higher representation theory*, Mem. Amer. Math. Soc. **250** (2017), no. 1191, v+141, DOI 10.1090/memo/1191. MR3709726

[Wed19] Paul Wedrich, *Exponential growth of colored HOMFLY-PT homology*, Adv. Math. **353** (2019), 471–525, DOI 10.1016/j.aim.2019.06.023. MR3982970

[Wit89] Edward Witten, *Quantum field theory and the Jones polynomial*, Comm. Math. Phys. **121** (1989), no. 3, 351–399. MR990772

[Wit18] Edward Witten, *Two lectures on gauge theory and Khovanov homology*, Modern geometry: a celebration of the work of Simon Donaldson, Proc. Sympos. Pure Math., vol. 99, Amer. Math. Soc., Providence, RI, 2018, pp. 393–415, DOI 10.1090/pspum/099/01747. MR3838889

[Wu12a] Hao Wu, *Equivariant colored $\mathfrak{sl}(N)$-homology for links*, J. Knot Theory Ramifications **21** (2012), no. 2, 1250012, 104, DOI 10.1142/S0218216511009558. MR2885476

[Wu12b] Hao Wu, *On the Kauffman-Vogel and the Murakami-Ohtsuki-Yamada graph polynomials*, J. Knot Theory Ramifications **21** (2012), no. 10, 1250098, 40, DOI 10.1142/S0218216512500988. MR2949230

[Wu14] Hao Wu, *A colored $\mathfrak{sl}(N)$ homology for links in S^3*, Dissertationes Math. **499** (2014), 217, DOI 10.4064/dm499-0-1. MR3234803

[Wu15] Hao Wu, *Equivariant Khovanov-Rozansky homology and Lee-Gornik spectral sequence*, Quantum Topol. **6** (2015), no. 4, 515–607, DOI 10.4171/QT/70. MR3392963

[WW15] Katrin Wehrheim, *Floer field philosophy*, Advances in the mathematical sciences, Assoc. Women Math. Ser., vol. 6, Springer, [Cham], 2016, pp. 3–90, DOI 10.1007/978-3-319-34139-2_1. MR3654490

[Yon11] Yasuyoshi Yonezawa, *Quantum $(\mathfrak{sl}_n, \wedge V_n)$ link invariant and matrix factorizations*, Nagoya Math. J. **204** (2011), 69–123, DOI 10.1215/00277630-1431840. MR2863366

Department of Mathematics, United States Naval Academy, Annapolis, Maryland 21402

Email address: meeseongim@gmail.com

Department of Mathematics, Johns Hopkins University, Baltimore, Maryland 21218; and Department of Mathematics, Columbia University, New York, New York 10027

Email address: khovanov@jhu.edu; khovanov@math.columbia.edu

Contemporary Mathematics
Volume **817**, 2025
https://doi.org/10.1090/conm/817/16330

Quantum cluster mutations and reduced word graphs

Ivan C. H. Ip

Abstract. We give an algebraic proof of the independence of Coxeter moves involved in the construction of positive representations of split-real quantum groups, thus completing a gap in the original construction. To do this, we propose a new quantized version of Lusztig's Injectivity Lemma in the language of quantum cluster algebra, the proof of which by Tits' Lemma reduces to calculations involving sequences of Coxeter moves forming rank 3 cycles. We give a new, constructive proof of Tits' Lemma, and provide the required explicit computation of the quantum cluster mutations under these rank 3 cycles using certain cluster algebraic tricks via universally Laurent polynomials.

1. Introduction

Positive representations were introduced in [**8**] to study the representation theory of *split real quantum groups* $\mathcal{U}_q(\mathfrak{g}_\mathbb{R})$ associated to semisimple Lie algebra $\mathfrak{g}$, as well as its *modular double* $\mathcal{U}_{qq^\vee}(\mathfrak{g}_\mathbb{R})$ introduced in [**3**,**4**] in the regime where $|q|=1$. These representations are natural generalizations of a special family of representations of $\mathcal{U}_q(\mathfrak{sl}_2(\mathbb{R}))$ classified in [**20**] and studied in detail by Teschner *et al.* [**17**,**18**] from the physics point of view of quantum Liouville theory, which is characterized by the actions of *positive (essentially) self-adjoint operators* on the Hilbert space $L^2(\mathbb{R})$.

Based on quantizing the regular action on smooth functions on the totally positive flag variety $(G/B)_{>0}$, we constructed, for the simply-laced cases in [**8**,**11**] and non-simply-laced cases in [**12**], a family of irreducible representations $\mathcal{P}_\lambda$ of $\mathcal{U}_{qq^\vee}(\mathfrak{g}_\mathbb{R})$ with the Chevalley generators acting on certain Hilbert space as positive self-adjoint operators, parametrized by $\lambda \in P_{\mathbb{R}+}$ in the positive real-span of the dominant weights.

The construction of $\mathcal{P}_\lambda$ depends on a choice of reduced expression $\mathbf{i}_0$ of the longest element of the Weyl group W of G, and a crucial claim is that the representations corresponding to different reduced expression $\mathbf{i}_0'$ are unitary equivalent, and these equivalence should be independent of the choice of Coxeter moves from $\mathbf{i}_0$ to $\mathbf{i}_0'$, see Section 3.3 for more details. In [**11**], we claimed that this is due to a quantized version of *Lusztig's Injectivity Lemma* [**16**, Proposition 2.7].

2020 *Mathematics Subject Classification.* Primary 17B37, 13F60.

The author is supported by the Hong Kong RGC General Research Funds ECS #26303319.

To explain in more elementary terms, consider the 3×3 matrices, where we have the following relation for $a, b, c > 0$:

$$\begin{pmatrix} 1 & a & 0 \\ 0 & 1 & 0 \\ 0 & 0 & 1 \end{pmatrix} \begin{pmatrix} 1 & 0 & 0 \\ 0 & 1 & b \\ 0 & 0 & 1 \end{pmatrix} \begin{pmatrix} 1 & c & 0 \\ 0 & 1 & 0 \\ 0 & 0 & 1 \end{pmatrix} = \begin{pmatrix} 1 & a+c & ab \\ 0 & 1 & b \\ 0 & 0 & 1 \end{pmatrix} \in U_{>0}^{+} \tag{1.1}$$

where $U_{>0}^{+}$ denotes the totally positive upper unipotent matrices. This can be rewritten using the notation of root subspaces as

$$(a, b, c) \in \mathbb{R}_{>0}^3 \mapsto x_1(a)x_2(b)x_1(c) \in U_{>0}^{+} \tag{1.2}$$

Lusztig's Injectivity Lemma (see Lemma 2.5 below) states that this map $\mathbb{R}_{>0}^3 \longrightarrow U_{>0}^{+}$ is injective. Furthermore, there exists a *Coxeter move* $(1, 2, 1) \sim (2, 1, 2)$:

(1.3)
$$\begin{pmatrix} 1 & a & 0 \\ 0 & 1 & 0 \\ 0 & 0 & 1 \end{pmatrix} \begin{pmatrix} 1 & 0 & 0 \\ 0 & 1 & b \\ 0 & 0 & 1 \end{pmatrix} \begin{pmatrix} 1 & c & 0 \\ 0 & 1 & 0 \\ 0 & 0 & 1 \end{pmatrix} = \begin{pmatrix} 1 & 0 & 0 \\ 0 & 1 & a' \\ 0 & 0 & 1 \end{pmatrix} \begin{pmatrix} 1 & b' & 0 \\ 0 & 1 & 0 \\ 0 & 0 & 1 \end{pmatrix} \begin{pmatrix} 1 & 0 & 0 \\ 0 & 1 & c' \\ 0 & 0 & 1 \end{pmatrix}$$

or equivalently

$$x_1(a)x_2(b)x_1(c) = x_2(a')x_1(b')x_2(c') \tag{1.4}$$

where

$$a' = \frac{bc}{a+c}, \qquad b' = a + c, \qquad c' = \frac{ab}{a+c} \tag{1.5}$$

is an involution between the variables.

The above index can clearly be generalized from $(1, 2, 1)$ to any reduced words $\mathbf{i}_0$ for general semisimple Lie types, and sequence of Coxeter moves relating different longest word $\mathbf{i}_0 \sim \mathbf{i}_0'$. As a consequence of Lusztig's Injectivity Lemma, if we have a sequence of Coxeter moves $\mathbf{i}_0 \sim \mathbf{i}_0' \sim \cdots \sim \mathbf{i}_0$ that returns to itself, which by (1.4) induces a sequence of equalities

$$x_{i_1}(a_1) \cdots x_{i_N}(a_N) = x_{i_1'}(a_1') \cdots x_{i_N'}(a_N') = \cdots = x_{i_1}(a_1'') \cdots x_{i_N}(a_N'') \in U_{>0}^{+} \tag{1.6}$$

then $a_i = a_i''$ for all $i = 1, \ldots, N$.

As explained, in [**11**] we proposed a quantized version of Lusztig's Injectivity Lemma and cited [**1**], where the variables $a_1, \ldots, a_N$ are non-commutative quantum variables in a quantum torus algebra $\mathbf{T}_q$, and stated that if a_i, a_i'' are as in (1.6), then they are the same elements in $\mathbf{T}_q$. However, in a private communication, Linhui Shen kindly pointed out that this reference is not accurate, since the equation (1.4) is not well-defined for general Lie types, as we do not really have a well-defined notion of "quantum matrices" other than type A_n. Furthermore, the classical proof using the Bruhat decomposition cannot carry over to the quantum case, as the definition of the *quantum Bruhat cells* (which belongs to the dual space) used by [**1**] does not match with the setup of our proposed quantum Lusztig transformation. Therefore, the quantum analogue of the injectivity Lemma is actually unclear and non-trivial. In fact, this result was also assumed implicitly in the construction of the non-simply-laced case [**12**] without proof. Therefore this posed a gap in the completeness of the general construction of positive representations, which was criticized by Goncharov–Shen in [**10**, Section 5.2.6].

In [**13**], based on the idea from [**6, 21**], a quantum cluster realization of the positive representations of $\mathcal{U}_q(\mathfrak{g}_\mathbb{R})$ was proposed, and the quantum Lusztig's Injectivity Lemma can be understood as the independence of choice of quantum cluster mutations associated to different Coxeter moves between reduced word $\mathbf{i}_0 \sim \mathbf{i}_0'$. Unfortunately, again the statement was not explicitly proved since the cluster realization solely relies on the explicit algebraic expressions of the Chevalley generators obtained from [**8, 11, 12**] directly without changes. In Goncharov–Shen [**10**], the positive representations are instead essentially obtained geometrically by canonically quantizing certain regular functions, known as the potential functions, of cluster varieties arising from the moduli spaces of framed G-local systems, which do not depend on the choice of reduced expression of w_0, and in particular on the choice of Coxeter moves.

In this paper, we attempt to provide an alternative, rigorous treatment to fill the gap of the original construction of the positive representations, using the cluster algebraic realization studied in [**13**] which will be reviewed in the Preliminaries Section. In brief, given a reduced word $\mathbf{i}_0$ of the longest element $w_0 \in W$ of the Weyl group, we can associate with it a special quantum torus algebra $\mathbf{T}_q^{\mathbf{i}_0}$, and for any Coxeter move relating $\mathbf{i}_0 \sim \mathbf{i}_0'$, there exists a birational involution called the *quantum cluster mutation* $\mu^q : \mathbf{T}_q^{\mathbf{i}_0'} \longrightarrow \mathbf{T}_q^{\mathbf{i}_0}$.

The main result (Theorem 3.11) of the paper is the following:

MAIN THEOREM. *If $\mathcal{C}_1, \mathcal{C}_2 : \mathbf{i}_0 \longrightarrow \mathbf{i}_0'$ are two sequences of Coxeter moves, and*

$$\mu^q_{\mathcal{C}_j} : \mathbf{T}_q^{\mathbf{i}_0'} \longrightarrow \mathbf{T}_q^{\mathbf{i}_0}, \qquad j = 1, 2$$

are the induced compositions of quantum cluster mutations corresponding to the moves $\mathcal{C}_j$, then

$$\mu^q_{\mathcal{C}_q} = \mu^q_{\mathcal{C}_2}.$$

As a consequence of this result, we propose in Corollary 3.12 the quantized analogue of Lusztig's Injectivity Lemma by relating the quantum Lusztig's coordinates a_i with the quantum cluster variables X_i of $\mathbf{T}_q^{\mathbf{i}_0}$ via a monomial transformation, which completes the gaps in the original construction of the positive representations of $\mathcal{U}_{q\tilde{q}}(\mathfrak{g}_\mathbb{R})$.

To stay chronicle to the construction, we will prove the Main Theorem using methods independent of [**10**]. This is based on the Tits' Lemma [**23**], restated in Theorem 4.10, which reduces the study of the fundamental group of the *reduced words graph* under Coxeter moves, to cycles arising only from rank 3 cases. This reduces the proof of the Main Theorem to checking images of quantum cluster variables under only those sequences of quantum cluster mutations forming rank 3 cycles, and this follows from an application of a well-known result, due to [**7**] and restated in Theorem 6.2, of σ-periods of quantum tropical Y-seed mutations.

However, both of these results were proved implicitly. In this paper, we wish to be more transparent, and instead give a new, constructive proof of Tits' Lemma, and furthermore provide an explicitly computation of the image of the quantum cluster variables under the required rank 3 cycles of quantum cluster mutations using several algebraic tricks. We also find it illuminating to demonstrate some general principles in calculation involving quantum cluster mutations, in particular the merit of using *universally Laurent polynomials*.

While the argument and the idea behind the fix mentioned above may be elementary to experts in cluster algebra, unfortunately the setup of the notations and preliminaries required to state our results are quite overwhelming. In order for the paper to be self-contained, we decide to spend a substantial portion of the text to provide explanation of the notations and preliminaries required to state and prove the Main Theorem. This also clarifies, in Section 3, certain relationship between the original construction in [**11,12**] with the newer language of quantum cluster algebra not realized back then. Particularly important is that we explain the *ad hoc* choice of orientations of Dynkin diagram used in the original construction [**11**, Definition 5.3] between the quantum Lusztig's coordinates is nothing but an unimportant choice of arrows between the frozen variables, and therefore unnecessary.

The outline of the paper is as follows. In Section 2, we fix notations and provide definitions and main properties of Lusztig's coordinates on totally positive semigroups and their transformations via Coxeter moves, the quantum group $\mathcal{U}_q(\mathfrak{g})$ and the Drinfeld's double $\mathfrak{D}_q(\mathfrak{g})$, and finally the quantum cluster algebra with the associated quivers required in this paper. In Section 3, we propose a new definition of quantum Lusztig's transformations, and relates our new definition to the one originally proposed in the construction of positive representations [**11**], which reduces the problem to proving the Main Theorem. In Section 4, we introduce the terminologies required to state the Tits' Lemma, and in Section 5, we give a new constructive proof by induction on the size of the reduced word graphs. Finally in Section 6, we complete the proof of the Main Theorem by explicitly calculating the quantum cluster mutations forming rank 3 cycles.

2. Notations and preliminaries

In this section, we recall some terminologies required in this paper, keeping the exposition minimal to what is needed. For example, we do not recall the full construction of the basic quiver, nor the Hopf algebra structures of the quantum groups. For more elaborated details see [**11,13**], where the choice of most notations in this paper are based on.

2.1. Coxeter moves and Lusztig's data. Let $\mathfrak{g}$ be a simple[1] Lie algebra over $\mathbb{C}$. Let G be the real simple Lie group corresponding to the split real form $\mathfrak{g}_{\mathbb{R}}$ of the Lie algebra $\mathfrak{g}$, and let B, B^- be two opposite Borel subgroups containing a split real maximal torus $T = B \cap B^-$. Let $U^+ \subset B, U^- \subset B^-$ be the corresponding unipotent subgroups.

Let I be the root index of the Dynkin diagram of $\mathfrak{g}$ such that

$$|I| = n = \mathrm{rank}(\mathfrak{g}). \tag{2.1}$$

Let Φ be the set of roots of $\mathfrak{g}$, $\Phi_+ \subset \Phi$ be the positive roots, and $\Delta_+ = \{\alpha_i\}_{i\in I} \subset \Phi_+$ be the positive simple roots. Let $W = \langle s_i\rangle_{i\in I}$ be the Weyl group of Φ generated by the simple reflections $s_i := s_{\alpha_i}$.

DEFINITION 2.1. Let $(-,-)$ be a W-invariant inner product on the root lattice $\mathbb{Z}\Phi$. We define

$$a_{ij} := \frac{2(\alpha_i,\alpha_j)}{(\alpha_i,\alpha_i)}, \qquad i,j \in I \tag{2.2}$$

such that $A := (a_{ij})$ is the *Cartan matrix*.

[1]The results of this paper naturally extends to the semisimple case by taking direct product.

We normalize $(-,-)$ as follows: we choose the symmetrization factors (also called the *multipliers*) $d_i \in \mathbb{Q}$ such that for any $i \in I$,

$$d_i := \frac{1}{2}(\alpha_i, \alpha_i) = \begin{cases} 1 & i \text{ is long root or in the simply-laced case,} \\ \frac{1}{2} & i \text{ is short root in type } B, C, F, \\ \frac{1}{3} & i \text{ is short root in type } G, \end{cases} \tag{2.3}$$

and $(\alpha_i, \alpha_j) = -1$ when $i, j \in I$ are adjacent in the Dynkin diagram, such that

$$d_i a_{ij} = d_j a_{ji}. \tag{2.4}$$

DEFINITION 2.2. Let $w \in W$. We call a sequence[2]

$$\mathbf{i} = (i_1, \ldots, i_m), \qquad i_k \in I \tag{2.5}$$

a *reduced word* of w if

$$w = s_{i_1} s_{i_2} \cdots s_{i_m}$$

is a reduced expression. The set of reduced words of w is denoted by $\mathcal{I}(w)$.

We let

$$l(w) := |\mathbf{i}| := m \tag{2.6}$$

to be the length function.

Let w_0 be the unique longest element of W. Throughout the paper we let

$$N := l(w_0) \tag{2.7}$$

and denote by $\mathbf{i}_0$ a reduced word of w_0.

Let $J \subset I$ and denote by $W_J \subset W$ the subgroup generated by $\langle s_i \rangle_{i \in J}$.[3]

DEFINITION 2.3. Let $i, j \in I$. We denote by $\mathbf{p}_{ij}$ the reduced word ending with j of the longest element $w_{ij} \in W_{\{i,j\}}$. i.e.

$$\mathbf{p}_{ij} = ij, \quad jij, \quad ijij, \text{ or } ijijij$$

respectively for $a_{ij}a_{ji} = 0, 1, 2$ or 3.

A *Coxeter move* C_{ij} of $\mathbf{i}$ transforms involutively a reduced word $\mathbf{i} \in \mathcal{I}(w)$ to another reduced word $\mathbf{i}' \in \mathcal{I}(w)$ by the rule

$$C_{ij} : (\ldots, \mathbf{p}_{ij}, \ldots) \longleftrightarrow (\ldots, \mathbf{p}_{ji}, \ldots).$$

at certain letter position, keeping the other letters fixed. More explicitly, we have

$$(\ldots, i, j, \ldots) \longleftrightarrow (\ldots, j, i, \ldots), \qquad a_{ij}a_{ji} = 0 \tag{2.8}$$

$$(\ldots, j, i, j, \ldots) \longleftrightarrow (\ldots, i, j, i, \ldots), \qquad a_{ij}a_{ji} = 1 \tag{2.9}$$

$$(\ldots, i, j, i, j, \ldots) \longleftrightarrow (\ldots, j, i, j, i, \ldots), \qquad a_{ij}a_{ji} = 2 \tag{2.10}$$

$$(i, j, i, j, i, j) \longleftrightarrow (j, i, j, i, j, i), \qquad a_{ij}a_{ji} = 3 \tag{2.11}$$

The move (2.8) will be referred to as the *commutative Coxeter move*. In this paper we will not need to consider the move (2.11).

Next we recall Lusztig's definition of the totally positive semigroup. Let $\mathbf{i}_0 = (i_1, \ldots, i_N)$ be a reduced word of w_0.

---

[2]We may sometimes omit the comma in $\mathbf{i}$ for clarity.

[3]By convention $W_I = W$ and $W_\emptyset = \{e\}$.

DEFINITION 2.4. For any $i \in I$, there exists a homomorphism $SL_2(\mathbb{R}) \longrightarrow G$ denoted by

$$\begin{pmatrix} 1 & a \\ 0 & 1 \end{pmatrix} \mapsto x_i(a) \in U_i^+, \tag{2.12}$$

$$\begin{pmatrix} b & 0 \\ 0 & b^{-1} \end{pmatrix} \mapsto \chi_i(b) \in T, \tag{2.13}$$

$$\begin{pmatrix} 1 & 0 \\ c & 1 \end{pmatrix} \mapsto y_i(c) \in U_i^-, \tag{2.14}$$

called the *pinning* of G, where U_i^+ and U_i^- are the simple root subgroups of the unipotent subgroup U^+ and U^- respectively. The *positive unipotent semigroup* $U_{>0}^+$ is defined as the image of the map $\iota : \mathbb{R}_{>0}^N \longrightarrow U^+$ given by

$$\iota : (a_1, a_2, \ldots, a_N) \mapsto x_{i_1}(a_1)x_{i_2}(a_2)\ldots x_{i_N}(a_N). \tag{2.15}$$

LEMMA 2.5 ([**16**, Proposition 2.7]). *The map $\iota : \mathbb{R}_{>0}^N \longrightarrow U^+$ is injective: if*

$$x_{i_1}(a_1)x_{i_2}(a_2)\ldots x_{i_N}(a_N) = x_{i_1}(a_1')x_{i_2}(a_2')\ldots x_{i_N}(a_N'), \tag{2.16}$$

then $a_k = a_k'$ for every $k = 1, \ldots, N$.

PROOF. By moving a root subgroup to the other side we obtain

$$x_{i_1}(a_1 - a_1')x_{i_2}(a_2)\ldots x_{i_N}(a_N) = x_{i_2}(a_2')\ldots x_{i_N}(a_N'),$$

hence both sides belong to different Bruhat cells unless $a_1 - a_1' = 0$. The claim then follows by induction. □

DEFINITION 2.6. We define the totally positive semigroup to be

$$G_{>0} := U_{>0}^- T_{>0} U_{>0}^+, \tag{2.17}$$

where $U_{>0}^{\pm}$ is as above, and $T_{>0}$ are generated by the images $\chi_i(b)$ with $b \in \mathbb{R}_{>0}$.

In this paper, we are only interested in the identities related to the Coxeter moves (2.9)–(2.10), namely

PROPOSITION 2.7. *We have the following identities in $U_{>0}^+$: for $a, b, c \in \mathbb{R}_{>0}$ and $i, j \in I$,*

- *In the simply-laced case (i.e. $a_{ij}a_{ji} = 1$), (2.9) corresponds to the identity*

$$x_i(a)x_j(b)x_i(c) = x_j(a')x_i(b')x_j(c') \tag{2.18}$$

  where

$$a' = \frac{bc}{a+c}, \qquad b' = a + c, \qquad c' = \frac{ab}{a+c}. \tag{2.19}$$

- *In the doubly-laced case (i.e. $a_{ij}a_{ji} = 2$), (2.10) corresponds to the identity*

$$x_i(a)x_j(b)x_i(c)x_j(d) = x_j(d')x_i(c')x_j(b')x_i(a'), \tag{2.20}$$

  where

$$a' = \frac{abc}{R}, \qquad b' = \frac{R^2}{S}, \qquad c' = \frac{S}{R}, \qquad d' = \frac{bc^2d}{S}, \tag{2.21}$$

  and

$$R = ab + ad + cd, \qquad S = a^2b + d(a+c)^2. \tag{2.22}$$

In both cases, the transformation $(a, b, c, (d)) \mapsto (a', b', c', (d'))$ is an involution.

In particular since the transformation of the variables is an involution, we conclude that the image of $\mathbb{R}^N_{>0}$ in (2.15) is independent of the choice of $\mathbf{i}_0 \in \mathcal{I}(w_0)$.

2.2. Quantum groups $\mathcal{U}_q(\mathfrak{g})$ and $\mathfrak{D}_q(\mathfrak{g})$. For any finite dimensional complex semisimple Lie algebra $\mathfrak{g}$, Drinfeld [**2**] and Jimbo [**14**] associated to it a remarkable Hopf algebra $\mathcal{U}_q(\mathfrak{g})$ known as *quantum group*, which is certain deformation of the universal enveloping algebra. We follow the notations used in [**13**] for $\mathcal{U}_q(\mathfrak{g})$ as well as the Drinfeld's double $\mathfrak{D}_q(\mathfrak{g})$ of its Borel part.

In the following, we assume again that $\mathfrak{g}$ is of simple Dynkin type.

DEFINITION 2.8. Let d_i be the multipliers (2.3). We define

$$q_i := q^{d_i}, \tag{2.23}$$

which we will also write as

$$q_l := q, \tag{2.24}$$

$$q_s := \begin{cases} q^{\frac{1}{2}} & \text{if } \mathfrak{g} \text{ is of type } B, C, F, \\ q^{\frac{1}{3}} & \text{if } \mathfrak{g} \text{ is of type } G, \end{cases} \tag{2.25}$$

for the parameters corresponding to long and short roots respectively.

DEFINITION 2.9. We define $\mathfrak{D}_q(\mathfrak{g})$ to be the $\mathbb{C}(q_s)$-algebra generated by the elements

$$\{\mathbf{E}_i, \mathbf{F}_i, \mathbf{K}_i^{\pm 1}, \mathbf{K}_i'^{\pm 1}\}_{i \in I}$$

subject to the following relations (including the obvious relations involving $\mathbf{K}_i^{-1}$ and $\mathbf{K}_i'^{-1}$ which we omit below for simplicity):

$$\mathbf{K}_i\mathbf{E}_j = q_i^{a_{ij}}\mathbf{E}_j\mathbf{K}_i, \qquad \mathbf{K}_i\mathbf{F}_j = q_i^{-a_{ij}}\mathbf{F}_j\mathbf{K}_i, \tag{2.26}$$

$$\mathbf{K}_i'\mathbf{E}_j = q_i^{-a_{ij}}\mathbf{E}_j\mathbf{K}_i', \qquad \mathbf{K}_i'\mathbf{F}_j = q_i^{a_{ij}}\mathbf{F}_j\mathbf{K}_i', \tag{2.27}$$

$$\mathbf{K}_i\mathbf{K}_j = \mathbf{K}_j\mathbf{K}_i, \qquad \mathbf{K}_i'\mathbf{K}_j' = \mathbf{K}_j'\mathbf{K}_i', \qquad \mathbf{K}_i\mathbf{K}_j' = \mathbf{K}_j'\mathbf{K}_i, \tag{2.28}$$

$$[\mathbf{E}_i, \mathbf{F}_j] = \delta_{ij}\frac{\mathbf{K}_i - \mathbf{K}_i'}{q_i - q_i^{-1}}, \tag{2.29}$$

together with the *Serre relations* for $i \neq j$:

$$\sum_{k=0}^{1-a_{ij}} (-1)^k \begin{bmatrix} 1-a_{ij} \\ k \end{bmatrix}_q \mathbf{E}_i^k\mathbf{E}_j\mathbf{E}_i^{1-a_{ij}-k} = 0, \tag{2.30}$$

$$\sum_{k=0}^{1-a_{ij}} (-1)^k \begin{bmatrix} 1-a_{ij} \\ k \end{bmatrix}_q \mathbf{F}_i^k\mathbf{F}_j\mathbf{F}_i^{1-a_{ij}-k} = 0, \tag{2.31}$$

where the *q-binomial* is given by

$$\begin{bmatrix} n \\ k \end{bmatrix}_q := \frac{[n]_q!}{[k]_q![n-k]_q!} \tag{2.32}$$

with $[k]_q := \dfrac{q^k - q^{-k}}{q - q^{-1}}$ the *q-number* and $[n]_q! := \prod_{k=1}^{n}[k]_q$ the *q-factorial.*

DEFINITION 2.10. The quantum group $\mathcal{U}_q(\mathfrak{g})$ is defined as the quotient

$$\mathcal{U}_g(\mathfrak{g}) := \mathfrak{D}_q(\mathfrak{g})/\langle \mathbf{K}_i\mathbf{K}_i' = 1\rangle_{i \in I}, \tag{2.33}$$

and it inherits a well-defined Hopf algebra structure from $\mathfrak{D}_q(\mathfrak{g})$.

REMARK 2.11. $\mathfrak{D}_q(\mathfrak{g})$ is the *Drinfeld's double* of the quantum Borel subalgebra $\mathcal{U}_q(\mathfrak{b})$ generated by $\mathbf{E}_i$ and $\mathbf{K}_i$.

DEFINITION 2.12. We define the rescaled generators

$$\mathbf{e}_i := \left(\frac{\sqrt{-1}}{q_i - q_i^{-1}}\right)^{-1} \mathbf{E}_i, \qquad \mathbf{f}_i := \left(\frac{\sqrt{-1}}{q_i - q_i^{-1}}\right)^{-1} \mathbf{F}_i. \tag{2.34}$$

Then we can also consider $\mathfrak{D}_q(\mathfrak{g})$ as the $\mathbb{C}(q_s)$-algebra generated by

$$\{\mathbf{e}_i, \mathbf{f}_i, \mathbf{K}_i, \mathbf{K}_i'\}_{i\in I} \tag{2.35}$$

and $\mathcal{U}_q(\mathfrak{g})$ as the corresponding quotient, where the generators satisfy all the defining relations (2.26)–(2.31) above, except (2.29) which is modified to be

$$[\mathbf{e}_i, \mathbf{f}_j] = \delta_{ij}(q_i - q_i^{-1})(\mathbf{K}_i' - \mathbf{K}_i). \tag{2.36}$$

2.3. Quantum torus algebra. In this section we recall some definitions and properties concerning quantum torus algebra and their cluster realizations. One of the main purposes of this section is to fix the normalization used when defining the quantum cluster mutations, which has a vast variety of conventions in the literature [**1**, **6**, **10**, **15**].

DEFINITION 2.13. A *cluster seed* is a datum

$$\mathbf{Q} = (Q, Q_0, B, D) \tag{2.37}$$

where Q is a finite set, $Q_0 \subset Q$ is a subset called the *frozen subset*, $B = (b_{ij})_{i,j\in Q}$ is a skew-symmetrizable $\frac{1}{2}\mathbb{Z}$-valued matrix called the *exchange matrix*, and $D = \mathrm{diag}(d_j)_{j\in Q}$ is a diagonal $\mathbb{Q}$-matrix called the *multiplier* with $d_j^{-1} \in \mathbb{Z}$, such that

$$W := DB = -B^T D \tag{2.38}$$

is skew-symmetric. We assume that $b_{ij} \in \mathbb{Z}$ unless both $i, j \in Q_0$.

The *rank* of $\mathbf{Q}$ is defined to be the rank of the exchange matrix B, or equivalently the rank of the skew-symmetric matrix W.

In the following, we will only consider the case when there exists a *decoration*

$$\eta : Q \longrightarrow I \tag{2.39}$$

to the set of simple roots of a simple Dynkin diagram, such that $D = \mathrm{diag}(d_{\eta(j)})_{j\in Q}$ where $(d_i)_{i\in I}$ are the multipliers given in (2.3).

Let $\Lambda_{\mathbf{Q}}$ be a $\mathbb{Z}$-lattice with basis $\{\overrightarrow{e_i}\}_{i\in Q}$, and let $d = \min_{j\in Q}(d_j)$ [4] Also let

$$w_{ij} = d_i b_{ij} = -w_{ji}. \tag{2.40}$$

We define a skew symmetric $d\mathbb{Z}$-valued form $(-,-)$ on $\Lambda_{\mathbf{Q}}$ by

$$(\overrightarrow{e_i}, \overrightarrow{e_j}) := w_{ij}. \tag{2.41}$$

DEFINITION 2.14. Let q be a formal parameter. We define the *quantum torus algebra* $\mathcal{X}_q^{\mathbf{Q}}$ associated to a cluster seed $\mathbf{Q}$ to be an associative algebra over $\mathbb{C}[q^d]$ generated by $\{X_i^{\pm 1}\}_{i\in Q}$ subject to the relations

$$X_i X_j = q^{-2w_{ij}} X_j X_i, \qquad i, j \in Q. \tag{2.42}$$

[4]For general skew-symmetrizable B we define $d = \left(l.c.m.(d_j^{-1})_{j\in Q}\right)^{-1}$.

The generators $X_i \in \mathcal{X}_q^{\mathbf{Q}}$ are called the *quantum cluster variables*, and they are *frozen* if $i \in Q_0$. We denote by $\mathbf{T}_q^{\mathbf{Q}}$ the non-commutative field of fractions of $\mathcal{X}_q^{\mathbf{Q}}$, which is well defined since $\mathcal{X}_q^{\mathbf{Q}}$ is an Noetherian domain [**9**].

Alternatively, $\mathcal{X}_q^{\mathbf{Q}}$ is generated by $\{X_\lambda\}_{\lambda\in\Lambda_{\mathbf{Q}}}$ with $X_0 := 1$ subject to the relations

$$q^{(\lambda,\mu)}X_\lambda X_\mu = X_{\lambda+\mu}, \qquad \mu,\lambda \in \Lambda_{\mathbf{Q}}. \tag{2.43}$$

The merit of this notation is that if we define a star structure on the quantum cluster variables by

$$X_i^* := X_i, \qquad i \in Q \tag{2.44}$$

$$q^* := q^{-1} \tag{2.45}$$

then for any $\lambda \in \Lambda_{\mathbb{Q}}$ the generator X_λ is self-adjoint under the anti-involution $*$.

NOTATION 2.15. Comparing both realizations of $\mathcal{X}_q^{\mathbf{Q}}$, we shall identify

$$X_i = X_{\overrightarrow{e_i}}, \tag{2.46}$$

and define the notation

$$X_{i_1,\ldots,i_k} := X_{\overrightarrow{e_{i_1}}+\cdots+\overrightarrow{e_{i_k}}}, \tag{2.47}$$

or more generally for $n_1, \ldots, n_k \in \mathbb{Z}$,

$$X_{i_1^{n_1},\ldots,i_k^{n_k}} := X_{n_1\overrightarrow{e_{i_1}}+\cdots+n_k\overrightarrow{e_{i_k}}}. \tag{2.48}$$

DEFINITION 2.16. We associate to each cluster seed $\mathbf{Q} = (Q, Q_0, B, D)$ with decoration η a quiver, denoted again by $\mathbf{Q}$, with vertices labeled by Q and adjacency matrix $C = (c_{ij})_{i,j\in Q}$, where

$$c_{ij} = \begin{cases} b_{ij} & \text{if } d_i = d_j, \\ w_{ij} & \text{otherwise.} \end{cases} \tag{2.49}$$

We call $i \in Q$ a *short* (resp. *long*) *node* if $q_i := q^{d_i} = q_s$ (resp. $q_i = q_l = q$). An arrow $i \longrightarrow j$ represents the algebraic relation in $\mathcal{X}_q^{\mathbf{Q}}$

$$X_iX_j = q_*^{-2}X_jX_i \tag{2.50}$$

where $q_* = q_s$ if both i, j are short nodes, or $q_* = q$ otherwise.

Obviously one can recover the cluster seed from the quiver and the multipliers.

NOTATION 2.17. We will use squares to denote frozen nodes $i \in Q_0$ and circles otherwise. We will also use dashed arrows if $c_{ij} = \frac{1}{2}$, which only occur between frozen nodes.

We will represent the algebraic relations (2.50) by thick or thin arrows (see Figure 1) for display conveniences (thickness is *not* part of the data of the quiver). Thin arrows only occur in the non-simply-laced case between two short nodes.

Next we recall two operations on cluster seeds and their induced actions on the quantum torus algebra. Let $\mathbf{Q} = (Q, Q_0, B, D)$ and $\mathbf{Q}' = (Q', Q_0', B', D')$ denote two cluster seeds with the same vertex sets $Q = Q'$ and $Q_0 = Q_0'$.

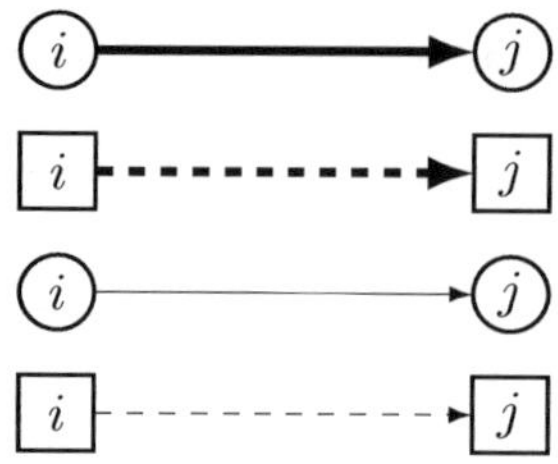

$$X_iX_j = q^{-2}X_jX_i$$
$$X_iX_j = q^{-1}X_jX_i$$
$$X_iX_j = q_s^{-2}X_jX_i$$
$$X_iX_j = q_s^{-1}X_jX_i$$

FIGURE 1. Arrows between nodes and their algebraic meaning.

DEFINITION 2.18. A *permutation* of a cluster seed $\sigma : \mathbf{Q} \longrightarrow \mathbf{Q}'$ is a bijection $\sigma : Q \longrightarrow Q'$ such that

$$\begin{aligned} \sigma(i) &= i, \qquad \forall i \in Q_0, \\ b'_{ij} &= b_{\sigma(i)\sigma(j)}, \\ d'_i &= d_{\sigma(i)}. \end{aligned} \tag{2.51}$$

It induces an isomorphism $\sigma^* : \mathbf{T}_q^{\mathbf{Q}'} \longrightarrow \mathbf{T}_q^{\mathbf{Q}}$ by

$$\sigma^*(X'_{\sigma(i)}) := X_i \tag{2.52}$$

where we denote by X'_i the quantum cluster variables of $\mathbf{T}_q^{\mathbf{Q}'}$.

DEFINITION 2.19. A *cluster mutation in direction* k, $\mu_k : \mathbf{Q} \longrightarrow \mathbf{Q}'$ is a bijection $\mu_k : Q \longrightarrow Q'$ such that $Q'_0 = Q_0$ and

$$\begin{aligned} b'_{ij} &= \begin{cases} -b_{ij} & \text{if } i = k \text{ or } j = k, \\ b_{ij} + \frac{b_{ik}|b_{kj}|+|b_{ik}|b_{kj}}{2} & \text{otherwise,} \end{cases} \\ d'_i &= d_i. \end{aligned} \tag{2.53}$$

DEFINITION 2.20. A cluster mutation $\mu_k : \mathbf{Q} \longrightarrow \mathbf{Q}'$ induces an isomorphism $\mu_k^q : \mathbf{T}_q^{\mathbf{Q}'} \longrightarrow \mathbf{T}_q^{\mathbf{Q}}$ called the *quantum cluster mutation*, defined by

$$\mu_k^q(X'_i) = \begin{cases} X_k^{-1} & \text{if } i = k, \\ X_i \prod_{r=1}^{|b_{ki}|}(1 + q_k^{2r-1}X_k) & \text{if } i \neq k \text{ and } b_{ki} \leq 0, \\ X_i \prod_{r=1}^{b_{ki}}(1 + q_k^{2r-1}X_k^{-1})^{-1} & \text{if } i \neq k \text{ and } b_{ki} \geq 0, \end{cases} \tag{2.54}$$

where again we denote by X'_i the quantum cluster variables of $\mathbf{T}_q^{\mathbf{Q}'}$.

REMARK 2.21. Sometimes we may just write X'_i for the element of $\mathbf{T}_q^{\mathbf{Q}}$ given by the formula of $\mu_k^q(X'_i)$ if no confusion arises.

REMARK 2.22. It is easy to see that μ_k^q preserves the $*$-structure (2.44). For example the second expression of (2.54) can be expanded and rewritten as

$$\mu_k^q(X'_i) = X_i + \sum_{m=1}^{n} \begin{bmatrix} n \\ m \end{bmatrix}_{q_k} X_{i,k^m}, \qquad n = |b_{ki}| \tag{2.55}$$

using the Notation 2.15, which is manifestly self-adjoint by (2.44) and the fact that the q-binomial (2.32) is invariant under $q \longleftrightarrow q^{-1}$. Such observation allows for easy consistency check of any formula involving quantum cluster mutations.

PROPOSITION 2.23 ([**6**]). *The quantum cluster mutation μ_k^q can be written as a composition of two homomorphisms*

$$\mu_k^q = \mu_k^\# \circ \mu_k^m, \tag{2.56}$$

where $\mu_k' : \mathbf{T}_q^{\mathbf{Q}'} \longrightarrow \mathbf{T}_q^{\mathbf{Q}}$ is a monomial transformation defined by

$$\mu_k^m(X_i') := \begin{cases} X_k^{-1} & \text{if } i = k, \\ X_i & \text{if } i \neq k \text{ and } b_{ki} \leq 0, \\ X_{i,k^{b_{ki}}} & \text{if } i \neq k \text{ and } b_{ki} \geq 0, \end{cases} \tag{2.57}$$

and $\mu_k^\# : \mathbf{T}_q^{\mathbf{Q}} \longrightarrow \mathbf{T}_q^{\mathbf{Q}}$ is a conjugation by the quantum dilogarithm function

$$\mu_k^\# := Ad_{\Psi^{q_k}(X_k)}, \tag{2.58}$$

where

$$\Psi^q := \prod_{r=0}^{\infty} (1 + q^{2r+1}x)^{-1}. \tag{2.59}$$

DEFINITION 2.24. A *coframed quiver* $\widehat{\mathbf{Q}}$ of $\mathbf{Q}$ consists of $\mathbf{Q}$ together with additional frozen vertex k' for each mutable vertex k of $\mathbf{Q}$ where the multiplier $d_{k'} := d_k$, and with additional arrow $k \longrightarrow k'$.

We will need the following simple observation about mutations of monomials.

PROPOSITION 2.25. *Let $\mu_{i_s} \circ \cdots \circ \mu_{i_1} : \mathbf{Q}' \longrightarrow \mathbf{Q}$ be a sequence of cluster mutations. Assume $X_\lambda \in \mathbf{T}_q^{\mathbf{Q}}$ is a monomial that commutes with any mutable variables X_k in $\mathbf{T}_q^{\mathbf{Q}}$. Then $\mu_{i_1}^q \circ \cdots \mu_{i_s}^q(X_\lambda)$ remains a monomial in $\mathbf{T}_q^{\mathbf{Q}'}$.*

Furthermore, if $\mu_{i_s} \circ \cdots \circ \mu_{i_1} = \sigma$ is a permutation of the coframed quiver $\widehat{\mathbf{Q}}$, then

$$\mu_{i_1}^q \circ \cdots \mu_{i_s}^q(X_\lambda) = \sigma^*(X_\lambda). \tag{2.60}$$

Here by Definition 2.18 if $\lambda = \sum \lambda_i \vec{e_i} \in \Lambda$, then

$$\sigma^*(X_\lambda) := X'_{\sum \lambda_i \vec{e}_{\sigma^{-1}(i)}}. \tag{2.61}$$

PROOF. If X_λ commutes with any mutable variables X_k, then the conjugation $\mu_k^\#$ is trivial, hence the only effect of a cluster mutation is given by the monomial transformation μ_k^m. Since μ_k^q is an algebra homomorphism, the image of the cluster mutation remains commutative with the mutable variables in the new seed.

Now as the image of X_λ remains a monomial, its mutation is equivalent to the mutation of its tropicalization. Notice that the monomial transformation μ_k^m coincides with the tropical Y-mutation [**15**], hence the tropicalization of the X-variables are just the c-vectors, with respect to the coframing in our setting. Hence if the sequence of cluster mutations is a permutation σ of mutable index keeping the coframing fixed, it induces a permutation of the c-vectors under the tropicalization, and the claim follows. □

2.4. Basic quivers. In this subsection, we recall the construction of the basic quiver $\mathbf{Q}(\mathbf{i})$ required for the quantum group embedding [10, 13], which helps clarifying the cluster nature of the quantum Lusztig's transformation studied in [11], to be explained in the next section. However, we do not require the full construction (which comes with extra vertices), but only the reduced part corresponding to a given reduced word.

DEFINITION 2.26. Let $i \in I$. The *elementary quiver* $\mathbf{J}(i)$ consists of

- The vertex set

$$Q = Q_0 = (I \setminus \{i\}) \cup \{i_l\} \cup \{i_r\} \tag{2.62}$$

which are all frozen;
- The multipliers $D = (d_j)_{j \in Q}$ which is the pull-back of the multipliers (2.3) from I under the natural projection $Q \longrightarrow I$ sending $\{i_l, i_r\}$ to i; and
- The adjacency matrix $C = (c_{ij})$ which is defined to be

$$c_{i_l,j} = c_{j,i_r} = \frac{d_i a_{ij}}{2}, \qquad j \in I \setminus \{i\}, \tag{2.63}$$

$$c_{i_l,i_r} = 1. \tag{2.64}$$

The vertices are organized in *levels*, such that the vertex $j \in I \setminus \{i\}$ is placed at level j, and $\{i_l, i_r\}$ are placed on the left and right hand side of level i respectively. The arrows between the vertex j and $\{i_l, i_r\}$ are dashed.

Intuitively, we call the set $(I\setminus\{i\})\cup\{i_l\}$ the *left frozen vertices*, and $(I\setminus\{i\})\cup\{i_r\}$ the *right frozen vertices*.

DEFINITION 2.27. Let $\mathbf{i} = (i_1, \ldots, i_m)$ be a reduced word. The *basic quiver* $\mathbf{Q}(\mathbf{i})$ is constructed by *amalgamating* the elementary quivers

$$\mathbf{Q}(\mathbf{i}) := \mathbf{J}(i_1) * \mathbf{J}(i_2) * \cdots * \mathbf{J}(i_m), \tag{2.65}$$

successively on each level, in the sense that the rightmost[5] frozen vertices of $\mathbf{J}(i_{k-1})$ are glued to the leftmost frozen vertices of $\mathbf{J}(i_k)$, with the weight of the corresponding arrows added together. Finally we remove all the vertices that are disjoint from the quiver, and redefine Q_0 so that the frozen vertices in the resulting quiver only consists of the left- and right-most vertices of each level.

We will write $\mathcal{X}_q^{\mathbf{i}} := \mathcal{X}_q^{\mathbf{Q}(\mathbf{i})}$ and $\mathbf{T}_q^{\mathbf{i}} := \mathbf{T}_q^{\mathbf{Q}(\mathbf{i})}$ to denote the quantum torus algebra associated to $\mathbf{Q}(\mathbf{i})$ and its field of fractions respectively.

DEFINITION 2.28. Following [13], we define the *canonical labeling* of $\mathbf{Q}(\mathbf{i})$ as

$$\{f_i^j : i \in I, j = 0, \ldots, n_i\} \tag{2.66}$$

where n_i is the number of occurrences of i in the reduced word $\mathbf{i}$, such that the vertex at level i and j-th position from the left is labeled by f_i^j.

EXAMPLE 2.29. Consider $\mathfrak{g} = \mathfrak{sl}_4$ and let $\mathbf{i}_0 = (1, 2, 1, 3, 2, 1)$ be a reduced word of $w_0 \in W$. The basic quiver is then the amalgamation of

$$\mathbf{Q}(\mathbf{i}_0) = \mathbf{J}(1) * \mathbf{J}(2) * \mathbf{J}(1) * \mathbf{J}(3) * \mathbf{J}(2) * \mathbf{J}(1) \tag{2.67}$$

see Figure 2.

[5] By definition, on $\mathbf{J}(i)$, except the i-th level, the leftmost and rightmost frozen vertices coincide.

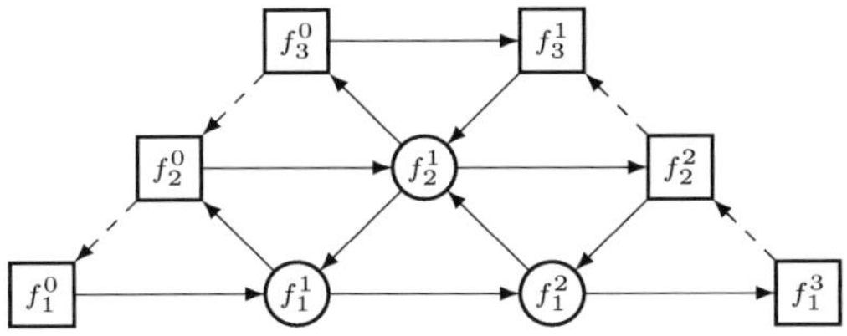

FIGURE 2. The basic quiver for $\mathbf{i}_0 = (1,2,1,3,2,1)$.

Recall [**13**, Section 7] that a Coxeter move $C : \mathbf{i} \mapsto \mathbf{i}'$ corresponds to a (sequence) of cluster mutations that transforms the quiver

$$\mathbf{Q}(\mathbf{i}) \longrightarrow \mathbf{Q}(\mathbf{i}'),$$

and its pullback induces an isomorphism $\mathbf{T}_q^{\mathbf{i}'} \longrightarrow \mathbf{T}_q^{\mathbf{i}}$.

To be more explicit, under the commutative Coxeter move $(\dots i, j \dots) \longrightarrow (\dots j, i, \dots)$ the quiver $\mathbf{Q}(\mathbf{i}) = \mathbf{Q}(\mathbf{i}')$ stays the same. For the other two cases (2.9)–(2.10), they correspond to the following mutations of the local pieces (2.65) inside $\mathbf{Q}(\mathbf{i})$ as

$$\mathbf{J}(i) * \mathbf{J}(j) * \mathbf{J}(i) \longrightarrow \mathbf{J}(j) * \mathbf{J}(i) * \mathbf{J}(j), \qquad a_{ij}a_{ji} = 1 \tag{2.68}$$

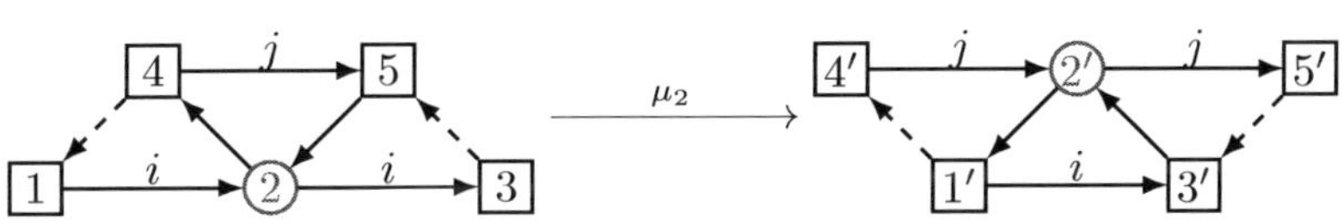

and

$$\mathbf{J}(i) * \mathbf{J}(j) * \mathbf{J}(i) * \mathbf{J}(j) \longrightarrow \mathbf{J}(j) * \mathbf{J}(i) * \mathbf{J}(j) * \mathbf{J}(i), \qquad a_{ij}a_{ji} = 2 \tag{2.69}$$

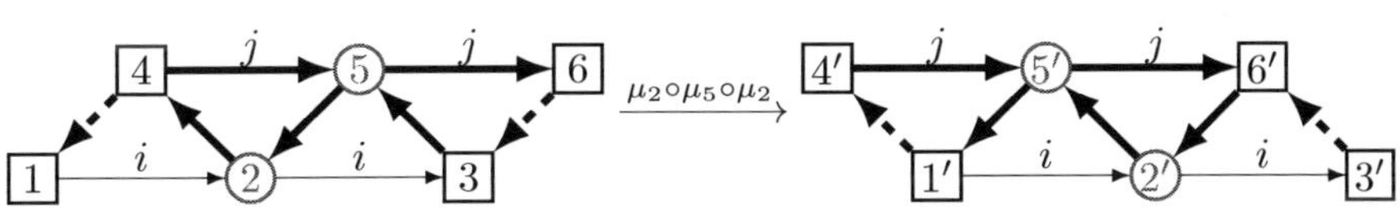

where nodes on level i are short.

NOTATION 2.30. Let

$$\mathcal{C} := \{C_1 \longrightarrow C_2 \longrightarrow C_3 \longrightarrow \cdots \longrightarrow C_k\} : \mathbf{i} \mapsto \mathbf{i}'$$

denotes a sequence of Coxeter moves, which corresponds to a sequence of cluster mutation $\mu_{i_m} \circ \cdots \circ \mu_{i_1} : \mathbf{Q}(\mathbf{i}) \longrightarrow \mathbf{Q}(\mathbf{i}')$. For simplicity we will write these as

$$\mu_{\mathcal{C}} := \mu_{i_m} \circ \cdots \circ \mu_{i_1} : \mathbf{Q}(\mathbf{i}) \longrightarrow \mathbf{Q}(\mathbf{i}')$$

$$\mu_{\mathcal{C}}^q := \mu_{i_1}^q \circ \cdots \circ \mu_{i_m}^q : \mathbf{T}_q^{\mathbf{i}'} \longrightarrow \mathbf{T}_q^{\mathbf{i}}$$

3. Quantum Lusztig's transformation

3.1. Rank 2 case. In [**11**], we proposed a quantization of the Lusztig's transformation (2.18) in order to construct the unitary transformation necessary for the definition of positive representations of $\mathcal{U}_q(\mathfrak{g}_\mathbb{R})$. It was stated as part of [**11**, Proposition 5.1], but we find it more logical to separately reproduce here as a Definition[6].

DEFINITION 3.1. Let $\mathbb{C}(q)\langle\alpha,\beta,\gamma\rangle$ be the noncommutative field of fractions generated by a triplet of quantum variables (α,β,γ) satisfying

$$\alpha\beta = q^{-2}\beta\alpha, \qquad \gamma\alpha = q^{-2}\alpha\gamma, \qquad \beta\gamma=\gamma\beta. \tag{3.1}$$

Then we define (α',β',γ') in $\mathbb{C}(q)\langle\alpha,\beta,\gamma\rangle$ by

$$\begin{aligned} \alpha' &:= (\alpha+\gamma)^{-1}\gamma\beta = \beta\gamma(\alpha+\gamma)^{-1} \\ \beta' &:= \alpha+\gamma \\ \gamma' &:= (\alpha+\gamma)^{-1}\alpha\beta = \beta\alpha(\alpha+\gamma)^{-1} \end{aligned} \tag{3.2}$$

such that they satisfy

$$\alpha'\beta'=\beta'\alpha', \qquad \beta'\gamma' = q^{-2}\gamma'\beta', \qquad \gamma'\alpha' = q^{-2}\alpha'\gamma'. \tag{3.3}$$

The same formula (3.2) applies to (α',β',γ') shows that it is an involution. Indeed we can check that

$$(\alpha'',\beta'',\gamma'') = (\alpha,\beta,\gamma). \tag{3.4}$$

In Notation 2.17, they can be presented as:

(3.5)
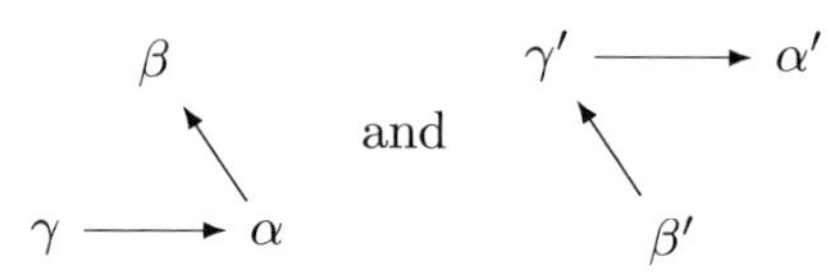

In other words, (3.2) defines an isomorphism

$$\mathbb{C}(q)\langle a',\beta',\gamma'\rangle \longrightarrow \mathbb{C}(q)\langle\alpha,\beta,\gamma\rangle$$

and serves as the quantum analogue of the Lusztig's transformation (2.18). These collection of non-commutative variables and the corresponding reduced words are written *formally* as

$$\mathbf{x}_i(\alpha)\mathbf{x}_j(\beta)\mathbf{x}_i(\gamma) = \mathbf{x}_j(\alpha')\mathbf{x}_i(\beta')\mathbf{x}_j(\gamma') \tag{3.6}$$

However, all these relations in fact suggest that the quantum Lusztig's transformation can be recast into the framework of quantum cluster mutation, which was not realized in [**11**]. Let

$$X_1 = \gamma, \qquad X_2 = q\alpha\gamma^{-1}, \qquad X_4=\beta, \tag{3.7}$$

$$X_4' = \gamma', \qquad X_2' = q\alpha'\gamma'^{-1}, \qquad X_1' = \beta'. \tag{3.8}$$

Then the q commutation relation coincides with the subquivers in (2.68), up to the arrows between the frozen nodes:

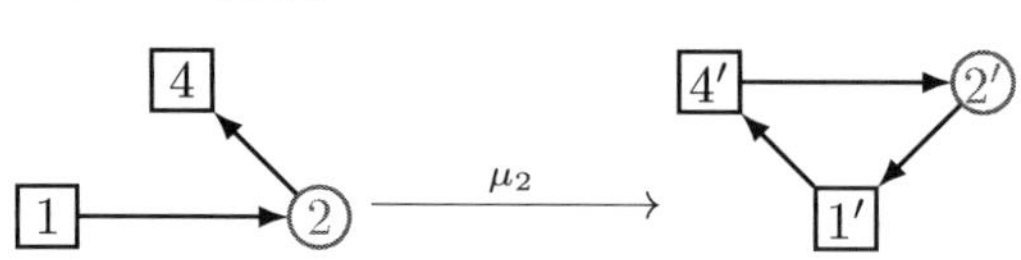

[6]We swap $q \longleftrightarrow q^{-1}$ compared with the definition used in [**11**].

Furthermore, one can rewrite (3.2) using (3.7) as

$$X_2' = X_2^{-1} \tag{3.9}$$

$$X_1' = X_1(1 + qX_2) \tag{3.10}$$

$$X_4' = X_4(1 + qX_2^{-1})^{-1} \tag{3.11}$$

which is exactly the quantum cluster mutation (2.54) at the vertex 2! (See Remark 2.21.)

Using Notation 2.15, note that (3.7)–(3.8) is equivalent to

$$\alpha = X_{1,2}, \qquad \beta = X_4, \qquad \gamma = X_1, \tag{3.12}$$

$$\alpha' = X_{2,4}', \qquad \beta' = X_1', \qquad \gamma' = X_4', \tag{3.13}$$

so that the variables are manifestly self-adjoint under (2.44).

Moreover, notice that the quantum cluster mutation formula does not depend on the adjacency between the frozen index (1 and 4) at all, so in principle we are free to choose *any* commutation relation between β and γ, which shows that the *ad hoc* choices of *orientation* arrows made in [**11**, Definition 5.3] was actually unnecessary.

Now the doubly-laced case follows the same strategy using the Coxeter move (2.10). Unfortunately this was not discussed in [**12**], so the following Definition completes its description. We define the analogue of (2.20) as follows.

DEFINITION 3.2. Define the field of fractions $\mathbb{C}(q)\langle\alpha,\beta,\gamma,\delta\rangle$ generated by $\{\alpha,\beta,\gamma,\delta\}$ with multipliers $\{1,2,1,2\}$ respectively, such that they satisfy the commutation relations given by the diagram

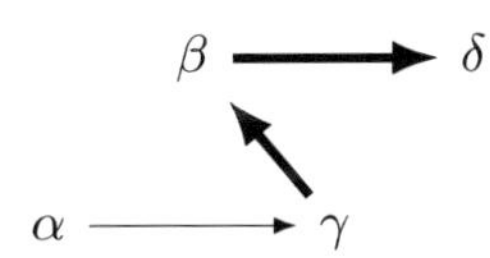

i.e.

$$\alpha\gamma = q^{-1}\gamma\alpha, \qquad \alpha\delta = q^2\delta\alpha, \qquad \beta\delta = q^{-2}\delta\beta \tag{3.14}$$

and commute otherwise. Then we define $\{\alpha',\beta',\gamma',\delta'\}$ in $\mathbb{C}(q)\langle\alpha,\beta,\gamma,\delta\rangle$ by

$$\alpha' := \gamma\beta\alpha R^{-1} \qquad \beta' := R^2S^{-1}$$

$$\gamma' := SR^{-1} \qquad \delta' := \beta\gamma\delta\gamma S^{-1}$$

where (written non-commutatively)

$$R := \beta\alpha + \delta\alpha + \delta\gamma, \qquad S := \alpha\beta\alpha + (\alpha+\gamma)\delta(\alpha+\gamma). \tag{3.15}$$

Note that the elements R and S commute.

PROPOSITION 3.3. *The transformation $(\alpha,\beta,\gamma,\delta) \mapsto (\alpha',\beta',\gamma',\delta')$ is an involution (with the same formula for the primed variables), and the new variables satisfy the commutation relation given by the diagram*

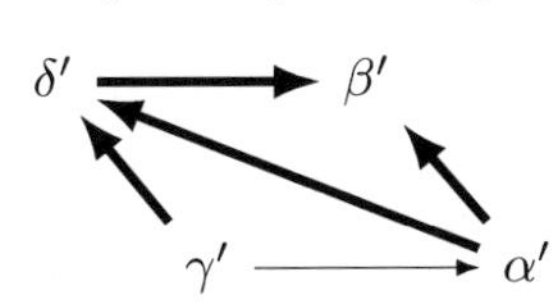

This will serve as the quantum analogue of the Lusztig transform, again written *formally* as

$$\mathbf{x}_i(\alpha)\mathbf{x}_j(\beta)\mathbf{x}_i(\gamma)\mathbf{x}_j(\delta) = \mathbf{x}_j(\delta')\mathbf{x}_i(\gamma')\mathbf{x}_j(\beta')\mathbf{x}_i(\alpha'). \tag{3.16}$$

We can again recast this as quantum cluster mutations (in the skew-symmetrizable case of (2.10)) as follows.

PROPOSITION 3.4. *Let*

$$X_1 = \alpha, \qquad X_2 = q^{-\frac{1}{2}}\alpha^{-1}\gamma, \qquad X_4 = \beta, \qquad X_5 = q^{-1}\beta^{-1}\delta, \tag{3.17}$$

$$X_1' = \gamma', \qquad X_2' = q^{\frac{1}{2}}\alpha'\gamma'^{-1}, \qquad X_4' = \delta', \qquad X_5' = q\beta'\delta'^{-1}. \tag{3.18}$$

Then we have the commutation relations given by the subquivers of (2.69) *(up to arrows between frozen nodes)*

and the new variables are given by the quantum mutations

$$(X_1, X_2, X_4, X_5) \xrightarrow{\mu_2\mu_5\mu_2} (X_1', X_2', X_4', X_5') \tag{3.19}$$

respectively.

Note that again (3.17)–(3.18) is equivalent to

$$\alpha = X_1, \qquad \beta = X_4, \qquad \gamma = X_{1,2}, \qquad \delta = X_{4,5} \tag{3.20}$$

$$\alpha' = X_{1,2}', \qquad \beta' = X_{4,5}', \qquad \gamma' = X_1', \qquad \delta' = X_4' \tag{3.21}$$

so that the variables are also self-adjoint under (2.44).

3.2. General case. Armed with the above observations, we can now restate properly the construction of [**11**] in terms of quantum cluster mutations.

Let $\mathbf{i}_0 = (i_1, \ldots, i_N) \in \mathcal{I}(w_0)$ be a longest reduced word, and consider the basic quiver $\mathbf{Q}(\mathbf{i}_0)$ constructed in Definition 2.27 equipped with the canonical labeling.

DEFINITION 3.5. Let $\mathbf{H}$ be an orientation of the Dynkin diagram with weights $\frac{1}{2}a_{ij}a_{ji}$ (presented as thin or thick dashed arrows with all the nodes frozen), and define the amalgamation

$$\widetilde{\mathbf{Q}}(\mathbf{i}_0) := \mathbf{H} * \mathbf{Q}(\mathbf{i}_0) \tag{3.22}$$

gluing $\mathbf{H}$ from the left matching the decoration, so that all the dashed arrows between the leftmost frozen nodes of $\mathbf{Q}(\mathbf{i}_0)$ are either canceled or replaced by a solid one.

DEFINITION 3.6. For $1 \le k \le N$, if $i_k = i$, define v_k such that i_k is the v_k-th occurrence of i from the left of $\mathbf{i}_0$, and define the monomials $\alpha_k \in \mathcal{X}_q^{\widetilde{\mathbf{Q}}(\mathbf{i}_0)}$ by

$$\alpha_k := X_{\sum_{j=1}^{v_k} f_i^j} \tag{3.23}$$

i.e. it is (up to a q factor) the product of all the X variables on the i-th level of $\widetilde{\mathbf{Q}}(\mathbf{i}_0)$ from the leftmost frozen node to the k-th position of $\mathbf{i}_0$.

Similarly define α'_k using the quantum cluster variables X'_i of the quiver $\mathbf{Q}(\mathbf{i}'_0)$ obtained from the mutation sequence corresponding to the Coxeter moves (2.68)–(2.69). Then we can generalize (3.6) and (3.16) as

$$\mathbf{x}_{i_1}(\alpha_1)\cdots\mathbf{x}_{i_N}(\alpha_N) = \mathbf{x}_{i'_1}(\alpha'_1)\cdots\mathbf{x}_{i'_N}(\alpha'_N). \tag{3.24}$$

We can now restate the construction of the quantum Lusztig transformation from [**11**], the proof of which follows directly from the explicit commutation relations between the α_k's defined in [**11**, Definition 5.3].

PROPOSITION 3.7. *In the simply-laced case,* $(\alpha_k)_{k=1}^N$ *coincides with* [**11**, Definition 5.3] *corresponding to the orientation* $\mathbf{H}$ *of the Dynkin diagram. Furthermore,* $(\alpha'_k)_{k=1}^N$ *coincides with the mutation rule in* [**11**, Proposition 5.4].

EXAMPLE 3.8. Consider type A_3 and an orientation of the Dynkin diagram by

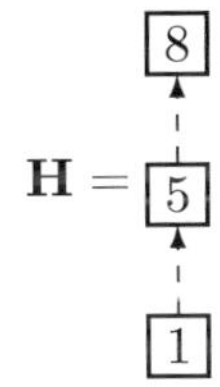

If the resulting quiver $\widetilde{\mathbf{Q}}(\mathbf{i}_0)$ corresponding to the initial seed $\mathbf{i}_0 = (121321)$ is labled by

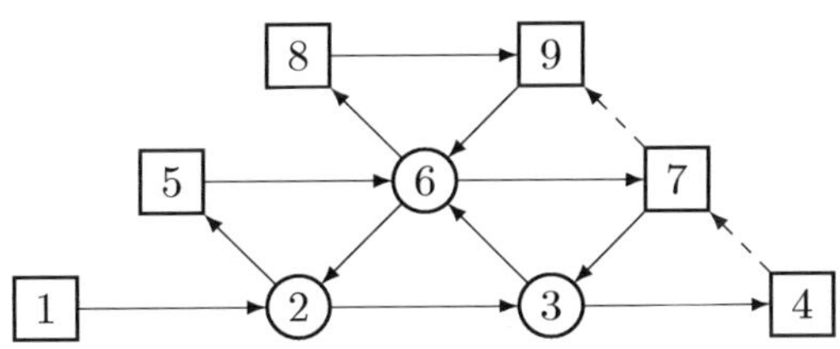

then

$$\alpha_4 = X_8,$$
$$\alpha_2 = X_5, \qquad \alpha_5 = X_{5,6},$$
$$\alpha_1 = X_1, \qquad \alpha_3 = X_{1,2}, \qquad \alpha_6 = X_{1,2,3}.$$

Notice that the rightmost frozen variables are not involved in the definition of α_k.

REMARK 3.9. In the original construction [**11**] where we swap $q^{-1} \longleftrightarrow q$ in (3.1), in order to be compatible with quantum mutation formula (2.54), an equivalent identification of the quantum Lusztig's variable with the quantum cluster variables is to remove the *leftmost frozen nodes* from $\mathbf{Q}(\mathbf{i}_0)$ instead, and also introduce an extra inverse for the X_i variables in (3.12)–(3.13). Similar modifications can be done for the non-simply laced and higher rank case.

REMARK 3.10. Up to a choice of the orientation $\mathbf{H}$ of the Dynkin diagram, we expect that the variables $(\alpha_k)_{k=1}^N$ can be identified with the canonical quantization of the *primary coordinates* P_i of [**10**] with a certain Poisson structure.

3.3. Positive representations. Given a longest reduced word $\mathbf{i}_0$, we give the construction of the positive representation $\mathcal{P}_\lambda^{\mathbf{i}_0}$ of $\mathcal{U}_q(\mathfrak{g}_\mathbb{R})$ in [**11**] in terms of self-adjoint operators induced from the transformation of the quantum Lusztig's coordinate defined in the previous section. Later in [**13**], we implicitly understood the relationship between quantum Lusztig transformation and quantum cluster mutation, and gave the construction in terms of quantum cluster algebra using the basic quiver $\mathbf{Q}(\mathbf{i}^{op}) * \mathbf{Q}(\mathbf{i})$.

In both construction, the representation depends on the choice of $\mathbf{i}_0$, but for any other longest reduced words $\mathbf{i}_0'$, we have a unitary equivalence

$$\mathcal{P}_\lambda^{\mathbf{i}_0} \simeq \mathcal{P}_\lambda^{\mathbf{i}_0'} \tag{3.25}$$

since Coxeter moves correspond to a sequence of mutations, which under any polarization becomes a unitary transformation induced by the *quantum dilogarithm function* [**11**].

Furthermore, the action of the Chevalley generators $\mathbf{f}_i, \mathbf{K}_i'$ can be written down explicitly as certain telescopic sums and monomials respectively, and is shown in both construction to depend only on the reduced word $\mathbf{i}_0$, and coincide with the doubling of the *Feigin's homomorphism* [**19**]. This was generalized to the Kac–Moody case in [**10**].

However, the problem lies with the $\mathbf{e}_i, \mathbf{K}_i$ generators. For each reduced words $\mathbf{i}_0'$ ending with letter i, there exists a finite difference operator known as the *canonical operator*. The action of the $\mathbf{e}_i, \mathbf{K}_i$ generators for $\mathbf{i}_0$ is then constructed by first taking any reduced word $\mathbf{i}_0'$ which ends with the letter i, take the canonical operator in this special case, and transform this operator back to the given $\mathbf{i}_0$ using a sequence of unitary transformations induced by the Coxeter moves. In the quantum cluster algebraic setting, the canonical operator is realized as a sum of two cluster monomials, and we perform the corresponding quantum cluster mutations in both the $\mathbf{Q}(\mathbf{i})$ and $\mathbf{Q}(\mathbf{i}^{op})$ copy of the basic quiver.

Therefore the crucial step is to show that the construction of the representation of $\mathbf{e}_i, \mathbf{K}_i$ does not depend on the choice of Coxeter moves from $\mathbf{i}_0$ to $\mathbf{i}_0'$. As pointed out in the Introduction, this statement is not explicit proved in both the simply-laced case [**11**] and the non-simply-laced case [**12**], in which the cluster realization in [**13**] solely depends on.

With the discussion in the previous subsections, we can now state the Main Theorem as follows, the proof of which will be done in Section 6 as a consequence of the Tits' Lemma.

THEOREM 3.11. *Let $\mathbf{i}_0, \mathbf{i}_0'$ be two longest reduced word, and let $C_1, C_2 : \mathbf{i}_0 \mapsto \mathbf{i}_0'$ be two sequences of Coxeter moves. Let*

$$\mu_{C_j}^q : \mathbf{T}_q^{\mathbf{i}_0'} \longrightarrow \mathbf{T}_q^{\mathbf{i}_0}, \qquad j = 1, 2 \tag{3.26}$$

be the induced sequences of quantum cluster mutations corresponding to the Coxeter moves given by (2.9)–(2.10)*. Then*

$$\mu_{C_1}^q = \mu_{C_2}^q. \tag{3.27}$$

Since the quantum Lusztig's coordinates are related to the quantum cluster variables by (3.23), we can now state the quantum version of the injectivity Lemma as follows, which completes the gap of the construction of positive representations described in [**11, 12**].

COROLLARY 3.12. *If* $\mathbf{i}_0 = (i_1, \ldots, i_N)$ *is a longest reduced word, and*

$$\mathbf{x}_{i_1}(\alpha_1) \cdots \mathbf{x}_{i_N}(\alpha_N) = \mathbf{x}_{i_1}(a'_1) \cdots \mathbf{x}_{i_N}(a'_N) \tag{3.28}$$

in the sense of (3.6) *or* (3.16)*, where the right hand side is obtained from the left hand side by a sequence of Coxeter moves* $\mathbf{i}_0 \mapsto \cdots \mapsto \mathbf{i}_0$ *that returns to itself, then*

$$a'_i = \alpha_i, \qquad i = 1, \ldots, N \tag{3.29}$$

as elements in the noncommutative field of fractions generated by α_i.

4. Tits' Lemma

In this section, we introduce the terminologies and preliminaries in order to state Tits' Lemma [**23**], which reduces the study of the fundamental group of the reduced words graph under Coxeter moves, to cycles arising only from rank 3 cases. This reduces the proof of Theorem 3.11 to explicitly checking only mutation sequences forming rank 3 cycles. We will give a new, constructive proof of this result in Section 5.

4.1. Definitions.

DEFINITION 4.1. A *reduced word graph* of $w \in W$ is a graph $\mathcal{G} := \mathcal{G}(w) := (\mathcal{G}_0, \mathcal{G}_1)$ where the vertex set $\mathcal{G}_0 := \mathcal{I}(w)$, and we have an edge joining $\mathbf{i}, \mathbf{i}' \in \mathcal{I}(w)$ if the two words are related by a Coxeter move. We will use dashed edge to indicate move (2.8), and thick edges for (2.10) or (2.11). It is well known that any two reduced word of $w \in W$ are related by a sequence of Coxeter moves, hence $\mathcal{G}$ is a connected graph.

In the construction of positive representations, Coxeter moves of the first type (2.8) amounts to relabeling of the index of variables, and are not important in the algebraic structures of the representations. Therefore it is useful to consider a contraction of the reduced word graph and identify vertices that are related by the commutative Coxeter moves (2.8).

DEFINITION 4.2. We define the equivalence classes $[\mathbf{i}]$ of *commutative words*, where $\mathbf{i} \sim \mathbf{i}'$ if they are related by a commutative Coxeter move (2.8).

DEFINITION 4.3. A *quotient reduced word graph* $\overline{\mathcal{G}}$ is a graph where the vertex set $\overline{\mathcal{G}}_0$ is the set of equivalence classes of commutative words $[\mathbf{i}]$, and an edge joins $[\mathbf{i}]$ and $[\mathbf{i}']$ if there exists representatives $\mathbf{i} \in [\mathbf{i}]$ and $\mathbf{i}' \in [\mathbf{i}']$ that are related by a non-commutative Coxeter move (2.9)–(2.11).

In other words $\overline{\mathcal{G}}$ is obtained from a contraction of the reduced word graph $\mathcal{G}$ where we contract any edges $\mathbf{i} - \mathbf{i}'$ corresponding to commutative Coxeter moves, and identify the vertices $\mathbf{i} = \mathbf{i}'$.

NOTATION 4.4. We will use the *stack notation* to denote the equivalence classes, such that commuting letter are written in the same position. For example in type A_3 consider $\mathbf{i} = 212321$, so that the position of $1, 3$ can be interchanged by the commutative Coxeter move. We then use the notation

$$[\mathbf{i}] = \{213213, 213231, 231213, 231231\} =: 2^3_1 2^3_1.$$

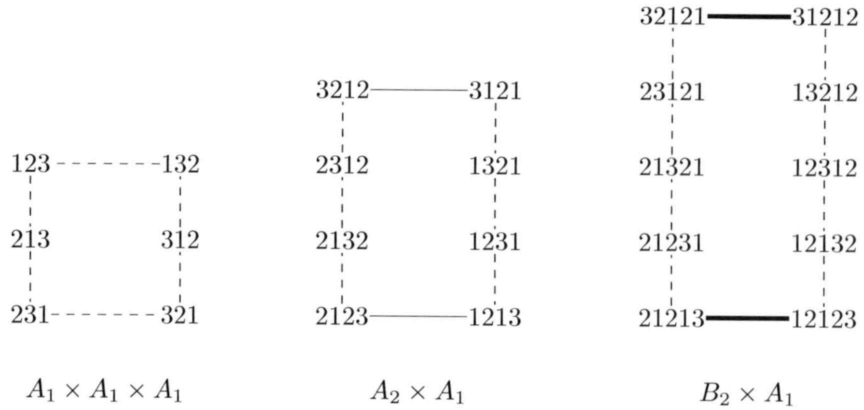

FIGURE 3. Reduced word graph of decomposable rank 3 types. Here the index $1, 2, 3$ correspond to the respective type factors indicated.

EXAMPLE 4.5. If $\mathfrak{g}$ is of rank 2, for the longest element $w_0 \in W$, obviously its reduced word graph just consists of a connected pair of vertices:

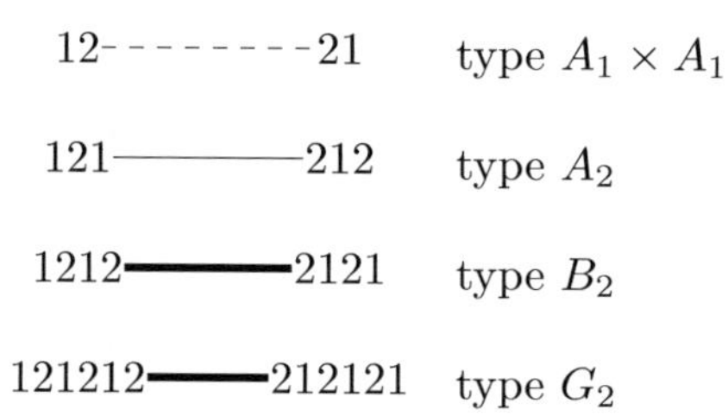

EXAMPLE 4.6. If $\mathfrak{g}$ is decomposable type of rank 3, i.e. of type $A_1 \times A_1 \times A_1$, $A_2 \times A_1$ or $B_2 \times A_1$, then $\mathcal{G}$ is a 6-, 8- or 10-gon respectively, see Figure 3.

The most important cases we need in this paper are the simple rank 3 cases.

THEOREM 4.7. *The quotient reduced word graph for the longest element $w_0 \in W$ is a*

- *8-gon if $\mathfrak{g}$ is of type A_3,*
- *14-gon if $\mathfrak{g}$ is of type B_3.*

PROOF. This can be done by computing the reduced word graph $\mathcal{G}$ explicitly. For type A_3 it is given by Figure 4 . The corresponding Coxeter moves among the vertices is clear by comparing the different expression of the reduced words.

On the other hand the quotient graph $\overline{\mathcal{G}}$ forms an 8-gon as in Figure 5.

The quotient reduced word graph $\overline{\mathcal{G}}$ in type B_3 is given in Figure 6. Here $a_{12}a_{21} = 2$ so that we need to consider Coxeter move (2.10): $1212 - 2121$.

The original reduced graph $\mathcal{G}$ can be obtained from $\overline{\mathcal{G}}$ by replacing the commutative words with the corresponding subgraph that involves only the commutative Coxeter move (2.8). For example, in $\mathcal{G}$ we have the subgraph as in Figure 7. In particular, while $\overline{\mathcal{G}}$ consists of 14 vertices, $\mathcal{G}$ consists of 42 vertices. $\square$

An important observation from Example 4.6 and Theorem 4.7 is the following

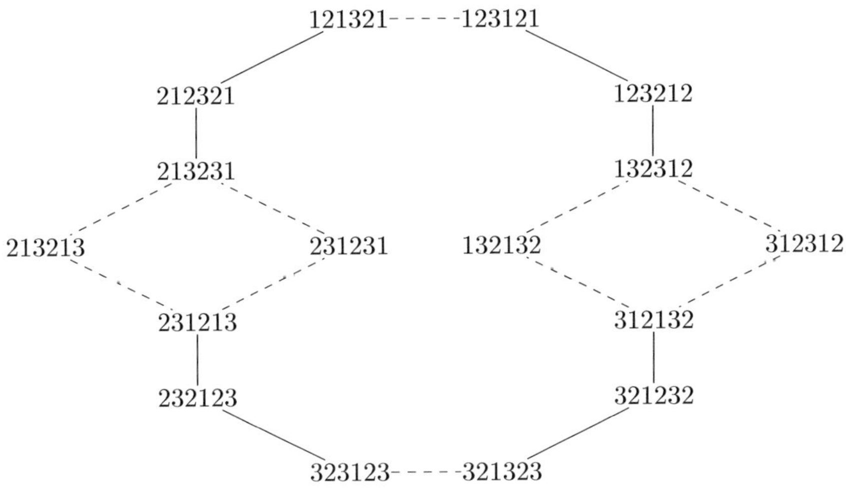

FIGURE 4. Reduced word graph in type A_3.

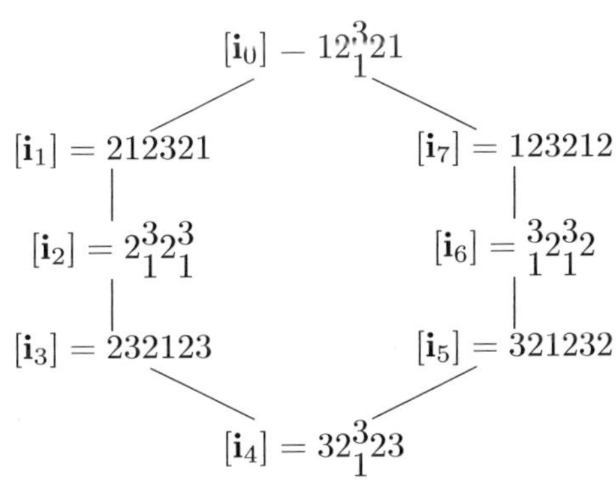

FIGURE 5. Quotient reduced word graph in type A_3.

COROLLARY 4.8. *In any reduced word graph $\mathcal{G}$ of rank 3, the subgraph consisting of words with the same last letter is connected.*

4.2. Statement.

DEFINITION 4.9. Let $\mathcal{G}$ be a connected graph. We say that the fundamental group $\pi_1(\mathcal{G})$ (without specified basepoint) is generated by closed paths $\alpha_1, \dots, \alpha_n$ if the fundamental group $\pi_1(\mathcal{G}, *)$ over any fixed based points $* \in \mathcal{G}_0$ is generated by closed paths of the form $\gamma^{-1} \circ \alpha_i \circ \gamma$ where γ is any path joining $*$ to any vertex of α_i.

We can now state one of the main results that is required for the construction of positive representations as follows.

THEOREM 4.10 (Tits' Lemma). *Let $w \in W$ with $l(w) \geq 3$. The fundamental group of the reduced word graph $\mathcal{G}(w)$ is generated by closed paths of the form*

(a) *Commuting Coxeter moves at different positions:*

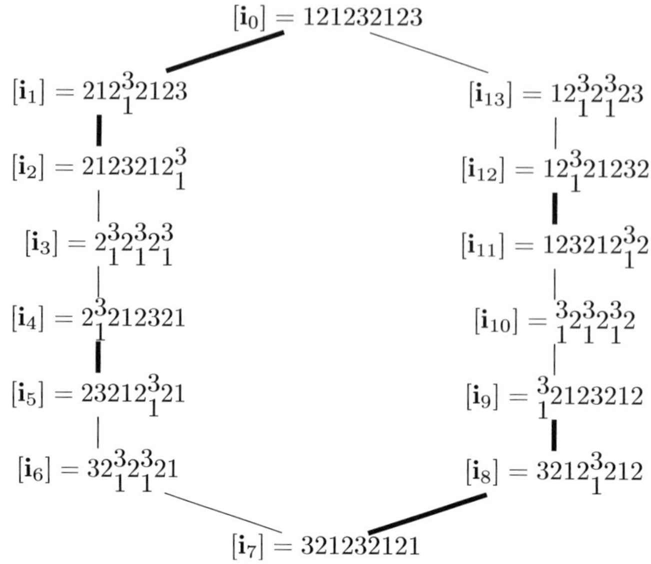

FIGURE 6. Quotient reduced word graph in type B_3.

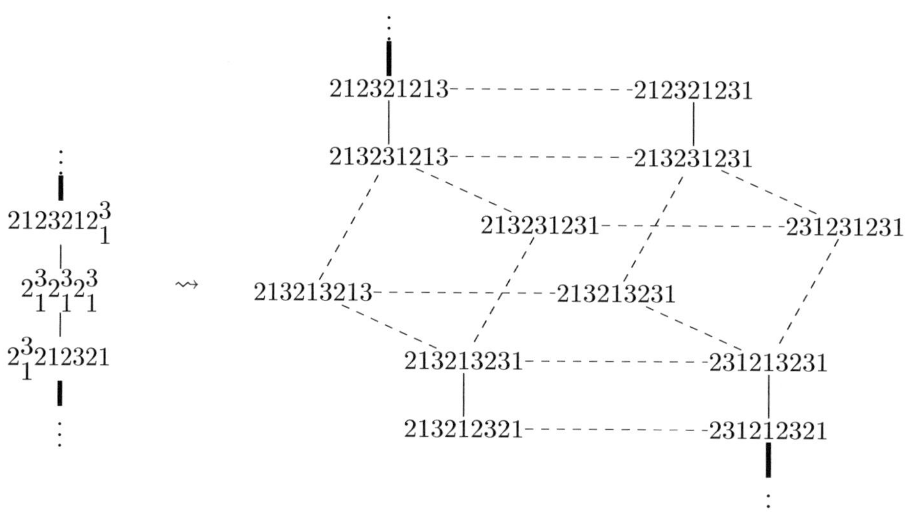

FIGURE 7. A subgraph of $\mathcal{G}$ obtained from $\overline{\mathcal{G}}$.

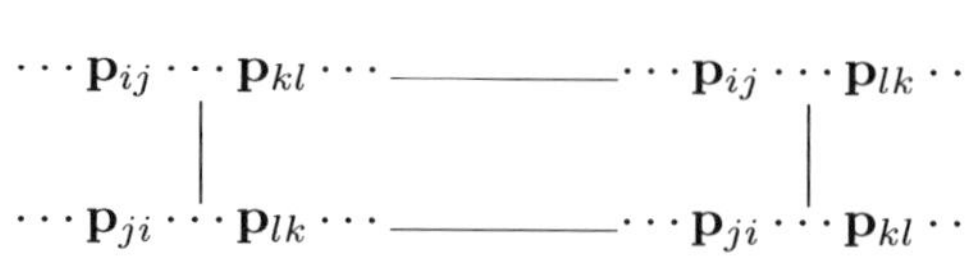

(b) *Rank 3 cycles:*

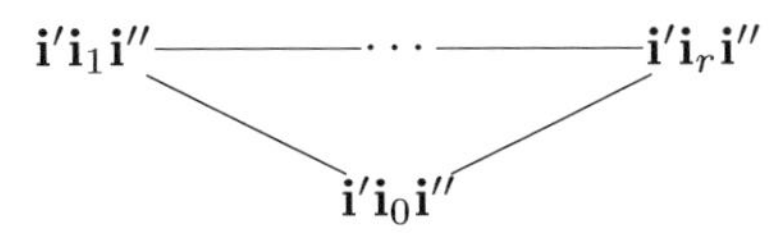

where $\mathbf{i}_0$, ..., $\mathbf{i}_r$ *and their Coxeter moves involve only three elements of* I.

By reducing the rank 3 cycles in type (b) to the quotient graph, we have

COROLLARY 4.11. *The fundamental group* $\pi_1(\overline{\mathcal{G}})$ *of the quotient reduced word graph is generated by squares, octagons and 14-gons, corresponding to commuting Coxeter moves, and rank 3 Coxeter cycles of type* A_3 *and* B_3 *respectively.*

PROOF. The only things to check is that for the rank 3 cycles, after quotienting the commutative Coxeter moves, the index forms the subtypes

- $A_1 \times A_1 \times A_1$: trivial
- $A_2 \times A_1$: single edge
- A_3: octagon
- $B_2 \times A_1$ single edge
- B_3: 14-gon

by Theorem 4.7. □

5. Proof of Theorem 4.10

We are ready to give a new, constructive proof of Tits' Lemma by producing an explicit decomposition of a given reduced word cycle. One can consider this argument as explicitly spelling out and simplifying the original abstract proof presented in [**23**, Section 4] using homotopies and morphisms of graphs.

REMARK 5.1. Throughout the proof, all the closed paths are not assumed to be simple, but self intersection naturally breaks the paths into unions of smaller cycles, as well as trivial cycles of the form $\gamma^{-1} \circ \gamma$ in the fundamental group $\pi_1(\mathcal{G})$. Therefore without loss of generality we may just present pictures of simple paths.

Let $\beta \in \pi_1(\mathcal{G}(w))$ be an oriented closed path in $\mathcal{G}(w)$. We first make two elementary observations:

(*) If there exists a path δ cutting β into two components β_1, β_2, then β is generated by $\{\alpha_1, \ldots, \alpha_n\}$ if both closed paths $\beta_1\delta$ and $\delta^{-1}\beta_2$ are generated by $\{\alpha_1, \ldots, \alpha_n\}$ (in the sense of Definition 4.9).

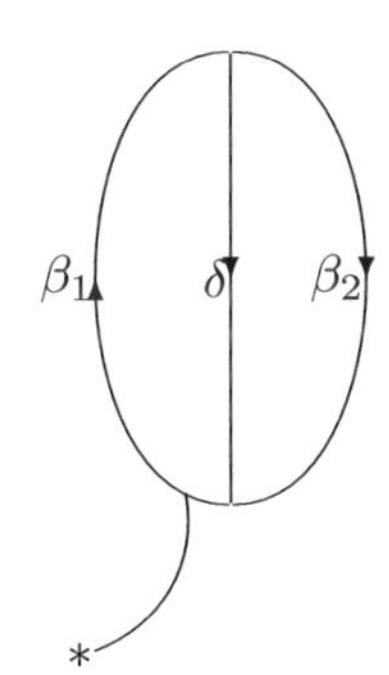

(**) If β is a closed path such that all the words (vertices) share the same letter $i \in I$ in the last position, then $l(ws_i) < l(w)$, and β is isomorphic to a closed path $\overline{\beta}$ of $\mathcal{G}(ws_i)$ where the vertex set consists of the same words with all the last letter removed.

Conversely if $\overline{\beta}$ is a closed path in $\mathcal{G}(w)$ and $l(ws_i) > l(w)$, then $\overline{\beta}$ is isomorphic to a closed path β in $\mathcal{G}(ws_i)$ by adding the letter i at the end of each word of $\overline{\beta}$.

Define the *last letter map* $\omega : \mathcal{I}(w) \longrightarrow I$ by

$$\mathbf{i} = (i_1, \ldots, i_N) \mapsto i_N. \tag{5.1}$$

For any path β of $\mathcal{G}$ we will also identify it as a subset of $\mathcal{G}_0$ with its collection of vertices (words). Define

$$m_i(\beta) = \#\{\mathbf{i} \in \beta : \omega(\mathbf{i}) = i\} \tag{5.2}$$

to be the number of words in β ending with the letter i.

We are now ready to explain the proof of Theorem 4.10 by induction. We will use a double induction on $n = l(w)$ and the number $m = |\omega(\beta)|$ of distinct last letter appearing in the words of β.

For the base case of $n = 3$, this is just a restatement of the type (b) cycles.

For the base case of $m = 1$, every word $\mathbf{i} \in \beta$ is of the form $\mathbf{i} = (\ldots, i)$ for some letter $i \in I$. In particular $l(ws_i) < l(w)$ and hence by (**) the cycle is isomorphic to a cycle $\overline{\beta}$ in $\mathcal{G}(ws_i)$, and we can use induction on n to decompose $\overline{\beta}$, which by the converse of (**) also decompose β.

Now assume $n > 3$ and $m > 1$. Fix an index $i \in \omega(\beta)$. Then by the definition of reduced word graph, there exists vertices $\mathbf{i}_1, \mathbf{i}_1', \mathbf{i}_2, \mathbf{i}_2'$ of β such that

- $\mathbf{i}_1$ is of the form $(\ldots, \mathbf{p}_{ji})$ and $\mathbf{i}_1'$ of the form $(\ldots, \mathbf{p}_{ij})$ for some letter $j \neq i$, and they are related by a single Coxeter move C_{ij} involving the last letter.
- $\mathbf{i}_2$ is of the form $(\ldots, \mathbf{p}_{ki})$ and $\mathbf{i}_2'$ of the form $(\ldots, \mathbf{p}_{ik})$ for some letter $k \neq i$, and they are related by a single Coxeter move C_{ik} involving the last letter.
- There exists a subpath β' of β joining $\mathbf{i}_1$ and $\mathbf{i}_2$, such that $\omega(\mathbf{i}) = i$ for any $\mathbf{i} \in \beta'$, see Figure 8. The remaining part of the loop β is denoted by β''.

In particular, $l(ws_i) = l(ws_j) = l(ws_k) < l(w)$.

Note that in the case $m = 2$, we allow $j = k$. The fact that $n > 3$ then guarantees that we can find $\mathbf{i}_1 \neq \mathbf{i}_2$ to apply the induction below.

If we let $J = \{i, j, k\}$ and consider the longest element $w_J \in W_J$, then by construction $l(ww_J) < l(w)$. Hence if $\overline{\mathbf{i}^k}\mathbf{p}_{ij}$ is a reduced word for w_J, then there exists $\mathbf{j}_j, \mathbf{j}_j' \in \mathcal{I}(w)$ which coincide with each other except for the last portion where they end differently with $\overline{\mathbf{i}^k}\mathbf{p}_{ji}$ and $\overline{\mathbf{i}^k}\mathbf{p}_{ij}$ respectively. Similarly for $\mathbf{j}_k, \mathbf{j}_k'$ which is the same as $\mathbf{j}_j, \mathbf{j}_j'$ except the last portion with $j \longleftrightarrow k$.

Note that we allow $j = k$ in which case $\overline{\mathbf{i}^j} = \overline{\mathbf{i}^k}$ is empty, and $\mathbf{j}_j, \mathbf{j}_j'$ coincides with $\mathbf{j}_k, \mathbf{j}_k'$ respectively.

Since $\mathbf{i}_1$ and $\mathbf{j}_j$ both ends with $\mathbf{p}_{ji}$, there exists a sequence of Coxeter moves $\mathcal{C} := \{C_l\}$ joining $\mathbf{i}_1$ to $\mathbf{j}_j$ which does not involve the $\mathbf{p}_{ji}$ portion and commute with C_{ij}. The same sequence $\mathcal{C}$ also transform $\mathbf{i}_1'$ into $\mathbf{j}_j'$, hence they form a sequence of commuting squares as in Figure 9. Similarly for $\mathbf{i}_2, \mathbf{i}_2'$ to $\mathbf{j}_k$, $\mathbf{j}_k'$.

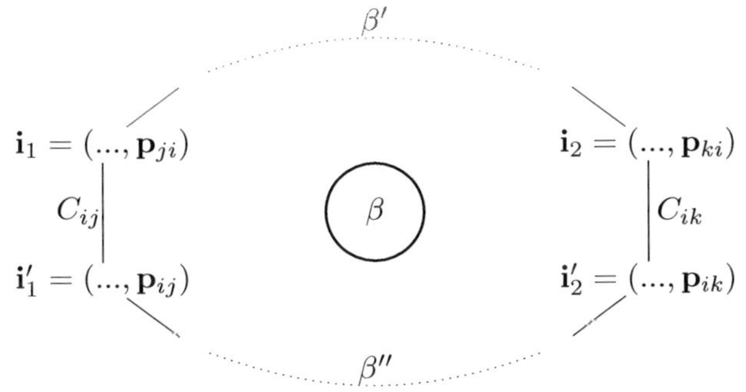

FIGURE 8. Decomposition of the cycle β. All the vertices appearing in the red path end with the letter i.

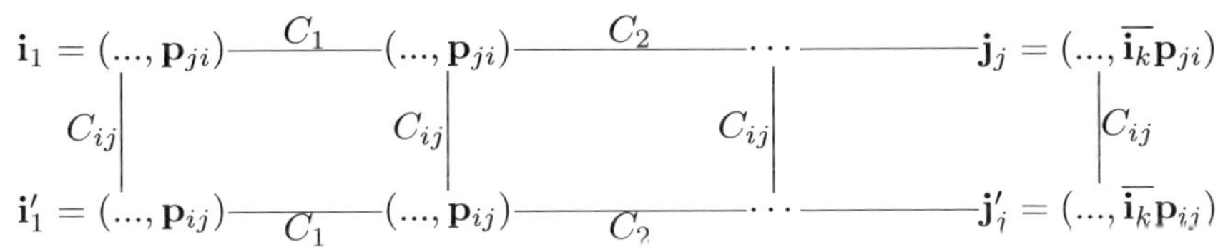

FIGURE 9. Commuting squares from $\mathbf{i}_1, \mathbf{i}_1'$ to $\mathbf{j}_j, \mathbf{j}_j'$.

If $j \neq k$, by Corollary 4.8, there exists a rank 3 cycle, that contains all $\mathbf{j}_j$, $\mathbf{j}_j'$, $\mathbf{j}_k$, $\mathbf{j}_k'$ in such a way that δ joins $\mathbf{j}_j$ and $\mathbf{j}_k$ with $\omega(\mathbf{i}) = i$ for any $\mathbf{i} \in \delta$, while δ' joins $\mathbf{j}_j'$ and $\mathbf{j}_k'$ with $\omega(\mathbf{i}) \neq i$ for any $\mathbf{i} \in \delta'$, see Figure 10. The cycle $C_{ij}\delta C_{ik}\delta'$ is then a rank 3 cycle of type (b). If $j = k$ then this part of the construction is trivial.

The other cycle $\beta_\#$ form by β' and $(\mathbf{i}_1 - \cdots - \mathbf{j}_j \overset{\delta}{-\cdots-} \mathbf{j}_k - \cdots - \mathbf{i}_2)$ all have words ending with the letter i, hence by (**) we can consider the last letter removed and apply induction on n to decompose $\beta_\#$. Therefore by (*), we are reduced to consider the remaining cycle β_* formed by β'' (which joins $\mathbf{i}_1'$ and $\mathbf{i}_2'$ in Figure 8) and $(\mathbf{i}_1' - \cdots - \mathbf{j}_j' \overset{\delta'}{-\cdots-} \mathbf{j}_k' - \cdots - \mathbf{i}_2')$, which does not involve i in the last letter.

Hence we now have

$$m_i(\beta_*) = m_i(\beta'') < m_i(\beta),$$

and so by applying the same argument on β_* repeatedly, we reduce to the case where $i \notin \omega(\beta_*)$ so that m decreases by at least 1, and we can apply induction on m to decompose β_* completely.

We remark that the $\mathbf{i}''$ part of the rank 3 cycles in type (b) of Theorem 4.10 comes from the letters removed during the induction step when $m = 1$ and $n \mapsto n - 1$. This conclude the proof of Theorem 4.10.

6. Proof of Theorem 3.11

We have the obvious property concerning operators defined over loops. Let $\{\mathcal{X}^{\mathbf{i}}\}_{\mathbf{i} \in \mathcal{G}_0}$ be a collection of algebras indexed by the vertices of $\mathcal{G}$. Assume that for any oriented edge $\mathbf{e} : \mathbf{i}_1 \longrightarrow \mathbf{i}_2$ of $\mathcal{G}$, we have a homomorphism $f_{\mathbf{e}} \in \mathrm{Hom}(\mathcal{X}^{\mathbf{i}_1}, \mathcal{X}^{\mathbf{i}_2})$. For any oriented path $\alpha = (\mathbf{e}_1, \ldots, \mathbf{e}_n)$ of $\mathcal{G}$, let $f_\alpha := f_{\mathbf{e}_n} \circ \cdots \circ f_{\mathbf{e}_1}$.

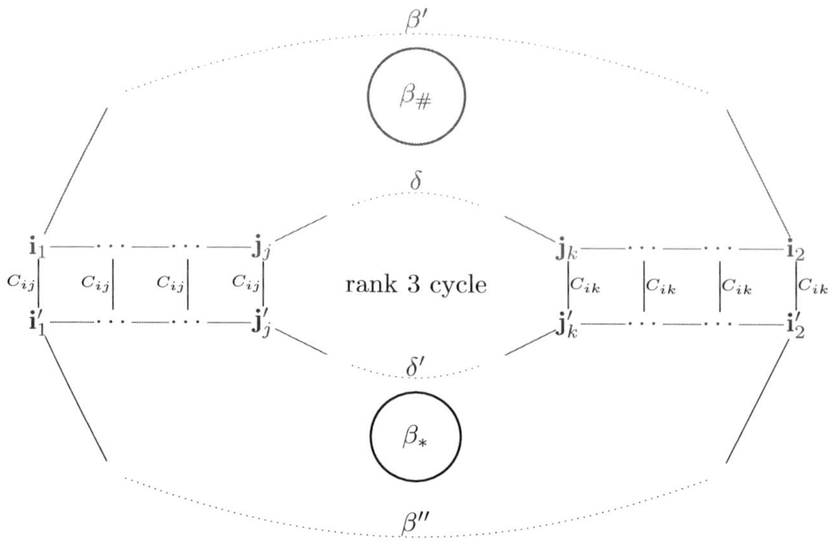

FIGURE 10. Decomposition of β,where all the vertices along the red cycle end with the letter i, and all the vertices along the blue path does not end with letter i.

LEMMA 6.1. *Assume that $f_{\mathbf{e}^{-1}} \circ f_{\mathbf{e}} = \mathrm{Id}_{\mathcal{X}_{\mathbf{i}_1}}$ for any oriented edge $\mathbf{e} : \mathbf{i}_1 \longrightarrow \mathbf{i}_2$. If $\pi_1(\mathcal{G})$ is generated by closed paths $\alpha_1, \ldots, \alpha_n$, and $f_{\alpha_i} = \mathrm{Id}$ for any i, then $f_\beta = \mathrm{Id}$ for any closed path β.*

PROOF. We have

$$f_{\gamma^{-1}\alpha_i\gamma} = f_{\gamma^{-1}} \circ f_{\alpha_i} \circ f_\gamma = f_{\gamma^{-1}} \circ f_\gamma = \mathrm{Id}.$$

Since any closed path β is generated by $\gamma^{-1} \circ \alpha_i \circ \gamma$, the claims follow. □

Therefore to complete the proof of the Main Theorem 3.11, we just need to check that the sequence of quantum cluster mutations along any commutative squares and any rank 3 cycles are trivial. The case of commutative squares is straightforward since it is just a change of index of the cluster variables. On the other hand, since the identity (3.27) is satisfied for the classical relations due to Lusztig's Injectivity Lemma applied to the classical Bruhat cells of type A_3 or B_3, the quantum version is a consequence of the following result due to [**7**], see also [**15**, Proposition 3.4].

THEOREM 6.2. *$(k_1, \ldots, k_n)$ is a σ-period of the quantum Y-seed $(B, \mathbf{Y})$ if and only if it is a σ-period of the classical y-seed (B, y).*

While this completes the proof of Theorem 3.11, in order to be self-contained (and to avoid another mis-citation since we could not find a direct proof of this Theorem), we find it illuminating to provide an alternative, direct computational proof of the quantum identity (3.27) over any 3-cycles.

6.1. Algorithms for explicit checking. More precisely, fix any reduced word $\mathbf{i}_0$ of w_0, and let

$$\mu_{i_m} \circ \cdots \circ \mu_{i_1} : \mathbf{Q}(\mathbf{i}_0) \longrightarrow \mathbf{Q}(\mathbf{i}_0) \tag{6.1}$$

be a mutation sequence corresponding to a rank 3 cycle. Then we have to check that

$$\mu_{i_1}^* \circ \cdots \circ \mu_{i_m}^*(X_i) = X_{\sigma(i)} \in \mathcal{X}_q^{\mathbf{i}_0} \tag{6.2}$$

where σ is a permutation of index corresponding to the overall effects involving commutative Coxeter moves.

In principle, the checking can be done explicitly, which involves *huge* rational expressions in the quantum cluster variables of $\mathbf{T}_q^{\mathbf{i}_0}$ and it is very hard to present. More precisely, during the sequence of cluster mutations, the initial variable $X_i \in \mathcal{X}_q^{\mathbf{i}_0}$ will be mutated to a series of rational expressions, followed by factorization of the numerators and denominators where some factors canceled and the expression simplified, finally ending again at a simple monomial $X_{\sigma(i)}$. While it is OK to write down all such expressions explicitly in type A_3, the expressions become *very complicated* in type B_3, and the factorization of non-commutative rational expressions could not be done effectively by simple computer program with symbolic computation.

To aid with the presentation in order to avoid putting the calculations into a huge appendix, we employ several tricks to simplify our verification. These tricks will also shed light onto some general principles in calculation involving quantum cluster mutations.

Trick 1: To aid the bookkeeping of index, while (6.2) is obtained from pullback of the mutation sequence so that the order of operators are reversed, we can make use of the fact that μ_k^* are involution, hence by taking inverse we can deduce that (6.2) holds if and only if the forward version hold:

$$\mu_{i_m}^* \circ \cdots \circ \mu_{i_1}^*(X_i) = X_{\sigma^{-1}(i)}, \qquad \forall X_i \in \mathcal{X}_q^{\mathbf{i}_0} \tag{6.3}$$

or in other words

$$\sigma^* \circ \mu_{i_m}^* \circ \cdots \circ \mu_{i_1}^*(X_i) = \mathrm{Id}_{\mathcal{X}_q^{\mathbf{i}_0}} \tag{6.4}$$

which is the identity that we will be checking. Also recall that if we use the *canonical labeling* of the quiver $\mathbf{Q}(\mathbf{i}_0)$, then σ^* becomes the identity map on $\mathcal{X}_q^{\mathbf{i}_0}$.

Trick 2: Let $P(X_{i_1}, \ldots, X_{i_k})$ be a Laurent polynomial which is linear in X_{i_k} or $X_{i_k}^{-1}$. If the identity (6.4) holds for $X_{i_1}, \ldots, X_{i_{k-1}}$ and $P(X_{i_1}, \ldots, X_{i_k})$, then the identity holds for X_{i_k} since it is a rational expression of the former and (6.4) is a homomorphism of $\mathbf{T}_q^{\mathbf{i}_0}$. In particular, the verification becomes very simple if the polynomial P arises from quantum group embedding $\mathfrak{D}_q \hookrightarrow \mathcal{X}_q^{\mathbf{i}_0}$ as in [**13**].

Trick 3: If X_λ is a *standard monomial*, i.e. on $\Lambda_{\mathbf{Q}}$ we have

$$(\lambda, \overrightarrow{e_k}) \leq 0, \qquad \forall k \in Q \setminus Q_0, \tag{6.5}$$

then its image is universally Laurent under any mutation sequence [**10**], and the calculation can be checked (by hand) and presented effectively.

We remark that even though Trick 3 utilizes the properties of standard monomials as universally Laurent polynomials, the calculation is just conveniently chosen to simplify the expression, and is independent of the proofs of those properties presented in [**10**].

6.2. Type A_3. Let $\mathfrak{g}$ be type A_3. We will use the labeling in Figure 5 for our seeds, so that the cluster variables $X_i^{(k)} \in \mathcal{X}_q^{\mathbf{i}_k}$. (We will omit the bracket for $\mathbf{i}_0$).

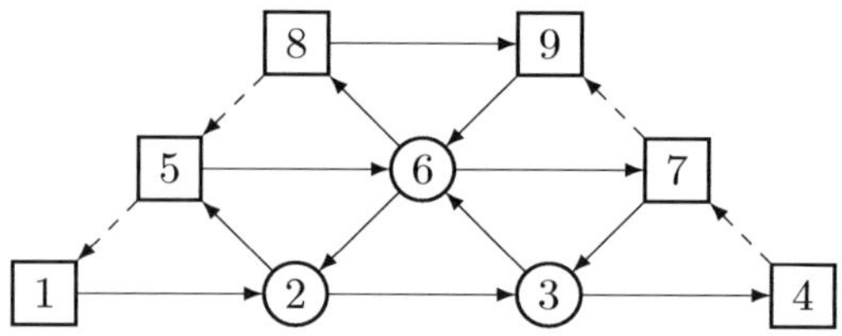

FIGURE 11. Basic quiver in type A_3.

Recall that if $\mathbf{i} \sim \mathbf{i}'$ in the commutative class, then the quiver $\mathbf{Q}(\mathbf{i}) = \mathbf{Q}(\mathbf{i}')$ up to relabeling the index.

The quiver corresponding to the initial seed $\mathbf{i}_0 = (121321)$ is labeled as in Figure 11. Then the rank 3 cycles corresponds to mutation at the vertices $2, 6, 3, 2, 6, 3, 2, 6$ sequentially, and we have

$$\mu_6 \circ \mu_2 \circ \mu_3 \circ \mu_6 \circ \mu_2 \circ \mu_3 \circ \mu_6 \circ \mu_2 = \sigma \tag{6.6}$$

where $\sigma = (263)$ is the permutation changing the index $(2,3,6)$ into $(6,2,3)$ in the final quiver.

REMARK 6.3. In fact the mutable part of $Q(\mathbf{i}_0)$ is cluster finite A_3 type, hence we know *a priori* that there are finitely many (quantum) cluster variables, and the periodicity of the mutation sequence is automatic from the general theory of quantum cluster algebra. Nonetheless, we will give a direct computational proof of the identity.

Recall [**13**, Theorem 4.14] that the image of the Chevalley generators of the lower Borel subalgebra $\mathcal{U}_q(\mathfrak{b}_-)$ under the Coxeter moves depend only on the reduced words $\mathbf{i} \in \mathcal{I}(w_0)$ and not on any choice of mutations. The images of the generators $\mathbf{f}_i$ are given by a telescopic sum along the horizontal arrows of the basic quivers $\mathbf{Q}(\mathbf{i})$ (with the right-most vertices removed), while $\mathbf{K}_i'$ are the corresponding cluster monomials. Hence the identity (6.3) holds automatically for these expressions. Explicitly they are given by:

$$\begin{aligned} \mathbf{f}_1 &= X_1 + X_{1,2} + X_{1,2,3}, & \mathbf{K}_1' &= X_{1,2,3,4} \\ \mathbf{f}_2 &= X_5 + X_{5,6}, & \mathbf{K}_2' &= X_{5,6,7} \\ \mathbf{f}_3 &= X_8, & \mathbf{K}_3' &= X_{8,9}. \end{aligned} \tag{6.7}$$

Also we observe that the monomials

$$C_1 = X_{2,3,6}, \qquad C_2 = X_{1,5,8} \tag{6.8}$$

commute with all the mutable variables $\{X_2, X_3, X_6\}$ and invariant under σ, therefore by Proposition 2.25 they remain monomials throughout the sequence of cluster mutations and the identity (6.3) holds automatically.

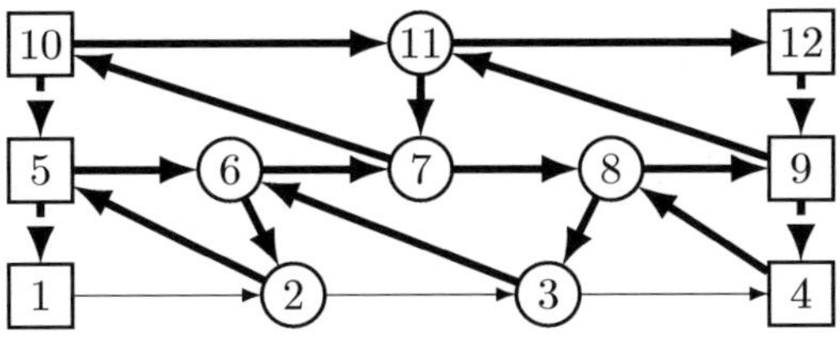

FIGURE 12. Basic quiver in type B_3.

By Trick 3, we claim that it is enough to check the identity (6.3) for the following *standard monomial* $X_{5,8}$:

$$\begin{aligned} X_{5,8} &\mapsto_{\mu_2^*} X_{5,8}^{(1)} + X_{2,5,8}^{(1)} \\ &\mapsto_{\mu_6^*} X_{5,8}^{(2)} + X_{5,6,8}^{(2)} + X_{2,5,6,8}^{(2)} \\ &\mapsto_{\mu_3^*} X_{5,8}^{(3)} + X_{5,6,8}^{(3)} + X_{2,5,6,8}^{(3)} + X_{2,3,5,6,8}^{(3)} \\ &\mapsto_{\mu_2^*} X_{2,5,8}^{(4)} + X_{2,5,6,8}^{(4)} + X_{2,3,5,6,8}^{(4)} \\ &\mapsto_{\mu_6^*} X_{2,5,8}^{(5)} + X_{2,3,5,8}^{(5)} \\ &\mapsto_{\mu_3^*} X_{2,5,8}^{(6)} \\ &\mapsto_{\mu_2^*} X_{5,8}^{(7)} \\ &\mapsto_{\mu_6^*} X_{5,8}. \end{aligned}$$

Now it follows by Trick 2 that in the following, if (6.3) holds for the elements on the left hand side, then it also holds for the right hand side due to linearity:

$$\begin{aligned} \mathbf{f}_3 &\Longrightarrow X_8 \\ \mathbf{K}_3' &\Longrightarrow X_9 \\ X_{5,8} &\Longrightarrow X_5 \\ \mathbf{f}_2 &\Longrightarrow X_{5,6} \Longrightarrow X_6 \\ \mathbf{K}_2' &\Longrightarrow X_7 \\ C_2 &\Longrightarrow X_1 \\ C_1 &\Longrightarrow X_{2,3} \Longrightarrow X_{1,2,3} \\ X_{1,2,3}, \mathbf{f}_1 &\Longrightarrow X_1 + X_{1,2} \Longrightarrow X_{1,2} \Longrightarrow X_2 \\ X_{2,3} &\Longrightarrow X_3 \\ \mathbf{K}_1' &\Longrightarrow X_4. \end{aligned}$$

Hence we conclude that (6.3) holds for all the cluster variables $X_i \in \mathcal{X}_q^{\mathbf{i}_0}$. This completes the discussion in type A_3.

6.3. Type B_3. The situation is much more complicated for type B_3, and as explained, huge non-commutative rational expressions appear if we just consider the images of any cluster variables.

To begin, we will choose the initial word $\mathbf{i}_0 = (121232123)$ with the corresponding quiver given in Figure 12.

REMARK 6.4. The mutable part of this quiver is of finite-mutation type $F_4^{(*,*)}$, but it is not finite type [**5**, Table 6]. In particular there will be infinitely many cluster variables, therefore the same argument in Remark 6.3 of type A_3 using finitely many cluster variables do not apply in this case.

The rank 3 cycle of Coxeter moves corresponds to a sequence of 26 mutations from $\mathbf{i}_0$ given sequentially by

$$\underline{2,6,2},\underline{3,8,3},7,11,\underline{2,8,2},6,7,\underline{3,11,3},\underline{2,8,2},7,6,\underline{3,8,3},11,7. \tag{6.9}$$

Here the cluster mutations of the underlined triplets correspond to the Coxeter moves involving the short index as in (2.69). Furthermore, in type B_3, it turns out that we do not need any extra permutation σ to complete the cycle.

To check the identity (6.3), again it is satisfied automatically by some special elements of $\mathcal{X}_q^{\mathbf{i}_0}$, including the image of the generators of $\mathcal{U}_q(\mathfrak{b}_-)$ expressed by the cluster variables along the horizontal arrows of $\mathbf{Q}(\mathbf{i})$:

$$\begin{aligned} \mathbf{f}_1 &= X_1 + X_{1,2} + X_{1,2,3}, & \mathbf{K}_1' &= X_{1,2,3,4}, \\ \mathbf{f}_2 &= X_5 + X_{5,6} + X_{5,6,7} + X_{5,6,7,8}, & \mathbf{K}_2' &= X_{5,6,7,8,9}, \\ \mathbf{f}_3 &= X_{10} + X_{10,11}, & \mathbf{K}_3' &= X_{10,11,12}, \end{aligned} \tag{6.10}$$

as well as some monomials that commute with all the mutable variables:

$$C_1 = X_{2^2,7^{-2},8^{-1},9^{-2},11^{-1}}, \qquad C_2 = X_{3^2,6,7^2,8,12}, \qquad C_3 = X_{4^2,8,9^2,11,12^2}, \tag{6.11}$$

which remain monomials during the sequence of cluster mutations and satisfy (6.3) automatically by Proposition 2.25.

We claim that it suffices to check that the identity (6.3) holds for the following monomials in $\mathcal{X}_q^{\mathbf{i}_0}$.

$$X_1^{-1}, \qquad X_{12}, \qquad X_{9,12}, \qquad X_{4,9,12}, \tag{6.12}$$

from which the following implications follow by Trick 2 due to linearity:

$$\begin{aligned} X_1^{-1} &\Longrightarrow X_1 \\ X_{12}, X_{9,12} &\Longrightarrow X_9 \\ X_{4,9,12} &\Longrightarrow X_4 \\ \mathbf{K}_1' &\Longrightarrow X_{1,2,3} \\ X_{1,2,3}, \mathbf{f}_1 &\Longrightarrow X_1 + X_{1,2} \Longrightarrow X_2 \Longrightarrow X_3 \\ X_{12}, \mathbf{K}_3' &\Longrightarrow X_{10,11} \\ \mathbf{f}_3, X_{10,11} &\Longrightarrow X_{10} \Longrightarrow X_{11} \\ C_3 &\Longrightarrow X_8 \\ C_1 &\Longrightarrow X_7 \\ C_2 &\Longrightarrow X_6 \\ \mathbf{f}_2 &\Longrightarrow X_5. \end{aligned}$$

Hence we conclude the identity (6.3) holds for all the cluster variables $X_i \in \mathcal{X}_q^{\mathbf{i}_0}$.

It remains to check the identity (6.3) holds for the 4 elements in (6.12), which are actually *standard monomials* in $\mathcal{X}_q^{\mathbf{i}_0}$ by checking (6.5) directly. Hence by Trick 3, they are *universally Laurent polynomials* and their mutations can be reasonably computed. In the following, to simplify notations we indicate with the bracket (k)

which seed $\mathcal{X}_q^{\mathbf{i}_k}$ the elements on the right belong to, following the indexing in Figure 6. Also we denote by $[2] := q^{\frac{1}{2}} + q^{-\frac{1}{2}}$.

$$X_1^{-1} \mapsto_{\mu_2^*\mu_6^*\mu_2^*} X_1^{-1}+X_{1,6}^{-1}+X_{1,2,6}^{-1} \tag{1}$$

$$\mapsto_{\mu_3^*\mu_8^*\mu_3^*} X_1^{-1}+X_{1,6}^{-1}+X_{1,2,6}^{-1}+X_{1,2,6,8}^{-1}+X_{1,2,3,6,8}^{-1} \tag{2}$$

$$\mapsto_{\mu_7^*} X_1^{-1}+X_{1,6}^{-1}+X_{1,2,6}^{-1}+X_{1,6,7}^{-1}+X_{1,2,6,7}^{-1}+X_{1,2,6,7,8}^{-1}+X_{1,2,3,6,7,8}^{-1} \tag{3}$$

$$\mapsto_{\mu_{11}^*} X_1^{-1}+X_{1,6}^{-1}+X_{1,2,6}^{-1}+X_{1,6,7}^{-1}+X_{1,2,6,7}^{-1}+X_{1,6,7,11}^{-1}+X_{1,2,6,7,11}^{-1} + X_{1,2,6,7,8,11}^{-1}+X_{1,2,3,6,7,8,11}^{-1} \tag{4}$$

$$\mapsto_{\mu_2^*\mu_8^*\mu_2^*} X_1^{-1}+X_{1,8}^{-1}+X_{1,2,8}^{-1}+X_{1,6,8}^{-1}+X_{1,2,6,8}^{-1}+X_{1,6,7,8}^{-1}+X_{1,2,6,7,8}^{-1} + X_{1,2,6,7,8,11}^{-1}+X_{1,2,3,6,7,8,11}^{-1} \tag{5}$$

$$\mapsto_{\mu_6^*} X_1^{-1}+X_{1,8}^{-1}+X_{1,2,8}^{-1}+X_{1,7,8}^{-1}+X_{1,2,7,8}^{-1}+X_{1,2,7,8,11}^{-1}+X_{1,2,3,7,8,11}^{-1} \tag{6}$$

$$\mapsto_{\mu_7^*} X_1^{-1}+X_{1,8}^{-1}+X_{1,2,8}^{-1}+X_{1,2,8,11}^{-1}+X_{1,2,3,8,11}^{-1} \tag{7}$$

$$\mapsto_{\mu_3^*\mu_{11}^*\mu_3^*} X_1^{-1}+X_{1,8}^{-1}+X_{1,2,8}^{-1} \tag{8}$$

$$\mapsto_{\mu_2^*\mu_8^*\mu_2^*} X_1^{-1} \tag{9}$$

$$\mapsto_{\mu_7^*} X_1^{-1} \tag{10}$$

$$\mapsto_{\mu_6^*} X_1^{-1} \tag{11}$$

$$\mapsto_{\mu_3^*\mu_8^*\mu_3^*} X_1^{-1} \tag{12}$$

$$\mapsto_{\mu_{11}^*} X_1^{-1} \tag{13}$$

$$\mapsto_{\mu_7^*} X_1^{-1}$$

$$X_{12} \mapsto_{\mu_2^*\mu_6^*\mu_2^*} X_{12} \tag{1}$$

$$\mapsto_{\mu_3^*\mu_8^*\mu_3^*} X_{12} \tag{2}$$

$$\mapsto_{\mu_7^*} X_{12} \tag{3}$$

$$\mapsto_{\mu_{11}^*} X_{12}+X_{11,12} \tag{4}$$

$$\mapsto_{\mu_2^*\mu_8^*\mu_2^*} X_{12}+X_{11,12}+[2]X_{2,11,12}+X_{2^2,11,12}+X_{2^2,8,11,12} \tag{5}$$

$$\mapsto_{\mu_6^*} X_{12}+X_{11,12}+[2]X_{2,11,12}+X_{2^2,11,12}+X_{2^2,8,11,12}+X_{2^2,6,8,11,12} \tag{6}$$

$$\mapsto_{\mu_7^*} X_{12}+X_{7,12}+X_{7,11,12}+[2]X_{2,7,11,12}+X_{2^2,7,11,12}+X_{2^2,7,8,11,12} + X_{2^2,7^2,8,11,12}+X_{2^2,6,7^2,8,11,12} \tag{7}$$

$$\mapsto_{\mu_3^*\mu_{11}^*\mu_3^*} X_{12}+X_{7,12}+[2]X_{2,7,12}+X_{2^2,7,12}+X_{2^2,7,8,12}+X_{7,11,12} + [2]X_{2,7,11,12}+X_{2^2,7,11,12}+[2]X_{2,3,7,11,12}+[2]X_{2^2,3,7,11,12} + X_{2^2,3^2,7,11,12}+X_{2^2,7,8,11,12}+[2]X_{2^2,3,7,8,11,12} + X_{2^2,3^2,7,8,11,12}+X_{2^2,3^2,7^2,8,11,12}+X_{2^2,3^2,6,7^2,8,11,12} \tag{8}$$

$$\mapsto_{\mu_2^*\mu_8^*\mu_2^*} X_{12}+X_{7,12}+X_{7,11,12}+[2]X_{3,7,11,12}+X_{3,7,11,12}+X_{3,7,8,11,12} + X_{3,7,8,11,12}+X_{3,6,7,8,11,12} \tag{9}$$

$$\mapsto_{\mu_7^*} X_{12}+X_{11,12}+[2]X_{3,11,12}+X_{3^2,11,12}+X_{3^2,8,11,12}+X_{3^2,6,8,11,12} \tag{10}$$

$$\mapsto_{\mu_6^*} X_{12}+X_{11,12}+[2]X_{3,11,12}+X_{3^2,11,12}+X_{3^2,8,11,12} \tag{11}$$

$$\mapsto_{\mu_3^*\mu_8^*\mu_3^*} X_{12}+X_{11,12} \tag{12}$$

$$\mapsto_{\mu_{11}^*} X_{12} \tag{13}$$

$$\mapsto_{\mu_7^*} X_{12}$$

$$
\begin{aligned}
X_{9,12} \mapsto_{\mu_2^*\mu_6^*\mu_2^*} \quad & X_{9,12} && (1)\\
\mapsto_{\mu_3^*\mu_8^*\mu_3^*} \quad & X_{9,12} + [2]X_{3,9,12} + X_{3^2,9,12} + X_{3^2,8,9,12} && (2)\\
\mapsto_{\mu_7^*} \quad & X_{9,12} + [2]X_{3,9,12} + X_{3^2,9,12} + X_{3^2,8,9,12} + X_{3^2,7,8,9,12} && (3)\\
\mapsto_{\mu_{11}^*} \quad & X_{9,11,12} + [2]X_{3,9,11,12} + X_{3^2,9,11,12} + X_{3^2,8,9,11,12} && (4)\\
& + X_{3^2,8,9,11^2,12} + X_{3^2,7,8,9,11^2,12}\\
\mapsto_{\mu_2^*\mu_8^*\mu_2^*} \quad & X_{9,11,12} + [2]X_{2,9,11,12} + X_{2^2,9,11,12} + [2]X_{2,3,9,11,12} && (5)\\
& + [2]X_{2^2,3,9,11,12} + X_{2^2,3^2,9,11,12} + X_{2^2,8,9,11,12}\\
& + [2]X_{2^2,3,8,9,11,12} + X_{2^2,3^2,8,9,11,12} + X_{2^2,3^2,8,9,11^2,12}\\
& + X_{2^2,3^2,7,8,9,11^2,12}\\
\mapsto_{\mu_6^*} \quad & X_{9,11,12} + [2]X_{2,9,11,12} + X_{2^2,9,11,12} + [2]X_{2,3,9,11,12} && (6)\\
& + [2]X_{2^2,3,9,11,12} + X_{2^2,3^2,9,11,12} + X_{2^2,8,9,11,12}\\
& + [2]X_{2^2,3,8,9,11,12} + X_{2^2,3^2,8,9,11,12} + X_{2^2,6,8,9,11,12}\\
& + [2]X_{2^2,3,6,8,9,11,12} + X_{2^2,3^2,6,8,9,11,12} + X_{2^2,3^2,8,9,11^2,12}\\
& + X_{2^2,3^2,6,8,9,11^2,12} + X_{2^2,3^2,6,7,8,9,11^2,12}\\
\mapsto_{\mu_7^*} \quad & X_{7,9,11,12} + [2]X_{2,7,9,11,12} + X_{2^2,7,9,11,12} + [2]X_{2,3,7,9,11,12} && (7)\\
& + [2]X_{2^2,3,7,9,11,12} + X_{2^2,3^2,7,9,11,12} + X_{2^2,7,8,9,11,12}\\
& + [2]X_{2^2,3,7,8,9,11,12} + X_{2^2,3^2,7,8,9,11,12} + X_{2^2,7^2,8,9,11,12}\\
& + [2]X_{2^2,3,7^2,8,9,11,12} + X_{2^2,3^2,7^2,8,9,11,12} + X_{2^2,6,7^2,8,9,11,12}\\
& + [2]X_{2^2,3,6,7^2,8,9,11,12} + X_{2^2,3^2,6,7^2,8,9,11,12} + X_{2^2,3^2,7^2,8,9,11^2,12}\\
& + X_{2^2,3^2,6,7^2,8,9,11^2,12}\\
\mapsto_{\mu_3^*\mu_{11}^*\mu_3^*} \quad & X_{7,9,11,12} + [2]X_{2,7,9,11,12} + X_{2^2,7,9,11,12} + [2]X_{2,3,7,9,11,12} && (8)\\
& + [2]X_{2^2,3,7,9,11,12} + X_{2^2,3^2,7,9,11,12} + X_{2^2,7,8,9,11,12}\\
& + [2]X_{2^2,3,7,8,9,11,12} + X_{2^2,3^2,7,8,9,11,12}\\
& + X_{2^2,3^2,7^2,8,9,11,12} + X_{2^2,3^2,6,7^2,8,9,11,12}\\
\mapsto_{\mu_2^*\mu_8^*\mu_2^*} \quad & X_{7,9,11,12} + [2]X_{3,7,9,11,12} + X_{3^2,7,9,11,12} + X_{3^2,7,8,9,11,12} && (9)\\
& + X_{3^2,7^2,8,9,11,12} + X_{3^2,6,7^2,8,9,11,12}\\
\mapsto_{\mu_7^*} \quad & X_{9,11,12} + [2]X_{3,9,11,12} + X_{3^2,9,11,12} + X_{3^2,8,9,11,12} && (10)\\
& + X_{3^2,6,8,9,11,12}\\
\mapsto_{\mu_6^*} \quad & X_{9,11,12} + [2]X_{3,9,11,12} + X_{3^2,9,11,12} + X_{3^2,8,9,11,12} && (11)\\
\mapsto_{\mu_3^*\mu_8^*\mu_3^*} \quad & X_{9,11,12} && (12)\\
\mapsto_{\mu_{11}^*} \quad & X_{9,12} && (13)\\
\mapsto_{\mu_7^*} \quad & X_{9,12}
\end{aligned}
$$

$$
\begin{aligned}
X_{4,9,12} \mapsto_{\mu_2^*\mu_6^*\mu_2^*} \quad & X_{4,9,12} && (1)\\
\mapsto_{\mu_3^*\mu_8^*\mu_3^*} \quad & X_{3,4,9,12} + X_{3^2,4,9,12} + X_{3^2,4,8,9,12} && (2)\\
\mapsto_{\mu_7^*} \quad & X_{3,4,9,12} + X_{3^2,4,9,12} + X_{3^2,4,8,9,12} + X_{3^2,4,7,8,9,12} && (3)\\
\mapsto_{\mu_{11}^*} \quad & X_{3,4,9,11,12} + X_{3^2,4,9,11,12} + X_{3^2,4,8,9,11,12} + X_{3^2,4,8,9,11^2,12} && (4)\\
& + X_{3^2,4,7,8,9,11^2,12} \\
\mapsto_{\mu_2^*\mu_8^*\mu_2^*} \quad & X_{2,3,4,9,11,12} + X_{2^2,3,4,9,11,12} + X_{2^2,3^2,4,9,11,12} + X_{2^2,3,4,8,9,11,12} && (5)\\
& + X_{2^2,3^2,4,8,9,11,12} + X_{2^2,3^2,4,8,9,11^2,12} + X_{2^2,3^2,4,7,8,9,11^2,12} \\
\mapsto_{\mu_6^*} \quad & X_{2,3,4,9,11,12} + X_{2^2,3,4,9,11,12} + X_{2^2,3^2,4,9,11,12} && (6)\\
& + X_{2^2,3,4,8,9,11,12} + X_{2^2,3^2,4,8,9,11,12} + X_{2^2,3,4,6,8,9,11,12} \\
& + X_{2^2,3^2,4,6,8,9,11,12} + X_{2^2,3^2,4,8,9,11^2,12} \\
& + X_{2^2,3^2,4,6,8,9,11^2,12} + X_{2^2,3^2,4,6,7,8,9,11^2,12} \\
\mapsto_{\mu_7^*} \quad & X_{2,3,4,7,9,11,12} + X_{2^2,3,4,7,9,11,12} + X_{2^2,3^2,4,7,9,11,12} && (7)\\
& + X_{2^2,3,4,7,8,9,11,12} + X_{2^2,3^2,4,7,8,9,11,12} + X_{2^2,3,4,7^2,8,9,11,12} \\
& + X_{2^2,3^2,4,7^2,8,9,11,12} + X_{2^2,3,4,6,7^2,8,9,11,12} \\
& + X_{2^2,3^2,4,6,7^2,8,9,11,12} + X_{2^2,3^2,4,7^2,8,9,11^2,12} \\
& + X_{2^2,3^2,4,6,7^2,8,9,11^2,12} \\
\mapsto_{\mu_3^*\mu_{11}^*\mu_3^*} \quad & X_{2,3,4,7,9,11,12} + X_{2^2,3,4,7,9,11,12} && (8)\\
& + X_{2^2,3^2,4,7,9,11,12} + X_{2^2,3,4,7,8,9,11,12} + X_{2^2,3^2,4,7,8,9,11,12} \\
& + X_{2^2,3^2,4,7^2,8,9,11,12} + X_{2^2,3^2,4,6,7^2,8,9,11,12} \\
\mapsto_{\mu_2^*\mu_8^*\mu_2^*} \quad & X_{3,4,7,9,11,12} + X_{3^2,4,7,9,11,12} + X_{3^2,4,7,8,9,11,12} + X_{3^2,4,7^2,8,9,11,12} && (9)\\
& + X_{3^2,4,6,7^2,8,9,11,12} \\
\mapsto_{\mu_7^*} \quad & X_{3,4,9,11,12} + X_{3^2,4,9,11,12} + X_{3^2,4,8,9,11,12} + X_{3^2,4,6,8,9,11,12} && (10)\\
\mapsto_{\mu_6^*} \quad & X_{3,4,9,11,12} + X_{3^2,4,9,11,12} + X_{3^2,4,8,9,11,12} && (11)\\
\mapsto_{\mu_3^*\mu_8^*\mu_3^*} \quad & X_{4,9,11,12} && (12)\\
\mapsto_{\mu_{11}^*} \quad & X_{4,9,12} && (13)\\
\mapsto_{\mu_7^*} \quad & X_{4,9,12}
\end{aligned}
$$

This completes the discussion in type B_3.

Acknowledgment

We thank Linhui Shen for pointing out the logical gap in the original construction and subsequent discussions related to the issue. We also thank Ronald Chun Wai Wong for aiding with the computations in the rank 3 cluster mutations.

References

[1] Arkady Berenstein and Andrei Zelevinsky, *Quantum cluster algebras*, Adv. Math. **195** (2005), no. 2, 405–455, DOI 10.1016/j.aim.2004.08.003. MR2146350

[2] V. G. Drinfel′d, *Hopf algebras and the quantum Yang-Baxter equation* (Russian), Dokl. Akad. Nauk SSSR **283** (1985), no. 5, 1060–1064. MR802128

[3] L. D. Faddeev, *Discrete Heisenberg-Weyl group and modular group*, Lett. Math. Phys. **34** (1995), no. 3, 249–254, DOI 10.1007/BF01872779. MR1345554

[4] Ludwig Faddeev, *Modular double of a quantum group*, Conférence Moshé Flato 1999, Vol. I (Dijon), Math. Phys. Stud., vol. 21, Kluwer Acad. Publ., Dordrecht, 2000, pp. 149–156. MR1805888

[5] Anna Felikson, Michael Shapiro, and Pavel Tumarkin, *Cluster algebras of finite mutation type via unfoldings*, Int. Math. Res. Not. IMRN **8** (2012), 1768–1804, DOI 10.1093/imrn/rnr072. MR2920830

[6] V. V. Fock and A. B. Goncharov, *Cluster $\mathcal{X}$-varieties, amalgamation, and Poisson-Lie groups*, Algebraic geometry and number theory, Progr. Math., vol. 253, Birkhäuser Boston, Boston, MA, 2006, pp. 27–68, DOI 10.1007/978-0-8176-4532-8_2. MR2263192

[7] V. V. Fock and A. B. Goncharov, *The quantum dilogarithm and representations of quantum cluster varieties*, Invent. Math. **175** (2009), no. 2, 223–286, DOI 10.1007/s00222-008-0149-3. MR2470108

[8] Igor B. Frenkel and Ivan C. H. Ip, *Positive representations of split real quantum groups and future perspectives*, Int. Math. Res. Not. IMRN **8** (2014), 2126–2164, DOI 10.1093/imrn/rns288. MR3194015

[9] K. R. Goodearl and R. B. Warfield Jr., *An introduction to noncommutative Noetherian rings*, 2nd ed., London Mathematical Society Student Texts, vol. 61, Cambridge University Press, Cambridge, 2004, DOI 10.1017/CBO9780511841699. MR2080008

[10] A. Goncharov, L. Shen, *Quantum geometry of moduli spaces of local systems and representation theory*, `arXiv:1904.10491v3`, 2019.

[11] Ivan C. H. Ip, *Positive representations of split real simply-laced quantum groups*, Publ. Res. Inst. Math. Sci. **56** (2020), no. 3, 603–646, DOI 10.4171/PRIMS/56-3-6. MR4116693

[12] Ivan C. H. Ip, *Positive representations of non-simply-laced split real quantum groups*, J. Algebra **425** (2015), 245–276, DOI 10.1016/j.jalgebra.2014.11.019. MR3295985

[13] Ivan C. H. Ip, *Cluster realization of $\mathcal{U}_q(\mathfrak{g})$ and factorizations of the universal R-matrix*, Selecta Math. (N.S.) **24** (2018), no. 5, 4461–4553, DOI 10.1007/s00029-018-0432-0. MR3874697

[14] Michio Jimbo, *A q-difference analogue of $U(\mathfrak{g})$ and the Yang-Baxter equation*, Lett. Math. Phys. **10** (1985), no. 1, 63–69, DOI 10.1007/BF00704588. MR797001

[15] Rinat M. Kashaev and Tomoki Nakanishi, *Classical and quantum dilogarithm identities*, SIGMA Symmetry Integrability Geom. Methods Appl. **7** (2011), Paper 102, 29, DOI 10.3842/SIGMA.2011.102. MR2861174

[16] G. Lusztig, *Total positivity in reductive groups*, Lie theory and geometry, Progr. Math., vol. 123, Birkhäuser Boston, Boston, MA, 1994, pp. 531–568, DOI 10.1007/978-1-4612-0261-5_20. MR1327548

[17] B. Ponsot, J. Teschner, *Liouville bootstrap via harmonic analysis on a noncompact quantum group*, `arXiv:hep-th/9911110`, (1999).

[18] B. Ponsot and J. Teschner, *Clebsch-Gordan and Racah-Wigner coefficients for a continuous series of representations of $\mathcal{U}_q(\mathfrak{sl}(2,\mathbb{R}))$*, Comm. Math. Phys. **224** (2001), no. 3, 613–655, DOI 10.1007/PL00005590. MR1871903

[19] Markus Reineke, *Feigin's map and monomial bases for quantized enveloping algebras*, Math. Z. **237** (2001), no. 3, 639–667, DOI 10.1007/PL00004885. MR1845343

[20] Konrad Schmüdgen, *Operator representations of $\mathcal{U}_q(\mathrm{sl}_2(\mathbf{R}))$*, Lett. Math. Phys. **37** (1996), no. 2, 211–222, DOI 10.1007/BF00416024. MR1391203

[21] Gus Schrader and Alexander Shapiro, *A cluster realization of $U_q(\mathfrak{sl}_\mathfrak{n})$ from quantum character varieties*, Invent. Math. **216** (2019), no. 3, 799–846, DOI 10.1007/s00222-019-00857-6. MR3955710

[22] A. Scott, *Classes of maximal-length reduced words in Coxeter groups*, Ph.D. Thesis, University of Warwick, (1996).

[23] J. Tits, *A local approach to buildings*, The geometric vein, Springer, New York-Berlin, 1981, pp. 519–547. MR661801

Department of Mathematics, Hong Kong University of Science and Technology, Clearwater Bay, Hong Kong, People's Republic of China

Email address: `ivanip@ust.hk`

Contemporary Mathematics
Volume **817**, 2025
https://doi.org/10.1090/conm/817/16326

A trilogy of mapping class group representations from three-dimensional quantum gravity

Hyun Kyu Kim

To Igor B. Frenkel on the occasion of his 70th birthday

Abstract. For a punctured surface $\mathfrak{S}$, the author and Scarinci have recently constructed a quantization of a moduli space of Lorentzian metrics on the 3-manifold $\mathfrak{S} \times \mathbb{R}$ of constant sectional curvature $\Lambda \in \{-1, 0, 1\}$. The invariance of this quantization under the action of the mapping class group $\mathrm{MCG}(\mathfrak{S})$ of $\mathfrak{S}$ yields families of unitary representations of $\mathrm{MCG}(\mathfrak{S})$ on a Hilbert space, with key ingredients being three versions of the quantum dilogarithm functions depending on Λ. In this survey article, we review and elaborate on this result.

Contents

1. Introduction

Quantization of Teichmüller spaces of surfaces appeared in 1990's as an approach to quantization of 3-dimensional gravity, as it had been expected in the mathematical physics literature since 1980's that the phase space of 3-dimensional pure gravity theory would be closely related to the Teichmüller space. Kashaev [**K98**], and independently Chekhov and Fock [**CF97, F97**], established the first major constructions of the quantum Teichmüller theory in late 1990's. What are

2020 *Mathematics Subject Classification.* Primary 57K20; Secondary 13F60, 81R60, 83C45, 20C35, 57K35, 20G42, 47B02.

Key words and phrases. Three-dimensional quantum gravity, quantum dilogarithm, mapping class group, quantum cluster variety, quantum Teichmüller theory.

This work was supported by the National Research Foundation of Korea (NRF) grant funded by the Korea government (MSIT) (No. 2020R1C1C1A01011151). The author has been supported by a KIAS Individual Grant (MG047203, MG047204) at Korea Institute for Advanced Study.

quantized in these works are certain versions of the *Teichmüller space* of an oriented punctured surface $\mathfrak{S}$. The Teichmüller space of $\mathfrak{S}$, which dates back at least to early 20th century, can be viewed as the space of all complex structures on $\mathfrak{S}$ considered up to pullback by self-diffeomorphisms of $\mathfrak{S}$ isotopic to identity, and can also be viewed as the space of all complete hyperbolic metrics on $\mathfrak{S}$ considered up to isotopy. There exist some variants considered in 1980's, depending on what kind of conditions or extra data one wants to put at the punctures. The Teichmüller space is equipped with a canonical Poisson structure called the Weil-Petersson Poisson structure, and there is a natural pullback action on the Teichmüller space, preserving the Weil-Petersson Poisson structure, by the *mapping class group* $\mathrm{MCG}(\mathfrak{S})$ of $\mathfrak{S}$, which is defined as the group of isotopy classes of orientation-preserving self-diffeomorphisms of $\mathfrak{S}$. The above mentioned works establish certain versions of quantization of the Teichmüller space with respect to the Weil-Petersson Poisson structure, equivariant under the action of $\mathrm{MCG}(\mathfrak{S})$. They build a complex Hilbert space $\mathscr{H}$, a set of self-adjoint operators on $\mathscr{H}$ that play roles of quantum counterparts of classical observable functions on the Teichmüller space, and a family of unitary projective representations of $\mathrm{MCG}(\mathfrak{S})$ on $\mathscr{H}$ such that the set of (quantum) self-adjoint operators on $\mathscr{H}$ satisfy certain equivariance under these representations of $\mathrm{MCG}(\mathfrak{S})$. It is often said that this family of representations of $\mathrm{MCG}(\mathfrak{S})$ is a main consequence of the quantum Teichmüller theory. These representations are interpreted as forming a family because they depend on the real quantum parameter $\hbar$, usually called the Planck constant.

These early works are based on special coordinate systems on the Teichmüller space developed in 1980's by Penner and Thurston [**P87, T80**], as well as on a special function called the *quantum dilogarithm* studied in 1990's by Faddeev and Kashaev [**F95, FK94**]. Especially, the framekwork of Chekhov and Fock [**F97, CF97**] later generalized by Fock and Goncharov to the theory of quantization of cluster varieties [**FG09**], a prominent application being the quantization of higher Teichmüller spaces [**FG06**]. So, the Chekhov-Fock quantization of the Teichmüller space can be viewed as a prototypical example of the Fock-Goncharov quantization of cluster varieties.

In the meantime, as the initial ideas of quantum Teichmüller theory were evolving to more general setting as mentioned above, the motivating original question of three dimensional quantum gravity seems to have been lost a bit in this process of generalization. Only relatively recently, Meusburger and Scarinci [**MS16**] found a direct interpretation of the moduli space of a certain version of the three dimensional gravity in terms of the language of cluster varieties. Building on this work, Scarinci and the author [**KS24**] developed a corresponding quantum theory, by suitably modifying Fock and Goncharov's quantum theory of cluster varieties. Roughly speaking, what Fock and Goncharov quantized in [**FG09**] is the Poisson manifold given as the set of positive real points of a cluster variety [**FG06, FG09**]. The work [**MS16**] realized a certain moduli space of 3d gravity as the set of points of a cluster variety valued at a special ring

$$\mathbb{R}_\Lambda = \mathbb{R}[\ell]/(\ell^2 = -\Lambda) \tag{1.1}$$

called the ring of *generalized complex numbers*, where Λ is a real parameter that appears in the theory of gravity, called the *cosmological constant*. In [**KS24**] we developed an $\mathbb{R}_\Lambda$-version of the Fock-Goncharov quantization, and consequently were able to establish a version of the 3d quantum gravity. Just as quantum Teichmüller theory, one of the output of our 3d quantum gravity result [**KS24**] would be a family of unitary projective representations of $\mathrm{MCG}(\mathfrak{S})$ on a Hilbert space, which this time depends not only on the quantum parameter $\hbar$, but also on the gravity parameter Λ. In fact, the formulation of [**KS24**] can be viewed as a construction of a family of representations of a certain groupoid version of $\mathrm{MCG}(\mathfrak{S})$, but not exactly $\mathrm{MCG}(\mathfrak{S})$. Hence, some extra work is needed if one wants to extract concrete $\mathrm{MCG}(\mathfrak{S})$ representations from the results of [**KS24**]. In [**KS24**], this task is not explicitly done, but only hints are given.

We undertake this task in the present article. Namely, we review part of the 3d quantum gravity result of [**KS24**] together with some background materials, and provide necessary extra algebraic and functional-analytic treatments to construct unitary projective representations of $\mathrm{MCG}(\mathfrak{S})$. We note that the way how the results of [**KS24**] are reviewed is adapted to the goal of the present article, and hence may contain somewhat different underlying viewpoints from those of [**KS24**], which could provide new insights to the subject. We also touch upon some applications and possible future research topics.

2. A three-dimensional quantum gravity

2.1. Basic definitions. Throughout the present article, we let $\mathfrak{S}$ be an oriented punctured surface of finite type; one can view it as being obtained from the closed oriented surface of genus g by removing n points, called *punctures*, such that $n > 1$ and $\chi(\mathfrak{S}) = 2 - 2g - n < 0$. The moduli space of 3d gravity to be quantized is $\mathcal{GH}_\Lambda(\mathfrak{S} \times \mathbb{R})$ (see [**MS16**]), depending on the parameter $\Lambda \in \mathbb{R}$ called the cosmological constant. As only the sign of Λ matters, it suffices to consider only three possible values -1, 0 and 1 for Λ. The space $\mathcal{GH}_\Lambda(\mathfrak{S} \times \mathbb{R})$ parametrizes isotopy classes of maximal globally hyperbolic Cauchy-complete Lorentzian metrics g on the 3-manifold $\mathfrak{S} \times \mathbb{R}$ of constant sectional curvature Λ having parabolic holonomy around punctures.

Let us briefly explain the notions just appeared for the metrics g. A metric g on $\mathfrak{S} \times \mathbb{R}$ of Lorentzian signature $(++-)$ being globally hyperbolic means that there is a Cauchy surface $\mathfrak{S}$, i.e. an embedded surface in $\mathfrak{S} \times \mathbb{R}$ diffeomorphic to $\mathfrak{S}$ that intersects each inextendible timelike curve exactly once. Such g being maximal means that the Lorentzian manifold $(\mathfrak{S}, g)$ is not properly contained via isometric embedding into a larger globally hyperbolic Lorentzian 3-manifold. Such g is Cauchy-complete if the induced metric on $\mathfrak{S}$ is geodesically complete. Such a Lorentzian manifold $(\mathfrak{S} \times \mathbb{R}, g)$ can be obtained as a quotient of the model geometry, which is the anti-de Sitter space AdS^3 for $\Lambda = -1$, the Minkowski space Mink^3 for $\Lambda = 0$, and the de Sitter space dS^3 for $\Lambda = 1$, by a discrete subgroup of the isometry group of the model geometry, which is $\mathrm{PSL}_2(\mathbb{R}_\Lambda)$, where $\mathbb{R}_\Lambda$ denotes the ring of generalized complex numbers $\mathbb{R}_\Lambda = \mathbb{R}[\ell]/(\ell^2 = -\Lambda)$ (see eq.(1.1)). Thus a point of $\mathcal{GH}_\Lambda(\mathfrak{S} \times \mathbb{R})$ yields an element of $\mathrm{Hom}(\pi_1(\mathfrak{S}), \mathrm{PSL}_2(\mathbb{R}_\Lambda)) / \mathrm{PSL}_2(\mathbb{R}_\Lambda)$, i.e. a group homomorphism from $\pi_1(\mathfrak{S})$ to $\mathrm{PSL}_2(\mathbb{R}_\Lambda)$, called the holonomy representation, defined up to conjugation in $\mathrm{PSL}_2(\mathbb{R}_\Lambda)$. The parabolicity condition is understood in terms of the holonomy.

An *ideal triangulation* Δ of $\mathfrak{S}$ means a collection of unoriented simple paths in $\mathfrak{S}$ called *ideal arcs* running between punctures, dividing $\mathfrak{S}$ into regions called *ideal triangles* bounded by three ideal arcs. In the present article we do not allow an ideal triangle to be 'self-folded'; this means that we require the three sides of an ideal triangle to be mutually distinct ideal arcs. An ideal arc is often considered up to isotopy, and an ideal triangulation up to simultaneous isotopy of its constituent ideal arcs. It is a simple exercise to show that Δ has $6g-6+3n$ constituent ideal arcs, and forms $4g-4+2n$ ideal triangles.

Per each choice of an ideal triangulation Δ of $\mathfrak{S}$, Meusburger and Scarinci [**MS16**] constructed a special coordinate system on the space $\mathcal{GH}_\Lambda(\mathfrak{S}\times\mathbb{R})$, with the coordinate functions x_i and y_i associated to ideal arcs i of Δ. In fact, these do not form a genuine coordinate system, but a constrained one. Namely, these coordinates satisfy the constraint equations

$$\sum_{i\in\Delta} v_{i,p}x_i = 0, \qquad \sum_{i\in\Delta} v_{i,p}y_i = 0 \tag{2.1}$$

for each puncture p of $\mathfrak{S}$, where $v_{i,p}\in\{0,1,2\}$ denotes the valence of Δ at p; we note that these constraints correspond to the parabolicity condition in the definition of $\mathcal{GH}_\Lambda(\mathfrak{S}\times\mathbb{R})$. What is useful to know is that for each puncture p, the vector $(v_{i,p})_{i\in\Delta}$, viewed as a column vector, belongs to the kernel of the exchange matrix ε, i.e.

$$\sum_{j\in\Delta} \varepsilon_{ij}v_{j,p} = 0, \qquad \forall i\in\Delta, \quad \forall\text{puncture } p. \tag{2.2}$$

It is well known that these vectors actually form a basis of the entire kernel of ε (see e.g. [**BL07**, Lem.8]). The natural gravitational Poisson structure $\{\cdot,\cdot\}$ of $\mathcal{GH}_\Lambda(\mathfrak{S}\times\mathbb{R})$ takes values on these coordinate functions as

$$\{x_i,x_j\}=0, \qquad \{y_i,y_j\}=0, \qquad \{x_i,y_i\}=\varepsilon_{ij}, \qquad \forall i,j\in\Delta.$$

We notice that there is a natural action on $\mathcal{GH}_\Lambda(\mathfrak{S}\times\mathbb{R})$ of the *mapping class group* $\mathrm{MCG}(\mathfrak{S}) = \mathrm{Diff}_+(\mathfrak{S})/\mathrm{Diff}(\mathfrak{S})_0$ of $\mathfrak{S}$, which is the group of orientation-preserving self-diffeomorphisms of $\mathfrak{S}$ considered up to isotopy, preserving the Poisson structure. See [**MS16, KS24**] for more details and references for the structure of the moduli space $\mathcal{GH}_\Lambda(\mathfrak{S}\times\mathbb{R})$.

2.2. A brief overview of a quantization program. We describe the formulation and results of the quantization program established in the work of the author and Scarinci [**KS24**]. For each ideal triangulation Δ, using the above mentioned coordinate system on $\mathcal{GH}_\Lambda(\mathfrak{S}\times\mathbb{R})$ a certain class of classical observables $\mathcal{A}_\Delta=\mathcal{A}_{\Delta;\Lambda}$ is suggested, which depends on the cosmological constant $\Lambda\in\{-1,0,1\}$. The best way to view this class is to formulate it as an algebra of certain functions on the space $\mathcal{GH}_\Lambda(\mathfrak{S}\times\mathbb{R})$ valued in the ring of generalized complex numbers $\mathbb{R}_\Lambda=\mathbb{R}[\ell]/(\ell^2=-\Lambda)$, or maybe in a suitable complexification of $\mathbb{R}_\Lambda$. In the present article we do not delve into a detailed discussion about it, and one can just view $\mathcal{A}_\Delta$ as a certain subset of $C^\infty(\mathcal{GH}_\Lambda(\mathfrak{S}\times\mathbb{R}))$, which is a collection of functions to be quantized. A drawback that we have to endure by giving up working with $\mathbb{R}_\Lambda$ is that the class $\mathcal{A}_\Delta\subset C^\infty(\mathcal{GH}_\Lambda(\mathfrak{S}\times\mathbb{R}))$ may not form an algebra.

Next, a quantum counterpart $\hat{\mathcal{A}}^\hbar_\Delta$ is introduced, which could be viewed as a one-real-parameter family of non-commutative algebras that deforms $\mathcal{A}_\Delta$ in the direction of the Poisson structure. A representation $\pi^\hbar_\Delta$ of $\hat{\mathcal{A}}^\hbar_\Delta$ is constructed on a Hilbert space $\mathscr{H}_\Delta$, in an appropriate sense. In particular, to each element $\hat{u}\in\hat{\mathcal{A}}^\hbar_\Delta$, is associated a densely-defined operator $\pi^\hbar_\Delta(\hat{u})$ on $\mathscr{H}_\Delta$. There are some subtle functional-analytic issues on the domains of definition, but we do not discuss them in this article.

The classes $\mathcal{A}_\Delta$ and $\mathcal{A}^\hbar_\Delta$ are constructed in such a way that for each pair of ideal triangulations Δ and Δ', there are natural bijections of sets (or of algebras)

$$\mu_{\Delta,\Delta'} : \mathcal{A}_{\Delta'} \to \mathcal{A}_\Delta \quad \text{and} \quad \mu^\hbar_{\Delta,\Delta'} : \mathcal{A}^\hbar_{\Delta'} \to \mathcal{A}^\hbar_\Delta,$$

that satisfy the consistency equations

$$\mu_{\Delta,\Delta'} \circ \mu_{\Delta',\Delta''} = \mu_{\Delta,\Delta''} \quad \text{and} \quad \mu^\hbar_{\Delta,\Delta'} \circ \mu^\hbar_{\Delta',\Delta''} = \mu^\hbar_{\Delta,\Delta''}$$

for each triple of ideal triangulations Δ, Δ' and Δ''. Hence, the classes $\mathcal{A}_\Delta$ for different Δ can be consistently identified with each other, and likewise for $\hat{\mathcal{A}}^\hbar_\Delta$.

For each pair of ideal triangulations Δ and Δ', a unitary map

$$\mathbf{K}^\hbar_{\Delta,\Delta'} : \mathscr{H}_{\Delta'} \to \mathscr{H}_\Delta$$

between the Hilbert spaces associated to Δ and Δ' is constructed, so that it intertwines the representations $\pi^\hbar_{\Delta'}$ and $\pi^\hbar_\Delta$ of $\hat{\mathcal{A}}^\hbar_{\Delta'}$ and $\hat{\mathcal{A}}^\hbar_\Delta$, which are related via the map $\mu^\hbar_{\Delta,\Delta'}$. To be more precise, for each $\hat{u} \in \hat{\mathcal{A}}^\hbar_{\Delta'}$ we would like the diagram

$$\begin{array}{ccc}
\mathscr{H}_{\Delta'} & \xrightarrow{\pi^\hbar_{\Delta'}(\hat{u})} & \mathscr{H}_{\Delta'} \\
\Big\downarrow \mathbf{K}^\hbar_{\Delta,\Delta'} & & \Big\downarrow \mathbf{K}^\hbar_{\Delta,\Delta'} \\
\mathscr{H}_\Delta & \xrightarrow{\pi^\hbar_\Delta(\mu^\hbar_{\Delta,\Delta'}(\hat{u}))} & \mathscr{H}_\Delta
\end{array}$$

to commute, that is, we would like the intertwining equations

$$\mathbf{K}^\hbar_{\Delta,\Delta'} \circ \pi^\hbar_{\Delta'}(\hat{u}) = \pi^\hbar_{\Delta,\Delta'}(\mu^\hbar_{\Delta,\Delta'}(\hat{u})) \circ \mathbf{K}^\hbar_{\Delta,\Delta'}$$

to hold. Another important stipulation is that these *unitary intertwiners* $\mathbf{K}^\hbar_{\Delta,\Delta'}$ should satisfy the consistency equations

$$\mathbf{K}^\hbar_{\Delta,\Delta'} \circ \mathbf{K}^\hbar_{\Delta',\Delta''} = \mathbf{K}^\hbar_{\Delta,\Delta''}$$

for each triple of triangulations Δ, Δ' and Δ'', perhaps up to multiplicative constants. So, the unitary maps $\mathbf{K}^\hbar_{\Delta,\Delta'}$ let us identify the representations $\pi^\hbar_\Delta$ of $\hat{\mathcal{A}}^\hbar_\Delta$ on the Hilbert spaces $\mathscr{H}_\Delta$ for different Δ, in a consistent manner.

An algebraic deformation quantization map

$$\hat{Q}^\hbar_\Delta : \mathcal{A}_\Delta \to \hat{\mathcal{A}}^\hbar_\Delta$$

is constructed for each Δ. The most important property of these maps $\hat{Q}^\hbar_\Delta$ is the compatibility with respect to the algebraic compatibility maps $\mu_{\Delta,\Delta'}$ and $\mu^\hbar_{\Delta,\Delta'}$. Namely, they make the diagram

$$\begin{array}{ccc}
\mathcal{A}_{\Delta'} & \xrightarrow{\hat{Q}^\hbar_{\Delta'}} & \hat{\mathcal{A}}^\hbar_{\Delta'} \\
\Big\downarrow \mu_{\Delta,\Delta'} & & \Big\downarrow \mu^\hbar_{\Delta,\Delta'} \\
\mathcal{A}_\Delta & \xrightarrow{\hat{Q}^\hbar_\Delta} & \hat{\mathcal{A}}^\hbar_\Delta
\end{array}$$

commute, that is, the following equation holds for each pair of ideal triangulations Δ and Δ':

$$\mu^\hbar_{\Delta,\Delta'} \circ \hat{Q}^\hbar_{\Delta'} = \hat{Q}^\hbar_\Delta \circ \mu_{\Delta,\Delta'}.$$

This means that the deformation quantization maps $\hat{Q}^\hbar_\Delta$ for different Δ can be identified with each other in a consistent manner. The construction of these maps $\hat{Q}^\hbar_\Delta$ is built on a previous joint work of the author and Allegretti [**AK17**], which in turn is based on the work of Bonahon and Wong [**BW11**].

Let us summarize. For each ideal triangulation Δ a quantization is constructed in the form of the following composition

$$\mathcal{A}_\Delta \xrightarrow{\hat{Q}^\hbar_\Delta} \hat{\mathcal{A}}^\hbar_\Delta \xrightarrow{\pi^\hbar_\Delta} \{\text{densely-defined operators on } \mathscr{H}_\Delta\}\,.$$

For each pair of ideal triangulations Δ and Δ', these quantizations are compatible with each other in the sense that the following diagram commutes:

$$\begin{array}{ccccc} \mathcal{A}_{\Delta'} & \xrightarrow{\hat{Q}^\hbar_{\Delta'}} & \hat{\mathcal{A}}^\hbar_{\Delta'} & \xrightarrow{\pi^\hbar_{\Delta'}} & \{\text{densely-defined operators on } \mathscr{H}_{\Delta'}\} \\ {\scriptstyle \mu_{\Delta,\Delta'}}\downarrow & & \downarrow{\scriptstyle \mu^\hbar_{\Delta,\Delta'}} & & \downarrow{\scriptstyle \text{conjugation by } \mathbf{K}^\hbar_{\Delta,\Delta'}} \\ \mathcal{A}_{\Delta} & \xrightarrow{\hat{Q}^\hbar_{\Delta}} & \hat{\mathcal{A}}^\hbar_{\Delta} & \xrightarrow{\pi^\hbar_{\Delta}} & \{\text{densely-defined operators on } \mathscr{H}_{\Delta}\} \end{array}$$

Hence, in the end, one can say that the quantization does not depend on the choice of Δ, and thus write the result as

$$\mathcal{A}_\mathfrak{S} \xrightarrow{\hat{Q}^\hbar_\mathfrak{S}} \hat{\mathcal{A}}^\hbar_\mathfrak{S} \xrightarrow{\pi^\hbar_\mathfrak{S}} \{\text{densely-defined operators on } \mathscr{H}_\mathfrak{S}\}\,,$$

which depends only on the surface $\mathfrak{S}$.

2.3. On the mapping class group action. An important stipulation we should impose on a quantization is an invariance or equivariance under the mapping class group $\mathrm{MCG}(\mathfrak{S})$, which acts naturally on the moduli space $\mathcal{GH}_\Lambda(\mathfrak{S}\times\mathbb{R})$ as Poisson automorphisms, hence on $C^\infty(\mathcal{GH}_\Lambda(\mathfrak{S}\times\mathbb{R}))$.

We first require that the class of classical observables $\mathcal{A}_\Delta$ should be invariant under the action of $\mathrm{MCG}(\mathfrak{S})$. That is, we should have $\mathcal{A}_\Delta = \mathcal{A}_{g.\Delta}$ for $g \in \mathrm{MCG}(\mathfrak{S})$, where $g.\Delta$ denotes the isotopy class of an ideal triangulation obtained by acting the mapping class g to Δ; the bijections between these classical classes associated to changes of ideal triangulations must also satisfy the invariance $\mu_{\Delta,\Delta'} = \mu_{g.\Delta,g.\Delta'}$. Upon quantization, invariance should hold for the classes of quantum observables $\mathcal{A}^\hbar_\Delta = \mathcal{A}^\hbar_{g.\Delta}$, the quantum bijections between them $\mu^\hbar_{\Delta,\Delta'} = \mu^\hbar_{g.\Delta,g.\Delta'}$, the quantum Hilbert spaces $\mathscr{H}_\Delta = \mathscr{H}_{g.\Delta}$, the representations $\pi^\hbar_\Delta = \pi^\hbar_{g.\Delta}$, the unitary intertwiners between them $\mathbf{K}^\hbar_{\Delta,\Delta'} = \mathbf{K}^\hbar_{g.\Delta,g.\Delta'}$, and the deformation quantization maps $\hat{Q}^\hbar_\Delta = \hat{Q}^\hbar_{g.\Delta}$.

In fact, one should carefully think of the meaning of the equalities appearing in the invariance condition, such as $\mathcal{A}_\Delta = \mathcal{A}_{g.\Delta}$ and $\mathscr{H}_\Delta = \mathscr{H}_{g.\Delta}$. For simplicity of discussion, at the moment, assume that $\mathcal{A}_\Delta$ is realized as a subalgebra of $C^\infty(\mathcal{GH}_\Lambda(\mathfrak{S}\times\mathbb{R}))$ for each Δ; this is indeed the case when $\Lambda = -1$. Thus both $\mathcal{A}_\Delta$ and $\mathcal{A}_{g.\Delta}$ are subalgebras of $C^\infty(\mathcal{GH}_\Lambda(\mathfrak{S}\times\mathbb{R}))$, hence the equality $\mathcal{A}_\Delta = \mathcal{A}_{g.\Delta}$ seems to have an unambiguous meaning. However, for a better control, we should discuss what is a correct identification map between $\mathcal{A}_\Delta$ and $\mathcal{A}_{g.\Delta}$. An obvious option is the restriction of the identity map on $C^\infty(\mathcal{GH}_\Lambda(\mathfrak{S}\times\mathbb{R}))$, sending each f to f. In fact, the identification $\mathcal{A}_\Delta \to \mathcal{A}_{g.\Delta}$ that suits better for the formulation of the invariance under the action of $\mathrm{MCG}(\mathfrak{S})$ is the map sending f to $g.f$, using the action of $\mathrm{MCG}(\mathfrak{S})$ on $C^\infty(\mathcal{GH}_\Lambda(\mathfrak{S}\times\mathbb{R}))$. All the equalities appearing in the formulation of the $\mathrm{MCG}(\mathfrak{S})$-invariance should be of the latter sort, taking into account the non-trivial natural action of $\mathrm{MCG}(\mathfrak{S})$; so, in a sense, maybe it is a better idea to denote them by the isomorphism sign $\cong$ rather than the equality sign $=$.

We claim that by a further investigation of the MCG($\mathfrak{S}$)-invariance along this line, one can obtain a version of MCG($\mathfrak{S}$)-equivariance of the quantization of $\mathcal{GH}_\Lambda(\mathfrak{S} \times \mathbb{R})$ as a consequence. In particular, this will yield unitary representations of MCG($\mathfrak{S}$) on Hilbert spaces, which are the principal objects of study of the present article. As mentioned in §1, the main goal of the present article is to present an explicit construction of these representations of MCG($\mathfrak{S}$), for it is only hinted but not fully displayed in [**KS24**].

3. Quantum dilogarithm functions

Here we explain some special functions called the *quantum dilogarithm* functions, being crucial ingredients of the constructions of [**KS24**]. Here we focus mostly just on the definitions and brief partial history, and refrain from giving much further survey; for more information, see [**FK94, Z07, KS24, K21a**]. Recall that the logarithm function with complex variable admits a power series expansion $-\log(1-z) = \sum_{m=1}^\infty \frac{z^m}{m}$ in the region $|z|<1$. Euler's dilogarithm is the function $\mathrm{Li}_2(z) = \sum_{m=1}^\infty \frac{z^m}{m^2}$, related to the usual logarithm by integration. Replacing one m in the denominator of the m-th summand by the 'quantum m', one obtains a crude version of what is called the quantum dilogarithm. It is more convenient to consider a variant $\mathrm{Li}_2(z;q) = \sum_{m=1}^\infty \frac{z^m}{m(1-q^m)}$, or better, its negative exponential $e^{-\mathrm{Li}_2(z;q)}$, which by a slight change of variables for q and z (namely, put q^2 and $-qz$ in place of q and z) yields the function

$$\psi^q(z) = \prod_{m=0}^\infty (1+q^{2m+1}z)^{-1}$$

which is sometimes referred to as the *compact quantum dilogarithm*. It enjoys several remarkable properties, the most fundamental one being the difference equation

$$\psi^q(q^2 z) = (1+qz)\psi^q(z),$$

and the most interesting one being the *pentagon identity*: for variables u, v such that $uv = q^2 vu$,

$$\psi^q(u)\psi^q(v) = \psi^q(v)\psi^q(qvu)\psi^q(v)$$

holds in an appropriate sense ([**FK94**]).

This function ψ^q has appeared in the literature for a long time in many different guises, and relatively recently in 1990's Faddeev and Kashaev started to establish some basics of it for use in mathematical physics [**F95, FK94**]. Then it was used crucially in the quantization of Teichmüller spaces by Kashaev [**K98**] and by Chekhov and Fock [**CF97, F97**], and in the quantization of cluster varieties by Fock and Goncharov [**FG09**].

One way of understanding ψ^q is to let q be a formal variable. However, for a 'physical' quantization, q should be an actual number. When q is a complex number satisfying $|q|<1$, the infinite product expression yields a well-defined meromorphic function on the complex plane. However, in quantization of Teichmüller spaces and cluster varieties [**K98, CF97, F97, FG09**], what was needed is the case of $|q|=1$, for which the infinite product does not make sense. Hence another version of a quantum dilogarithm function had to be used, which would share some similar properties with ψ^q. It is the *non-compact quantum dilogarithm* function $\Phi^\hbar(z)$ [**F95, FK94**], which is a function in the complex variable z, associated to a nonzero real parameter $\hbar$. Assuming $\hbar > 0$ at the moment for convenience, it is given by the following contour integral formula which goes back to Barnes [**B01**] 100 years ago in the study of 'double Gamma function':

$$\Phi^\hbar(z) = \exp\left(-\frac{1}{4}\int_\Omega \frac{e^{-\mathrm{i}pz}}{\sinh(\pi p)\sinh(\pi\hbar p)}\frac{dp}{p}\right) \tag{3.1}$$

for z living in the strip $|\,\mathrm{Im}(z)| < \pi(1+\hbar)$, where Ω is the contour along the real line that avoids the origin by a small half-circle above the origin. One can show that this formula defines a non-vanishing complex analytic function on this strip, and that this function has an analytic continuation to a meromorphic function on the whole complex plane with an explicitly described sets of zeroes and poles. A fundamental property is the pair of difference equations

$$\begin{cases} \Phi^\hbar(z+2\pi\mathrm{i}\hbar) = (1+e^{\pi\mathrm{i}\hbar}e^z)\Phi^\hbar(z), \\ \Phi^\hbar(z+2\pi\mathrm{i}) = (1+e^{\pi\mathrm{i}/\hbar}e^{z/\hbar})\Phi^\hbar(z), \end{cases}$$

which play crucial roles in solving the intertwining equations. It also satisfies $|\Phi^\hbar(x)| = 1$ for $x \in \mathbb{R}$; hence, for any self-adjoint operator $\mathbf{A}$ on a Hilbert space, the operator $\Phi^\hbar(\mathbf{A})$, obtained by applying the functional calculus of $\mathbf{A}$ to the function $\Phi^\hbar$, is unitary. Another important property of $\Phi^\hbar$ is the pentagon identity: for self-adjoint operators $\mathbf{P}$ and $\mathbf{Q}$ on a separable Hilbert space satisfying the Heisenberg relation $[\mathbf{P},\mathbf{Q}] = 2\pi\mathrm{i}\hbar\cdot\mathrm{id}$ (more precisely, its 'Weyl-relations version' as in eq.(4.2)), one has the following operator identity among unitary maps ([**FK94, FKV01, G08, W00**]):

$$\Phi^\hbar(\mathbf{P})\Phi^\hbar(\mathbf{Q}) = \Phi^\hbar(\mathbf{Q})\Phi^\hbar(\mathbf{P}+\mathbf{Q})\Phi^\hbar(\mathbf{P}).$$

Meanwhile, we will also need the case $\hbar < 0$. The formula in eq.(3.1) actually yields

$$\Phi^{-\hbar}(z) = (\Phi^\hbar(z))^{-1},$$

which is a way of understanding the case of a negative real parameter $\hbar$.

Comparing the difference equation of ψ^q and the first of the two difference equations of $\Phi^\hbar$, keeping in mind that $q = e^{\pi\mathrm{i}\hbar}$, one can immediately expect that ψ^q and $\Phi^\hbar$ should be closely related. To see this more clearly, let h be any nonzero complex number with $\mathrm{Re}(h) \geq 0$. Notice that we are not using the symbol $\hbar$ here. Consider the integral expression

$$\Phi^h(z) = \exp\left(-\frac{1}{4}\int_{\Omega'} \frac{e^{-\mathrm{i}pz}}{\sinh(\pi p)\sinh(\pi h p)}\frac{dp}{p}\right), \tag{3.2}$$

which is referred to as the Barnes integral in [**KS24**], for some contour Ω' to be explained. Let us first focus on the case when $\mathrm{Re}(h) > 0$, for which we can take the contour Ω' to be same as before, i.e. as Ω. This expression defines a non-vanishing complex analytic function on the strip $|\,\mathrm{Im}(z)| < \pi(1+\mathrm{Re}(h))$, and is analytically continued to a meromorphic function. When $\mathrm{Im}(h) > 0$ is also satisfied, then by the residue computation, one can show that

$$\Phi^h(z) = \frac{\psi^{\exp(\pi\mathrm{i}h)}(e^z)}{\psi^{\exp(-\pi\mathrm{i}/h)}(e^{z/h})} \tag{3.3}$$

holds, which is a straightforward exercise in complex analysis. That is, the function Φ^h is a certain ratio of two compact quantum dilogarithm functions. Notice that both quantum parameters $e^{\pi\mathrm{i}h}$ and $e^{-\pi\mathrm{i}/h}$ for the two compact quantum dilogarithms are complex numbers of modulus less than 1, due to the assumption

$\mathrm{Im}(h) > 0$; one also notes that h and $-1/h$ are related to each other by the 'modular transformation'. However, notice that what we referred to as the non-compact quantum dilogarithm falls into the case when $h = \hbar$ belongs to $\mathbb{R}_{>0}$, i.e. $\mathrm{Im}(h) = 0$. In this case, the parameters $e^{\pi i h}$ and $e^{-\pi i/h}$ are of modulus 1, hence the associated compact quantum dilogarithm functions do not make sense, therefore nor does the equality in eq.(3.3). In a sense, the non-compact quantum dilogarithm can be thought of as being a limit of the ratio of compact quantum dilogarithms as in the right hand side of eq.(3.3) as $\mathrm{Im}(h) \to 0$. Anyways, the function $\Phi^h(z)$ defined in eq.(3.2) can also be referred to as a quantum dilogarithm function as well.

The case $\mathrm{Re}(h) > 0$ has been studied in the literature, but for the case $\mathrm{Re}(h) = 0$, i.e. when $h = \mathrm{i}\hbar \in \mathrm{i}\mathbb{R}$ is a pure imaginary number, the integral expression in eq.(3.2) seems to have been somewhat elusive. In [**KS24**] we undertook the task of studying this case. The contour integral in eq.(3.2) does not work if we use the previous Ω as the contour Ω', because in case $\mathrm{Re}(h) = 0$ the integrand has poles along real and imaginary axes, hence also on Ω. Thus we used a 'slanted' contour $\Omega' = e^{\mathrm{i}\theta}\Omega$ for some angle $\theta \in \mathbb{R}$ of small absolute value, which is taken to be a negative angle when $\mathrm{Im}(h) \ge 0$ and a positive angle when $\mathrm{Im}(h) \le 0$. We then still get a well-defined meromorphic function $\Phi^h = \Phi^{\mathrm{i}\hbar}$, and when $\mathrm{Im}(h) > 0$ the ratio expression in eq.(3.3) still holds; in particular, when $\mathrm{Im}(h) > 0$ and $\mathrm{Re}(h) = 0$, i.e. $h = \mathrm{i}\hbar \in \mathrm{i}\mathbb{R}_{>0}$, we have:

$$\Phi^{\mathrm{i}\hbar}(z) = \frac{\psi^{\exp(-\pi\hbar)}(e^z)}{\psi^{\exp(-\pi/\hbar)}(e^{z/(\mathrm{i}\hbar)})}.$$

Notice that the quantum parameters $e^{-\pi\hbar}$ and $e^{-\pi/\hbar}$ for the two compact quantum dilogarithms in this case are positive real numbers less than 1. Similar arguments show that for $\mathrm{Im}(h) < 0$, i.e. when $h = -\mathrm{i}\hbar$ with $\hbar > 0$, the function $\Phi^{-\mathrm{i}\hbar}(z)$ can be computed to be

$$\Phi^{-\mathrm{i}\hbar}(z) = \frac{\psi^{\exp(-\pi/\hbar)}(e^{-z/(\mathrm{i}\hbar)})}{\psi^{\exp(-\pi\hbar)}(e^z)} = \overline{\Phi^{\mathrm{i}\hbar}(\overline{z})}^{-1}.$$

One can view each of $\Phi^{\mathrm{i}\hbar}$ and $\Phi^{-\mathrm{i}\hbar}$ as a 'modular double' version of a compact quantum dilogarithm function $\psi^{\mathbf{q}}$, with a real quantum parameter $\mathbf{q} = e^{-\pi\hbar}$. As mentioned, although these functions $\Phi^{\mathrm{i}\hbar}$ and $\Phi^{-\mathrm{i}\hbar}$ are genuine ratios of certain compact quantum dilogarithm functions, they do not seem to have been considered before in the literature.

What is actually introduced and used in [**KS24**] for the quantization of $\mathcal{GH}_\Lambda(\mathfrak{S}\times\mathbb{R})$ is a two-real-variable function $F_\Lambda^\hbar : \mathbb{R}^2 \to \mathbb{C}$, depending on the cosmological constant $\Lambda \in \{-1, 0, 1\}$, defined by

$$F_\Lambda^\hbar(x,y) = \begin{cases} \Phi^\hbar(x+\hbar y)\Phi^{-\hbar}(x-\hbar y) & \text{if } \Lambda = -1, \\ \Phi^{\mathrm{i}\hbar}(x+\mathrm{i}\hbar y)\Phi^{-\mathrm{i}\hbar}(x-\mathrm{i}\hbar y) & \text{if } \Lambda = 1, \\ (1+e^x)^{y/(\pi \mathrm{i})} = \exp\left(\frac{y}{\pi \mathrm{i}}\log(1+e^x)\right) & \text{if } \Lambda = 0. \end{cases} \tag{3.4}$$

The three formulas may not look so uniform at first glance. We note that in fact they all come from the heuristic expression

$$\Phi^{\ell\hbar}(x+\ell\hbar y)\Phi^{-\ell\hbar}(x-\ell\hbar y)$$

in terms of the ring $\mathbb{R}_\Lambda = \mathbb{R}[\ell]/(\ell^2 = -\Lambda)$ of generalized complex numbers, or in terms of a suitable complexification of $\mathbb{R}_\Lambda$. Note also that the second factor $\Phi^{-\ell\hbar}(x-\ell\hbar y)$ is obtained by applying the $\mathbb{R}_\Lambda$-conjugation on the first factor $\Phi^{\ell\hbar}(x+$

$\ell\hbar y$), so the first and the second factors can be thought of as $\mathbb{R}_\Lambda$-holomorphic and $\mathbb{R}_\Lambda$-anti-holomorphic parts. But this is indeed just at a heuristic level, as the $\mathbb{R}_\Lambda$-*version of the quantum dilogarithm* $\Phi^{\ell\hbar}$ cannot be defined using the contour integral on the nose. Anyways, the function $F_0^\hbar$ for $\Lambda = 0$ 'comes from' an $\mathbb{R}_0$-version of the quantum dilogarithm, hence can be viewed as a degenerate, or flat, version of the quantum dilogarithm.

Unlike $\Phi^{\pm\ell\hbar}$, the defining expressions in the right hand side of eq.(3.4) for the function $F_\Lambda^\hbar$ do make good senses. One useful property of the function $F_\Lambda^\hbar$ we shall use is its 'unitarity':

$$|F_\Lambda^\hbar(x,y)| = 1, \qquad \forall x, y \in \mathbb{R}. \tag{3.5}$$

If $\mathbf{A}$ and $\mathbf{B}$ are self-adjoint operators on a Hilbert space that strongly commute with each other, i.e. if they satisfy the Weyl-relations version (eq.(4.2)) of the Heisenberg relation $[\mathbf{A},\mathbf{B}] = 0$, then one can apply the two-variable functional calculus for $\mathbf{A}$ and $\mathbf{B}$ to this function $F_\Lambda^\hbar$ to obtain an operator $F_\Lambda^\hbar(\mathbf{A},\mathbf{B})$. For the multi-variable functional calculus for a strongly commuting collection of self-adjoint operators, see [**S12**]. The unitarity as in eq.(3.5) implies that $F_\Lambda^\hbar(\mathbf{A},\mathbf{B})$ is unitary. Another important property of $F_\Lambda^\hbar$ is the pentagon operator identity, as shown in [**KS24**, Prop.6.16]. It says that, if $\mathbf{x}_1$, $\mathbf{y}_1$, $\mathbf{x}_2$ and $\mathbf{y}_2$ are self-adjoint operators satisfying the (Weyl-relations versions of the) Heisenberg relations $[\mathbf{x}_1,\mathbf{x}_2] = [\mathbf{y}_1,\mathbf{y}_2] = [\mathbf{x}_1,\mathbf{y}_1] = [\mathbf{x}_2,\mathbf{y}_2] = 0$, $[\mathbf{x}_1,\mathbf{y}_2] = [\mathbf{y}_1,\mathbf{x}_2] = \pi\mathrm{i}\cdot\mathrm{id}$, then

$$F_\Lambda^\hbar(\mathbf{x}_1,\mathbf{y}_1)F_\Lambda^\hbar(\mathbf{x}_2,\mathbf{y}_2) = F_\Lambda^\hbar(\mathbf{x}_2,\mathbf{y}_2)F_\Lambda^\hbar(\mathbf{x}_1+\mathbf{x}_2,\mathbf{y}_1+\mathbf{y}_2)F_\Lambda^\hbar(\mathbf{x}_1,\mathbf{y}_1) \tag{3.6}$$

holds as an equality of unitary operators. For other properties of $F_\Lambda^\hbar$, see [**KS24**].

4. The unitary interwiners

Among the aspects of the quantization program of [**KS24**] as spelled out in §2, in the present article we focus only on the system of Hilbert spaces $\mathscr{H}_\Delta$ of quantum states associated to ideal triangulations Δ, together with the unitary maps $\mathbf{K}^\hbar_{\Delta,\Delta'}$ which identify these Hilbert spaces with each other. For the remaining aspects of the quantization program, see [**KS24**].

4.1. The Hilbert space and basic operators. Fix any ideal triangulation Δ. Let $\mathscr{H}_\Delta$ be a separable complex Hilbert space, and let $\mathbf{x}_i$ and $\mathbf{y}_i$ ($i \in \Delta$) be self-adjoint operators on $\mathscr{H}_\Delta$ satisfying the Heisenberg relations

$$[\mathbf{x}_i,\mathbf{x}_j] = 0, \qquad [\mathbf{y}_i,\mathbf{y}_j] = 0, \qquad [\mathbf{x}_i,\mathbf{y}_j] = \pi\mathrm{i}\varepsilon_{ij}\cdot\mathrm{id}, \qquad \forall i,j \in \Delta. \tag{4.1}$$

One might wonder why there is no quantum parameter $\hbar$ in these Heisenberg relations. Indeed, what should be quantum counterparts of the coordinate functions x_i and y_i of $\mathcal{GH}_\Lambda(\mathfrak{S}\times\mathbb{R})$ are the operators $\mathbf{x}_i$ and $\hbar\mathbf{y}_i$, as we have $[\mathbf{x}_i,\hbar\mathbf{y}_j] = \pi\mathrm{i}\hbar\varepsilon_{ij}\cdot\mathrm{id}$. In [**KS24**] we decided to take out $\hbar$, which made it easier to '$\mathbb{R}_\Lambda$-complexify' $\hbar\in\mathbb{R}$ to $\ell\hbar\in\mathbb{R}_\Lambda$.

When we say that two self-adjoint operators $\mathbf{A}$ and $\mathbf{B}$ satisfy a Heisenberg relation

$$[\mathbf{A},\mathbf{B}] = \mathrm{i}c\cdot\mathrm{id}$$

for some $c\in\mathbb{R}$, unless otherwise stated we mean that they satisfy the *Weyl-relations* version of it, i.e. that the family of equations

$$e^{\mathrm{i}\alpha\mathbf{A}}e^{\mathrm{i}\beta\mathbf{B}} = e^{-\mathrm{i}c\alpha\beta}e^{\mathrm{i}\beta\mathbf{B}}e^{\mathrm{i}\alpha\mathbf{A}} \tag{4.2}$$

of unitary operators are satisfied for all $\alpha, \beta \in \mathbb{R}$. This is what makes the solution of the Heisenberg relations more rigid, and is required as a hypothesis for a (generalized) Stone-von Neumann theorem. Let us also consider the quantum constraint equations

$$\sum_{i\in\Delta} v_{i,p}\mathbf{x}_i = 0 \qquad \text{and} \qquad \sum_{i\in\Delta} v_{i,p}\mathbf{y}_i = 0, \qquad \forall \text{puncture } p. \tag{4.3}$$

One can observe that the Stone-von Neumann theorem [**vN31**] [**H13**, §14] implies that the collection of self-adjoint operators $\{\mathbf{x}_i, \mathbf{y}_i\}_{i\in\Delta}$ on a separable complex Hilbert space $\mathscr{H}$ satisfying eq.(4.1) and eq.(4.3) is uniquely determined up to unitary equivalence, if we also stipulate that these operators should act in an *irreducible* manner, meaning that the only closed subspaces of $\mathscr{H}$ invariant under $e^{\mathrm{i}\alpha\mathbf{x}_i}$ and $e^{\mathrm{i}\alpha\mathbf{y}_i}$ for all $\alpha \in \mathbb{R}$ and i are 0 and $\mathscr{H}$ itself.

As just mentioned, by using the existence and uniqueness result of the Stone-von Neumann theorem, it is enough to just write the equations eq.(4.1) and eq.(4.3), without writing down the actual Hilbert space and the operators explicitly; see [**K21b**] [**H13**, §14]. However, instead of doing so, we shall explicitly present a particular collection of operators as a solution. For convenience of our presentation, we will at the moment forget the constraint equations in eq.(4.3) and work with a certain solution to just the Heisenberg relations in eq.(4.1), based on which we will build our unitary intertwiners. Then afterwards in §5.3, we will give a recipe how to actually obtain a solution satisfying both eq.(4.1) and eq.(4.3), and the corresponding intertwiners.

The Hilbert space is set to be

$$\mathscr{H}_\Delta := L^2(\mathbb{R}^\Delta, \bigwedge_{i\in\Delta} dt_i),$$

with respect to the standard Lebesgue measure on $\mathbb{R}^\Delta = \mathbb{R}^{|\Delta|} = \mathbb{R}^{6g-6+3n}$. Consider the position and the momentum operators

$$t_i \qquad \text{and} \qquad \mathrm{i}\frac{\partial}{\partial t_i}, \qquad \text{for } i \in \Delta,$$

which are self-adjoint operators on $\mathscr{H}_\Delta$ satisfying the Heisenberg relations

$$[t_i, t_j] = 0, \qquad [\mathrm{i}\frac{\partial}{\partial t_i}, \mathrm{i}\frac{\partial}{\partial t_j}] = 0, \qquad [t_i, \mathrm{i}\frac{\partial}{\partial t_j}] = -\mathrm{i}\delta_{i,j}\cdot \mathrm{id},$$

where $\delta_{i,j}$ is the Kronecker delta. It is well known that any $\mathbb{R}$-linear combinations of the position and the momentum operators yield self-adjoint operators, and any two of them satisfy the Heisenberg relations that can be deduced from the bilinearity of the commutator bracket; see [**H13**]. Using this fact, we consider the operators

$$\mathbf{x}_i = -\pi \mathrm{i}\frac{\partial}{\partial t_i}, \qquad \mathbf{y}_i = \sum_{j\in\Delta} \varepsilon_{ij} t_j, \qquad \forall i \in \Delta, \tag{4.4}$$

which are self-adjoint operators on $\mathscr{H}_\Delta$ satisfying the Heisenberg relations in eq.(4.1).

4.2. The intertwiner for a flip at an arc. Suppose that Δ and Δ' are ideal triangulations of $\mathfrak{S}$ that are related to each other by *flip* at an arc k, that is, they differ exactly at one arc denoted by k. In this case, there is a natural set bijection between Δ and Δ', allowing us to enumerate the members of these two triangulations by the same symbols like i, j and k. We aim to describe the solution to the unitary intertwiner

$$\mathbf{K}^{\hbar}_{\Delta,\Delta'} : \mathscr{H}_{\Delta'} \to \mathscr{H}_{\Delta}$$

constructed in [**KS24**]. Since Δ' is obtained from Δ by flip at k, we may write $\Delta' = \mu_k(\Delta)$ using the symbol μ_k which hints at the connection to the 'cluster mutations' [**FZ02, FG06, FG09**], and denote this intertwiner as

$$\mathbf{K}^{\hbar}_k = \mathbf{K}^{\hbar}_{\Delta,\Delta'} = \mathbf{K}^{\hbar}_{\Delta,\mu_k(\Delta)} : \mathscr{H}_{\mu_k(\Delta)} = \mathscr{H}_{\Delta'} \to \mathscr{H}_{\Delta}.$$

One has to be careful when dealing with the symbol $\mathbf{K}^{\hbar}_k$ because it is ambiguous, as it does not indicate the underlying ideal triangulations.

Following [**FG09**], the intertwiner $\mathbf{K}^{\hbar}_k = \mathbf{K}^{\hbar}_{\Delta,\Delta'}$ for a flip μ_k at k is obtained as the composition

$$\mathbf{K}^{\hbar}_k = \mathbf{K}^{\hbar\sharp}_k \circ \mathbf{K}'_k : \mathscr{H}_{\Delta'} \xrightarrow{\mathbf{K}'_k} \mathscr{H}_{\Delta} \xrightarrow{\mathbf{K}^{\hbar\sharp}_k} \mathscr{H}_{\Delta}$$

of two unitary operators $\mathbf{K}'_k$ and $\mathbf{K}^{\hbar\sharp}_k$, where

$$\mathbf{K}'_k = \mathbf{K}'_{\Delta,\Delta'} : \mathscr{H}_{\Delta'} = L^2(\mathbb{R}^{\Delta'}, \bigwedge_{i\in\Delta'} dt'_i) \to L^2(\mathbb{R}^{\Delta}, \bigwedge_{i\in\Delta} dt_i) = \mathscr{H}_{\Delta}$$

is called the 'monomial transformation part' and

$$\mathbf{K}^{\hbar\sharp}_k = \mathbf{K}^{\hbar\sharp}_{\Delta,\Delta'} : \mathscr{H}_{\Delta} = L^2(\mathbb{R}^{\Delta}, \bigwedge_{i\in\Delta} dt_i) \to L^2(\mathbb{R}^{\Delta}, \bigwedge_{i\in\Delta} dt_i) = \mathscr{H}_{\Delta}$$

is called the 'automorphism part'.

The monomial transformation part $\mathbf{K}'_k : L^2(\mathbb{R}^{\Delta'}, \bigwedge_{i\in\Delta'} dt'_i) \to L^2(\mathbb{R}^{\Delta}, \bigwedge_{i\in\Delta} dt_i)$ is the operator induced by the map

$$\chi_k : \mathbb{R}^{\Delta} \to \mathbb{R}^{\Delta'}$$

between the Euclidean spaces $\mathbb{R}^{\Delta}$ and $\mathbb{R}^{\Delta'}$ defined as follows: the pullback of each coordinate function of the codomain is given by

$$\chi_k^* t'_i = \begin{cases} t_i & \text{if } i \neq k, \\ -t_k + \sum_{j\in\Delta}[-\varepsilon_{kj}]_+ t_j & \text{if } i = k, \end{cases}$$

where $\varepsilon = (\varepsilon_{ij})_{i,j\in\Delta}$ is the *exchange matrix* for the triangulation Δ defined as

$$\begin{aligned} \varepsilon_{ij} &= a_{ij} - a_{ji}, \\ a_{ij} &= \text{the number of corners of ideal triangles of } \Delta \\ &\quad\ \text{delimited by } i \text{ at the left and } j \text{ at the right,} \end{aligned}$$

and $[a]_+$ is the positive part of a real number a, i.e.

$$[a]_+ = \frac{a + |a|}{2}.$$

In [**K21a**], the author studied the operators on $L^2(\mathbb{R}^N)$ induced by invertible linear maps on $\mathbb{R}^N$, precisely to the extent we need in this setting; in particular, as observed in [**K21a**], the operator $\mathbf{K}'_k$ is unitary.

The automorphism part $\mathbf{K}_k^{\hbar\sharp} : L^2(\mathbb{R}^\Delta, \bigwedge_{i\in\Delta} dt_i) \to L^2(\mathbb{R}^\Delta, \bigwedge_{i\in\Delta} dt_i)$ is the operator given by the following formula, depending on the cosmological constant Λ:

$$\mathbf{K}_k^{\hbar\sharp} = F_\Lambda^\hbar(\mathbf{x}_k, \mathbf{y}_k) = \begin{cases} \Phi^\hbar(\mathbf{x}_k + \hbar\mathbf{y}_k)\Phi^{-\hbar}(\mathbf{x}_k - \hbar\mathbf{y}_k) & \text{if } \Lambda = -1, \\ \Phi^{\mathrm{i}\hbar}(\mathbf{x}_k + \mathrm{i}\hbar\mathbf{y}_k)\Phi^{-\mathrm{i}\hbar}(\mathbf{x}_k - \mathrm{i}\hbar\mathbf{y}_k) & \text{if } \Lambda = 1, \\ (1+e^{\mathbf{x}_k})^{\mathbf{y}_k/(\pi\mathrm{i})} & \text{if } \Lambda = 0, \end{cases} \tag{4.5}$$

where the function $F_\Lambda^\hbar : \mathbb{R}^2 \to \mathbb{C}$ is as defined in eq.(3.4) in §3. Here $\Phi^{\pm\hbar}$ and $\Phi^{\pm\mathrm{i}\hbar}$ are some versions of the quantum dilogarithm functions described in §3. The formula $F_\Lambda^\hbar(\mathbf{x}_k, \mathbf{y}_k)$, i.e. each of the three lines in the right of eq.(4.5), is defined by the two-variable functional calculus for the self-adjoint operators $\mathbf{x}_k$ and $\mathbf{y}_k$ on $\mathscr{H}_\Delta = L^2(\mathbb{R}^\Delta, \bigwedge_{i\in\Delta} dt_i)$ in eq.(4.4), which makes sense because $\mathbf{x}_k$ and $\mathbf{y}_k$ strongly commute with each other; indeed, recall that they satisfy the Weyl-relations version of the Heisenberg relation $[\mathbf{x}_k, \mathbf{y}_k] = 0$, as in eq.(4.1). From eq.(3.5), one observes that $\mathbf{K}_k^{\hbar\sharp}$ is unitary.

As said in §3, we have $\Phi^{-\hbar}(z) = (\Phi^\hbar(z))^{-1}$, which entails that the above described intertwiner $\mathbf{K}_k^\hbar = \mathbf{K}_k^{\hbar\sharp} \circ \mathbf{K}'_k$ for the case $\Lambda = -1$ coincides with Fock and Goncharov's intertwiner associated to a cluster mutation μ_k [**FG09**], for the quantization of their 'cluster symplectic double variety', or 'cluster $\mathscr{D}$-variety' [**FG09**]. The construction of [**FG09**] applies to *any* skew-symmetric integer matrix ε. If we restrict ourselves to the case when the exchange matrix ε comes from an ideal triangulation of a punctured surface, what Fock and Goncharov quantized is a certain double of the 'enhanced Teichmüller space' $\mathcal{T}^+(\mathfrak{S})$ [**BL07, F97**], which in a sense can be viewed as the cotangent bundle of $\mathcal{T}^+(\mathfrak{S})$. Anyhow, we note that the functions used in this case of $\Lambda = -1$ are the non-compact quantum dilogarithm functions $\Phi^{\pm\hbar}$ of Faddeev and Kashaev [**F95, FK94, B01**], which are also what are used in the original works of quantum Teichmüller theory by Kashaev [**K98**] and by Chekhov and Fock [**CF97, F97**] for the construction of unitary intertwiners; in these works, the flip intertwiner involves just one copy of $\Phi^\hbar$, instead of both $\Phi^\hbar$ and $\Phi^{-\hbar} = (\Phi^\hbar)^{-1}$.

4.3. The intertwiner for an arbitrary change of ideal triangulations. Now let Δ and Δ' be any two ideal triangulations of $\mathfrak{S}$. It is well known that they can be connected to each other by a sequence of flips; see Labardini-Fragoso [**L-F09**] for such a statement when we restrict only to the ideal triangulations without self-folded triangles. So one can find a sequence of ideal triangulations $\Delta_0, \Delta_1, \ldots, \Delta_m$, such that $\Delta_0 = \Delta$, $\Delta_m = \Delta'$, and $\forall i = 1, \ldots, m$, Δ_{i-1} and Δ_i are related by a flip, say $\Delta_i = \mu_{k_i}(\Delta_{i-1})$ for some arc k_i. We then define the intertwiner $\mathbf{K}_{\Delta,\Delta'}^\hbar$ associated to the change of ideal triangulations $\Delta \leadsto \Delta'$ as the composition of the intertwiners for flips in the above sequence. That is, we let

$$\begin{aligned} \mathbf{K}_{\Delta,\Delta'}^\hbar &:= \mathbf{K}_{\Delta_0,\Delta_1}^\hbar \circ \mathbf{K}_{\Delta_1,\Delta_2}^\hbar \circ \cdots \circ \mathbf{K}_{\Delta_{m-1},\Delta_m}^\hbar \\ &= \mathbf{K}_{k_1}^\hbar \circ \mathbf{K}_{k_2}^\hbar \circ \cdots \circ \mathbf{K}_{k_m}^\hbar : \mathscr{H}_{\Delta'} \to \mathscr{H}_\Delta. \end{aligned} \tag{4.6}$$

As the sequence of flips connecting Δ and Δ' is not unique, one has to make sure that the resulting operator $\mathbf{K}_{\Delta,\Delta'}^\hbar$ does not depend on the choice of such a sequence. This well-definedness is guaranteed if we show that the flip operators $\mathbf{K}_k^\hbar$ satisfy all possible relations satisfied by the flips of ideal triangulations. The first step is to classify all relations of flips of triangulations. The following result is well known; see [**FST08**, Thm.3.10] and references therein.

PROPOSITION 4.1. *Any relation satisfied by flips of ideal triangulations without self-folded triangles is generated by the following ones:*

(1) *(twice-flip identity)* $\mu_k \circ \mu_k = \mathrm{id}$ *when applied to any* Δ*;*
(2) *(quadrilateral identity)* $\mu_i \circ \mu_j \circ \mu_i \circ \mu_j = \mathrm{id}$ *when applied to any* Δ *with* $\varepsilon_{ij} = 0$*;*
(3) *(pentagon identity)* $P_{(ij)} \circ \mu_i \circ \mu_j \circ \mu_i \circ \mu_j \circ \mu_i = \mathrm{id}$ *when applied to any* Δ *with* $\varepsilon_{ij} = \pm 1$*, where* $P_{(ij)}$ *means re-labeling* $i \leftrightarrow j$.

The last relation actually holds without the re-labeling process. However, we find it useful to keep track of the labels of ideal arcs, in which case we are dealing in fact with labeled ideal triangulations in effect. We fix an index set I, say $\{1, 2, \ldots, 6g - 6 + 3n\}$, and by a labeled ideal triangulation we mean an ideal triangulation Δ together with a set bijection $\Delta \to I$, which is the labeling map. By a slight abuse of notation, we still denote just by the same symbol Δ a labeled ideal triangulation. Each element of I corresponds to an ideal arc of a labeled ideal triangulation; so the symbols i, j and k which appeared so far can be understood as elements of I. The Hilbert space $\mathscr{H}_\Delta = L^2(\mathbb{R}^\Delta, \bigwedge_{i\in\Delta} dt_i)$ can now be written as

$$\mathscr{H}_\Delta = L^2(\mathbb{R}^I, \bigwedge_{i\in I} dt_i).$$

When we deal with several labeled ideal triangulations Δ at the same time, we will denote the variables t_i for $\mathscr{H}_\Delta$ differently for different Δ, if necessary.

Now, in addition to the flip μ_k at an arc $k \in I$, we add one more kind of elementary transformation of labeled ideal triangulations, which is the label permutation P_σ associated to any permutation σ of the index set I. We write $\Delta' = P_\sigma(\Delta)$ if Δ and Δ' are the same ideal triangulations with their labeling maps related as follows: if the constituent ideal arc is labeled by i in Δ then it is labeled by $\sigma(i)$ in Δ'. To this change of ideal triangulations $\Delta \rightsquigarrow \Delta' = P_\sigma(\Delta)$ we associate the index-permutation sintertwiner

$$\mathbf{P}_\sigma = \mathbf{K}^\hbar_{\Delta, P_\sigma(\Delta)} = \mathbf{K}^\hbar_{\Delta,\Delta'} : \mathscr{H}_{\Delta'} = L^2(\mathbb{R}^I, \bigwedge_{i\in I} dt'_i) \to L^2(\mathbb{R}^I, \bigwedge_{i\in I} dt_i) = \mathscr{H}_\Delta$$

defined as the operator induced by the re-labeling map $L_\sigma : \mathbb{R}^I = \mathbb{R}^\Delta \to \mathbb{R}^{\Delta'} = \mathbb{R}^I$, whose pullback action on the coordinate functions is given by

$$L^*_\sigma t'_{\sigma(i)} = t_i, \qquad \forall i \in I.$$

It's easy to see that $\mathbf{P}_\sigma$ is unitary.

Now that we decided to work with labeled ideal triangulations, eq.(4.6) has to be modified because now a change of ideal triangulation $\Delta \rightsquigarrow \Delta'$ is expressed as a sequence of mutations and label permutations; hence $\mathbf{K}^\hbar_{\Delta,\Delta'}$ should be defined as composition of corresponding flip intertwiners $\mathbf{K}^\hbar_k$ and index permutation operators $\mathbf{P}_\sigma$. To ensure the well-definedness of $\mathbf{K}^\hbar_{\Delta,\Delta'}$, we now need to classify all relations satisfied by flips and label permutations.

PROPOSITION 4.2. *Any relation satisfied by flips and label permutations of labeled ideal triangulations without self-folded triangles is generated by the relations in Prop.4.1, together with the following:*

$$P_{\mathrm{id}} = \mathrm{id}, \qquad P_{(\sigma\circ\gamma)^{-1}} \circ P_\sigma \circ P_\gamma = \mathrm{id}, \qquad P_\sigma \circ \mu_i \circ P_{\sigma^{-1}} \circ \mu_{\sigma(i)} = \mathrm{id},$$

applied to any labeled ideal triangulation, for all permutations σ and γ of I and for all $i \in I$.

This proposition would follow from Prop.4.1 by an argument analogous to the one in [**K16b**, §4.1], using a lemma by Bakalov and Kirillov [**BK00**].

As desired, all relations satisfied by flips and index permutations are indeed shown to be satisfied by the corresponding intertwining operators. Note that due to the contravariant nature of the construction of the intertwining operators, we need to reverse the order of composition in the relations when translating to relations of operators. For example, the operator identity corresponding to the pentagon identity in Prop.4.1(3) reads

$$\mathbf{K}_i^\hbar \circ \mathbf{K}_j^\hbar \circ \mathbf{K}_i^\hbar \circ \mathbf{K}_j^\hbar \circ \mathbf{K}_i^\hbar \circ \mathbf{P}_{(ij)} = \mathrm{id} : \mathscr{H}_\Delta \to \mathscr{H}_\Delta \tag{4.7}$$

which holds when the 'initial' triangulation Δ satisfies $\varepsilon_{ij} = \pm 1$; the core of a proof of this identity is eq.(3.6), the pentagon operator identity of the function $F_\Lambda^\hbar$ which is used in the definition of the flip intertwiners.

THEOREM 4.3 (consistency equations of intertwiners [**KS24**]). *The intertwining operators $\mathbf{K}_k^\hbar$ and $\mathbf{P}_\sigma$ corresponding to the flip μ_k and label permutation P_σ of labeled ideal triangulations satisfy all relations appearing in Prop.4.1–4.2, with the order of composition reversed.*

This implies the well-definedness of the intertwiner $\mathbf{K}_{\Delta,\Delta'}^\hbar$ associated to an arbitrary change of labeled ideal triangulations $\Delta \rightsquigarrow \Delta'$; that is, the operator $\mathbf{K}_{\Delta,\Delta'}^\hbar$ does not depend on the choice of a sequence of flips and label permutations connecting Δ and Δ'. Moreover, this result also implies the consistency equations of the resulting intertwiners. Namely,

$$\mathbf{K}_{\Delta,\Delta'}^\hbar \circ \mathbf{K}_{\Delta',\Delta''}^\hbar = \mathbf{K}_{\Delta,\Delta''}^\hbar \tag{4.8}$$

holds for any triple of labeled ideal triangulations Δ, Δ' and Δ''.

This finishes the description of the system of Hilbert spaces $\mathscr{H}_\Delta$ of quantum states associated to labeled ideal triangulations Δ, together with a consistent system of unitary intertwiners $\mathbf{K}_{\Delta,\Delta'}^\hbar$ among these Hilbert spaces, for the quantization program for the moduli space $\mathcal{GH}_\Lambda(\mathfrak{S} \times \mathbb{R})$ established in [**KS24**].

5. Mapping class group representations

We explain how the unitary intertwiners can be used to construct representations of the mapping class group MCG($\mathfrak{S}$) of $\mathfrak{S}$. As a matter of fact, perhaps a better viewpoint is to interpret the intertwiners as forming a representation of a groupoid rather than that of a group, as pointed out to the author by Dennis Sullivan at Igor Frenkel's 70th birthday conference. Indeed, in [**FG09**] Fock and Goncharov first formulate their quantization result in terms of a groupoid, and as a consequence obtains a representation of a group. Here we do likewise. We note that the constructions presented in this section are not explicitly written down in [**KS24**].

5.1. The Ptolemy groupoid, and the Ptolemy class groupoid. Following [**P87**] with some modification, one may define the *Ptolemy groupoid* Pt($\mathfrak{S}$) of the surface $\mathfrak{S}$ as the category whose set of objects Ob(Pt($\mathfrak{S}$)) is the set of all labeled ideal triangulations Δ without self-folded triangles, such that for each pair of objects Δ and Δ' there is exactly one morphism from Δ to Δ', denoted by

$(\Delta \rightsquigarrow \Delta')$. Here, an ideal triangulation Δ is considered up to simultaneous isotopy of its constituent ideal arcs.

One can formulate the results reviewed in §4 as a contravariant functor

$$\mathrm{Pt}(\mathfrak{S}) \to \mathrm{Hilb},$$

where Hilb denotes the category of Hilbert spaces whose morphisms are unitary maps. The object Δ of $\mathrm{Pt}(\mathfrak{S})$ goes to the Hilbert space $\mathscr{H}_\Delta$, and the morphism $(\Delta \rightsquigarrow \Delta')$ of $\mathrm{Pt}(\mathfrak{S})$ goes to the unitary map $\mathbf{K}^\hbar_{\Delta,\Delta'} : \mathscr{H}_{\Delta'} \to \mathscr{H}_\Delta$.

An important property of this functor is the invariance under the action of the mapping class group $\mathrm{MCG}(\mathfrak{S}) = \mathrm{Diff}_+(\mathfrak{S})/\mathrm{Diff}(\mathfrak{S})_0$, which we now describe. First, we note that $\mathrm{MCG}(\mathfrak{S})$ naturally acts on the Ptolemy groupoid $\mathrm{Pt}(\mathfrak{S})$. The element $\varphi \in \mathrm{MCG}(\mathfrak{S})$ naturally sends the object Δ to $\varphi.\Delta$ because φ is an isotopy class of homeomorphisms or diffeomorphisms from $\mathfrak{S}$ to itself, and accordingly sends the morphism $(\Delta \rightsquigarrow \Delta')$ to $\varphi.(\Delta \rightsquigarrow \Delta') := (\varphi.\Delta \rightsquigarrow \varphi.\Delta')$. Recall that an object Δ of $\mathrm{Pt}(\mathfrak{S})$ is not just an ideal triangulation, but is equipped with a labeling map $\Delta \to I$. The object $\varphi.\Delta$ then is given the natural labeling map given by the composition $\varphi.\Delta \to \Delta \to I$; that is, if α is an ideal arc of Δ labeled by $i \in I$, then the ideal arc $\varphi.\alpha$ of $\varphi.\Delta$ is labeled by i. This induces a natural identification between the Hilbert spaces

$$\mathscr{H}_{\varphi.\Delta} = L^2(\mathbb{R}^I, \bigwedge_{i\in I} dt_i) \to L^2(\mathbb{R}^I, \bigwedge_{i\in I} dt_i) = \mathscr{H}_\Delta$$

which is the identity map $L^2(\mathbb{R}^I, \bigwedge_{i\in I} dt_i) \to L^2(\mathbb{R}^I, \bigwedge_{i\in I} dt_i)$. Via this identification, one can regard

$$\mathscr{H}_{\varphi.\Delta} \equiv \mathscr{H}_\Delta.$$

Now, in case $(\Delta \rightsquigarrow \Delta')$ is a flip μ_k at $k \in I$, recall that the construction of the corresponding intertwiner $\mathbf{K}^\hbar_{\Delta,\Delta'} = \mathbf{K}^\hbar_k$ uses only the exchange matrix ε (and k), rather than the full topological information of Δ. Notice that the exchange matrix $\varepsilon = \varepsilon_\Delta$ of a labeled ideal triangulation Δ is insensitive to the mapping class group action, namely

$$\varepsilon_\Delta = \varepsilon_{\varphi.\Delta}.$$

That is, for any $i, j \in I$, the (i,j)-th entry of ε_Δ coincides with the (i,j)-th entry of $\varepsilon_{\varphi.\Delta}$. From this we deduce the invariance

$$\mathbf{K}^\hbar_{\Delta,\Delta'} = \mathbf{K}^\hbar_{\varphi.\Delta,\varphi.\Delta'}$$

in the case when $(\Delta \rightsquigarrow \Delta')$ is a flip. In the case when $(\Delta \rightsquigarrow \Delta')$ is a label permutation, i.e. $\Delta' = P_\sigma(\Delta)$ for some permutation σ of I, it is easy to see that this invariance again holds. It then follows that the invariance holds for an arbitrary change of labeled ideal triangulations $(\Delta \rightsquigarrow \Delta')$.

Consider the quotient groupoid

$$\mathrm{PCG}(\mathfrak{S}) := \mathrm{Pt}(\mathfrak{S})/\mathrm{MCG}(\mathfrak{S})$$

and call it the *Ptolemy class groupoid*. An object of $\mathrm{PCG}(\mathfrak{S})$ representing $\Delta \in \mathrm{Ob}(\mathrm{Pt}(\mathfrak{S}))$ is denoted by $[\Delta]$, and a morphism of $\mathrm{PCG}(\mathfrak{S})$ representing a morphism $(\Delta \rightsquigarrow \Delta')$ of $\mathrm{Pt}(\mathfrak{S})$ is denoted by $[\Delta \rightsquigarrow \Delta']$. In particular, we have

$$[\Delta] = [\varphi.\Delta] \quad \text{and} \quad [\Delta \rightsquigarrow \Delta'] = [\varphi.\Delta \rightsquigarrow \varphi.\Delta']$$

for each $\varphi \in \mathrm{MCG}(\mathfrak{S})$. Using this notation, one can define

$$\mathscr{H}_{[\Delta]} := \mathscr{H}_\Delta \quad \text{and} \quad \mathbf{K}^\hbar_{[\Delta \rightsquigarrow \Delta']} := \mathbf{K}^\hbar_{\Delta,\Delta'},$$

which are well-defined by the invariance statements we have above. Thus the results of §4 yield a contravariant functor

$$\text{PCG}(\mathfrak{S}) \to \text{Hilb} \tag{5.1}$$

which sends an object $[\Delta]$ to the Hilbert space $\mathscr{H}_{[\Delta]}$ and a morphism $[\Delta \rightsquigarrow \Delta']$ to $\mathbf{K}^{\hbar}_{[\Delta\rightsquigarrow\Delta']}$.

5.2. A trilogy of representations of the mapping class group. Pick any labeled ideal triangulation Δ of $\mathfrak{S}$, and consider the corresponding object $[\Delta]$ of the Ptolemy class groupoid $\text{PCG}(\mathfrak{S}) = \text{Pt}(\mathfrak{S})/\text{MCG}(\mathfrak{S})$. Note that for each mapping class $\varphi \in \text{MCG}(\mathfrak{S})$, the morphism $(\Delta \rightsquigarrow \varphi.\Delta)$ of the Ptolemy groupoid $\text{Pt}(\mathfrak{S})$ yields a morphism $[\Delta \rightsquigarrow \varphi.\Delta]$ of $\text{PCG}(\mathfrak{S})$ from $[\Delta]$ to $[\varphi.\Delta] = [\Delta]$. Hence $[\Delta \rightsquigarrow \varphi.\Delta]$ is an automorphism of $[\Delta]$. This gives a map

$$\mathcal{F}_\Delta : \text{MCG}(\mathfrak{S}) \to \text{Aut}([\Delta]), \qquad \varphi \mapsto [\Delta \rightsquigarrow \varphi.\Delta]$$

For $\varphi_1, \varphi_2 \in \text{MCG}(\mathfrak{S})$ note

$$\begin{aligned}
\mathcal{F}_\Delta(\varphi_1) \circ \mathcal{F}_\Delta(\varphi_2) &= [\Delta \rightsquigarrow \varphi_1.\Delta] \circ [\Delta \rightsquigarrow \varphi_2.\Delta] \\
&= [\varphi_2.\Delta \rightsquigarrow \varphi_2.(\varphi_1.\Delta)] \circ [\Delta \rightsquigarrow \varphi_2.\Delta] \\
&= [\Delta \rightsquigarrow \varphi_2.(\varphi_1.\Delta)] \\
&= [\Delta \rightsquigarrow (\varphi_2\varphi_1).\Delta] \\
&= \mathcal{F}_\Delta(\varphi_2\varphi_1),
\end{aligned}$$

so $\mathcal{F}_\Delta$ is an anti-homomorphism of groups. It is easy to see that unless $\mathfrak{S}$ is a once-punctured torus, $\mathcal{F}_\Delta$ is an isomorphism, since the action of $\text{MCG}(\mathfrak{S})$ on $\text{Ob}(\text{Pt}(\mathfrak{S}))$ is free. When $\mathfrak{S}$ is a once-punctured torus, the kernel $\mathcal{F}_\Delta$ is generated by the 'hyperelliptic' involution which is of order 2.

Composing $\mathcal{F}_\Delta$ with the functor in eq.(5.1), one obtains a map

$$\begin{aligned}
&\rho^{\hbar}_\Delta : \text{MCG}(\mathfrak{S}) \to \text{U}(\mathscr{H}_{[\Delta]}) = \{\text{unitary maps } \mathscr{H}_{[\Delta]} \to \mathscr{H}_{[\Delta]}\}, \\
&\qquad \varphi \mapsto \mathbf{K}^{\hbar}_{\mathcal{F}_\Delta(\varphi)} = \mathbf{K}^{\hbar}_{[\Delta\rightsquigarrow\varphi.\Delta]} = \mathbf{K}^{\hbar}_{\Delta,\varphi.\Delta} : \mathscr{H}_{[\Delta]} \equiv \mathscr{H}_{\varphi.\Delta} \to \mathscr{H}_\Delta = \mathscr{H}_{[\Delta]}.
\end{aligned}$$

It is easy to see that $\rho^{\hbar}_\Delta$ is a group homomorphism. Just to write down explicitly, note that for $\varphi_1, \varphi_2 \in \text{MCG}(\mathfrak{S})$, one has

$$\begin{aligned}
\rho^{\hbar}_\Delta(\varphi_1) \circ \rho^{\hbar}_\Delta(\varphi_2) &= \mathbf{K}^{\hbar}_{[\Delta\rightsquigarrow\varphi_1.\Delta]} \circ \mathbf{K}^{\hbar}_{[\Delta\rightsquigarrow\varphi_2.\Delta]} \\
&= \mathbf{K}^{\hbar}_{[\Delta\rightsquigarrow\varphi_1.\Delta]} \circ \mathbf{K}^{\hbar}_{[\varphi_1.\Delta\rightsquigarrow\varphi_1.(\varphi_2.\Delta)]} \\
&= \mathbf{K}^{\hbar}_{\Delta,\varphi_1.\Delta} \circ \mathbf{K}^{\hbar}_{\varphi_1.\Delta,(\varphi_1\varphi_2).\Delta} \\
&= \mathbf{K}^{\hbar}_{\Delta,(\varphi_1\varphi_2).\Delta} \qquad\qquad (\because \text{eq.}(4.8)) \\
&= \mathbf{K}^{\hbar}_{[\Delta\rightsquigarrow(\varphi_1\varphi_2).\Delta]} \\
&= \rho^{\hbar}_\Delta(\varphi_1\varphi_2).
\end{aligned}$$

This $\rho^{\hbar}_\Delta$ is the promised unitary representation of $\text{MCG}(\mathfrak{S})$ on the Hilbert space $\mathscr{H}_{[\Delta]} = L^2(\mathbb{R}^I)$ coming from the quantization of the moduli space $\mathcal{GH}_\Lambda(\mathfrak{S} \times \mathbb{R})$ of three-dimensional gravity. Roughly summarizing, for each mapping class $\varphi \in \text{MCG}(\mathfrak{S})$ and a labeled ideal triangulation Δ, one finds a sequence of flips μ_k and label permutations P_σ connecting Δ and $\varphi.\Delta$, and takes the composition of the intertwiners $\mathbf{K}^{\hbar}_k$ and $\mathbf{P}_\sigma$ corresponding to these flips and label permutations, to construct a unitary operator $\rho^{\hbar}_\Delta(\varphi)$. Recall that the flip intertwiner $\mathbf{K}^{\hbar}_k$ is a composition of a simple operator on $L^2(\mathbb{R}^I)$ induced by a linear automorphism of $\mathbb{R}^I$ and an operator expressed by the three kinds of quantum dilogarithm functions, depending on the cosmological constant Λ.

Let us briefly discuss the dependence of this representation $\rho^\hbar_\Delta$ on the choice of a labeled ideal triangulation Δ. To any ordered pair of labeled ideal triangulations Δ and Δ', is associated a unique unitary intertwiner $\mathbf{K}^\hbar_{\Delta,\Delta'} : \mathscr{H}_{\Delta'} \to \mathscr{H}_\Delta$, which can be interpreted as the unitary map $\mathbf{K}^\hbar_{[\Delta\rightsquigarrow\Delta']} : \mathscr{H}_{[\Delta']} \to \mathscr{H}_{[\Delta]}$. It is easy to observe that this map $\mathbf{K}^\hbar_{[\Delta\rightsquigarrow\Delta']}$ gives an intertwiner between the representations $\rho^\hbar_{\Delta'}$ and $\rho^\hbar_\Delta$, in the sense that the diagram

$$\begin{array}{ccc} \mathscr{H}_{[\Delta']} & \xrightarrow{\rho^\hbar_{\Delta'}(\varphi)} & \mathscr{H}_{[\Delta']} \\ {\scriptstyle \mathbf{K}^\hbar_{[\Delta\rightsquigarrow\Delta']}}\downarrow & & \downarrow{\scriptstyle \mathbf{K}^\hbar_{[\Delta\rightsquigarrow\Delta']}} \\ \mathscr{H}_{[\Delta]} & \xrightarrow{\rho^\hbar_{\Delta}(\varphi)} & \mathscr{H}_{[\Delta]} \end{array}$$

commutes for all $\varphi \in \mathrm{MCG}(\mathfrak{S})$, i.e. the following equation holds:

$$\mathbf{K}^\hbar_{[\Delta\rightsquigarrow\Delta']} \circ \rho^\hbar_{\Delta'}(\varphi) = \rho^\hbar_\Delta(\varphi) \circ \mathbf{K}^\hbar_{[\Delta\rightsquigarrow\Delta']}.$$

Hence, one can conclude that the representations $\rho^\hbar_\Delta$ on $\mathscr{H}_{[\Delta]}$ for different labeled ideal triangulations Δ are isomorphic to each other through a consistent system of isomorphisms $\mathbf{K}^\hbar_{[\Delta\rightsquigarrow\Delta']}$ between them.

5.3. Irreducible representations satisfying quantum constraints. The representation of $\mathrm{MCG}(\mathfrak{S})$ constructed above is built on the unitary intertwiners $\mathbf{K}^\hbar_{\Delta,\Delta'}$, for which we used a particular self-adjoint operators $\mathbf{x}_i$ and $\mathbf{y}_i$ in eq.(4.4) on the Hilbert space $\mathscr{H}_\Delta$ for each Δ. They were taken as forming a standard solution to the Heisenberg relations in eq.(4.1). But they did not satisfy the quantum constraint equations in eq.(4.3), so they cannot be regarded as honest quantizations of the coordinate functions x_i and y_i on the space $\mathcal{GH}_\Lambda(\mathfrak{S}\times\mathbb{R})$. Another related fact is that they do not form an irreducible representation. In the present subsection we give a recipe to obtain an irreducible system of self-adjoint operators $\mathbf{x}_i$ and $\mathbf{y}_i$ (for $i \in \Delta$) on some Hilbert space satisfying both eq.(4.1) and eq.(4.3). We note that this recipe is only hinted in [**KS24**], but details are not given there.

The problem of finding such operators can be viewed as a problem of constructing representations of a generalized Heisenberg algebra, which was thoroughly studied in the literature; for example, one can find a version useful for us stated in [**L77**]. The construction of such representations as necessary for our purposes is reviewed in [**K21b**], which we apply to our situation here. Consider the vector space $(\mathbb{R}^\Delta)^2 = (\mathbb{R}^I)^2 = \mathbb{R}^{12g-12+6n}$ equipped with the skew-symmetric bilinear form $\langle\cdot,\cdot\rangle$ given by $\left(\begin{smallmatrix} 0 & \varepsilon \\ -\varepsilon & 0 \end{smallmatrix}\right)$ on the basis $\{v_i, w_i\}_{i\in\Delta}$, i.e. $\langle v_i, v_j\rangle = 0 = \langle w_i, w_j\rangle$, $\langle v_i, w_j\rangle = \varepsilon_{ij} = \varepsilon_{i,j}$. Choose any Lagrangian subspace $\mathscr{L}$ of this space $((\mathbb{R}^\Delta)^2, \langle\cdot,\cdot\rangle)$, i.e. a maximal isotropic subspace; here, isotropic means $\langle v, w\rangle = 0$, $\forall v, w \in \mathscr{L}$. Since $\dim(\ker\varepsilon) = n$ as mentioned in §2.2, one can observe that $\mathscr{L}$ is of dimension $2n + \frac{1}{2}(12g - 12 + 4n)$. Upon a choice of $\mathscr{L}$ and a basis $\mathcal{B}$ of the complement of $\mathscr{L}$ in $(\mathbb{R}^\Delta)^2$, using a basic harmonic analysis on the Heisenberg group for the Heisenberg Lie algebra defined as a 1-dimensional central extension of the abelian

Lie algebra $(\mathbb{R}^\Delta)^2$ constructed by using the form $\langle\cdot,\cdot\rangle$, an irreducible system of self-adjoint operators $\mathbf{x}_i$ and $\mathbf{y}_i$ (for $i\in\Delta$) satisfying eq.(4.1) and eq.(4.3) can be constructed on a Hilbert space isomorphic to $L^2(\mathbb{R}^\Delta/\mathscr{L}) = L^2(\mathbb{R}^\mathcal{B}) = L^2(\mathbb{R}^{6g-6+2n})$ with respect to the usual Lebesgue measure on $\mathbb{R}^{6g-6+2n}$. The resulting operators $\mathbf{x}_i$ and $\mathbf{y}_i$ are $\mathbb{R}$-linear combinations of the position and the momentum operators on $L^2(\mathbb{R}^{6g-6+2n}, \bigwedge_{j=1}^{6g-6+2n} ds_j)$.

Using the last comment, one can just try to look, by hand, for a system of $\mathbb{R}$-linear combinations of the position and the momentum operators that satisfy both eq.(4.1) and eq.(4.3). But we could also use the full power of the construction we discussed above (reviewed in [**K21b**]), which gives an explicit solution. For this, we first recall from §2.2 that the (column) vectors $(v_{i,p})_{i\in\Delta}$ for punctures p form a basis of the kernel of ε; in particular, $\sum_{j\in\Delta}\varepsilon_{ij}v_{j,p}=0$ for each puncture p and each $i\in\Delta$. This means that $\{\sum_{i\in\Delta}v_{i,p}v_i, \sum_{i\in\Delta}v_{i,p}w_i\}_{p:\text{punctures}}$ forms a basis of the kernel (or, radical) of $\langle\cdot,\cdot\rangle$. Since the $(6g-6+3n)\times n$ matrix $(v_{i,p})_{i\,\in\,\Delta,\ p:\text{punctures}}$ is of full rank n, one can find n pivot rows, say row numbers $i_1,i_2,\ldots,i_n\in\Delta$. Then one can transform this matrix into a column reduced echelon form using column operations, with leading 1 at the pivot rows, i.e. at i_j-th rows for $j=1,\ldots,n$. One can further assume that at each i_j-th row, there is only one non-zero entry, say at p_j-th place where the entry is 1. Here $p_1,\ldots,p_n$ is an enumeration of (or, labels of) punctures. Denote this reduced matrix by $(w_{i,p})_{i\,\in\,\Delta,\ p:\text{punctures}}$; by the condition we imposed, for each $j=1,\ldots,n$, we have $w_{i_j,p_j}=1$, and $w_{i_j,p}=0$ for each $p\neq p_j$. Hence, if we write

$$\mathring{\Delta} := \Delta\setminus\{i_1,\ldots,i_n\},$$

which is a collection of $6g-6+2n$ arcs, then for each $j=1,\ldots,n$, in the p_j-th column we have $w_{i_{j'},p_j}=0$ whenever $j'\neq j$. Each column of the reduced matrix $(w_{i,p})_{i,p}$ is still in the kernel of the matrix ε, so $\sum_{k\in\Delta}\varepsilon_{ik}w_{k,p_j}=0$, and hence

$$\varepsilon_{i,i_j} + \textstyle\sum_{k\in\mathring{\Delta}}\varepsilon_{ik}w_{k,p_j} = 0 \tag{5.2}$$

holds for each $j=1,\ldots,n$ and $i\in\Delta$. The linear relations among the rows of $(v_{i,p})_{i,p}$ are the same as those of $(w_{i,p})_{i,p}$. For the latter matrix, for each $j=1,\ldots,n$, the i_j-th row is p_j-th standard basis row vector; hence, for each $i\in\mathring{\Delta}$, the i-th row $(w_{i,p})_{p:\text{punctures}}$ is the sum of w_{i,p_j} times i_j-th row, where j runs in $\{1,\ldots,n\}$. The same relation among the rows hold for the former matrix $(v_{i,p})$, too; for each $i\in\mathring{\Delta}$, one has

$$v_{i,p} = \textstyle\sum_{j=1}^n w_{i,p_j}v_{i_j,p} \tag{5.3}$$

for each puncture p.

Now, for an example of the construction of Heisenberg algebra representation, let's choose the Lagrangian $\mathscr{L}$ of $(\mathbb{R}^\Delta)^2$ to be spanned by w_i for $i\in\Delta$, together with $\sum_{i\in\Delta}v_{i,p}v_i$ for punctures p. As a basis $\mathcal{B}$ of the complement of $\mathscr{L}$ we take v_i for $i\in\mathring{\Delta}$. Then, the construction reviewed in [**K21b**] (especially, Examples 2.9 and 2.10 there are useful), with a slight scaling applied at the end for our purposes, yields the following self-adjoint operators

(5.4)

$$\begin{cases} \mathbf{x}_i = -\pi \mathrm{i} \dfrac{\partial}{\partial s_i}, & \forall i \in \mathring{\Delta} = \Delta \setminus \{i_1, \ldots, i_n\}, \\ \mathbf{x}_{i_j} = \sum_{k \in \mathring{\Delta}} \pi \mathrm{i} w_{k,p_j} \dfrac{\partial}{\partial s_k}, & \forall j = 1, \ldots, n, \\ \mathbf{y}_i = \sum_{j \in \mathring{\Delta}} \varepsilon_{ij} s_j, & \forall i \in \Delta, \end{cases}$$

on the Hilbert space

$$\mathscr{H}_\Delta = L^2(\mathbb{R}^{\mathring{\Delta}}, \textstyle\bigwedge_{i \in \mathring{\Delta}} ds_i) = L^2(\mathbb{R}^{6g-6+2n})$$

One can now forget about the construction reviewed in [**K21b**] and just take the above final result. One can check that they indeed satisfy the desired equations, as follows. It is easy to verify $[\mathbf{x}_i, \mathbf{x}_j] = 0 = [\mathbf{y}_i, \mathbf{y}_j]$ for all $i, j \in \Delta$, as well as $[\mathbf{x}_i, \mathbf{y}_j] = \pi \mathrm{i} \varepsilon_{ij}$ for all $i \in \mathring{\Delta}$ and $j \in \Delta$. For $j = 1, \ldots, n$ and $k \in \Delta$, note that

$$\begin{aligned} [\mathbf{x}_{i_j}, \mathbf{y}_k] &= [\textstyle\sum_{a \in \mathring{\Delta}} \pi \mathrm{i} w_{a,p_j} \frac{\partial}{\partial s_a}, \sum_{b \in \mathring{\Delta}} \varepsilon_{kb} s_b] \\ &= \textstyle\sum_{a \in \mathring{\Delta}} \sum_{b \in \mathring{\Delta}} \pi \mathrm{i} w_{a,p_j} \varepsilon_{kb} \delta_{a,b} \\ &= \textstyle\sum_{a \in \mathring{\Delta}} \pi \mathrm{i} w_{a,p_j} \varepsilon_{ka} \\ &= \pi \mathrm{i} \varepsilon_{i_j,k}, \end{aligned}$$

where the last equality holds by eq.(5.2). So the Heisenberg relations in eq.(4.1) are satisfied. Moreover, for each puncture p, note that

$$\textstyle\sum_{i \in \Delta} v_{i,p} \mathbf{y}_i = \sum_{i \in \Delta} v_{i,p} \sum_{j \in \mathring{\Delta}} \varepsilon_{ij} s_j = \sum_{j \in \mathring{\Delta}} (\sum_{i \in \Delta} v_{i,p} \varepsilon_{ij}) s_j = 0$$

holds by eq.(2.2), and that

$$\begin{aligned} \textstyle\sum_{i \in \Delta} v_{i,p} \mathbf{x}_i &= \textstyle\sum_{i \in \mathring{\Delta}} v_{i,p} (-\pi \mathrm{i} \frac{\partial}{\partial s_i}) + \sum_{j=1}^n v_{i_j,p} \sum_{k \in \mathring{\Delta}} (\pi \mathrm{i} w_{k,p_j} \frac{\partial}{\partial s_k}) \\ &= \textstyle\sum_{i \in \mathring{\Delta}} (-\pi \mathrm{i}) (v_{i,p} - \sum_{j=1}^n v_{i_j,p} w_{i,p_j}) \frac{\partial}{\partial s_i} \\ &= 0, \end{aligned}$$

where the last equality holds by eq.(5.3). Hence the quantum constraint equations in eq.(4.3) are also satisfied. One can also verify that these operators forms an irreducible system, and that any irreducible system of self-adjoint operators satisfying eq.(4.1) and eq.(4.3) is unitarily equivalent to the above one, using (generalized) Stone-von Neumann theorem; see e.g. [**H13**, §14].

The choice of a pair of Lagrangian subspace $\mathscr{L}$ and a complementary basis $\mathcal{B}$ made above is not being claimed to be canonical. A different choice yields in general a different set of operators on a Hilbert space isomorphic to $L^2(\mathbb{R}^{6g-6+2n})$. The construction reviewed in [**K21b**] in fact also gives an explicit unitary equivalence between different solutions, using basic harmonic analysis. Due to the irreducibility, the unitary equivalence is unique up to scalar, and is what is referred to as the *(Segal-)Shale-Weil intertwiner* [**S62, S63, W64, L77**], which can be thought of as a generalization of the Fourier transformation. The monomial transformation part operator $\mathbf{K}'_k$ we described in the present article can also be thought of as a special example of a Shale-Weil intertwiner for a certain setting.

If one decides to use the operators $\mathbf{x}_i$ and $\mathbf{y}_i$ as in eq.(5.4) instead of eq.(4.4), one should also modify the subsequent constructions in §4–§5 accordingly, in particular the unitary intertwiners $\mathbf{K}^\hbar_{\Delta,\Delta'}$. The automorphism part operator $\mathbf{K}^{\hbar\sharp}_k$ for a flip μ_k is given by the same formula in eq.(4.5), this time using new $\mathbf{x}_k$ and $\mathbf{y}_k$.

The monomial transformation part operator $\mathbf{K}'_k$ for a flip μ_k, as well as the index-permutation operator $\mathbf{P}_\sigma$, must be constructed afresh. Due to the irreducibility of the new set of operators $\mathbf{x}_i$ and $\mathbf{y}_i$, these sought-for operators $\mathbf{K}'_k$ and $\mathbf{P}_\sigma$ are completely determined, up to scalar, by their conjugation actions as follows (and these equations for conjugation actions are what we really need from the intertwiners):

$$\mathbf{K}'_k \mathbf{x}'_i (\mathbf{K}'_k)^{-1} = \begin{cases} -\mathbf{x}_k & \text{if } i = k, \\ \mathbf{x}_i + [\varepsilon_{ik}]_+ \mathbf{x}_k & \text{if } i \neq k, \end{cases} \qquad \mathbf{K}'_k \mathbf{y}'_i (\mathbf{K}'_k)^{-1} = \begin{cases} -\mathbf{y}_k & \text{if } i = k, \\ \mathbf{y}_i + [\varepsilon_{ik}]_+ \mathbf{y}_k & \text{if } i \neq k, \end{cases}$$

$$\mathbf{P}_\sigma \mathbf{x}'_{\sigma(i)} \mathbf{P}_\sigma^{-1} = \mathbf{x}_i, \qquad \mathbf{P}_\sigma \mathbf{y}'_{\sigma(i)} \mathbf{P}_\sigma^{-1} = \mathbf{y}_i, \quad \forall i \in I.$$

In order to explicitly construct such operators, one could again use the above construction reviewed in [**K21b**], this time for the Shale-Weil intertwiners. To say briefly, take the case of $\mathbf{K}'_k : \mathscr{H}_{\Delta'} \to \mathscr{H}_\Delta$, where $\Delta' = \mu_k(\Delta)$. The construction of a Heisenberg representation for Δ has required us to choose some data $(\mathscr{L}, \mathcal{B})$ of a Lagrangian subspace $\mathscr{L}$ of $((\mathbb{R}^\Delta)^2, \langle \cdot, \cdot \rangle)$ and a complementary basis $\mathcal{B}$. Likewise, that for Δ' needed a choice $(\mathscr{L}', \mathcal{B}')$ for the space $((\mathbb{R}^{\Delta'})^2, \langle \cdot, \cdot \rangle')$. One then takes a pullback of $(\mathscr{L}', \mathcal{B}')$ along the map $C : (\mathbb{R}^\Delta)^2 \to (\mathbb{R}^{\Delta'})^2$ given on the basis by $C(v_k) = -v'_k$, $C(w_k) = -w'_k$, $C(v_i) = v'_i + [\varepsilon_{ik}]_+ v'_k$ and $C(w_i) = w'_i + [\varepsilon_{ik}]_+ w'_k$ for $i \neq k$. This pullback $(C^{-1}(\mathscr{L}'), C^{-1}(\mathcal{B}'))$ forms a data of a Lagrangian and a complementary basis for $(\mathbb{R}^\Delta)^2$, hence yields a Heisenberg algebra representation. Then $\mathbf{K}'_k$ is essentially defined as the Shale-Weil intertwiner between this representation for $(C^{-1}(\mathscr{L}'), C^{-1}(\mathcal{B}'))$ and that for $(\mathscr{L}, \mathcal{B})$. For more details about how to do this, we refer the readers to [**K21b**], where such a task was carefully done for quantization of the Teichmüller space. It will be a good exercise for a reader to explicitly construct $\mathbf{K}'_k$ and $\mathbf{P}_\sigma$ for our setting, mimicking [**K21b**].

To conclude, the quantization result deduced from the (irreducible) system of operators satisfying eq.(4.1) and eq.(4.3), as appearing in eq.(5.4) for example which we discussed in the present subsection, instead of the reducible solution in eq.(4.4) which does not satisfy eq.(4.3), is what could really be regarded as a sought-for honest quantization of the moduli space $\mathcal{GH}_\Lambda(\mathfrak{S} \times \mathbb{R})$ for 3d gravity.

One worthwhile comment is that the intertwiners $\mathbf{K}^\hbar_{\Delta,\Delta'}$ constructed in the present subsection for the irreducible representations may satisfy the consistency equations in eq.(4.8) only up to multiplicative constants. This means that the corresponding MCG($\mathfrak{S}$) representations $\rho^\hbar_\Delta$ constructed using the methods of §5.1–§5.2 a priori would only be *projective* representations, i.e. $\rho^\hbar_\Delta(\varphi_1)\rho^\hbar_\Delta(\varphi_2) = \rho^\hbar_\Delta(\varphi_1\varphi_2)$ holds only up to multiplicative constants. See [**K21b**] for a method of explicit computation of these constants as well as for their meaning.

5.4. Meaning of the results. As mentioned in [**KS24**], it has essentially been shown in the literature that the moduli space $\mathcal{GH}_\Lambda(\mathfrak{S} \times \mathbb{R})$ of 3d spacetimes, with the gravitational symplectic structure, is symplectomorphic to the cotangent bundle $T^*\mathcal{T}(\mathfrak{S})$ with the canonical (contangent-bundle) symplectic structure, where $\mathcal{T}(\mathfrak{S})$ is the *(cusped) Teichmüller space* of $\mathfrak{S}$, which is the set of all finite-area complete hyperbolic metrics on $\mathfrak{S}$, considered up to pullback by self-diffeomorphisms of $\mathfrak{S}$ isotopic to identity. This is true regardless of the value of the cosmological constant $\Lambda \in \{-1, 0, 1\}$.

So, one could say that the problem of quantization of $\mathcal{GH}_\Lambda(\mathfrak{S} \times \mathbb{R})$ boils down to quantization of the cotangent bundle $T^*\mathcal{T}(\mathfrak{S})$. As quantization of the cotangent bundle of a manifold has been studied extensively, one could just think of applying the known results. For example, once we choose a coordinate system on the base manifold, which is $\mathcal{T}(\mathfrak{S})$ in our case, there is a standard canonical quantization of the coordinate functions on the base manifold and the corresponding coordinates of the cotangent fiber, as self-adjoint operators on the Hilbert space $L^2(\mathcal{T}(\mathfrak{S}))$; these operators are just the position and the momentum operators. Even if one takes this well-known construction as a partial solution to the quantization problem, one has to do more. Namely, what are other classical observable functions on $T^*\mathcal{T}(\mathfrak{S})$ that one should and could quantize, and what operators should be assigned to them, in a canonical manner?

The quantization constructed in [**KS24**] takes a completely different route. It does not make use of the identification of $\mathcal{GH}_\Lambda(\mathfrak{S}\times\mathbb{R})$ with the cotangent bundle $T^*\mathcal{T}(\mathfrak{S})$ and its canonical quantization (of its Darboux coordinate functions). Instead, based on the special coordinate systems on $\mathcal{GH}_\Lambda(\mathfrak{S}\times\mathbb{R})$ valued in the ring $\mathbb{R}_\Lambda = \mathbb{R}[\ell]/(\ell^2 = -\Lambda)$ studied by Meusburger and Scarinci [**MS16**] using 3d geometry, a certain set of classical observables to be quantized is suggested, mimicking the upper cluster algebra from the theory of cluster algebras [**FZ02**] and cluster varieties [**FG06, FG09**]. An algebraic deformation quantization of these observables is constructed, by translating the results from quantization of cluster varieties from surfaces [**AK17**] with suitable modifications. A corresponding operator representation on a Hilbert space is constructed, by developing an $\mathbb{R}_\Lambda$-version of Fock and Goncharov's representation of quantum cluster variety at $\mathbb{R}_{>0}$ [**FG09**]. By 'composing' the algebraic deformation quantization and the operator representation, one obtains a quantization of the special classical observables we chose by operators on a Hilbert space.

What is reviewed in the present article is the Hilbert space representation (without the algebraic deformation quantization part). Although this representation is described in [**KS24**] using a version of functional calculus valued in $\mathbb{R}_\Delta$, here we avoided working with the $\mathbb{R}_\Lambda$-version, and translated the final results without using $\mathbb{R}_\Lambda$. An advantage of avoiding $\mathbb{R}_\Lambda$ is that the results can be described using usual theory of self-adjoint operators and their functional calculus. A slight disadvantage is that the answers for different values of the cosmological constant Λ look different from each other, not a priori directly related. If one is willing to work with the $\mathbb{R}_\Lambda$-version as in [**KS24**], then one can understand the quantization results in a uniform manner for all values of Λ, maybe with some parts remaining only heuristic. Especially, the formula of eq.(4.5) for the automorphism part operator $\mathbf{K}_k^{\hbar\sharp}$ for the unitary intertwiner $\mathbf{K}_k^\hbar$ for a flip μ_k can be described by a single formula

$$\Phi^{\ell\hbar}(\mathbf{x}_k + \hat{\ell}\hbar\mathbf{y}_k)\Phi^{-\ell\hbar}(\mathbf{x}_k + \hat{\ell}\hbar\mathbf{y}_k)$$

using the somewhat heuristic $\mathbb{R}_\Lambda$-version $\Phi^{\ell\hbar}$ of the quantum dilogarithm, where $\hat{\ell}$ is a suitable operator that quantizes the element ℓ of $\mathbb{R}_\ell$; in particular, $\hat{\ell}^2 = -\Lambda\cdot\mathrm{id}$. We refer the readers to [**KS24**] for more details.

In any case, the quantization of $\mathcal{GH}_\Lambda(\mathfrak{S}\times\mathbb{R})$ obtained in [**KS24**], which we reviewed in the present article, is not based on the canonical quantization of the cotangent bundle $T^*\mathcal{T}(\mathfrak{S})$. It uses 3d geometry and does depend on the value of Λ. As an important consequence, it also yields a trilogy of unitary representations of the mapping class group $\mathrm{MCG}(\mathfrak{S})$ for the three values of Λ, as we are focusing in the present article. As a matter of fact, the natural classical actions of $\mathrm{MCG}(\mathfrak{S})$ on $\mathcal{GH}_\Lambda(\mathfrak{S}\times\mathbb{R})$ for different Λ yield different actions on the cotangent bundle

$T^*\mathcal{T}(\mathfrak{S})$. So, in fact, our quantizations for different Λ can be interpreted as giving quantizations of three different quotients of $T^*\mathcal{T}(\mathfrak{S})$ by MCG($\mathfrak{S}$).

When $\Lambda \neq 0$, the quantum dilogarithm function used in the quantization, $\Phi^{\pm\hbar}$ or $\Phi^{\pm i\hbar}$, has the real parameter $\hbar$ in it, so the resulting MCG($\mathfrak{S}$)-representation $\rho^{\hbar}_{\Delta}$, based on the unitary intertwiners $\mathbf{K}^{\hbar}_{\Delta,\Delta'}$, can be thought of as forming a one-real-parameter family of representations. However, for the case $\Lambda = 0$, there is no dependence on the quantum parameter $\hbar$, as observed in eq.(4.5); so we get a single representation, not a family, although this is still a 'quantum' representation.

As mentioned in §4.2, for $\Lambda = -1$ which corresponds to the 3d anti-de Sitter geometry, the intertwiners and the resulting unitary MCG($\mathfrak{S}$)-representations obtained in [**KS24**] as written in the present article coincide with Fock and Goncharov's representation for the quantum cluster symplectic double variety of the Teichmüller space [**FG09**]. The crucial technical ingredient is the non-compact quantum dilogarithm $\Phi^{\pm\hbar}$ of Faddeev and Kashaev [**F95, FK94**], which actually goes back to Barnes 100 years ago [**B01**], and which is used already in the quantization of Teichmüller spaces by Kashaev [**K98**] and by Chekhov and Fock [**CF97, F97**] in 1990's. We note that the Chekhov-Fock quantization of Teichmüller spaces [**CF97, F97**] is roughly 'half' of the quantization described in the present article for $\Lambda = -1$, or equivalently of the quantization of [**FG09**]; for Chekhov and Fock's intertwiner $\mathbf{K}^{\hbar}_k$ for a flip μ_k, the automorphism part $\mathbf{K}^{\hbar\sharp}_k$ is given roughly by a single non-compact quantum dilogarithm $\Phi^{\hbar}(\mathbf{x}_k + \hbar\mathbf{y}_k)$, which is a 'half' of our answer for $\Lambda = -1$, and the monomial transformation part $\mathbf{K}'_k$ is given by a suitable Shale-Weil intertwiner, as explicitly written in [**K21b**] for the irreducible representation.

For $\Lambda = 1$ which corresponds to the 3d de Sitter geometry, the intertwiners and unitary MCG($\mathfrak{S}$)-representations are based on the version of a quantum dilogarithm function denoted by $\Phi^{\pm i\hbar}$ [**KS24**], which is a certain 'modular double' version of the compact quantum dilogarithm $\psi^q(z) = \prod_{n\geq 0}(1 + q^{2n+1}z)^{-1}$. We note that the compact quantum dilogarithm ψ^q (of Faddeev and Kashaev [**F95, FK94**]) is not suitable for quantization of Teichmüller spaces because it does not make sense when $|q| = 1$. This is the reason why the non-compact quantum dilogarithm $\Phi^{\pm\hbar}$ was used in quantum Teichmüller theory, which can be viewed as a *limit* of a certain ratio of two compact quantum dilogarithm functions, but not a genuine ratio of such. Meanwhile, our function $\Phi^{\pm i\hbar}$, which does not seem to have been particularly considered in the literature before [**KS24**], is a genuine ratio of two compact quantum dilogarithm functions, not a limit of such. A single copy of $\Phi^{\pm i\hbar}$ does not provide a unitary operator, and only a suitable product of the two versions $\Phi^{i\hbar}$ and $\Phi^{-i\hbar}$ does, where these two copies represent the $\mathbb{R}_\Lambda$-holomorphic and the $\mathbb{R}_\Lambda$-anti-holomorphic parts respectively. In the end, the intertwiner for a flip is a unitary intertwiner made up by four genuine compact quantum dilogarithm functions $\psi^{\mathbf{q}}$ with real quantum parameters $\mathbf{q}$ related to each other in a specific way. As far as the author knows, this was the first time that honest compact quantum dilogarithm functions (without taking their limits) were used directly to give unitary representations of MCG($\mathfrak{S}$) on a Hilbert space.

The intertwiners and unitary MCG($\mathfrak{S}$)-representations for $\Lambda = 0$ which corresponds to the 3d Minkowski ('flat') geometry can be thought of as being built on the classical logarithm function, not quantum nor dilogarithm. They can also be regarded as coming from a 'derivative' of the family of quantum dilogarithm functions, due to the nature of the ring $\mathbb{R}_\Lambda$ when $\Lambda = 0$; indeed, notice that for an 'analytic' function f, observe that $f(x + \ell y) = f(x) + \ell f'(x) y$ holds in the ring $\mathbb{R}_0 = \mathbb{R}[\ell]/(\ell^2 = 0)$, for $x, y \in \mathbb{R}$.

6. Intertwiners and quantum group representations

Here we discuss a relationship with the quantum group representations. For this we will first review the results of the joint work of the author with Igor Frenkel [**FK12**].

6.1. Quantum Teichmüller theory and quantum plane algebra. Recall that to each labeled ideal triangulation Δ is associated a Hilbert space $\mathscr{H}_\Delta$, and to each change of labeled ideal triangulations $(\Delta \leadsto \Delta')$ is associated a unitary intertwining operator $\mathbf{K}^\hbar_{\Delta,\Delta'}$. An elementary case is when Δ' is obtained from Δ by a flip μ_k at an arc k, and we denoted the unitary intertwiner as $\mathbf{K}^\hbar_k$. One property of this flip intertwiner is that it satisfies the pentagon identity, as in eq.(4.7) of Thm.4.3; a slight nuisance is that this version of the pentagon identity involves not just the flip intertwiners, but also an index permutation operator $\mathbf{P}_\sigma$. Analogous results hold for the Chekhov-Fock quantization of Teichmüller spaces [**CF97**, **F97**], which is also based on labeled ideal triangulations. The unitary intertwiner for a flip in this quantization also satisfies the pentagon identity as in eq.(4.7), up to a multiplicative scalar. Any reader who is interested in this scalar can consult [**K21b**, **FS10**].

Meanwhile, Kashaev's quantization of Teichmüller spaces [**K98**] is based not on labeled ideal triangulations, but on ideal triangulations together with a different kind of extra data. It uses a *dotted ideal triangulation*, which consists of an ideal triangulation Δ, a labeling of ideal triangles of Δ, and the choice of a distinguished corner for each ideal triangle of Δ which is depicted by a dot $\bullet$ in pictures. To each dotted ideal triangulation, is associated a Hilbert space $\bigotimes_{t:\text{ triangles}} L^2(\mathbb{R}) \cong L^2(\mathbb{R}^{\{\text{triangles}\}}) = L^2(\mathbb{R}^{4g-4+2n})$ (here, the symbol $\bigotimes$ denotes the Hilbert space tensor product), and to each change of dotted ideal triangulations is associated a unitary intertwiner. There are three kinds of elementary changes: 1) T_{ts} which is a flip at an arc adjacent to triangles t and s with a special dot configuration on these triangles, 2) A_t which is a counterclockwise change of a dot in the triangle t, and 3) P_σ which is a label permutation of triangles. The corresponding unitary intertwiners can be denoted by $\mathbf{T}^\hbar_{ts}$, $\mathbf{A}^\hbar_t$ and $\mathbf{P}_\sigma$. One of the consistency equations satisfied by these operators is another version of a pentagon identity, which reads

$$\mathbf{T}^\hbar_{ts}\mathbf{T}^\hbar_{rt} = \mathbf{T}^\hbar_{rt}\mathbf{T}^\hbar_{rs}\mathbf{T}^\hbar_{ts}.$$

The subscripts indicate which tensor factors of $\bigotimes_{\text{triangles}} L^2(\mathbb{R})$ are involved; more precisely, there is a single operator $\mathbf{T}^\hbar : L^2(\mathbb{R}) \otimes L^2(\mathbb{R}) \to L^2(\mathbb{R}) \otimes L^2(\mathbb{R})$ such that $\mathbf{T}^\hbar_{ij}$ means that this operator $\mathbf{T}^\hbar$ is being applied to the i-th and the j-th tensor factors $L^2(\mathbb{R})$. The above pentagon identity involves only three tensor factors, and its prototypical version can be written as

$$\mathbf{T}^\hbar_{23}\mathbf{T}^\hbar_{12} = \mathbf{T}^\hbar_{12}\mathbf{T}^\hbar_{13}\mathbf{T}^\hbar_{23} \tag{6.1}$$

which is on $L^2(\mathbb{R}) \otimes L^2(\mathbb{R}) \otimes L^2(\mathbb{R})$. We note that this version of the pentagon identity involves only $\mathbf{T}^\hbar$, but not other operators like index permutation operators.

The pentagon identity naturally arises in several areas of mathematics, although maybe in slightly different forms. For example, the version as in eq.(6.1) acting on a triple tensor product is observed in a tensor (or, monoidal) category as a consistency axiom for the associativity morphism $(A \otimes B) \otimes C \to A \otimes (B \otimes C)$. Thus, it is natural to ask whether there exists a tensor category whose associativity morphisms recover the above operator $\mathbf{T}^\hbar$ appearing in Kashaev's quantum Teichmüller theory. This question is answered affirmative in the author's joint work with Igor Frenkel [**FK12**], where we consider the category of certain well-behaved representations of a rather simple Hopf $*$-algebra called the *quantum plane* $\mathcal{B}_q$, which is an associative algebra generated by $X^{\pm 1}$ and Y, modulo the relation

$$XY = q^2 YX,$$

where the associated quantum parameter $q = e^{\pi i \hbar}$ is a complex number of modulus 1. This algebra can be viewed as a quantum observable algebra of the plane $\mathbb{R}^2 = \{(x,y) : x, y \in \mathbb{R}\}$ with a standard symplectic structure $dx \wedge dy$, where X and Y are quantum counterparts of the classical functions e^x and e^y. What may distinguish this algebra $\mathcal{B}_q$ from what is called the quantum torus is the $*$-structure, which is given by $X^* = X$ and $Y^* = Y$ (for the quantum plane), rather than $X^* = X^{-1}$ and $Y^* = Y^{-1}$ (for the quantum torus); this means that when we consider a representation of $\mathcal{B}_q$ on a Hilbert space, we would like X and Y to be represented by self-adjoint operators, rather than unitary operators.

An irreducible 'well-behaved' representation π of $\mathcal{B}_q$ on a Hilbert space $\mathscr{H}$ is uniquely determined up to equivalence [**S92**]; it is given by $\mathscr{H} = L^2(\mathbb{R})$, where X and Y are represented by real exponentials of the position and the momentum operators. The algebra $\mathcal{B}_q$ can be viewed as a Borel subalgebra of the split real quantum group $\mathcal{U}_q(\mathfrak{sl}(2,\mathbb{R}))$, and in particular is a Hopf algebra whose coproduct is given by $\Delta(X) = X \otimes X$ and $\Delta(Y) = Y \otimes X + 1 \otimes Y$ (here, Δ is the coproduct, and has nothing to do with an ideal triangulation!). Using the coproduct, one defines action of $\mathcal{B}_q$ on the tensor product of representations. We show that the tensor square $\mathscr{H} \otimes \mathscr{H}$ decomposes into direct integral $\int_{\mathbb{R}}^{\oplus} \mathscr{H}$ of the unique irreducible representations $\mathscr{H}$; we show this in the form of an isomorphism $\mathbf{F}^\hbar : \mathscr{H} \otimes \mathscr{H} \xrightarrow{\sim} M \otimes \mathscr{H}$, where $M = L^2(\mathbb{R})$ is the trivial $\mathcal{B}_q$-module called the multiplicity module. We realize the decomposition isomorphism $\mathbf{F}^\hbar$ as a map $L^2(\mathbb{R}) \otimes L^2(\mathbb{R}) \to L^2(\mathbb{R}) \otimes L^2(\mathbb{R})$, given by an integral transformation expressed using the non-compact quantum dilogarithm function $\Phi^\hbar$; we note that it is also possible to express this operator $\mathbf{F}^\hbar$ as a composition of a Shale-Weil type operator and an operator given by functional calculus applied to the function $\Phi^\hbar$, just as we became familiar with through the present article.

The tensor cube of $\mathscr{H}$ has two decompositions, $(\mathscr{H} \otimes \mathscr{H}) \otimes \mathscr{H}$ and $\mathscr{H} \otimes (\mathscr{H} \otimes \mathscr{H})$. The coassociatity of the coproduct implies that the 'identity' map $(\mathscr{H} \otimes \mathscr{H}) \otimes \mathscr{H} \to \mathscr{H} \otimes (\mathscr{H} \otimes \mathscr{H})$ is an isomorphism of representations. Meanwhile, the parentheses tell us how to decompose each of these representations into direct integral of $\mathscr{H}$, using the decomposition maps $\mathbf{F}^\hbar$. More precisely, we obtain the diagram

$$\begin{array}{ccccc} (\mathscr{H} \otimes \mathscr{H}) \otimes \mathscr{H} & \xrightarrow{\mathbf{F}^\hbar_{12}} & M \otimes \mathscr{H} \otimes \mathscr{H} & \xrightarrow{\mathbf{F}^\hbar_{23}} & M \otimes M \otimes \mathscr{H} \\ \mathrm{id} \big\downarrow & & & & \big\downarrow \\ \mathscr{H} \otimes (\mathscr{H} \otimes \mathscr{H}) & \xrightarrow{\mathbf{F}^\hbar_{23}} & \mathscr{H} \otimes M \otimes \mathscr{H} & \xrightarrow{\mathbf{F}^\hbar_{13}} & M \otimes M \otimes \mathscr{H} \end{array}$$

By composing the four $\mathbf{F}^\hbar$ in the diagram, one gets the rightmost vertical map $\mathbf{F}^\hbar_{13}\mathbf{F}^\hbar_{23}(\mathbf{F}^\hbar_{12})^{-1}(\mathbf{F}^\hbar_{23})^{-1} : M \otimes M \otimes \mathscr{H} \to M \otimes M \otimes \mathscr{H}$ which we prove to be of the form $\mathbf{T}^\hbar \otimes \mathrm{id}$ for some $\mathbf{T}^\hbar : M \otimes M \to M \otimes M$. So, the 'identity' associativity morphism $(\mathscr{H} \otimes \mathscr{H}) \otimes \mathscr{H} \to \mathscr{H} \otimes (\mathscr{H} \otimes \mathscr{H})$ is encoded via $\mathbf{F}^\hbar$ as a map $\mathbf{T}^\hbar$: $M \otimes M \to M \otimes M$ between the multiplicity modules. We note that, by using the pentagon operator identity satisfied by the non-compact quantum dilogarithm $\Phi^\hbar$, we computed $\mathbf{T}^\hbar$ to be again of a similar form as $\mathbf{F}^\hbar$, namely composition of a Shale-Weil type operator and an operator expressed by $\Phi^\hbar$.

Consider the five different possible decompositions of the quadruple tensor product $\mathscr{H} \otimes \mathscr{H} \otimes \mathscr{H} \otimes \mathscr{H}$, represented by the choices of paranthesizing such as $(((\mathscr{H} \otimes \mathscr{H}) \otimes \mathscr{H}) \otimes \mathscr{H}$. The identity maps between them can be translated into the maps $\mathbf{T}^\hbar$ between the multiplicity modules $M \otimes M \otimes M$. The identity maps all commute with each other, and this commutativity is translated into the pentagon identity of five copies of $\mathbf{T}^\hbar$, which is exactly of the form as in eq.(6.1).

Part of the main theorem of [**FK12**] says that this operator $\mathbf{T}^\hbar$ coming from the representation theory of the quantum plane algebra $\mathcal{B}_q$, or in fact more precisely of its modular double $\mathcal{B}_q \otimes \mathcal{B}_{q^\vee}$ with $q^\vee = e^{\pi \mathrm{i}/\hbar}$, coincides with the unitary intertwiner $\mathbf{T}^\hbar$ for a flip T_{ts} of dotted ideal triangulations appearing in Kashaev's quantum Teichmüller theory [**K98**], up to conjugation by a unitary map. Moreover, the rigid tensor category structure for the representations of $\mathcal{B}_q$ is also studied. By making use of the identification $M \cong \mathrm{Hom}_{\mathcal{B}_q}(\mathscr{H}, \mathscr{H} \otimes \mathscr{H})$ of M with the space of intertwiners, and by permuting the roles of the three copies of $\mathscr{H}$ appearing in the right hand side of this isomorphism using various natural isomorphisms involving the dual representations, we construct a map $\mathbf{A}^\hbar : M \to M$ of order three, and we have shown that it essentially coincides with the intertwiner $\mathbf{A}^\hbar$ for a dot-change A_t in Kashaev's quantum Teichmüller theory.

6.2. 3d quantum gravity and quantum pseudo-Kähler plane algebra. Let's come back now to 3d quantum gravity. Inspired by the flip intertwiners $\mathbf{K}^\hbar_k$ for the quantization of $\mathcal{GH}_\Lambda(\mathfrak{S} \times \mathbb{R})$ constructed in [**KS24**], the author has developed results analogous to [**FK12**] suited to 3d quantum gravity, especially for the case of positive cosmological constant $\Lambda = 1$, in [**K23**]. First, a new Hopf $*$-algebra $\mathcal{C}_\mathbf{q}$ is introduced, where the associated quantum parameter $\mathbf{q} = e^{-\pi\hbar}$ is real; $\mathcal{C}_\mathbf{q}$ is generated by $Z_1^{\pm 1}$ and Z_2, with the relation

$$Z_1 Z_2 = \mathbf{q}^2 Z_2 Z_1.$$

This algebra structure seems just like the quantum torus algebra or the quantum plane algebra, this time with a real quantum parameter. What distinquishes $\mathcal{C}_\mathbf{q}$ vastly with these previous algebras is the $*$-structure. In fact, being more precise, $\mathcal{C}_\mathbf{q}$ is generated by $Z_1^{\pm 1}$, Z_2, $(Z_1^*)^{\pm 1}$, Z_2^*, modulo the above relation, and also additional relations saying that each of Z_1 and Z_2 commutes with each of Z_1^* and Z_2^*, and that $Z_1^* Z_2^* = \mathbf{q}^{-2} Z_2^* Z_1^*$. The notations of the generators immediately suggest the $*$-structure. The coproduct is given as before by $\Delta Z_1 = Z_1 \otimes Z_1$, $\Delta Z_2 = Z_2 \otimes Z_1 + 1 \otimes Z_2$, and similarly for Z_1^* and Z_2^*. One can view this algebra as being a quantum observable algebra for the complex 'plane' $\mathbb{C}^2 = \{(z_1, z_2) : z_1, z_2 \in \mathbb{C}\}$, equipped with the pseudo-Kähler metric $dz_1 \otimes d\overline{z}_2 + dz_2 \otimes d\overline{z}_1$. Hence this algebra is called the *quantum pseudo-Kähler plane* in [**K23**].

A basic representation of $\mathcal{C}_{\mathbf{q}}$, or more precisely of its modular double $\mathcal{C}_{\mathbf{q}} \otimes \mathcal{C}_{1/\mathbf{q}^\vee}$ (with $\mathbf{q}^\vee = e^{-\pi/\hbar}$), is constructed on $\mathscr{H} = L^2(\mathbb{R}^2)$, where the generators are represented by normal operators given by exponentials of simple $\mathbb{C}$-linear combinations of the position and the momentum operators. Then all the constructions as in [**FK12**] are mimicked by studying the representation theory of $\mathcal{C}_{\mathbf{q}}$, yielding the new corresponding versions of the operators $\mathbf{F}^\hbar$, $\mathbf{T}^\hbar$ and $\mathbf{A}^\hbar$, which satisfy the same consistency relations as before, including the pentagon identity in eq.(6.1). Notice that $\mathbf{F}^\hbar$ and $\mathbf{T}^\hbar$ this time are given in terms of the modular-double version of the compact quantum dilogarithm, i.e $\Phi^{\pm i\hbar}$.

For the case of the quantum plane algebra $\mathcal{B}_q$ as in [**FK12**], these operators $\mathbf{T}^\hbar$ and $\mathbf{A}^\hbar$ are what constitute Kashaev's version of quantization of Teichmüller spaces. Note that Kashaev's quantum Teichmüller theory [**K98**] is based Kashaev's special coordinate functions which are ratios of Penner's lambda length coordinates on the 'decorated' Teichmüller space $\mathcal{T}^d(\mathfrak{S})$ [**P87**]. Kashaev's functions do not really form a coordinate system of a Teichmüller space, and can be viewed as being a coordinate system of some related space, given by the product of Penner's projectivized decorated Teichmüller space $\mathcal{T}^d(\mathfrak{S})/\mathbb{R}_{>0}$ and the first cohomology $H^1(\mathfrak{S};\mathbb{R})$ [**K98**, Prop.5].

Similarly for the case of the quantum pseudo-Kähler plane algebra $\mathcal{C}_{\mathbf{q}}$ as in [**K23**], the operators $\mathbf{T}^\hbar$ and $\mathbf{A}^\hbar$ should be what constitute a Kashaev-type quantization of 3d gravity for $\Lambda = 1$, which is never considered. One should define a version of 'decorated' moduli space of 3d gravity, establish 3d complexified version of lambda lengths there (one may want to see [**NN04**]), consider certain ratios of them, and so on. Such a construction is still yet to be found and left open.

Note that in [**K23**], only the case $\Lambda = 1$ is dealt with in detail, which inspired the new Hopf algebra as above. One may continue investigating the cases for the remaining two values of Λ. Most likely, the case of $\Lambda = -1$, which is treated in [**K23**, §14] without proof, would just be a 'symplectic double' version of what was obtained in [**FK12**] already. The case $\Lambda = 0$ might be interesting. The first step would be to define a certain new version of 'quantum plane' algebra, probably over $\mathbb{R}_0$ or over the suitable complexification of $\mathbb{R}_0$ (see [**KS24**]), which is without the quantum parameter at all! The representation theory of this algebra should yield the 'flat' versions of the operators $\mathbf{T}$ and $\mathbf{A}$ as well, where $\mathbf{T}$ is given in terms of the usual logarithm function. Then again, the Kashaev-type 3d quantum gravity theory should also be sought for.

7. Some applications and future directions

Modular functor conjecture

Notice that the quantum Teichmüller theory associates to each punctured surface $\mathfrak{S}$ a family of unitary (projective) representations of MCG($\mathfrak{S}$) on a Hilbert space $\mathscr{H}_{\mathfrak{S}}$; the construction a priori depends on the choice of an ideal triangulation Δ of $\mathfrak{S}$, but the results for different Δ are consistently identified with each other. The *(SL$_2$) modular functor conjecture* speculates that this assignment of Hilbert spaces to different surfaces is organized in such a way that it forms a so-called 2-dimensional 'modular functor' corresponding to some 2-dimensional conformal field theory. A crucial axiom to be checked is, for each punctured surface $\mathfrak{S}$ and a simple loop γ in $\mathfrak{S}$ that does not contract to an interior point or a puncture, if $\mathfrak{S}_\gamma$ denotes the new surface obtained from $\mathfrak{S}$ by cutting along γ (and shrinking the two holes thus created from γ to punctures), then the Hilbert space representations $\mathscr{H}_{\mathfrak{S}}$ and $\mathscr{H}_{\mathfrak{S}_\gamma}$ associated to $\mathfrak{S}$ and $\mathfrak{S}_\gamma$ by quantum Teichmüller theory should be related through a special way. For details, see [**T07**] and [**FG09**]. A modified version is verified in [**T07**], based on a novel adaptation of Kashaev's quantum Teichmüller. Still, a more direct version using the Chekhov-Fock-Goncharov quantum Teichmüller theory [**CF97,F97,FG09,K21b**] has not been fully established in a rigorous manner, and some of the first steps toward it are performed in [**K24**].

Meanwhile, the version of 3d quantum gravity constructed in [**KS24**] also associates a Hilbert space representation of $\mathrm{MCG}(\mathfrak{S})$ to each punctured surface $\mathfrak{S}$. Hence one can formulate and try to solve the corresponding '$\mathrm{SL}_2(\mathbb{R}_\Lambda)$' version of the modular functor conjecture. In the end, it would yield a 2-dimensional modular functor corresponding to some 2-dimensional conformal field theory.

Meanwhile, notice that there is a general correspondence between 2-dimensional conformal field theories (2d CFT) and 3-dimensional topological quantum field theories (3d TQFT); see [**F95**]. So the above modular functor conjectures hint at certain 3d TQFT lurking behind the quantum Teichmüller theory and 3d quantum gravity theory. On the other hand, Kashaev and Andersen approach 3d TQFT more directly from quantum Teichmüller theory [**AK14**]. One could try an analog of this using the 3d quantum gravity theory of [**KS24**].

The universal case

The universal Teichmüller space has been known to be related to the universal phase space of 3d anti-de Sitter gravity; see [**BH86,SK13**]. Thus, quantization of the universal Teichmüller space, which can be viewed as the Teichmüller space of the open unit disc, with certain restriction on the boundary behavior [**P93**], should be related to quantization of 3d anti-de Sitter gravity. Besides, the mapping class group for a version of the universal Teichmüller space is the 'Ptolemy-Thompson group' T, which is one of the three infinite groups of R. Thompson [**CFP96**]. In particular, quantization of universal Teichmüller space yields representations of T; see [**FS10**] and [**K16a**]. On the other hand, Igor Frenkel and Bob Penner have suggested in [**FP22**] that the study of universal Teichmüller space might lead to the Monster 2d CFT which was constructed by I. Frenkel, J. Lepowsky and A. Meurman in 1980's [**FLM88**]. Keeping in mind Edward Witten's conjectural relationship between 3d quantum gravity and Monster 2d CFT [**W07**], it should be worth investigating the universal case of the construction of [**KS24**], and thinking how it could be connected to all these conjectures. We refer the readers to [**FK12,K16a,FS10,P93**] for more details on how to approach a quantization of universal Teichmüller space.

Going to higher rank

The Teichmüller space of a surface $\mathfrak{S}$ can be thought of as some version (or, some component) of the representation variety $\mathrm{Hom}(\pi_1(\mathfrak{S}),\mathrm{G})/\mathrm{G}$, when $\mathrm{G} = \mathrm{PSL}_2(\mathbb{R})$. The moduli space $\mathcal{GH}_\Lambda(\mathfrak{S}\times\mathbb{R})$ of 3d gravity on $\mathfrak{S}\times\mathbb{R}$ corresponds to the case when $\mathrm{G} = \mathrm{PSL}_2(\mathbb{R}_\Lambda)$ (see [**KS24**]). Meanwhile, the study of the representation variety for higher rank (real) Lie groups G, as being a higher rank analog of the Teichmüller theory, is referred to as the *higher Teichmüller theory* [**FG06**]. In some case like $\mathrm{G} = \mathrm{PGL}_3(\mathbb{R})$, the corresponding space can be interpreted as the moduli space of certain geometric structures on the surface $\mathfrak{S}$, such as convex projective structures [**G90**]. One might try to look for such a geometric description when we $\mathbb{R}_\Lambda$-complexify, e.g. in the case when $\mathrm{G} = \mathrm{PGL}_3(\mathbb{R}_\Lambda)$. Quantization of such $\mathbb{R}_\Lambda$-complexified moduli spaces, or their suitable cluster variety relatives, can in fact be obtained already using the results of [**KS24**], because most of the constructions of [**KS24**] are formulated to work for general cluster varieties, not just for the space $\mathcal{GH}_\Lambda(\mathfrak{S} \times \mathbb{R})$.

More general topology

After the author's talk on [**KS24**] at Igor Frenkel's 70th birthday conference, Nikita Nekrasov in the audience asked what would happen to the case of 3d gravity when the topology of the surface changes ('in time'). The quantization methods of [**KS24**] depend on the $\mathbb{R}_\Lambda$-valued cluster coordinates developed in [**MS16**], which work only on the space $\mathcal{GH}_\Lambda(\mathfrak{S} \times \mathbb{R})$ of 3d gravity on the 3-manifold $\mathfrak{S} \times \mathbb{R}$. For a moduli space of 3d gravity on a more general 3-manifold which does not look like a surface times an interval, the cluster variety structure has not been observed as far as the author knows, although somewhat similar looking coordinate change maps may have been found [**GTZ15**]. It seems that one might need to develop a new machinery to deal with these (e.g. to quantize them), instead of trying to apply the known theory of cluster varieties.

Acknowledgments

The author thanks his advisor Igor B. Frenkel for giving continuous support and inspirations. The author thanks the organizers of Igor's 70th birthday conference for providing a wonderful experience at the conference, and also for putting together this proceedings.

References

[AK17] Dylan G. L. Allegretti and Hyun Kyu Kim, *A duality map for quantum cluster varieties from surfaces*, Adv. Math. **306** (2017), 1164–1208, DOI 10.1016/j.aim.2016.11.007. MR3581328

[AK14] Jørgen Ellegaard Andersen and Rinat Kashaev, *A TQFT from quantum Teichmüller theory*, Comm. Math. Phys. **330** (2014), no. 3, 887–934, DOI 10.1007/s00220-014-2073-2. MR3227503

[BK00] B. Bakalov and A. Kirillov Jr., *On the lego-Teichmüller game*, Transform. Groups **5** (2000), no. 3, 207–244, DOI 10.1007/BF01679714. MR1780935

[B01] E. W. Barnes, *Theory of the double Gamma function*, Phil. Trans. Roy. Soc. A **196** (1901), 265–388.

[BL07] Francis Bonahon and Xiaobo Liu, *Representations of the quantum Teichmüller space and invariants of surface diffeomorphisms*, Geom. Topol. **11** (2007), 889–937, DOI 10.2140/gt.2007.11.889. MR2326938

[BW11] Francis Bonahon and Helen Wong, *Quantum traces for representations of surface groups in* $\mathrm{SL}_2(\mathbb{C})$, Geom. Topol. **15** (2011), no. 3, 1569–1615, DOI 10.2140/gt.2011.15.1569. MR2851072

[BH86] J. D. Brown and Marc Henneaux, *Central charges in the canonical realization of asymptotic symmetries: an example from three-dimensional gravity*, Comm. Math. Phys. **104** (1986), no. 2, 207–226. MR836000

[CFP96] J. W. Cannon, W. J. Floyd, and W. R. Parry, *Introductory notes on Richard Thompson's groups*, Enseign. Math. (2) **42** (1996), no. 3-4, 215–256. MR1426438

[CF97] V. V. Fock and L. O. Chekhov, *Quantum Teichmüller spaces* (Russian, with Russian summary), Teoret. Mat. Fiz. **120** (1999), no. 3, 511–528, DOI 10.1007/BF02557246; English transl., Theoret. and Math. Phys. **120** (1999), no. 3, 1245–1259. MR1737362

[F95] L. D. Faddeev, *Discrete Heisenberg-Weyl group and modular group*, Lett. Math. Phys. **34** (1995), no. 3, 249–254, DOI 10.1007/BF01872779. MR1345554

[FK94] L. D. Faddeev and R. M. Kashaev, *Quantum dilogarithm*, Modern Phys. Lett. A **9** (1994), no. 5, 427–434, DOI 10.1142/S0217732394000447. MR1264393

[FKV01] L. D. Faddeev, R. M. Kashaev, and A. Yu. Volkov, *Strongly coupled quantum discrete Liouville theory. I. Algebraic approach and duality*, Comm. Math. Phys. **219** (2001), no. 1, 199–219, DOI 10.1007/s002200100412. MR1828812

[F97] V. V. Fock, *Dual Teichmüller spaces*, arXiv:dg-ga/9702018

[FG06] V. V. Fock and Alexander Goncharov, *Moduli spaces of local systems and higher Teichmüller theory*, Publ. Math. Inst. Hautes Études Sci. **103** (2006), 1–211, DOI 10.1007/s10240-006-0039-4. MR2233852

[FG09] V. V. Fock and A. B. Goncharov, *The quantum dilogarithm and representations of quantum cluster varieties*, Invent. Math. **175** (2009), no. 2, 223–286, DOI 10.1007/s00222-008-0149-3. MR2470108

[FST08] Sergey Fomin, Michael Shapiro, and Dylan Thurston, *Cluster algebras and triangulated surfaces. I. Cluster complexes*, Acta Math. **201** (2008), no. 1, 83–146, DOI 10.1007/s11511-008-0030-7. MR2448067

[FZ02] Sergey Fomin and Andrei Zelevinsky, *Cluster algebras. I. Foundations*, J. Amer. Math. Soc. **15** (2002), no. 2, 497–529, DOI 10.1090/S0894-0347-01-00385-X. MR1887642

[FK12] Igor B. Frenkel and Hyun Kyu Kim, *Quantum Teichmüller space from the quantum plane*, Duke Math. J. **161** (2012), no. 2, 305–366, DOI 10.1215/00127094-1507390. MR2876932

[FLM88] Igor Frenkel, James Lepowsky, and Arne Meurman, *Vertex operator algebras and the Monster*, Pure and Applied Mathematics, vol. 134, Academic Press, Inc., Boston, MA, 1988. MR996026

[FP22] Igor Frenkel and Robert Penner, *Sketch of a program for automorphic functions from universal Teichmüller theory to capture monstrous moonshine*, Comm. Math. Phys. **389** (2022), no. 3, 1525–1567, DOI 10.1007/s00220-021-04268-8. MR4381179

[F95] Louis Funar, *$(2+1)$-D topological quantum field theory and 2-D conformal field theory*, Comm. Math. Phys. **171** (1995), no. 2, 405–458. MR1344733

[FS10] Louis Funar and Vlad Sergiescu, *Central extensions of the Ptolemy-Thompson group and quantized Teichmüller theory*, J. Topol. **3** (2010), no. 1, 29–62, DOI 10.1112/jtopol/jtp033. MR2608476

[GTZ15] Stavros Garoufalidis, Dylan P. Thurston, and Christian K. Zickert, *The complex volume of $\mathrm{SL}(n,\mathbb{C})$-representations of 3-manifolds*, Duke Math. J. **164** (2015), no. 11, 2099–2160, DOI 10.1215/00127094-3121185. MR3385130

[G90] William M. Goldman, *Convex real projective structures on compact surfaces*, J. Differential Geom. **31** (1990), no. 3, 791–845. MR1053346

[G08] Alexander B. Goncharov, *Pentagon relation for the quantum dilogarithm and quantized $\mathcal{M}_{0,5}^{\rm cyc}$*, Geometry and dynamics of groups and spaces, Progr. Math., vol. 265, Birkhäuser, Basel, 2008, pp. 415–428, DOI 10.1007/978-3-7643-8608-5_10. MR2402412

[H13] Brian C. Hall, *Quantum theory for mathematicians*, Graduate Texts in Mathematics, vol. 267, Springer, New York, 2013, DOI 10.1007/978-1-4614-7116-5. MR3112817

[K98] R. M. Kashaev, *Quantization of Teichmüller spaces and the quantum dilogarithm*, Lett. Math. Phys. **43** (1998), no. 2, 105–115, DOI 10.1023/A:1007460128279. MR1607296

[K16a] Hyun Kyu Kim, *The dilogarithmic central extension of the Ptolemy-Thompson group via the Kashaev quantization*, Adv. Math. **293** (2016), 529–588, DOI 10.1016/j.aim.2016.02.016. MR3474328

[K16b] Hyun Kyu Kim, *Ratio coordinates for higher Teichmüller spaces*, Math. Z. **283** (2016), no. 1-2, 469–513, DOI 10.1007/s00209-015-1607-4. MR3489077

[K21a] Hyun Kyu Kim, *Phase constants in the Fock-Goncharov quantum cluster varieties*, Anal. Math. Phys. **11** (2021), no. 1, Paper No. 2, 66, DOI 10.1007/s13324-020-00439-3. MR4179968

[K21b] Hyun Kyu Kim, *Irreducible self-adjoint representations of quantum Teichmüller space and the phase constants*, J. Geom. Phys. **162** (2021), Paper No. 104103, 36, DOI 10.1016/j.geomphys.2021.104103. MR4202190

[K23] Hyun Kyu Kim, *Three-dimensional quantum gravity from the quantum pseudo-Kähler plane*, Comm. Math. Phys. **402** (2023), no. 3, 2715–2763, DOI 10.1007/s00220-023-04783-w. MR4630484

[K24] Hyun Kyu Kim, *Quantized geodesic lengths for Teichmüller spaces: algebraic aspects*, arXiv:2405.14727, 2024.

[KS24] Hyun Kyu Kim and Carlos Scarinci, *A quantization of moduli spaces of 3-dimensional gravity*, Comm. Math. Phys. **405** (2024), no. 6, Paper No. 144, 62, DOI 10.1007/s00220-024-05012-8. MR4751726

[L-F09] D. Labardini-Fragoso, *Quivers with potentials associated to triangulated surfaces, part II: arc representations*, `arXiv:0909.4100`, 2009

[L77] Gérard Lion, *Intégrales d'entrelacement sur des groupes de Lie nilpotents et indices de Maslov* (French), Non-commutative harmonic analysis (Actes Colloq., Marseille-Luminy, 1976), Lecture Notes in Math., Vol. 587, Springer, Berlin-New York, 1977, pp. 160–176. MR486311

[MS16] Catherine Meusburger and Carlos Scarinci, *Generalized shear coordinates on the moduli spaces of three-dimensional spacetimes*, J. Differential Geom. **103** (2016), no. 3, 425–474. MR3523528

[NN04] Toshihiro Nakanishi and Marjatta Näätänen, *Complexification of lambda length as parameter for* $\mathrm{SL}(2,\mathbb{C})$ *representation space of punctured surface groups*, J. London Math. Soc. (2) **70** (2004), no. 2, 383–404, DOI 10.1112/S0024610704005435. MR2078900

[P87] R. C. Penner, *The decorated Teichmüller space of punctured surfaces*, Comm. Math. Phys. **113** (1987), no. 2, 299–339. MR919235

[P93] R. C. Penner, *Universal constructions in Teichmüller theory*, Adv. Math. **98** (1993), no. 2, 143–215, DOI 10.1006/aima.1993.1015. MR1213724

[SK13] Carlos Scarinci and Kirill Krasnov, *The universal phase space of* AdS_3 *gravity*, Comm. Math. Phys. **322** (2013), no. 1, 167–205, DOI 10.1007/s00220-012-1655-0. MR3073162

[S12] Konrad Schmüdgen, *Unbounded self-adjoint operators on Hilbert space*, Graduate Texts in Mathematics, vol. 265, Springer, Dordrecht, 2012, DOI 10.1007/978-94-007-4753-1. MR2953553

[S92] Konrad Schmüdgen, *Operator representations of* $\mathbf{R}_q^2$, Publ. Res. Inst. Math. Sci. **28** (1992), no. 6, 1029–1061, DOI 10.2977/prims/1195167736. MR1203759

[S63] I. E. Segal, *Transforms for operators and symplectic automorphisms over a locally compact abelian group*, Math. Scand. **13** (1963), 31–43, DOI 10.7146/math.scand.a-10686. MR163183

[S62] David Shale, *Linear symmetries of free boson fields*, Trans. Amer. Math. Soc. **103** (1962), 149–167, DOI 10.2307/1993745. MR137504

[T07] Jörg Teschner, *An analog of a modular functor from quantized Teichmüller theory*, Handbook of Teichmüller theory. Vol. I, IRMA Lect. Math. Theor. Phys., vol. 11, Eur. Math. Soc., Zürich, 2007, pp. 685–760, DOI 10.4171/029-1/17. MR2349683

[T80] W.P. Thurston, *The Geometry and Topology of three-manifolds* Lecture Notes, Princeton University, 1980, available at `http://library.msri.org/books/gt3m`

[vN31] J. v. Neumann, *Die Eindeutigkeit der Schrödingerschen Operatoren* (German), Math. Ann. **104** (1931), no. 1, 570–578, DOI 10.1007/BF01457956. MR1512685

[W64] André Weil, *Sur certains groupes d'opérateurs unitaires* (French), Acta Math. **111** (1964), 143–211, DOI 10.1007/BF02391012. MR165033

[W07] E. Witten, *Three-dimensional gravity reconsidered*, `arXiv:0706.3359`, 2007.

[W00] S. L. Woronowicz, *Quantum exponential function*, Rev. Math. Phys. **12** (2000), no. 6, 873–920, DOI 10.1142/S0129055X00000344. MR1770545

[Z07] Don Zagier, *The dilogarithm function*, Frontiers in number theory, physics, and geometry. II, Springer, Berlin, 2007, pp. 3–65, DOI 10.1007/978-3-540-30308-4_1. MR2290758

School of Mathematics, Korea Institute for Advanced Study (KIAS), 85 Hoegi-ro Dongdaemun-gu, Seoul 02455, Republic of Korea

Email address: `hkim@kias.re.kr`

Contemporary Mathematics
Volume **817**, 2025
https://doi.org/10.1090/conm/817/16327

Quantum loop groups for arbitrary quivers

Andrei Neguţ

ABSTRACT. We study the dual constructions of quantum loop groups and Feigin-Odesskii type shuffle algebras for an arbitrary quiver, for which the arrow parameters are arbitrary non-zero elements of any field. Examples of our setup include K-theoretic Hall algebras of quivers with 0 potential, quantum loop groups of Kac-Moody type and quiver quantum toroidal algebras.

1. Introduction

1.1. In the present paper, we develop a systematic treatment of quantum loop groups, which generalizes the particular cases treated in [**7**–**10**]. Specifically, we fix a finite set I, a field $\mathbb{K}$ of characteristic 0, and a collection of rational functions

$$\zeta_{ij}(x) \in \frac{\mathbb{K}[x^{\pm 1}]}{(1-x)^{\delta_{ij}}} \tag{1.1}$$

for all $i, j \in I$. To this datum, one can associate two objects

- The (half) **quadratic quantum loop groups**

$$\widetilde{\mathbf{U}}^+ = \mathbb{K}\Big\langle e_{i,d} \Big\rangle_{i \in I, d \in \mathbb{Z}} \Big/ \text{relation (2.3)} \tag{1.2}$$

$$\widetilde{\mathbf{U}}^- = \mathbb{K}\Big\langle f_{i,d} \Big\rangle_{i \in I, d \in \mathbb{Z}} \Big/ \text{relation (2.7)} \tag{1.3}$$

- The **big shuffle algebras** (following ideas of [**1**, **2**])

$$\mathcal{V}^{\pm} = \bigoplus_{\boldsymbol{n}=(n_i)_{i\in I} \in \mathbb{N}^I} \mathbb{K}[z_{i1}, z_{i1}^{-1}, \dots, z_{in_i}, z_{in_i}^{-1}]^{\text{sym}}_{i \in I} \tag{1.4}$$

  which are endowed with the product (2.5) and its opposite, respectively.

The constructions above are related by $\mathbb{K}$-algebra homomorphisms

$$\widetilde{\mathbf{U}}^{\pm} \xrightarrow{\widetilde{\Upsilon}^{\pm}} \mathcal{V}^{\pm}, \qquad e_{i,d}, f_{i,d} \mapsto z_{i1}^d \tag{1.5}$$

and pairings

$$\widetilde{\mathbf{U}}^+ \otimes \mathcal{V}^- \xrightarrow{\langle \cdot, \cdot \rangle} \mathbb{K} \tag{1.6}$$

$$\mathcal{V}^+ \otimes \widetilde{\mathbf{U}}^- \xrightarrow{\langle \cdot, \cdot \rangle} \mathbb{K} \tag{1.7}$$

2020 *Mathematics Subject Classification.* Primary 17B37.

The author was supported by NSF grant DMS-1845034. The author was also supported by the Alfred P. Sloan Foundation and the MIT Research Support Committee.

whose formulas will be recalled in (2.14) and (2.15). With this in mind, define

- The (half) **quantum loop groups** as

$$\mathbf{U}^{\pm} = \widetilde{\mathbf{U}}^{\pm} \Big/ K^{\pm} \tag{1.8}$$

where $K^{\pm} = \text{Ker}\, \widetilde{\Upsilon}^{\pm}$.

- The **shuffle algebras** as

$$\mathring{\mathcal{S}}^{\pm} = \text{Im}\, \widetilde{\Upsilon}^{\pm} \subset \mathcal{V}^{\pm} \tag{1.9}$$

Finally, let

$$\mathcal{S}^{\pm} \subset \mathcal{V}^{\pm} \tag{1.10}$$

denote the set of elements which pair trivially with $K^{\pm}$ under the pairings (1.6) and (1.7), respectively. The following is our first main result.

THEOREM 1.1 (Theorem 2.5). *We have* $\mathcal{S}^{\pm} = \mathring{\mathcal{S}}^{\pm}$, *thus* (1.5) *yields isomorphisms*

$$\Upsilon^{\pm} : \mathbf{U}^{\pm} \xrightarrow{\sim} \mathcal{S}^{\pm} \tag{1.11}$$

Moreover, the pairings (1.6)–(1.7) *descend to pairings*

$$\mathbf{U}^{+} \otimes \mathcal{S}^{-} \xrightarrow{\langle \cdot, \cdot \rangle} \mathbb{K} \tag{1.12}$$

$$\mathcal{S}^{+} \otimes \mathbf{U}^{-} \xrightarrow{\langle \cdot, \cdot \rangle} \mathbb{K} \tag{1.13}$$

which are non-degenerate in both arguments, and coincide under (1.11).

If the rational functions (1.1) have the property that

$$\frac{\zeta_{ij}(x)}{\zeta_{ji}(x^{-1})} \tag{1.14}$$

is regular and non-zero at $x = \infty$ for all $i, j \in I$, then we recall in Subsection 2.14 the usual construction of gluing the halves $\mathbf{U}^{\pm}$ into a quantum loop group

$$\mathbf{U} = \mathbf{U}^{+} \otimes \mathbf{U}^{0} \otimes \mathbf{U}^{-} \tag{1.15}$$

using the Drinfeld double construction (above, $\mathbf{U}^0$ is a commutative subalgebra generated by symbols $\{h_{i,d}, h'_{i,-d}\}_{i \in I, d \geq 0}$, modulo the relation $h_{i,0}h'_{i,0} = 1, \forall i \in I$).

1.2. Before we move on to the other main results of the present paper, let us recall two important particular cases of Theorem 1.1, which have already appeared in the literature. Let I be a finite set, assumed to be totally ordered in Definition 1.2.

DEFINITION 1.2. Consider $\mathbb{K} = \mathbb{Q}(q)$ and

$$\zeta_{ij}(x) = \frac{(q^{-d_{ij}} - x)(-x)^{-\delta_{i>j}}}{(1 - x)^{\delta_{ij}}} \tag{1.16}$$

where $\{d_{ij}\}$ is the Cartan matrix of a simply-laced[1] Kac-Moody Lie algebra $\mathfrak{g}$. Then

$$\mathbf{U} = U_q(L\mathfrak{g})$$

is the quantum loop group associated to $\mathfrak{g}$ (see [**8**] for details).

When $U_q(L\mathfrak{g})$ is replaced by the usual quantum group $U_q(\mathfrak{g})$, constructions analogous to those of Subsection 1.1 were performed in [**4, 14, 15**], although they are technically substantially different from those in the present paper.

DEFINITION 1.3. Let Q be a quiver with vertex set I and edge set E. Consider $\mathbb{K} = \mathbb{Q}(q, t_e)_{e \in E}$ and

$$\zeta_{ij}(x) = \left(\frac{1 - xq^{-1}}{1 - x}\right)^{\delta_{ij}} \prod_{e = \overrightarrow{ij}} \left(\frac{1}{t_e} - x\right) \prod_{e = \overrightarrow{ji}} \left(1 - \frac{t_e}{xq}\right) \tag{1.17}$$

Then it was shown in [**9**] that $\mathbf{U}^+$ is isomorphic to the localized preprojective K-theoretic Hall algebra of Q (defined as in [**16**]).

When Q is the quiver with one vertex and g loops, the algebra $\mathbf{U}^+$ from Definition 1.3 also matches the Hall algebra of a genus g curve over the finite field $\mathbb{F}_{q^{-1}}$ ([**10**]).

1.3. To complete the picture of Subsection 1.1, one would like an explicit presentation of the defining relations of $\mathbf{U}^\pm$, i.e. a collection of generators for the two-sided ideals $K^\pm$. In the setting of Definition 1.2 for $\mathfrak{g}$ of finite type, such a system of generators is given by the Drinfeld-Serre relations for all $i \neq j$ in I. However, for arbitrary symmetric Cartan matrices, one needs the more general expressions constructed in [**8**, formula (1.15)] for any $i \neq j$ in I and integers $k, l \geq 0$ such that

$$k + l = -d_{ij} \tag{1.18}$$

Similarly, in the setting of Definition 1.3, the ideals $K^\pm$ are generated by the cubic expressions defined in [**10**, formula (1.6)] for every arrow of the doubled quiver. Rather than present the aforementioned expressions explicitly, we will invoke the following principle that follows from the non-degeneracy of the pairings (1.6)–(1.7)

(1.19) **elements of the ideals $K^\pm \subset \widetilde{\mathbf{U}}^\pm$ are dual to**

linear conditions cutting out the inclusions $\mathcal{S}^\mp \subset \mathcal{V}^\mp$

With this in mind, we will find it more informative to present the latter linear conditions. In the setting of Definition 1.2, for any $i \neq j$ in I and k, l as in (1.18), Laurent polynomials $R \in \mathcal{S}^\mp$ must satisfy the condition

$$R\Big|^{z_{i1} = xq^k, z_{i2} = xq^{k-2}, \dots, z_{i,k+1} = xq^{-k}}_{z_{j1} = xq^l, z_{j2} = xq^{l-2}, \dots, z_{j,l+1} = xq^{-l}} = 0 \tag{1.20}$$

In the setting of Definition 1.3, Laurent polynomials $R \in \mathcal{S}^\mp$ satisfy the condition

$$R\Big|_{z_{ia} = \frac{qz_{jb}}{t_e} = qz_{ic}} = R\Big|_{z_{ja} = t_e z_{ib} = qz_{jc}} = 0 \tag{1.21}$$

[1]Recall that this means that $\{d_{ij}\}_{i,j \in I}$ are integers such that

$$\begin{cases} d_{ij} = d_{ji} \leq 0 & \text{if } i \neq j \\ d_{ii} = 2 & \text{for all } i \in I \end{cases}$$

for any arrow e from i to j, and for any $a \neq c$ (and moreover $a \neq b \neq c$ if $i = j$). Comparing (1.20) to (1.21) makes it apparent that the linear conditions that cut out $\mathcal{S}^{\mp} \subset \mathcal{V}^{\mp}$ (and dually, the generators of the ideals $K^{\pm}$) strongly depend on the particular ζ functions that define our quantum loop groups and shuffle algebras.

1.4. The common feature of the rational functions (1.16) and (1.17) is that their numerators are completely split. Therefore, we will henceforth assume that

$$\zeta_{ij}(x) = \frac{\alpha_{ij} x^{s_{ij}}}{(1-x)^{\delta_{ij}}} \prod_{e=1}^{\#_{ij}} (1 - x q_e^{ij}) \tag{1.22}$$

for certain $\alpha_{ij}, q_e^{ij} \in \mathbb{K}^{\times}$, integers s_{ij} and non-negative integers $\#_{ij}$.[2] Recall from Subsection 1.3 that in the special cases studied in Definitions 1.2 and 1.3, the inclusion $\mathcal{S}^{\pm} \subset \mathcal{V}^{\pm}$ can be presented as the intersection of ideals of Laurent polynomials satisfying conditions such as (1.20) and (1.21), respectively. Our next main Theorem shows that this kind of behavior holds in general. For any $\boldsymbol{n} \in \mathbb{N}^I$, we will write $\mathcal{V}_{\pm \boldsymbol{n}}$ for the $\boldsymbol{n}$-th direct summand of (1.4), and set $\mathcal{S}_{\pm \boldsymbol{n}} = \mathcal{S}^{\pm} \cap \mathcal{V}_{\pm \boldsymbol{n}}$.

THEOREM 1.4. *For any $\boldsymbol{n} = (n_i)_{i \in I} \in \mathbb{N}^I$ with $n = \sum_{i \in I} n_i$, and any*

$$p = (p_{ia})_{i \in I, a \in \{1, \dots, n_i\}} \in (\mathbb{K}^{\times})^n \Big/ \mathbb{K}^{\times} \tag{1.23}$$

there exists a homogeneous ideal

$$\mathcal{I}_{p, \boldsymbol{n}} \subseteq \mathbb{K}[z_{i1}, z_{i1}^{-1}, \dots, z_{in_i}, z_{in_i}^{-1}]_{i \in I} \tag{1.24}$$

supported on the one-dimensional locus $(z_{ia})_{i \in I, a \in \{1, \dots, n_i\}} \in p\mathbb{K}^{\times}$, such that

$$\mathcal{S}_{\pm \boldsymbol{n}} = \mathcal{V}_{\pm \boldsymbol{n}} \bigcap_{\substack{\boldsymbol{0} \leq \boldsymbol{k} \leq \boldsymbol{n} \\ p \in (\mathbb{K}^{\times})^k / \mathbb{K}^{\times}}} \mathcal{I}_{p, \boldsymbol{k}}^{(\boldsymbol{n})} \tag{1.25}$$

(see (3.28) and the discussion preceding it for the notation in the right-hand side).

1.5. According to the principle (1.19), dualizing Theorem 1.4 will yield a collection of generators for the ideals $K^{\pm}$ (and thus complete the generators-and-relations presentation of the quantum loop groups $\mathbf{U}$). Let us write

$$\widetilde{\mathbf{U}}^{\pm} = \bigoplus_{\boldsymbol{n} \in \mathbb{N}^I} \widetilde{\mathbf{U}}_{\pm \boldsymbol{n}} = \bigoplus_{(\boldsymbol{n}, d) \in \mathbb{N}^I \times \mathbb{Z}} \widetilde{\mathbf{U}}_{\pm \boldsymbol{n}, \pm d}$$

for the components with respect to the grading defined by $\deg e_{i,d} = (\boldsymbol{\varsigma}^i, d)$, $\deg f_{i,d} = (-\boldsymbol{\varsigma}^i, d)$ for all $i \in I$ and $d \in \mathbb{Z}$, where $\boldsymbol{\varsigma}^i = \underbrace{(0, \dots, 0, 1, 0, \dots, 0)}_{1 \text{ on } i\text{-th position}} \in \mathbb{N}^I$.

THEOREM 1.5. *For any $(\boldsymbol{n}, d) \in \mathbb{N}^I \times \mathbb{Z}$ and $p \in (\mathbb{K}^{\times})^n / \mathbb{K}^{\times}$, there are finite sets*

$$W_{p, \pm \boldsymbol{n}, \pm d} \subset \widetilde{\mathbf{U}}_{\pm \boldsymbol{n}, \pm d} \tag{1.26}$$

which generate the two-sided ideal $K^{\pm}$ as p and $(\boldsymbol{n}, d)$ vary. Thus, we have

$$\mathbf{U}^{\pm} = \widetilde{\mathbf{U}}^{\pm} \Big/ \left(W_{p, \pm \boldsymbol{n}, \pm d} \right)_{p \in (\mathbb{K}^{\times})^n / \mathbb{K}^{\times}, \boldsymbol{n} \in \mathbb{N}^I, d \in \mathbb{Z}} \tag{1.27}$$

[2] This is not a major restriction, as completely general rational functions $\zeta_{ij}(x)$ can be written as in (1.22), at the cost of replacing the ground field $\mathbb{K}$ by an algebraic field extension.

As we will show in the proof of Proposition 3.9, one can compute the ideals $\mathcal{I}_{p,\boldsymbol{n}}$ recursively in $\boldsymbol{n}$. In more detail, we emphasize the fact that each ideal (1.24) is homogeneous and supported on the one-dimensional locus $p\mathbb{K}^\times \cong \operatorname{Spec}\mathbb{K}[x^{\pm 1}]$. Therefore, in every homogeneous degree $d \in \mathbb{Z}$, the inclusion

$$\mathcal{I}_{p,\boldsymbol{n},d} \subseteq \mathbb{K}[z_{i1}, z_{i1}^{-1} \dots, z_{in_i}, z_{in_i}^{-1}]_{i\in I}^{\text{homogeneous degree d}}$$

is cut out as a $\mathbb{K}$-vector space by finitely many linear conditions. On $\mathcal{V}_{\pm\boldsymbol{n}}$, these linear conditions are realized by pairing with finitely many elements of $\widetilde{\mathbf{U}}_{\mp\boldsymbol{n},\mp d}$ under the pairings (1.6) (1.7), as shown in the proof of Proposition 3.12. The aforementioned elements are, by definition, the elements of the finite sets (1.26).

1.6. Because of the arbitrariness of the point (1.23), it might seem like one must consider a great multitude of ideals (1.24) and elements (1.26). However, this is not the case: as we will see in Subsection 3.2, for every $\boldsymbol{n} \in \mathbb{N}^I$ and every $p = (p_{ia})_{i\in I, a\in\{1,\dots,n_i\}}$ the corresponding ideals/elements only depend on the collection

$$\left\{ q_e^{ij} \middle| q_e^{ij} = \frac{p_{jb}}{p_{ia}} \text{ for some } a, b \right\}_{i,j\in I, e\in\{1,\dots,\#_{ij}\}} \subset \left\{ \text{roots of the } \zeta_{ij}\text{'s} \right\}$$

As there are only finitely many choices of such collections, one only needs to consider finitely many ideals/elements in (1.24)/(1.26) for every $\boldsymbol{n} \in \mathbb{N}^I$. Moreover, the ideal (1.24) is non-trivial (respectively the set (1.26) is non-empty) only if the point p is a wheel, i.e. there exists a cycle $(i_1, a_1), \dots, (i_k, a_k), (i_{k+1}, a_{k+1}) = (i_1, a_1)$ s.t.

$$\frac{p_{i_{\bullet+1}a_{\bullet+1}}}{p_{i_\bullet a_\bullet}} \in \left\{ q_e^{i_\bullet i_{\bullet+1}} \right\}_{e\in\{1,\dots,\#_{i_\bullet i_{\bullet+1}}\}}, \qquad \forall \bullet \in \{1, \dots, k\} \tag{1.28}$$

For such a wheel, the requirement that a Laurent polynomial $R(z_{ia})_{i\in I, a\in\{1,\dots,n_i\}}$ lies in $\mathcal{I}_{p,\boldsymbol{n}}$ is a linear condition on the derivatives of R at the points of the form

$$(z_{ia})_{i\in I, a\in\{1,\dots,n_i\}} \in p\mathbb{K}^\times$$

thus generalizing the **wheel conditions** discovered in [**2**] in the setting of quantum loop groups of finite and affine type (which involved certain special points p). In special cases (such as (1.20), (1.21) or (1.31) below), we see that wheel conditions are rather simple vanishing conditions, but in general they can be quite complicated.

1.7. Although we do not explicitly describe the wheel conditions for arbitrary ζ functions, in Subsection 3.2 we will explain that they are built out of the special cases when

$$\zeta_{ij}(x) = (1-x)^{\#_{ij} - \delta_{ij}} \tag{1.29}$$

for various non-negative integers $\#_{ij}$. In this setting, the shuffle algebra $\mathcal{V}^+$ is none other than the K-theoretic Hall algebra with 0 potential associated to the quiver Q with vertex set I and $\#_{ij}$ arrows from the vertex i to the vertex j, for all $i, j \in I$ (see [**5, 12**]). These algebras are very actively studied in connection with Donaldson-Thomas invariants for Calabi-Yau categories, and the upshot of our theorems is that the "spherical" subalgebra $\mathring{\mathcal{S}}^+ = \mathcal{S}^+ \subset \mathcal{V}^+$ admits a presentation as (half of) a quantum loop group.

As for K-theoretic Hall algebras equivariant with respect the torus

$$T = \prod_{i,j\in I} (\mathbb{C}^*)^{\#_{ij}}$$

we let q_e^{ij} denote the standard character corresponding to the e-th factor of $\mathbb{C}^*$ in the i, j term of the product above. Then let us work over the ring

$$\mathbb{L} = \mathbb{Q}\left[\left(q_e^{ij}\right)^{\pm 1}\right]_{i,j \in I, e \in \{1,\dots,\#_{ij}\}}$$

and define the $\mathbb{L}$-algebra

$$\mathcal{V}^+_{\text{int}} = \bigoplus_{\boldsymbol{n}=(n_i)_{i\in I} \in \mathbb{N}^I} \mathbb{L}[z_{i1}, z_{i1}^{-1}, \dots, z_{in_i}, z_{in_i}^{-1}]^{\text{sym}}_{i\in I}$$

endowed with the shuffle multiplication (2.5) associated to the rational functions (1.22).[3] Similarly, one lets

$$\mathcal{S}^+_{\text{int}} \subseteq \mathcal{V}^+_{\text{int}} \tag{1.30}$$

denote the $\mathbb{L}$-subalgebra generated by $\{z_{i1}^d\}_{i\in I, d\in\mathbb{Z}}$. The methods in the present paper allow to describe the fibers of the subalgebra (1.30) above various $\overline{\mathbb{Q}}$-points of $\operatorname{Spec}\mathbb{L}$, which corresponds to various specializations of the arrow parameters q_e^{ij}. If the specialization is generic, there are no wheels (1.28) and thus

$$\mathcal{S}^+_{\text{generic}} = \mathcal{V}^+_{\text{generic}}$$

(dually, the quantum loop group associated to generic parameters is $\mathbf{U}^- = \widetilde{\mathbf{U}}^-$). At the opposite extreme, if the arrow parameters are all specialized to 1, we recover the situation of (1.29), which as we will see is the most interesting and important one.

1.8. Our construction also applies to quiver quantum toroidal algebras. These are trigonometric versions (introduced in [**3, 11**]) of the quiver Yangians (introduced in [**6**], see also [**13**] for a related mathematical construction) that act on the vector spaces of BPS states for non-compact toric Calabi-Yau threefolds X. More specifically, one associates to such an X a quiver Q endowed with arrow parameters

$$\{q_e^{ij}\}_{i,j\in I, e\in\{1,\dots,\#_{ij}\}} \in \text{Rep}_{\mathbb{C}^*\times\mathbb{C}^*}$$

See [**11**, Section B] for the precise construction; the two-dimensional torus $\mathbb{C}^* \times \mathbb{C}^*$ should be interpreted as the kernel of the Calabi-Yau form. When the quiver is symmetric (i.e. $\#_{ij} = \#_{ji}$, which is known as “non-chiral” in the physics literature), the corresponding **quiver quantum toroidal algebra** is none other than our

$$\widetilde{\mathbf{U}} = \widetilde{\mathbf{U}}^+ \otimes \widetilde{\mathbf{U}}^0 \otimes \widetilde{\mathbf{U}}^-$$

defined with respect to the ζ functions (1.22), for certain α_{ij} and s_{ij}.[4] One of the main features of these quiver quantum toroidal algebras is that they act on the vector spaces of BPS crystal configurations

$$M = \bigoplus_{\Lambda \text{ 3d crystal configuration}} \mathbb{K} \cdot |\Lambda\rangle$$

[3]One can define α_{ij} and s_{ij} arbitrarily in (1.22), and various choices correspond to various twists of the K-theoretic Hall product; the customary choices are $\alpha_{ij} = 1$ and $s_{ij} = 0$ for all i, j.

[4]For non-symmetric (i.e. chiral) quivers, our construction describes the positive/negative parts of quiver quantum toroidal algebras, but the Hopf algebra structure is not well-defined.

where $\mathbb{K} = \text{Frac}(\text{Rep}_{\mathbb{C}^* \times \mathbb{C}^*})$ is the ground field in the present setup (see [**11**, Section 5] for a review of 3d crystal configurations, which are generalizations of plane partitions). As noted in [**3**, Section 5], the action

$$\widetilde{\mathbf{U}}^{\pm} \curvearrowright M$$

factors through the homomorphism $\widetilde{\mathbf{U}}^{\pm} \xrightarrow{\widetilde{\Upsilon}^{\pm}} \mathcal{S}^{\pm} \overset{(1.11)}{\cong} \mathbf{U}^{\pm}$. Thus, there is an action

$$\mathbf{U} \curvearrowright M$$

where $\mathbf{U}$ is the Drinfeld double (1.15). In other words, all the elements (1.26) act by 0 in the representation M, and so the quotient $\mathbf{U}$ plays the role of a "reduced" version of the quiver quantum toroidal algebra $\widetilde{\mathbf{U}}$. In [**7**, formula (1.13)], we will give a generators-and-relations presentation of the quantum loop group $\mathbf{U}$ in the setting at hand, by giving an explicit set of generators for the ideals $K^{\pm}$. For now let us mention the dual statement, in accordance with the principle (1.19): a Laurent polynomial $R \in \mathcal{V}^{\mp}$ lies in $\mathcal{S}^{\mp}$ if and only if satisfies the conditions

$$R\Big|_{z_a = z_{a-1} q_{e_a}^{i_a i_{a-1}},\ \forall a \in \{1,\dots,k\}} = 0 \tag{1.31}$$

for any face $F = \{i_0, i_1, i_2, \dots, i_{k-1}, i_k = i_0\}$ of the quiver Q (in the context of quiver quantum toroidal algebras, the quiver Q is naturally drawn on the torus with polygonal faces), whose boundary edges are denoted by $e_1, \dots, e_k$.

1.9. The outline of the present paper is the following.

- In Section 2, we prove Theorem 1.1 for arbitrary ζ functions (1.1)
- In Section 3, we prove Theorems 1.4 and 1.5 for factored ζ functions, as in (1.22)

2. Shuffle algebras

We will develop the basic theory of shuffle algebras, in the trigonometric setting studied by [**1**, **2**], which depend on a choice of rational functions $\zeta_{ij}(x)$. In the present Section we will deal mostly with generalities that apply to arbitrary ζ_{ij}'s.

2.1. We will work over a base field $\mathbb{K}$ of characteristic 0. Consider a finite set I, and let us fix a non-zero Laurent polynomial for any $i \neq j$ in I

$$\zeta_{ij}(x) \in \mathbb{K}[x^{\pm 1}] \tag{2.1}$$

and a non-zero rational function with at most a simple pole at $x = 1$ for any $i \in I$

$$\zeta_{ii}(x) \in \frac{\mathbb{K}[x^{\pm 1}]}{1 - x} \tag{2.2}$$

DEFINITION 2.1. The (positive part of the) **quadratic quantum loop group** $\widetilde{\mathbf{U}}^{+}$ associated to the datum $\{\zeta_{ij}\}_{i,j \in I}$ is the $\mathbb{K}$-algebra generated by symbols

$$\{e_{i,d}\}_{i \in I, d \in \mathbb{Z}}$$

modulo the following relations for all $i, j \in I$

$$e_i(z) e_j(w) \zeta_{ji}\left(\frac{w}{z}\right) = e_j(w) e_i(z) \zeta_{ij}\left(\frac{z}{w}\right) \tag{2.3}$$

Above and henceforth, we use the notation:

$$e_i(z) = \sum_{d \in \mathbb{Z}} \frac{e_{i,d}}{z^d}$$

for all $i \in I$, and relation (2.3) is interpreted as an infinite collection of relations obtained by equating the coefficients of all $\{z^a w^b\}_{a,b\in\mathbb{Z}}$ in the left and right-hand sides (if $i = j$, one clears the denominators $z - w$ from (2.3) before equating coefficients).

The algebra $\widetilde{\mathbf{U}}^+$ is graded by $\mathbb{N}^I \times \mathbb{Z}$ (in this paper, $\mathbb{N}$ is assumed to contain 0), via

$$\deg e_{i,d} = (\boldsymbol{\varsigma}^i, d)$$

for all $i \in I$ and $d \in \mathbb{Z}$. Above and throughout the present paper, $\boldsymbol{\varsigma}^i \in \mathbb{N}^I$ denotes the I-tuple of integers with a 1 on position i, and 0 everywhere else. We will write

$$\widetilde{\mathbf{U}}^+ = \bigoplus_{\boldsymbol{n}\in\mathbb{N}^I} \widetilde{\mathbf{U}}_{\boldsymbol{n}} = \bigoplus_{(\boldsymbol{n},d)\in\mathbb{N}^I\times\mathbb{Z}} \widetilde{\mathbf{U}}_{\boldsymbol{n},d}$$

for the graded components. We have the **shift automorphism**

$$\widetilde{\mathbf{U}}^+ \xrightarrow{\tau_{\boldsymbol{k}}} \widetilde{\mathbf{U}}^+, \qquad e_{i,d} \mapsto e_{i,d+k_i} \tag{2.4}$$

for any $\boldsymbol{k} = (k_i)_{i\in I} \in \mathbb{Z}^I$.

2.2. Let us consider an infinite collection of variables $z_{i1}, z_{i2}, \dots$ for all $i \in I$. For any $\boldsymbol{n} = (n_i)_{i\in I} \in \mathbb{N}^I$, we will write $|\boldsymbol{n}| = \sum_{i\in I} n_i$ and $\boldsymbol{n}! = \prod_{i\in I} n_i!$. The following construction is a straightforward generalization of that of [**1, 2**].

DEFINITION 2.2. The **big shuffle algebra** associated to the datum $\{\zeta_{ij}\}_{i,j\in I}$ is

$$\mathcal{V}^+ = \bigoplus_{\boldsymbol{n}\in\mathbb{N}^I} \mathbb{K}[z_{i1}^{\pm 1}, \dots, z_{in_i}^{\pm 1}]^{\text{sym}}_{i\in I}$$

endowed with the multiplication

$$R(\dots, z_{i1}, \dots, z_{in_i}, \dots) * R'(\dots, z_{i1}, \dots, z_{in'_i}, \dots) = \tag{2.5}$$

$$\text{Sym}\left[\frac{R(\dots, z_{i1}, \dots, z_{in_i}, \dots)R'(\dots, z_{i,n_i+1}, \dots, z_{i,n_i+n'_i}, \dots)}{\boldsymbol{n}!\boldsymbol{n}'!} \prod_{\substack{1\le a\le n_i \\ n_j<b\le n_j+n'_j}}^{i,j\in I} \zeta_{ij}\left(\frac{z_{ia}}{z_{jb}}\right)\right]$$

Above and henceforth, "sym" (resp. "Sym") denotes symmetric functions (resp. symmetrization) with respect to the variables $z_{i1}, z_{i2}, \dots$ for each $i \in I$ separately.[5]

Note that the algebra $\mathcal{V}^+$ is graded by $\mathbb{N}^I \times \mathbb{Z}$, via

$$\deg R(\dots, z_{i1}, \dots, z_{in_i}, \dots) = (\boldsymbol{n}, \text{hom deg } R)$$

where "hom deg R" denotes the homogeneous degree of R in all its variables. Let

$$\mathcal{V}^+ = \bigoplus_{\boldsymbol{n}\in\mathbb{N}^I} \mathcal{V}_{\boldsymbol{n}} = \bigoplus_{(\boldsymbol{n},d)\in\mathbb{N}^I\times\mathbb{Z}} \mathcal{V}_{\boldsymbol{n},d}$$

denote the graded pieces. The analogue of the shift automorphism (2.4) is

$$\mathcal{V}^+ \xrightarrow{\tau_{\boldsymbol{k}}} \mathcal{V}^+, \qquad R(\dots, z_{ia}, \dots) \mapsto R(\dots, z_{ia}, \dots) \prod_{i\in I, a\ge 1} z_{ia}^{k_i} \tag{2.6}$$

for any $\boldsymbol{k} = (k_i)_{i\in I} \in \mathbb{Z}^I$.

[5] Although the ζ functions might seem to contribute simple poles at $z_{ia} - z_{ib}$ for $a \neq b$ to the right-hand side of (2.5), these poles disappear when taking the symmetrization (the poles in question can only have even order in any symmetric rational function).

2.3. We will also encounter the opposite algebra

$$\widetilde{\mathbf{U}}^- = \mathbb{K}\Big\langle f_{i,d} \Big\rangle_{i \in I, d \in \mathbb{Z}} \Big/ \text{relation (2.7)}$$

with the defining relation

$$f_i(z) f_j(w) \zeta_{ij}\left(\frac{z}{w}\right) = f_j(w) f_i(z) \zeta_{ji}\left(\frac{w}{z}\right) \tag{2.7}$$

for all $i, j \in I$. Also let $\mathcal{V}^- = \mathcal{V}^+$ as a vector space, but in the multiplication (2.5) we replace

$$\zeta_{ij}\left(\frac{z_{ia}}{z_{jb}}\right) \quad \text{with} \quad \zeta_{ji}\left(\frac{z_{jb}}{z_{ia}}\right)$$

The algebra $\widetilde{\mathbf{U}}^-$ is $(-\mathbb{N}^I) \times \mathbb{Z}$ graded, via

$$\deg f_{i,d} = (-\boldsymbol{\varsigma}^i, d)$$

for all $i \in I$, $d \in \mathbb{Z}$, while $\mathcal{V}^-$ is $(-\mathbb{N}^I) \times \mathbb{Z}$ graded, via

$$\deg R(\dots, z_{i1}, \dots, z_{in_i}, \dots) = (-\boldsymbol{n}, \text{hom deg } R)$$

for all $R \in \mathcal{V}^-$. We will denote the graded pieces of $\widetilde{\mathbf{U}}^-$ and $\mathcal{V}^-$ by

$$\widetilde{\mathbf{U}}^- = \bigoplus_{\boldsymbol{n} \in \mathbb{N}^I} \widetilde{\mathbf{U}}_{-\boldsymbol{n}} = \bigoplus_{(\boldsymbol{n},d) \in \mathbb{N}^I \times \mathbb{Z}} \widetilde{\mathbf{U}}_{-\boldsymbol{n},d}$$

$$\mathcal{V}^- = \bigoplus_{\boldsymbol{n} \in \mathbb{N}^I} \mathcal{V}_{-\boldsymbol{n}} = \bigoplus_{(\boldsymbol{n},d) \in \mathbb{N}^I \times \mathbb{Z}} \mathcal{V}_{-\boldsymbol{n},d}$$

The shift automorphisms on the negative algebras are defined as

$$\widetilde{\mathbf{U}}^- \xrightarrow{\tau_{\boldsymbol{k}}} \widetilde{\mathbf{U}}^-, \qquad e_{i,d} \mapsto e_{i,d-k_i} \tag{2.8}$$

$$\mathcal{V}^- \xrightarrow{\tau_{\boldsymbol{k}}} \mathcal{V}^-, \qquad R(\dots, z_{ia}, \dots) \mapsto R(\dots, z_{ia}, \dots) \prod_{i \in I, a \geq 1} z_{ia}^{-k_i} \tag{2.9}$$

for any $\boldsymbol{k} = (k_i)_{i \in I} \in \mathbb{Z}^I$.

2.4. There are homomorphisms of $(\pm\mathbb{N}^I) \times \mathbb{Z}$ graded $\mathbb{K}$-algebras

$$\widetilde{\mathbf{U}}^\pm \xrightarrow{\widetilde{\Upsilon}^\pm} \mathcal{V}^\pm, \qquad e_{i,d}, f_{i,d} \mapsto z_{i1}^d \tag{2.10}$$

Indeed, this claim only entails checking the fact that relations (2.3) are respected by the shuffle product (2.5), which is straightforward. The map (2.10) is neither injective nor surjective, and one of the main goals of the present paper is to describe

$$\mathring{\mathcal{S}}^\pm = \operatorname{Im} \widetilde{\Upsilon}^\pm$$

(i.e. $\mathring{\mathcal{S}}^\pm$ is the $\mathbb{K}$-subalgebra of $\mathcal{V}^\pm$ generated by $\{z_{i1}^d\}_{i \in I, d \in \mathbb{Z}}$) and

$$K^\pm = \operatorname{Ker} \widetilde{\Upsilon}^\pm$$

as a two-sided ideal of $\widetilde{\mathbf{U}}^\pm$.

2.5. A key role in our study of the algebras $\widetilde{\mathbf{U}}^{\pm}$ and $\mathcal{V}^{\pm}$ is played by certain bilinear pairings that we will introduce shortly. Let us consider the following notation for all rational functions $F(z_1, \dots, z_n)$. If $Dz_a = \frac{dz_a}{2\pi i z_a}$, then we will write

$$\int_{|z_1| \gg \dots \gg |z_n|} F(z_1, \dots, z_n) \prod_{a=1}^{n} Dz_a \tag{2.11}$$

for the constant term in the expansion of F as a power series in

$$\frac{z_2}{z_1}, \dots, \frac{z_n}{z_{n-1}}$$

The notation in (2.11) is motivated by the fact that if $\mathbb{K} = \mathbb{C}$, one could compute this constant term as a contour integral (with the contours being concentric circles, situated very far from each other compared to the absolute values of the coefficients of F). We define $\int_{|z_1| \ll \dots \ll |z_n|} F(z_1, \dots, z_n) \prod_{a=1}^{n} Dz_a$ by analogy with (2.11).

DEFINITION 2.3. There exist bilinear pairings

$$\widetilde{\mathbf{U}}^{+} \otimes \mathcal{V}^{-} \xrightarrow{\langle \cdot, \cdot \rangle} \mathbb{K} \tag{2.12}$$

$$\mathcal{V}^{+} \otimes \widetilde{\mathbf{U}}^{-} \xrightarrow{\langle \cdot, \cdot \rangle} \mathbb{K} \tag{2.13}$$

given for all $R^{\pm} \in \mathcal{V}_{\pm \boldsymbol{n}}$ and all $i_1, \dots, i_n \in I$, $d_1, \dots, d_n \in \mathbb{Z}$ by

(2.14)

$$\Big\langle e_{i_1,d_1} \cdots e_{i_n,d_n}, R^- \Big\rangle = \int_{|z_1| \gg \dots \gg |z_n|} \frac{z_1^{d_1} \dots z_n^{d_n} R^-(z_1, \dots, z_n)}{\prod_{1 \le a < b \le n} \zeta_{i_b i_a}\left(\frac{z_b}{z_a}\right)} \prod_{a=1}^{n} Dz_a$$

(2.15)

$$\Big\langle R^+, f_{i_1,-d_1} \cdots f_{i_n,-d_n} \Big\rangle = \int_{|z_1| \ll \dots \ll |z_n|} \frac{z_1^{-d_1} \dots z_n^{-d_n} R^+(z_1, \dots, z_n)}{\prod_{1 \le a < b \le n} \zeta_{i_a i_b}\left(\frac{z_a}{z_b}\right)} \prod_{a=1}^{n} Dz_a$$

if $\varsigma^{i_1} + \dots + \varsigma^{i_n} = \boldsymbol{n}$, and 0 otherwise.

In the right-hand sides of (2.14) and (2.15), we implicitly identify

$$z_a \quad \text{with} \quad z_{i_a \bullet_a}, \qquad \forall a \in \{1, \dots, n\} \tag{2.16}$$

where $\bullet_a \in \{1, 2, \dots, n_{i_a}\}$ may be chosen arbitrarily (however, we require $\bullet_a \neq \bullet_b$ if $a \neq b$ and $i_a = i_b$) due to the symmetry of $R^{\pm}$. We call (2.16) a **relabeling** of the variables of $R^{\pm}(z_{i1}, \dots, z_{in_i})_{i \in I}$ in accordance with $i_1, \dots, i_n$.

2.6. The pairings (2.12)–(2.13) are non-degenerate in the $\mathcal{V}^{\pm}$ argument, i.e.

$$\Big\langle \widetilde{\mathbf{U}}^{+}, R^- \Big\rangle = 0 \quad \Rightarrow \quad R^- = 0 \tag{2.17}$$

$$\Big\langle R^+, \widetilde{\mathbf{U}}^{-} \Big\rangle = 0 \quad \Rightarrow \quad R^+ = 0 \tag{2.18}$$

for any $R^{\pm} \in \mathcal{V}^{\pm}$. This is simply because any rational function whose power series expansion (with respect to any order of variables) vanishes must be identically 0.

DEFINITION 2.4. Let $\mathcal{S}^{\mp} \subset \mathcal{V}^{\mp}$ denote the dual of $K^{\pm} = \text{Ker}\,\widetilde{\Upsilon}^{\pm}$ under the pairings (2.12)–(2.13), respectively, i.e.

$$R^- \in \mathcal{S}^- \quad \Leftrightarrow \quad \left\langle K^+, R^- \right\rangle = 0 \tag{2.19}$$

$$R^+ \in \mathcal{S}^+ \quad \Leftrightarrow \quad \left\langle R^+, K^- \right\rangle = 0 \tag{2.20}$$

We will call $\mathcal{S}^{\pm}$ the positive/negative **shuffle algebra**.[6]

As a consequence of Definition 2.4, the pairings (2.12)–(2.13) descend to pairings

$$\mathring{\mathcal{S}}^+ \otimes \mathcal{S}^- \xrightarrow{\langle \cdot, \cdot \rangle} \mathbb{K} \tag{2.21}$$

$$\mathcal{S}^+ \otimes \mathring{\mathcal{S}}^- \xrightarrow{\langle \cdot, \cdot \rangle} \mathbb{K} \tag{2.22}$$

which are non-degenerate in the $\mathcal{S}^{\pm}$ argument. The main result of the present Section is the following.

THEOREM 2.5. *We have $\mathcal{S}^{\pm} = \mathring{\mathcal{S}}^{\pm}$, and the pairings* (2.21) *and* (2.22) *coincide.*

2.7. Before we delve into proving Theorem 2.5, let us note that it states that

$$R \in \mathring{\mathcal{S}}^- \qquad \Leftrightarrow \qquad R \in \mathcal{S}^-$$

for an arbitrary $R \in \mathcal{V}^-$. The first $\in$ above is an issue of the Laurent polynomial R lying in a certain ideal, while the second $\in$ above involves checking that R is annihilated by a collection of linear maps. Explicitly, if $\boldsymbol{n} \in \mathbb{N}^I$ denotes the (negative of the) degree of R, let us choose a linear spanning set

$$\sum_{\substack{i_1,\dots,i_n \in I \\ \varsigma^{i_1}+\dots+\varsigma^{i_n}=\boldsymbol{n}}} \Big[p_{i_1,\dots,i_n}(z_1,\dots,z_n) e_{i_1}(z_1)\dots e_{i_n}(z_n)\Big]_{\text{ct}} \in K^+ \cap \widetilde{\mathbf{U}}_{\boldsymbol{n}} \tag{2.23}$$

(where $p_{i_1,\dots,i_n}$ are Laurent polynomials, and $[\dots]_{\text{ct}}$ denotes the constant term in the variables $z_1, \dots, z_n$). If we relabel the variables according to (2.16), then (2.23) is equivalent to the symmetric Laurent polynomial identity

$$\sum_{\substack{i_1,\dots,i_n \in I \\ \varsigma^{i_1}+\dots+\varsigma^{i_n}=\boldsymbol{n}}} \text{Sym}\left[p_{i_1,\dots,i_n}(z_{i_1\bullet_1},\dots,z_{i_n\bullet_n}) \prod_{1\leq a<b\leq n} \zeta_{i_a i_b}\left(\frac{z_{i_a\bullet_a}}{z_{i_b\bullet_b}}\right)\right] = 0 \tag{2.24}$$

Then $R \in \mathcal{V}_{-\boldsymbol{n}}$ lies in $\mathcal{S}_{-\boldsymbol{n}} = \mathcal{V}_{-\boldsymbol{n}} \cap \mathcal{S}^-$ if and only if for any $\{p_{i_1,\dots,i_n}\}_{i_1,\dots,i_n\in I}$ as in (2.23), we have

$$0 = \left\langle \sum_{\substack{i_1,\dots,i_n \in I \\ \varsigma^{i_1}+\dots+\varsigma^{i_n}=\boldsymbol{n}}} \Big[p_{i_1,\dots,i_n}(z_1,\dots,z_n) e_{i_1}(z_1)\dots e_{i_n}(z_n)\Big]_{\text{ct}}, R \right\rangle$$

$$= \sum_{\substack{i_1,\dots,i_n \in I \\ \varsigma^{i_1}+\dots+\varsigma^{i_n}=\boldsymbol{n}}} \int_{|z_1|\gg\dots\gg|z_n|} \frac{p_{i_1,\dots,i_n}(z_1,\dots,z_n) R(z_1,\dots,z_n)}{\prod_{1\leq a<b\leq n} \zeta_{i_b i_a}\left(\frac{z_b}{z_a}\right)} \prod_{a=1}^n Dz_a \tag{2.25}$$

Formula (2.25) yields a collection of equations, linear in R, which determine the $\mathbb{K}$-vector subspace $\mathcal{S}_{-\boldsymbol{n}} \subset \mathcal{V}_{-\boldsymbol{n}}$ for any given $\boldsymbol{n} \in \mathbb{N}^I$. The situation of the $\mathbb{K}$-vector subspace $\mathcal{S}_{\boldsymbol{n}} \subset \mathcal{V}_{\boldsymbol{n}}$ is treated analogously, with f's instead of e's.

[6]The fact that $\mathcal{S}^{\pm}$ is closed under shuffle product will be established as part of Theorem 2.5.

EXAMPLE 2.6. Assume $n = 3$, $i_1, i_2, i_3 \in I$ are distinct and

$$\zeta_{i_1 i_2}(x) = \zeta_{i_2 i_3}(x) = \zeta_{i_3 i_1}(x) = 1$$
$$\zeta_{i_2 i_1}(x) = \zeta_{i_3 i_2}(x) = \zeta_{i_1 i_3}(x) = 1 - x$$

Then every coefficient of the formal series

$$e_{i_1}(z_1)e_{i_2}(z_2)e_{i_3}(z_3) + \frac{z_1}{z_3} \cdot e_{i_2}(z_2)e_{i_3}(z_3)e_{i_1}(z_1) + \frac{z_2}{z_3} \cdot e_{i_3}(z_3)e_{i_1}(z_1)e_{i_2}(z_2) \tag{2.26}$$

lies in K^+, and property (2.25) is equivalent to

$$\delta\left(\frac{z_1}{z_2}\right)\delta\left(\frac{z_1}{z_3}\right) R(z_1, z_2, z_3) = 0 \quad \Leftrightarrow \quad R(x, x, x) = 0 \tag{2.27}$$

where $\delta(x) = \sum_{d \in \mathbb{Z}} x^d$ is the formal delta series.

2.8. Given $\boldsymbol{n} \in \mathbb{N}^I$, let us now fix $i_1, \dots, i_n \in I$ such that $\boldsymbol{n} = \varsigma^{i_1} + \dots + \varsigma^{i_n}$. Let

$$S(\boldsymbol{n}) \subset S(n)$$

be the subset of permutations σ such that $\sigma(a) < \sigma(b)$ if $a < b$ and $i_a = i_b$. In the present Subsection, we will relabel our variables according to (2.16). It is easy to see that (2.25) is equivalent to

$$0 = \sum_{\sigma \in S(\boldsymbol{n})} \int_{|z_{\sigma(1)}| \gg \dots \gg |z_{\sigma(n)}|} \frac{p_\sigma(z_{\sigma(1)}, \dots, z_{\sigma(n)}) R(z_1, \dots, z_n)}{\prod_{1 \leq a < b \leq n} \zeta_{i_{\sigma(b)} i_{\sigma(a)}}\left(\frac{z_{\sigma(b)}}{z_{\sigma(a)}}\right)} \prod_{a=1}^{n} Dz_a \tag{2.28}$$

where we choose to index the Laurent polynomial p by an arbitrary permutation $\sigma \in S(\boldsymbol{n})$ which permutes the fixed indices $i_1, \dots, i_n$. In the previous Subsection, we showed that $R \in \mathcal{V}_{-\boldsymbol{n}}$ lies in $\mathcal{S}_{-\boldsymbol{n}}$ if and only if (2.28) holds for any collection of Laurent polynomials $\{p_\sigma\}_{\sigma \in S(\boldsymbol{n})}$ that belongs to the kernel of the following map

$$\bigoplus_{\sigma \in S(\boldsymbol{n})} \mathbb{K}[z_1^{\pm 1}, \dots, z_n^{\pm 1}] \xrightarrow{\Phi} \mathbb{K}[z_1^{\pm 1}, \dots, z_n^{\pm 1}]^{\text{sym}} \tag{2.29}$$

$$\Phi\Big(\{p_\sigma\}_{\sigma \in S(\boldsymbol{n})}\Big) = \sum_{\sigma \in S(\boldsymbol{n})} \text{Sym}\left[p_\sigma(z_{\sigma(1)}, \dots, z_{\sigma(n)}) \prod_{1 \leq a < b \leq n} \zeta_{i_{\sigma(a)} i_{\sigma(b)}}\left(\frac{z_{\sigma(a)}}{z_{\sigma(b)}}\right)\right]$$

of $\mathbb{K}[z_1^{\pm 1}, \dots, z_n^{\pm 1}]^{\text{sym}}$-modules (in the present Subsection, "sym" and "Sym" refer to symmetric polynomials with respect to any variables z_a and z_b for which $i_a = i_b$).

PROPOSITION 2.7. *If $\{p_\sigma\}_{\sigma \in S(\boldsymbol{n})}$ lies in the image of the map*

$$\bigoplus_{\sigma \neq \sigma' \in S(\boldsymbol{n})} \mathbb{K}[z_1^{\pm 1}, \dots, z_n^{\pm 1}] \xrightarrow{\Psi} \bigoplus_{\sigma \in S(\boldsymbol{n})} \mathbb{K}[z_1^{\pm 1}, \dots, z_n^{\pm 1}] \tag{2.30}$$

$$\Psi\Big(\{f_{\sigma\sigma'}(z_1, \dots, z_n)\}_{\sigma \neq \sigma' \in S(\boldsymbol{n})}\Big) = \{p_\sigma(z_1, \dots, z_n)\}_{\sigma \in S(\boldsymbol{n})}$$

where each $p_\sigma(z_1, \dots, z_n)$ is given by

$$\sum_{\sigma' \neq \sigma} f_{\sigma\sigma'}(z_{\sigma^{-1}(1)}, \dots, z_{\sigma^{-1}(n)}) \prod_{\sigma'^{-1}(\sigma(a)) > \sigma'^{-1}(\sigma(b))}^{a<b} \widetilde{\zeta}_{i_b i_a}\left(\frac{z_b}{z_a}\right)$$
$$- \sum_{\sigma' \neq \sigma} f_{\sigma'\sigma}(z_{\sigma^{-1}(1)}, \dots, z_{\sigma^{-1}(n)}) \prod_{\sigma'^{-1}(\sigma(a)) > \sigma'^{-1}(\sigma(b))}^{a<b} \widetilde{\zeta}_{i_b i_a}\left(\frac{z_b}{z_a}\right)\left(-\frac{z_a}{z_b}\right)^{\delta_{i_a i_b}}$$

then the right-hand side of (2.28) *is 0 for all* $R \in \mathcal{V}_{-\boldsymbol{n}}$. *In the formulas above,* $\widetilde{\zeta}_{ij} = \zeta_{ij}$ *if* $i \neq j$, *but* $\widetilde{\zeta}_{ii}$ *denotes the Laurent polynomial in the numerator of* (2.2).

It is easy to see that $\Phi \circ \Psi = 0$. Then Proposition 2.7 tells us that $R \in \mathcal{V}_{-\boldsymbol{n}}$ actually lies in $\mathcal{S}_{-\boldsymbol{n}}$ if and only if (2.28) holds for a set of representatives

$$\{p_\sigma\}_{\sigma \in S(\boldsymbol{n})} \in \operatorname{Ker} \Phi \Big/ \operatorname{Im} \Psi$$

The quotient $\operatorname{Ker}\Phi / \operatorname{Im}\Psi$ can be thought of as (a stronger version of) a first Koszul homology module, and it controls the linear relations that cut out $\mathcal{S}_{-\boldsymbol{n}} \subset \mathcal{V}_{-\boldsymbol{n}}$.

PROOF OF PROPOSITION 2.7. It suffices to fix $\sigma \neq \sigma'$. Then the claim of Proposition 2.7 boils down to the fact that

$$\int_{|z_{\sigma(1)}| \gg \dots \gg |z_{\sigma(n)}|} f_{\sigma\sigma'}(z_1, \dots, z_n) R(z_1, \dots, z_n) \frac{\prod_{\sigma'^{-1}(a) > \sigma'^{-1}(b)}^{\sigma^{-1}(a) < \sigma^{-1}(b)} \left(1 - \frac{z_b}{z_a}\right)}{\prod_{\sigma'^{-1}(a) < \sigma'^{-1}(b)}^{\sigma^{-1}(a) < \sigma^{-1}(b)} \widetilde{\zeta}_{i_b i_a}\left(\frac{z_b}{z_a}\right)} \prod_{a=1}^n Dz_i$$

$$= \int_{|z_{\sigma'(1)}| \gg \dots \gg |z_{\sigma'(n)}|} f_{\sigma\sigma'}(z_1, \dots, z_n) R(z_1, \dots, z_n) \frac{\prod_{\sigma'^{-1}(a) > \sigma'^{-1}(b)}^{\sigma^{-1}(a) < \sigma^{-1}(b)} \left(1 - \frac{z_b}{z_a}\right)}{\prod_{\sigma'^{-1}(a) < \sigma'^{-1}(b)}^{\sigma^{-1}(a) < \sigma^{-1}(b)} \widetilde{\zeta}_{i_b i_a}\left(\frac{z_b}{z_a}\right)} \prod_{a=1}^n Dz_i$$

Indeed, the integrand is the same in both sides of the expression above, and there are no poles prohibiting us from moving the contours from $|z_a| \gg |z_b|$ if $\sigma^{-1}(a) > \sigma^{-1}(b)$ (as in the LHS) to $|z_a| \gg |z_b|$ if ${\sigma'}^{-1}(a) > {\sigma'}^{-1}(b)$ (as in the RHS). □

2.9. We will now set up the proof of Theorem 2.5. Consider the set of **letters**

$$\left\{ i^{(d)} \right\}_{i \in I, d \in \mathbb{Z}}$$

A **word** is simply a sequence of letters

$$w = \left[i_1^{(d_1)} \dots i_n^{(d_n)} \right] \tag{2.31}$$

We will call n the **length** of a word as above, and call

$$\overline{w} = (d_1, \dots, d_n) \tag{2.32}$$

the sequence of exponents of w. The **degree** of the word (2.31) is defined as

$$\deg w = (\boldsymbol{\varsigma}^{i_1} + \dots + \boldsymbol{\varsigma}^{i_n}, d_1 + \dots + d_n) \in \mathbb{N}^I \times \mathbb{Z} \tag{2.33}$$

Denote the set of all words by $\mathcal{W}$. To $w \in \mathcal{W}$ as above, we associate the elements

$$e_w = e_{i_1, d_1} \dots e_{i_n, d_n} \in \widetilde{\mathbf{U}}^+ \tag{2.34}$$

$$f_w = e_{i_1, -d_1} \dots e_{i_n, -d_n} \in \widetilde{\mathbf{U}}^- \tag{2.35}$$

It is clear that linear generating sets of $\mathring{\mathcal{S}}^{\pm}$ are given by

$$E_w := \widetilde{\Upsilon}^+(e_w) = \operatorname{Sym}\left[z_{i_1 \bullet_1}^{d_1} \dots z_{i_n \bullet_n}^{d_n} \prod_{1 \leq a < b \leq n} \zeta_{i_a i_b}\left(\frac{z_{i_a \bullet_a}}{z_{i_b \bullet_b}}\right) \right] \in \mathcal{V}^+ \tag{2.36}$$

$$F_w := \widetilde{\Upsilon}^-(f_w) = \operatorname{Sym}\left[z_{i_1 \bullet_1}^{-d_1} \dots z_{i_n \bullet_n}^{-d_n} \prod_{1 \leq a < b \leq n} \zeta_{i_b i_a}\left(\frac{z_{i_b \bullet_b}}{z_{i_a \bullet_a}}\right) \right] \in \mathcal{V}^- \tag{2.37}$$

(in the formulas above, we choose $\{\bullet_a\}_{a \in \{1,\dots,n\}} \in \{1, \dots, n_{i_a}\}$ such that $\bullet_a < \bullet_b$ whenever $a < b$ and $i_a = i_b$) as w runs over the set $\mathcal{W}$ of all words.

LEMMA 2.8. *We have $\mathring{\mathcal{S}}^{\pm} \subseteq \mathcal{S}^{\pm}$.*

PROOF. We will prove the required statement for $\pm = -$, as the case of $\pm = +$ is analogous. The following formula will come in handy repeatedly

$$\left\langle e_v, F_w \right\rangle = \left\langle E_v, f_w \right\rangle \tag{2.38}$$

$$= \int_{|z_1| \gg \dots \gg |z_n|} \sum_{\substack{\sigma \in S(n) \\ i_a = j_{\sigma(a)}, \forall a}} z_1^{d_1 - k_{\sigma(1)}} \dots z_n^{d_n - k_{\sigma(n)}} \prod_{\sigma(a) > \sigma(b)}^{a<b} \frac{\zeta_{i_a i_b}\left(\frac{z_a}{z_b}\right)}{\zeta_{i_b i_a}\left(\frac{z_b}{z_a}\right)} \prod_{a=1}^n Dz_a$$

for any pair of words

$$v = \left[i_1^{(d_1)} \dots i_n^{(d_n)}\right] \qquad \text{and} \qquad w = \left[j_1^{(k_1)} \dots j_n^{(k_n)}\right] \tag{2.39}$$

(the proof of (2.38) is almost word-for-word as that of [**9**, Remark 3.16], so we will not repeat it here). Thus, the pairing $\langle \text{LHS of (2.23)}, F_w \rangle$ equals

$$\sum_{\sigma \in S(n)} \int_{|z_1| \gg \dots \gg |z_n|} \frac{p_{j_{\sigma(1)}, \dots, j_{\sigma(n)}}(z_1, \dots, z_n)}{z_1^{k_{\sigma(1)}} \dots z_n^{k_{\sigma(n)}}} \prod_{\sigma(a) > \sigma(b)}^{a<b} \frac{\zeta_{j_{\sigma(a)} j_{\sigma(b)}}\left(\frac{z_a}{z_b}\right)}{\zeta_{j_{\sigma(b)} j_{\sigma(a)}}\left(\frac{z_b}{z_a}\right)} \prod_{a=1}^n Dz_a$$

In every summand over σ in the formula above, let us change the variable according to $z_a = y_{\sigma(a)}$. Then the formula above reads

$$\sum_{\sigma \in S(n)} \int_{|y_{\sigma(1)}| \gg \dots \gg |y_{\sigma(n)}|} \frac{p_{j_{\sigma(1)}, \dots, j_{\sigma(n)}}(y_{\sigma(1)}, \dots, y_{\sigma(n)})}{y_1^{k_1} \dots y_n^{k_n}} \prod_{\sigma^{-1}(a) > \sigma^{-1}(b)}^{a<b} \frac{\zeta_{j_b j_a}\left(\frac{y_b}{y_a}\right)}{\zeta_{j_a j_b}\left(\frac{y_a}{y_b}\right)} \prod_{a=1}^n Dy_a$$

As one moves the contours of integration from $|y_a| \gg |y_b| \Leftrightarrow \sigma^{-1}(a) < \sigma^{-1}(b)$ to $|y_a| \ll |y_b| \Leftrightarrow a < b$, one does not encounter any poles, so we conclude that

$$\left\langle \text{LHS of (2.23)}, F_w \right\rangle$$

$$= \int_{|y_1| \ll \dots \ll |y_n|} \sum_{\sigma \in S(n)} \frac{p_{j_{\sigma(1)}, \dots, j_{\sigma(n)}}(y_{\sigma(1)}, \dots, y_{\sigma(n)})}{y_1^{k_1} \dots y_n^{k_n}} \prod_{\sigma^{-1}(a) > \sigma^{-1}(b)}^{a<b} \frac{\zeta_{j_b j_a}\left(\frac{y_b}{y_a}\right)}{\zeta_{j_a j_b}\left(\frac{y_a}{y_b}\right)} \prod_{a=1}^n Dy_a$$

$$= \int_{|y_1| \ll \dots \ll |y_n|} \frac{\sum_{\sigma \in S(n)} p_{j_{\sigma(1)}, \dots, j_{\sigma(n)}}(y_{\sigma(1)}, \dots, y_{\sigma(n)}) \prod_{a<b} \zeta_{j_{\sigma(a)} j_{\sigma(b)}}\left(\frac{y_{\sigma(a)}}{y_{\sigma(b)}}\right)}{y_1^{k_1} \dots y_n^{k_n} \prod_{a<b} \zeta_{j_a j_b}\left(\frac{y_a}{y_b}\right)} \prod_{a=1}^n Dy_a$$

The numerator of the fraction above is precisely the LHS of (2.24), which vanishes by assumption. We conclude that $\langle \text{LHS of (2.23)}, F_w \rangle = 0$. Since the LHS of (2.23) was an arbitrary element of K^+, we conclude that $F_w \in \mathcal{S}^-$, as required. ☐

REMARK 2.9. Lemma 2.8 implies that (2.21) and (2.22) restrict to two pairings

$$\mathring{\mathcal{S}}^+ \otimes \mathring{\mathcal{S}}^- \xrightarrow{\langle \cdot, \cdot \rangle} \mathbb{K} \tag{2.40}$$

which coincide due to the equality on the first line of (2.38). Moreover, because (2.21) and (2.22) are non-degenerate in the $\mathcal{S}^{\pm}$ argument, we conclude that (2.40) is non-degenerate in both arguments.

2.10. Let us fix a total order $<$ on the finite set I, and associate to it the following total order on the set of letters

$$i^{(d)} < j^{(e)} \quad \text{if} \quad \begin{cases} d > e \\ \quad \text{or} \\ d = e \text{ and } i < j \end{cases}$$

Then we have the corresponding total lexicographic order on the set of words

$$\left[i_1^{(d_1)} \dots i_n^{(d_n)}\right] < \left[j_1^{(e_1)} \dots j_m^{(e_m)}\right]$$

if there exists k such that $i_1^{(d_1)} = j_1^{(e_1)}, \dots, i_k^{(d_k)} = j_k^{(e_k)}$ and either $i_{k+1}^{(d_{k+1})} < j_{k+1}^{(e_{k+1})}$ or $k = n < m$.

DEFINITION 2.10. Consider $\{s_{ij} \in \mathbb{Z}, \#_{ij} \geq 0\}_{i,j \in I}$ uniquely defined so that

$$\zeta_{ij}(x) = \alpha_{ij} \cdot x^{s_{ij}} + \dots + \beta_{ij} \cdot x^{s_{ij} + \#_{ij}} \tag{2.41}$$

for all $i \neq j$, while

$$\zeta_{ii}(x) = \frac{\alpha_{ii} \cdot x^{s_{ii}} + \dots + \beta_{ii} \cdot x^{s_{ii} + \#_{ii}}}{1 - x} \tag{2.42}$$

for all $i \in I$, where $\alpha_{ij}, \beta_{ij} \in \mathbb{K}^\times$. Then a word (2.31) is called **non-increasing** if

$$\begin{cases} d_a < d_b - \sum_{a \leq x < b} (s_{i_x i_b} + s_{i_b i_x}) \\ \quad \text{or} \\ d_a = d_b - \sum_{a \leq x < b} (s_{i_x i_b} + s_{i_b i_x}) \text{ and } i_a \geq i_b \end{cases} \tag{2.43}$$

for all $1 \leq a < b \leq n$. Let $\mathcal{W}_{\text{non-inc}}$ denote the set of non-increasing words.

LEMMA 2.11. *There are finitely many non-increasing words of given degree, which are bounded above by any given word v.*

PROOF. Let us assume we are counting non-increasing words $[i_1^{(d_1)} \dots i_n^{(d_n)}]$ with $d_1 + \dots + d_n = d$ for fixed n and d. The fact that such words are bounded above implies that d_1 is bounded below. But then the inequality (2.43) implies that $d_2, \dots, d_n$ are also bounded below. The fact that $d_1 + \dots + d_n$ is fixed implies that there can only be finitely many choices for the exponents $d_1, \dots, d_n$. Since there are also finitely many choices for $i_1, \dots, i_n \in I$, this concludes the proof. □

2.11. Our main interest in the notion of non-increasing words is the following.

PROPOSITION 2.12. *The set $\{e_w\}_{w \in \mathcal{W}_{\text{non-inc}}}$ is a $\mathbb{K}$-linear basis of $\widetilde{\mathbf{U}}^+$.*

PROOF. Relation (2.3) reads

$$e_i(z) e_j(w) \left[\alpha_{ji} \left(\frac{w}{z}\right)^{s_{ji}} + \dots + \beta_{ji} \left(\frac{w}{z}\right)^{s_{ji} + \#_{ji}}\right]$$
$$= e_j(w) e_i(z) \left[\alpha_{ij} \left(\frac{z}{w}\right)^{s_{ij}} + \dots + \beta_{ij} \left(\frac{z}{w}\right)^{s_{ij} + \#_{ij}}\right]$$

if $i \neq j$, and

$$e_i(z)e_i(w)\left[\alpha_{ii}\left(\frac{w}{z}\right)^{s_{ii}} + \dots + \beta_{ii}\left(\frac{w}{z}\right)^{s_{ii}+\#_{ii}}\right]$$
$$= e_i(w)e_i(z)\left[-\alpha_{ii}\left(\frac{z}{w}\right)^{s_{ii}-1} - \dots - \beta_{ii}\left(\frac{z}{w}\right)^{s_{ii}+\#_{ii}-1}\right]$$

for all $i \in I$. Foiling out the formulas above and equating the coefficients of various monomials in z and w yields the equalities

$$\alpha_{ji} \cdot e_{i,d}e_{j,k} + \text{larger} = \alpha_{ij} \cdot e_{j,k-s_{ij}-s_{ji}}e_{i,d+s_{ij}+s_{ji}} + \text{larger}$$

if $i \neq j$, and

$$\alpha_{ii} \cdot e_{i,d}e_{i,k} + \text{larger} = -\alpha_{ii} \cdot e_{i,k-2s_{ii}+1}e_{i,d+2s_{ii}-1} + \text{larger}$$

where in either of the formulas above, the word "larger" in the phrase "e_w+larger" stands for linear combinations of e_v's with $v > w$. Using the relations above, the proof of [**9**, Proposition 3.11 in Subsection 4.4] explains how to write any e_w for $w \notin \mathcal{W}_{\text{non-inc}}$ as a linear combination of e_v's with $v > w$ and $v \in \mathcal{W}_{\text{non-inc}}$. Following *loc. cit.*, one actually obtains the stronger fact that there exists a constant $\beta(n)$ for all $n \in \mathbb{N}$ (which only depends on the integers $\{s_{ij}, \#_{ij}\}_{i,j\in I}$) such that

$$e_w \in \sum_{\min \overline{w}-\beta(\text{length}(w)) \leq \min \overline{v} \leq \max \overline{v} \leq \max \overline{w}+\beta(\text{length}(w))}^{v\in\mathcal{W}_{\text{non-inc}} \text{ such that } v\geq w \text{ and}} \mathbb{K} \cdot e_v \tag{2.44}$$

for all $w \in \mathcal{W}$. In other words, when writing any e_w as a linear combination of e_v's with $v \geq w$ and $v \in \mathcal{W}_{\text{non-inc}}$, the exponents of any v which appears in the linear combination will always remain within a global constant from the exponents of w.

Let us now prove that $\{e_w\}_{w\in\mathcal{W}_{\text{non-inc}}}$ are linearly independent in $\widetilde{\mathbf{U}}^+$. Given a monomial in variables $\{z_{i\bullet}\}_{i\in I,\bullet\in\mathbb{N}}$, we will consider all ways to order its variables

$$\mu = z_{i_1\bullet_1}^{-k_1} \dots z_{i_n\bullet_n}^{-k_n} \tag{2.45}$$

The **leading word** of μ as above is defined as the (lexicographically) largest word

$$w_\mu = \left[i_1^{(d_1)} \dots i_n^{(d_n)}\right] \tag{2.46}$$

among all ways to order the variables in (2.45), where we write

$$d_a = k_a - \sum_{x>a} s_{i_x i_a} + \sum_{y<a} s_{i_a i_y}, \qquad \forall a \in \{1, \dots, n\} \tag{2.47}$$

LEMMA 2.13 ([**9**, Lemma 4.8]). *The leading word of any monomial is non-increasing in the sense of Definition* 2.10.

More generally, the leading word lead(R) of any $R \in \mathcal{V}^-$ is defined as the lexicographically largest of the leading words (2.46) for all the monomials which appear in R with non-zero coefficient (the corresponding monomial will be called the **leading monomial** of R). Conversely, any non-increasing word w is the leading word

$$w = \text{lead}(\text{Sym}\,\mu) \tag{2.48}$$

where the monomial μ is chosen such that $w_\mu = w$ in accordance with (2.46).

Analogously to [**9**, formula (4.18)], one can show by direct inspection that

$$\Big\langle e_w, R \Big\rangle \text{ is } \begin{cases} \neq 0 & \text{if } w = \text{lead}(R) \\ = 0 & \text{if } w > \text{lead}(R) \end{cases} \tag{2.49}$$

The formula above immediately shows the linear independence of the elements e_w, as w runs over non-increasing words. Indeed, if one were able to write such an element e_w as a linear combination of elements $\{e_v\}_{w<v\in\mathcal{W}_{\text{non-inc}}}$, then this would contradict (2.49) for $R = \text{Sym}\,\mu$ with μ as in (2.48). □

2.12. We are now ready to conclude the proof of Theorem 2.5.

DEFINITION 2.14. A word w is called **standard** if

$$E_w \notin \sum_{v>w} \mathbb{K} \cdot E_v \tag{2.50}$$

Let $\mathcal{W}_{\text{stan}}$ denote the set of standard words.

Because of the surjective algebra homomorphism $\widetilde{\Upsilon}^+ : \widetilde{\mathbf{U}}^+ \to \mathring{\mathcal{S}}^+$, $e_w \mapsto E_w$, Proposition 2.12 implies that any standard word is non-increasing.

PROOF OF THEOREM 2.5. We will prove the case of $\pm = -$, as that of $\pm = +$ is analogous. Fix any $n \in \mathbb{N}$; all words in the present proof will have length n. As

$$\frac{\zeta_{ij}(x)}{\zeta_{ji}(x^{-1})} \in \mathbb{K}^\times \cdot x^{s_{ij}+s_{ji}+\#_{ij}-\delta_{ij}} + (\text{smaller deegree as } x \to \infty) \tag{2.51}$$

for all $i, j \in I$, formula (2.38) implies that $\langle e_v, F_w\rangle \neq 0$ (for words v and w as in (2.39)) only if there exists $\sigma \in S(n)$ and natural numbers

$$\Big\{c_{ab} \leq s_{i_a i_b} + s_{i_b i_a} + \#_{i_a i_b} - \delta_{i_a i_b}\Big\}_{a<b, \sigma(a)>\sigma(b)}$$

such that we have for all $a \in \{1, \dots, n\}$

$$d_a + \sum_{\sigma(y)<\sigma(a)}^{y>a} c_{ay} - \sum_{\sigma(x)>\sigma(a)}^{x<a} c_{xa} = k_{\sigma(a)} \tag{2.52}$$

Thus, let us fix a natural number $m \gg$ the finitely many integers $\{s_{ij}, \#_{ij}\}_{i,j\in I}$ and n. Consider the infinite graph $G_{m,n}$ whose vertices are collections of integers

$$(d_1, \dots, d_n) \text{ such that } d_a \leq d_{a+1} + m$$

and whose edges are

$$(d_1, \dots, d_n) \to (k_1, \dots, k_n) \tag{2.53}$$

if and only if there exists $\sigma \in S(n)$ and natural numbers

$$\Big\{c_{ab} \leq m\Big\}_{a<b, \sigma(a)>\sigma(b)}$$

such that formula (2.52) holds (although the graph $G_{m,n}$ is defined to be directed, it is easy to see that an edge (2.53) exists if and only if the opposite edge exists). The preceding discussion implies that

$$\Big\langle e_v, F_w \Big\rangle \neq 0 \quad \Rightarrow \quad \text{exists edge } \overline{v} \to \overline{w} \tag{2.54}$$

in $G_{m,n}$, for any two non-increasing words v and w.

LEMMA 2.15 ([**9**, Lemma 3.18]). *All connected components of $G_{m,n}$ are finite.*

With the Lemma above in mind, let us define the following finite-dimensional $\mathbb{K}$-vector subspaces for any connected component $H \subset G_{m,n}$:

$$\mathring{\mathcal{S}}^+_H = \sum_{w \in \mathcal{W}_{\text{non-inc}}, \ \overline{w} \in H} \mathbb{K} \cdot E_w \subset \mathring{\mathcal{S}}^+$$

$$\mathring{\mathcal{S}}^-_H = \sum_{w \in \mathcal{W}_{\text{non-inc}}, \ \overline{w} \in H} \mathbb{K} \cdot F_w \subset \mathring{\mathcal{S}}^-$$

Because of (2.54), the restriction of the pairing (2.40) satisfies

$$\left\langle \mathring{\mathcal{S}}^+_H, \mathring{\mathcal{S}}^-_{H'} \right\rangle = 0 \tag{2.55}$$

for any distinct connected components $H \neq H'$. The following is an immediate consequence of (2.55) and the non-degeneracy of the pairing (2.40).

PROPOSITION 2.16. *The restriction of the pairing* (2.40) *to*

$$\mathring{\mathcal{S}}^+_H \otimes \mathring{\mathcal{S}}^-_H \xrightarrow{\langle \cdot, \cdot \rangle} \mathbb{K} \tag{2.56}$$

is non-degenerate for any connected component $H \subset G_{m,n}$.

PROPOSITION 2.17. *For any $n \in \mathbb{N}$, we have*

$$\bigoplus_{|\boldsymbol{n}|=n} \mathring{\mathcal{S}}^\pm_{\boldsymbol{n}} = \bigoplus_{H \subset G_{m,n}}^{\textit{connected components}} \mathring{\mathcal{S}}^\pm_H \tag{2.57}$$

PROOF. Let us prove (2.57) for $\pm = +$. Because the E_w's span $\mathring{\mathcal{S}}^+$ as w runs over all non-increasing words, all that we need to do to prove (2.57) is to show that there are no linear relations among the various direct summands of the RHS. To this end, assume that we had a relation

$$\sum_{H \text{ a connected component of } G_{m,n}} \alpha_H = 0$$

for various $\alpha_H \in \mathring{\mathcal{S}}^+_H$. Pairing the relation above with a given $\mathring{\mathcal{S}}^-_H$ implies that

$$\left\langle \alpha_H, \mathring{\mathcal{S}}^-_H \right\rangle = 0$$

Because the pairing (2.56) is non-degenerate, this implies that $\alpha_H = 0$. □

PROPOSITION 2.18. *For any $n \in \mathbb{N}$, we have*

$$\mathring{\mathcal{S}}^+_H = \bigoplus_{w \textit{ standard}, \ \overline{w} \in H} \mathbb{K} \cdot E_w \tag{2.58}$$

$$\mathring{\mathcal{S}}^-_H = \bigoplus_{w \textit{ standard}, \ \overline{w} \in H} \mathbb{K} \cdot F_w \tag{2.59}$$

for any connected component $H \subset G_{m,n}$.[7]

PROOF. Since $\mathring{\mathcal{S}}^\pm_H$ are finite-dimensional vector spaces, the Proposition boils down to the general claim that any vector space spanned by vectors $\alpha_1, \dots, \alpha_k$ has a basis consisting of those α_i's which cannot be written as linear combinations of $\{\alpha_j\}_{j>i}$. □

[7] A priori, the notion of "standard" words defined as in (2.50) might differ from the analogous notion with F's instead of E's. The indexing sets defining $\mathring{\mathcal{S}}^+_H$ and $\mathring{\mathcal{S}}^-_H$ in (2.58) and (2.59) must therefore run over the notion of standard words defined with respect to E's and F's, respectively.

To complete the proof of Theorem 2.5, consider any $R \in \mathcal{S}_{-\boldsymbol{n}}$ with $|\boldsymbol{n}| = n$. From (2.14), it is easy to see that:

$$\left\langle E_{[i_1^{(d_1)} \dots i_n^{(d_n)}]}, R \right\rangle = 0$$

if d_1 is small enough. However, by Lemma 2.11, there are only finitely many non-increasing words w of given degree with d_1 bounded below. This implies that

$$\left\langle E_w, R \right\rangle \neq 0$$

only for finitely many non-increasing words w. Let $H_1, \dots, H_t \subset G_{m,n}$ denote the connected components which contain the sequences of exponents of the aforementioned finitely many non-increasing words. The non-degeneracy of the pairings (2.56) of finite-dimensional vector spaces implies that there exists an element

$$R' \in \mathring{\mathcal{S}}^-_{H_1} \oplus \dots \oplus \mathring{\mathcal{S}}^-_{H_t} \subset \mathring{\mathcal{S}}^-$$

such that $\langle E_w, R \rangle = \langle E_w, R' \rangle$ for all non-increasing words w. Then the non-degeneracy of (2.21) in the second argument implies that $R = R'$, as we needed to establish $\mathcal{S}^- = \mathring{\mathcal{S}}^-$. The statement about the pairings (2.21) and (2.22) being identical therefore follows from Remark 2.9. $\square$

REMARK 2.19. Reversing the argument in the proof of Theorem 2.5 allows us to find (algorithmically) for any $R \in \mathcal{V}^- \backslash \mathcal{S}^-$ some element $\phi \in K^+$ such that

$$\left\langle \phi, R \right\rangle \neq 0 \tag{2.60}$$

Indeed, one need only isolate the finitely many non-increasing words w such that

$$\left\langle e_w, R \right\rangle \neq 0$$

and let $H_1, \dots, H_t \subset G_{m,n}$ denote the connected components which contain the sequences of exponents of the aforementioned finitely many non-increasing words. For any $w \in \mathcal{W}_{\text{non-inc}} \backslash \mathcal{W}_{\text{stan}}$, we have by definition

$$E_w = \sum_{v > w} \gamma_w^v \cdot E_v$$

for certain $\gamma_w^v \in \mathbb{K}$. Therefore,

$$e_w - \sum_{v>w} \gamma_w^v \cdot e_v \in K^+$$

and we claim that there exists $w \in \mathcal{W}_{\text{non-inc}} \backslash \mathcal{W}_{\text{stan}}$ such that

$$\left\langle e_w - \sum_{v>w} \gamma_w^v \cdot e_v, R \right\rangle \neq 0 \tag{2.61}$$

Indeed, if (2.61) failed for all $w \in \mathcal{W}_{\text{non-inc}} \backslash \mathcal{W}_{\text{stan}}$ then R would pair trivially with the whole of K^+, and this would violate the assumption $R \in \mathcal{V}^- \backslash \mathcal{S}^-$. Moreover, one can achieve (2.61) for one of those finitely many w's whose sequence of exponents $\overline{w}$ lies in the connected components $H_1, \dots, H_t \subset G_{m,n}$, because all other e_w's (as well as all e_v's with $v > w$) pair trivially with R by construction.

2.13. The quadratic quantum group of Definition 2.1 is merely the starting point of our study. The object we are actually interested in is the following.

DEFINITION 2.20. The (positive/negative part of the) **quantum loop group** associated to the datum $\{\zeta_{ij}\}_{i,j\in I}$ is the $\mathbb{K}$-algebra

$$\mathbf{U}^{\pm} = \widetilde{\mathbf{U}}^{\pm} \Big/ K^{\pm} \tag{2.62}$$

where $K^{\pm}$ denotes the kernel of the homomorphism $\widetilde{\Upsilon}^{\pm} : \widetilde{\mathbf{U}}^{\pm} \longrightarrow \mathcal{V}^{\pm}$.

As a direct consequence of the definition, the homomorphisms $\widetilde{\Upsilon}^{\pm}$ descend to isomorphisms

$$\Upsilon^{\pm} : \mathbf{U}^{\pm} \xrightarrow{\sim} \mathring{\mathcal{S}}^{\pm} = \mathcal{S}^{\pm} \tag{2.63}$$

with the last equality due to Theorem 2.5. The descended pairings (2.21)–(2.22) therefore yield pairings

$$\mathbf{U}^{+} \otimes \mathcal{S}^{-} \xrightarrow{\langle \cdot,\cdot \rangle} \mathbb{K} \tag{2.64}$$

$$\mathcal{S}^{+} \otimes \mathbf{U}^{-} \xrightarrow{\langle \cdot,\cdot \rangle} \mathbb{K} \tag{2.65}$$

which both coincide under the isomorphisms (2.63)

$$\mathbf{U}^{+} \otimes \mathbf{U}^{-} \xrightarrow{\langle \cdot,\cdot \rangle} \mathbb{K} \tag{2.66}$$

as shown in Remark 2.9.

2.14. In the present Subsection, let us assume that

$$\#_{ij} = -s_{ij} - s_{ji} + \delta_{ij} \tag{2.67}$$

for all $i, j \in I$, in the notation of (2.41)–(2.42). In particular, this implies $\#_{ij} = \#_{ji}$ for all i, j. We will refer to the collection $\{\zeta_{ij}\}_{i,j\in I}$ as being **symmetric**. Then all algebras studied in the present Section can be made into topological bialgebras, through the following procedure. First, we consider the extended algebras

$$\mathbf{U}^{\geq} = \frac{\mathbf{U}^{+}[h_{i,0}^{\pm 1}, h_{i,1}, h_{i,2}, \dots]_{i\in I}}{\left(h_i(z)e_j(w) = e_j(w)h_i(z)\frac{\zeta_{ij}\left(\frac{z}{w}\right)}{\zeta_{ji}\left(\frac{w}{z}\right)}\right)} \tag{2.68}$$

$$\mathbf{U}^{\leq} = \frac{\mathbf{U}^{-}[h'^{\pm 1}_{i,0}, h'_{i,-1}, h'_{i,-2}, \dots]_{i\in I}}{\left(h'_i(z)f_j(w) = f_j(w)h'_i(z)\frac{\zeta_{ji}\left(\frac{w}{z}\right)}{\zeta_{ij}\left(\frac{z}{w}\right)}\right)} \tag{2.69}$$

where

$$h_i(z) = \sum_{d=0}^{\infty} \frac{h_{i,d}}{z^d} \qquad \text{and} \qquad h'_i(z) = \sum_{d=0}^{\infty} h'_{i,-d} z^d$$

Note that the assumption (2.67) implies that

$$\frac{\zeta_{ij}(x)}{\zeta_{ji}(x^{-1})}$$

is regular and non-zero at $x = \infty$. With this in mind, one interprets the quotient relations in (2.68) (respectively (2.69)) by expanding them as power series in negative (respectively positive) powers of $\frac{z}{w}$. We can make $\mathbf{U}^{\geq}$ and $\mathbf{U}^{\leq}$ into topological bialgebras via the coproduct

$$\Delta(h_i(z)) = h_i(z) \otimes h_i(z) \tag{2.70}$$

$$\Delta(h'_i(z)) = h'_i(z) \otimes h'_i(z) \tag{2.71}$$

$$\Delta(e_i(z)) = h_i(z) \otimes e_i(z) + e_i(z) \otimes 1 \tag{2.72}$$

$$\Delta(f_i(z)) = 1 \otimes f_i(z) + f_i(z) \otimes h'_i(z) \tag{2.73}$$

It is straightforward to check that the pairing (2.66) extends to a bialgebra pairing

$$\mathbf{U}^{\geq} \otimes \mathbf{U}^{\leq} \xrightarrow{\langle \cdot, \cdot \rangle} \mathbb{K} \tag{2.74}$$

via

$$\Big\langle h_i(z), h'_i(w) \Big\rangle = \frac{\zeta_{ij}\left(\frac{z}{w}\right)}{\zeta_{ji}\left(\frac{w}{z}\right)} \qquad \text{expanded as } |z| \gg |w| \tag{2.75}$$

and properties

$$\Big\langle x, yy' \Big\rangle = \Big\langle \Delta(x), y \otimes y' \Big\rangle$$

$$\Big\langle xx', y \Big\rangle = \Big\langle x' \otimes x, \Delta(y) \Big\rangle$$

for all $x, x' \in \mathbf{U}^{\geq}$ and $y, y' \in \mathbf{U}^{\leq}$. There are also antipode maps on $\mathbf{U}^{\geq}$ and $\mathbf{U}^{\leq}$ satisfying the usual properties in a topological Hopf algebra.

DEFINITION 2.21. The **quantum loop group** is defined as

$$\mathbf{U} = \mathbf{U}^{\geq} \otimes \mathbf{U}^{\leq} \Big/ \Big(h_{i,0} \otimes h'_{i,0} = 1 \otimes 1 \Big) \tag{2.76}$$

with the multiplication governed by the Drinfeld double relation

$$(x \otimes y)(x' \otimes y') = \sum_a \sum_b \Big\langle S(x'_{1,a}), y_{1,b} \Big\rangle xx'_{2,a} \otimes y_{2,b} y' \Big\langle x'_{3,a}, y_{3,b} \Big\rangle \tag{2.77}$$

for any $x, x' \in \mathbf{U}^{\geq}$ and $y, y' \in \mathbf{U}^{\leq}$ whose twice iterated coproducts satisfy

$$\Delta^{(2)}(x') = \sum_a x'_{1,a} \otimes x'_{2,a} \otimes x'_{3,a} \qquad \text{and} \qquad \Delta^{(2)}(y) = \sum_b y_{1,b} \otimes y_{2,b} \otimes y_{3,b}$$

The algebra $\mathbf{U}$ inherits a Hopf algebra structure from its subalgebras $\mathbf{U}^{\geq}$ and $\mathbf{U}^{\leq}$.

Using relation (2.77), it is straightforward to deduce commutation relations between $e_i(z), h_i(z)$ and $f_j(w), h'_j(w)$. The most notable of these is the relation

$$[e_i(z), f_j(w)] = \delta_{ij} \delta\left(\frac{z}{w}\right) \cdot \frac{h_i(z) - h'_i(w)}{q - q^{-1}} \tag{2.78}$$

where $\delta(x) = \sum_{d \in \mathbb{Z}} x^d$ is a formal series, and δ_{ij} is the Kronecker delta function.

REMARK 2.22. Choosing to have the scalar $q - q^{-1}$ in the denominator of (2.78) is simply a matter of convention, to match the analogous constructions for usual quantum groups. However, to have (2.78) indeed follow from (2.77), one needs to rescale the pairings (2.14) and (2.15) by $(q^{-1} - q)^{-n}$. This modification does not substantially change any of the contents of the present Section, so we will ignore it.

3. The roots of the ζ functions

In the present Section, we will work over the splitting field of the Laurent polynomials $\zeta_{ij}(x)(1-x)^{\delta_{ij}}$ of (2.1)–(2.2), and determine the structure of the resulting shuffle algebras in terms of the pattern of zeroes of these Laurent polynomials.

3.1. In the present Section, we assume that

$$\zeta_{ij}(x) = \frac{\widetilde{\zeta}_{ij}(x)}{(1-x)^{\delta_{ij}}} \tag{3.1}$$

with

$$\widetilde{\zeta}_{ij}(x) = \alpha_{ij} x^{s_{ij}} \prod_{e=1}^{\#_{ij}} (1 - x q_e^{ij}) \tag{3.2}$$

for various scalars $q_e^{ij} \in \mathbb{K}^\times$ (in other words, we extend the ground field of Section 2 to the splitting field of the Laurent polynomials that appear in (2.1)–(2.2)).

EXAMPLE 3.1. When $\mathbb{K} = \mathbb{Q}$ and

$$\zeta_{ij}(x) = (1-x)^{\#_{ij} - \delta_{ij}}$$

for some matrix of non-negative integers $A = (\#_{ij})_{i,j \in I}$, we will use the notation

$$\widetilde{\mathbf{U}}_A^{\pm}, \mathbf{U}_A^{\pm}, \mathcal{V}_A^{\pm}, \mathcal{S}_A^{\pm}$$

for the $\mathbb{Q}$-algebras defined in Section 2. Note that $\mathcal{V}_A^+$ is the K-theoretic Hall algebra with 0 potential ([**5, 12**]) of the quiver with vertex set I and $\#_{ij}$ arrows from i to j.

We will soon show that the shuffle algebras from Example 3.1 are the building blocks of shuffle algebras for general zeta functions (3.2). Lemmas 3.2 and 3.4 hold in the generality of (2.1)–(2.2), but we will only use them in the setting of (3.2).

LEMMA 3.2. *For any $\boldsymbol{n} = (n_i)_{i \in I} \in \mathbb{N}^I$, we have*

$$\mathcal{S}_{\pm \boldsymbol{n}} = \frac{\mathcal{J}_{\boldsymbol{n}}}{\Delta_{\boldsymbol{n}}} \cap \mathcal{V}_{\pm \boldsymbol{n}} \tag{3.3}$$

where $\mathcal{J}_{\boldsymbol{n}} \subset \mathbb{K}[z_{i1}^{\pm 1}, \dots, z_{in_i}^{\pm 1}]_{i \in I}$ denotes the ideal

$$\mathcal{J}_{\boldsymbol{n}} = \left(\prod_{(i,a)<(j,b)} \widetilde{\zeta}_{ij} \left(\frac{z_{ia}}{z_{jb}} \right) \right)_{\text{total order } < \text{ on } \{(i,a) | i \in I, a \in \{1,\dots,n_i\}\}} \tag{3.4}$$

and

$$\Delta_{\boldsymbol{n}} = \prod_{i \in I} \prod_{1 \le a < b \le n_i} (z_{ia} - z_{ib}) \tag{3.5}$$

In (3.3) and henceforth, $\frac{\mathcal{J}_{\boldsymbol{n}}}{\Delta_{\boldsymbol{n}}}$ denotes the ideal quotient $(\mathcal{J}_{\boldsymbol{n}} : (\Delta_{\boldsymbol{n}}))$.

PROOF. The inclusion $\subseteq$ in (3.3) is an immediate consequence of $\mathcal{S}_{\pm \boldsymbol{n}} = \mathring{\mathcal{S}}_{\pm \boldsymbol{n}}$, which was established in Theorem 2.5. To conclude the proof, we need to show that any

$$R \in \frac{\mathcal{J}_{\boldsymbol{n}}}{\Delta_{\boldsymbol{n}}} \cap \mathcal{V}_{\pm \boldsymbol{n}}$$

actually lies in $\mathcal{S}_{\pm\boldsymbol{n}}$. The formula above implies that

$$R(z_{i1},\dots,z_{in_i})_{i\in I} = \sum_{<} p_<(z_{i1},\dots,z_{in_i})_{i\in I} \prod_{(i,a)<(j,b)} \zeta_{ij}\left(\frac{z_{ia}}{z_{jb}}\right) \tag{3.6}$$

for some Laurent polynomials $p_<$, where the sum goes over all total orders on the set $\{(i,a)|i\in I, a\in\{1,\dots,n_i\}\}$. However, the fact that $R\in\mathcal{V}_{\pm\boldsymbol{n}}$ implies that symmetrization (i.e. summing over all permutations of $z_{i1},\dots,z_{in_i}$, for each $i\in I$ separately) of R has the effect of multiplying R by a positive integer. As the symmetrization of the right-hand side of (3.6) clearly lies in $\mathring{\mathcal{S}}_{\pm\boldsymbol{n}} = \mathcal{S}_{\pm\boldsymbol{n}}$, then so does R. □

EXAMPLE 3.3. In the setting of Example 3.1, Lemma 3.2 implies that we can have $\mathcal{S}_{A,\pm\boldsymbol{n}} \neq \mathcal{V}_{A,\pm\boldsymbol{n}}$ only if there exist $i_1,\dots,i_k\in I$ such that $\#_{i_\bullet i_{\bullet+1}}>0$ for all $\bullet\in\{1,\dots,k\}$ (let $i_{k+1}=i_1$), which is precisely the existence of a **wheel** in the language of Subsection 1.6. Indeed, the absence of such a wheel would imply that $\#_{ii}=0,\forall i$ and that there exists a total order $<$ on I such that $\#_{ij}=0$ if $i<j$. Then the total order on $\{(i,a)|i\in I, 1\le a\le n_i\}$ which has $(i,a)<(j,b)$ if $i<j$ or if $i=j$ and $a<b$ would imply $1\in\mathcal{J}_{\boldsymbol{n}}$, and (3.3) would give us $\mathcal{S}_{A,\pm\boldsymbol{n}}=\mathcal{V}_{A,\pm\boldsymbol{n}}$.

LEMMA 3.4. *Assume that one can partition the set*

$$I = I_1 \sqcup \dots \sqcup I_k \tag{3.7}$$

such that $\#_{ij}=0$ (i.e. $\widetilde{\zeta}_{ij}(x)$ is a unit in the ring $\mathbb{K}[x^{\pm1}]$) whenever $i\in I_s$ and $j\in I_t$ for $s\neq t$. Then for any $\boldsymbol{n}\in\mathbb{N}^I$, we have

$$\mathcal{S}_{\pm\boldsymbol{n}} = \bigotimes_{t=1}^{k} \mathcal{S}^{(t)}_{\pm\boldsymbol{n}^t} \tag{3.8}$$

as $\mathbb{K}$-vector subspaces of

$$\mathbb{K}[z_{i1}^{\pm1},\dots,z_{in_i}^{\pm1}]^{\text{sym}}_{i\in I} = \bigotimes_{t=1}^{k} \mathbb{K}[z_{i1}^{\pm1},\dots,z_{in_i}^{\pm1}]^{\text{sym}}_{i\in I_t}$$

where $\mathcal{S}^{\pm(t)}$ denotes the shuffle algebra defined with respect to $\{\zeta_{ij}\}_{i,j\in I_t}$, and $\boldsymbol{n}^t\in\mathbb{N}^{I_t}$ denotes the projection of $\boldsymbol{n}\in\mathbb{N}^I$ corresponding to the subset $I_t\subset I$.

PROOF. By analogy with (3.4), for any $t\in\{1,\dots,k\}$ consider

$$\mathcal{J}^{(t)}_{\boldsymbol{n}^t} = \left(\prod_{(i,a)<(j,b)} \widetilde{\zeta}_{ij}\left(\frac{z_{ia}}{z_{jb}}\right)\right)_{\text{total order } < \text{ on } \{(i,a)|i\in I^t, a\in\{1,\dots,n_i\}\}}$$

as an ideal in $\mathbb{K}[z_{i1}^{\pm1},\dots,z_{in_i}^{\pm1}]_{i\in I_t}$. The equality

$$\mathcal{J}_{\boldsymbol{n}} = \bigotimes_{t=1}^{k} \mathcal{J}^{(t)}_{\boldsymbol{n}^t} \tag{3.9}$$

is an easy consequence of the fact that every generator of the ideal in the left-hand side is a product of generators of the ideals in the right-hand side, and vice versa (this uses the fact that $\widetilde{\zeta}_{ij}$ is a unit if i and j lie in different parts of the partition (3.7)). Since $\Delta_{\boldsymbol{n}} = \prod_{t=1}^{k} \Delta_{\boldsymbol{n}^t}$, it is an easy exercise to show that (3.9) implies

$$\frac{\mathcal{J}_{\boldsymbol{n}}}{\Delta_{\boldsymbol{n}}} = \bigotimes_{t=1}^{k} \frac{\mathcal{J}^{(t)}_{\boldsymbol{n}^t}}{\Delta_{\boldsymbol{n}^t}} \tag{3.10}$$

(indeed, the $\supseteq$ inclusion is trivial, while $\subseteq$ follows from the fact that the polynomials $\Delta_{\boldsymbol{n}^1}, \dots, \Delta_{\boldsymbol{n}^k}$ do not have any variables in common; see Claim 3.7 for a more complicated instance of this argument). Formulas (3.3) and (3.10) imply (3.8). □

3.2. In the setting of the zeta functions (3.2), consider any $\boldsymbol{n} \in \mathbb{N}^I$. For any

$$p = (p_{ia})_{i \in I, a \in \{1,\dots,n_i\}} \in (\mathbb{K}^\times)^n \Big/ \mathbb{K}^\times \tag{3.11}$$

(where $n = |\boldsymbol{n}|$ and $\mathbb{K}^\times$ acts on $(\mathbb{K}^\times)^n$ by simultaneous rescaling) we define the set of equivalence classes

$$I^p = \Big\{(i,a) \Big| i \in I, a \in \{1,\dots,n_i\}\Big\} \Big/ (i,a) \sim (j,b) \text{ if } i = j \text{ and } p_{ia} = p_{jb} \tag{3.12}$$

and the polynomial

$$\widetilde{\zeta}^p_{(i,a)(j,b)}(x) = \prod_{1 \leq e \leq \#_{ij} \text{ such that } q_e^{ij} = \frac{p_{jb}}{p_{ia}}} (1 - x q_e^{ij}) \tag{3.13}$$

The definition above is designed to achieve two goals. Firstly

$$\widetilde{\zeta}^p_{(i,a)(j,b)} \text{ divides } \widetilde{\zeta}_{ij} \tag{3.14}$$

in $\mathbb{K}[x^{\pm 1}]$. Secondly, define the matrix

$$A^p = \left\{\#^p_{(i,a)(j,b)}\right\}_{(i,a),(j,b) \in I^p}$$

where $\#^p_{(i,a)(j,b)}$ denotes the number of $e \in \{1,\dots,\#_{ij}\}$ such that

$$q_e^{ij} = \frac{p_{jb}}{p_{ia}}$$

(i.e. the number of linear factors in (3.13)). If we let $\boldsymbol{n}^p \in \mathbb{N}^{I^p}$ denote the vector whose (i,a) entry is the cardinality of the equivalence class of (i,a) in (3.12), then we have an isomorphism of vector spaces

$$\mathcal{S}_{p,\pm\boldsymbol{n}^p} \xrightarrow{\sim} \mathcal{S}_{A^p,\pm\boldsymbol{n}^p}, \qquad R(z_{ia}) \mapsto R(z_{ia} p_{ia}) \tag{3.15}$$

where $\mathcal{S}_p^\pm$ is the shuffle algebra associated to I^p and the functions (3.13), and $\mathcal{S}_A^\pm$ is the shuffle algebra defined in Example 3.1. The upshot is that, in degree $\boldsymbol{n}^p$, the shuffle algebra defined with respect to the divisors (3.13) is simply a rescaling of the shuffle algebra in Example 3.1 (which only depends on the choice of a quiver).

3.3. The following result explains our earlier statement that all shuffle algebras are built out of those from Example 3.1.

PROPOSITION 3.5. *For any $\boldsymbol{n} = (n_i)_{i \in I} \in \mathbb{N}^I$, we have*

$$\mathcal{S}_{\pm\boldsymbol{n}} = \bigcap_{p \in (\mathbb{K}^\times)^n / \mathbb{K}^\times} \mathcal{S}_{p,\pm\boldsymbol{n}^p} \tag{3.16}$$

where both sides are $\mathbb{K}$-vector subspaces of $\mathbb{K}[z_{i1}^{\pm 1}, \dots, z_{in_i}^{\pm 1}]_{i \in I}$.

Thus, formulas (3.15) and (3.16) reduce the study of the shuffle algebra $\mathcal{S}_{\pm \boldsymbol{n}}$ to the understanding of the shuffle algebras of Example 3.1. In fact, it is clear that the right-hand side of (3.16) is essentially a finite intersection, as $\mathcal{S}_{p,\pm \boldsymbol{n}^p}$ only depends on the equivalence relation defining (3.12) and on the collection of linear factors that appear in (3.13), both of which entail finitely many choices for each $\boldsymbol{n} \in \mathbb{N}^I$.

Proof of Proposition 3.5. Let us recall the ideal (3.4) and the polynomial (3.5), and define similarly for any $p \in (\mathbb{K}^\times)^n/\mathbb{K}^\times$ the ideal

$$\mathcal{J}_{p,\boldsymbol{n}^p} = \left(\prod_{(i,a)<(j,b)} \widetilde{\zeta}^p_{(i,a)(j,b)} \left(\frac{z_{ia}}{z_{jb}} \right) \right)_{\text{total order } < \text{ on } \{(i,a)|i\in I, a\in\{1,\dots,n_i\}\}} \tag{3.17}$$

and the polynomial

$$\Delta_{\boldsymbol{n}^p} = \prod_{i\in I} \prod_{\substack{1\leq a<b\leq n_i \\ p_{ia}=p_{ib}}} (z_{ia} - z_{ib}) \tag{3.18}$$

both in $\mathbb{K}[z_{i1}^{\pm 1}, \dots, z_{in_i}^{\pm 1}]_{i\in I}$. Formula (3.3) yields

$$\bigcap_{p\in(\mathbb{K}^\times)^n/\mathbb{K}^\times} \mathcal{S}_{p,\pm \boldsymbol{n}^p} = \bigcap_{p\in(\mathbb{K}^\times)^n/\mathbb{K}^\times} \left(\frac{\mathcal{J}_{p,\boldsymbol{n}^p}}{\Delta_{\boldsymbol{n}^p}} \cap \mathbb{K}[z_{i1}^{\pm 1}, \dots, z_{in_i}^{\pm 1}]_{i\in I}^{\text{sym}_p} \right) \tag{3.19}$$

where sym_p refers to polynomials which are symmetric in all z_{ia} and z_{ib} such that $p_{ia} = p_{ib}$. Thus, formula (3.16) is a consequence of

$$\frac{\mathcal{J}_{\boldsymbol{n}}}{\Delta_{\boldsymbol{n}}} = \bigcap_{p\in(\mathbb{K}^\times)^n/\mathbb{K}^\times} \frac{\mathcal{J}_{p,\boldsymbol{n}^p}}{\Delta_{\boldsymbol{n}^p}} \tag{3.20}$$

which we will now prove.

Claim 3.6. For any $p \in (\overline{\mathbb{K}}^\times)^n/\overline{\mathbb{K}}^\times$, there exists $p' \in (\mathbb{K}^\times)^n/\mathbb{K}^\times$ such that

$$\mathcal{J}_{p',\boldsymbol{n}^{p'}} = \mathcal{J}_{p,\boldsymbol{n}^p} \quad \text{and} \quad \Delta_{\boldsymbol{n}^{p'}} = \Delta_{\boldsymbol{n}^p} \tag{3.21}$$

where $\overline{\mathbb{K}}$ denotes the algebraic closure of $\mathbb{K}$.

We will prove Claim 3.6 by considering the partition

$$\{(i,a)|i \in I, a \in \{1,\dots,n_i\}\} = \bigsqcup_{\text{cosets } H \text{ of } \overline{\mathbb{K}}^\times/\mathbb{K}^\times} \{(i,a)|p_{ia} \in H\}$$

For any coset H of $\overline{\mathbb{K}}^\times/\mathbb{K}^\times$, let us simultaneously rescale $\{p_{ia} \in H\}$ to be generic elements of $\{p'_{ia} \in \mathbb{K}^\times\}$. We claim that

$$\widetilde{\zeta}^p_{(i,a)(j,b)}(x) = \widetilde{\zeta}^{p'}_{(i,a)(j,b)}(x) \tag{3.22}$$

for all (i,a) and (j,b), which would imply that $p' \in (\mathbb{K}^\times)^n/\mathbb{K}^\times$ satisfies (3.21). Indeed, the equality above is obvious for (i,a) and (j,b) lying in the same coset. Meanwhile, when (i,a) and (j,b) are in different cosets, both sides of (3.22) are equal to 1: for the LHS, this is because $p_{jb}/p_{ia} \notin \mathbb{K}^\times$, and thus cannot be equal to any of the roots $q_e^{ij} \in \mathbb{K}^\times$. For the RHS, this is because the field $\mathbb{K}$ of characteristic 0 is infinite, so the fact that p'_{jb} and p'_{ia} are generically chosen means that their ratio would not be equal to any one of the finitely many q_e^{ij}'s.

CLAIM 3.7. We have

$$\frac{\mathcal{J}_{\boldsymbol{n}}}{\Delta_{\boldsymbol{n}}} = \bigcap_{p \in (\overline{\mathbb{K}}^\times)^n / \overline{\mathbb{K}}^\times} \frac{\mathcal{J}_{p,\boldsymbol{n}^p}}{\Delta_{\boldsymbol{n}^p}} \tag{3.23}$$

It is clear that Claims 3.6 and 3.7 imply (3.20), so it remains to prove the latter claim. To show that two ideals of $\mathbb{K}[z_{i1}^{\pm 1}, \dots, z_{in_i}^{\pm 1}]_{i \in I}$ are equal, it suffices to show that they are equal in the localization at every closed point

$$p = (p_{ia})_{i \in I, a \in \{1, \dots, n_i\}} \in (\overline{\mathbb{K}}^\times)^n$$

In the case at hand, this follows from (3.24) and (3.25) below. Firstly, we need

$$\left(\frac{\mathcal{J}_{\boldsymbol{n}}}{\Delta_{\boldsymbol{n}}}\right)_p = \left(\frac{\mathcal{J}_{p,\boldsymbol{n}^p}}{\Delta_{\boldsymbol{n}^p}}\right)_p \tag{3.24}$$

which is a straightforward consequence of the fact that

$$\widetilde{\zeta}_{ij}\left(\frac{z_{ia}}{z_{jb}}\right) = \widetilde{\zeta}^p_{(i,a)(j,b)}\left(\frac{z_{ia}}{z_{jb}}\right) \cdot (\text{unit in the localization at } p)$$

and

$$\Delta_{\boldsymbol{n}} = \Delta_{\boldsymbol{n}^p} \cdot (\text{unit in the localization at } p)$$

Secondly, we need to show that

$$\left(\frac{\mathcal{J}_{p,\boldsymbol{n}^p}}{\Delta_{\boldsymbol{n}^p}}\right)_p \subseteq \left(\frac{\mathcal{J}_{p',\boldsymbol{n}^{p'}}}{\Delta_{\boldsymbol{n}^{p'}}}\right)_p \tag{3.25}$$

for any $p' \neq p$. To prove (3.25), note the obvious fact that

$$\widetilde{\zeta}^p_{(i,a)(j,b)}\left(\frac{z_{ia}}{z_{jb}}\right) \quad \text{is divisible by} \quad \widetilde{\zeta}^{p'}_{(i,a)(j,b)}\left(\frac{z_{ia}}{z_{jb}}\right)$$

in the localization at p. Therefore, we have

$$\left(\mathcal{J}_{p,\boldsymbol{n}^p}\right)_p \subseteq \left(\mathcal{J}_{p',\boldsymbol{n}^{p'}}\right)_p$$

Thus, to prove (3.25) it suffices to show that

$$\left(\frac{\mathcal{J}_{p',\boldsymbol{n}^{p'}}}{\Delta_{\boldsymbol{n}^p}}\right)_p = \left(\frac{\mathcal{J}_{p',\boldsymbol{n}^{p'}}}{\Delta_{\boldsymbol{n}^{p'}}}\right)_p \tag{3.26}$$

Equality (3.26) can only fail due to linear factors $z_{ia} - z_{ib}$ which appear in $\Delta_{\boldsymbol{n}^p}$ but do not appear in $\Delta_{\boldsymbol{n}^{p'}}$, i.e. whenever $p_{ia} = p_{ib}$ but $p'_{ia} \neq p'_{ib}$. It suffices to show that for any such linear factor, we have

$$\left(\frac{\mathcal{J}_{p',\boldsymbol{n}^{p'}}}{z_{ia} - z_{ib}}\right)_p = \left(\mathcal{J}_{p',\boldsymbol{n}^{p'}}\right)_p$$

or in other words, that

$$(z_{ia} - z_{ib})x \in \left(\mathcal{J}_{p',\boldsymbol{n}^{p'}}\right)_p \quad \Rightarrow \quad x \in \left(\mathcal{J}_{p',\boldsymbol{n}^{p'}}\right)_p \tag{3.27}$$

Let us construct the graph G with vertex set $\{(i', a') | i' \in I, a' \in \{1, \dots, n_{i'}\}\}$ and an edge between (i', a') and (j', b') if there exist linear factors in

$$\widetilde{\zeta}^{p'}_{(i',a')(j',b')}\left(\frac{z_{i'a'}}{z_{j'b'}}\right)$$

which are non-units in the localization at p. There can only be such an edge if

$$\frac{p'_{i'a'}}{p'_{j'b'}} = \frac{p_{i'a'}}{p_{j'b'}}$$

Just like in (3.9), we have

$$\left(\mathcal{J}_{p',\boldsymbol{n}^{p'}}\right)_p = \mathcal{J}_1 \cdots \mathcal{J}_k$$

where each $\mathcal{J}_\bullet$ is an ideal whose generators only involve the variables z_{ia} for (i,a) lying in the $\bullet$-th connected component of G. The following result is an easy exercise, whose proof we leave to the reader.

Lemma 3.8. *Assume that we have ideals $\mathcal{J}_1, \ldots, \mathcal{J}_k$ in $\mathbb{K}[w_1^{\pm 1}, \ldots, w_n^{\pm 1}]$, such that we can partition the variables $w_1, \ldots, w_n$ into disjoint sets $V_1, \ldots, V_k$ with the generators of each $\mathcal{J}_\bullet$ being Laurent polynomials in the variables of $V_\bullet$. Then*

$$\mathcal{J}_1 \ldots \mathcal{J}_k = \mathcal{J}_1 \cap \cdots \cap \mathcal{J}_k$$

Lemma 3.8 applies equally well to ideals in the localization of $\mathbb{K}[z_{i1}^{\pm 1}, \ldots, z_{in_i}^{\pm 1}]_{i\in I}$ at p, so we have

$$\left(\mathcal{J}_{p',\boldsymbol{n}^{p'}}\right)_p = \mathcal{J}_1 \cap \cdots \cap \mathcal{J}_k$$

As $p_{ia} = p_{ib}$ and $p'_{ia} \neq p'_{ib}$, this implies that (i,a) and (i,b) are in different connected components of the graph G. Thus, to prove (3.27) it suffices to show that

$$(z_{ia} - z_{ib})x \in \mathcal{J}_\bullet \quad \Rightarrow \quad x \in \mathcal{J}_\bullet$$

where $\mathcal{J}_\bullet$ is an ideal generated by polynomials that involve at most one of the variables z_{ia} and z_{ib}. This is an obvious fact, whose proof we leave to the reader. □

3.4. We will now consider the inclusion $\mathcal{S}_{A,\pm\boldsymbol{n}} \subset \mathcal{V}_{A,\pm\boldsymbol{n}}$ of Example 3.1, for any matrix A with non-negative entries and any $\boldsymbol{n} \in \mathbb{N}^I$. Given elements $\boldsymbol{k}, \boldsymbol{n} \in \mathbb{N}^I$, we will write $\mathbf{0} \leq \boldsymbol{k} \leq \boldsymbol{n}$ if $0 \leq k_i \leq n_i$ for all $i \in I$. In this case, if we have an ideal

$$\mathcal{I}_{\boldsymbol{k}} \subseteq \mathbb{Q}[z_{i1}^{\pm 1}, \ldots, z_{ik_i}^{\pm 1}]_{i\in I}$$

then we will write

$$\mathcal{I}_{\boldsymbol{k}}^{(\boldsymbol{n})} = \bigcap_{C=(\text{distinct } c_1^{(i)},\ldots,c_{k_i}^{(i)} \in \{1,\ldots,n_i\})_{i\in I}} \text{ideal generated by } \varphi_C(\mathcal{I}_{\boldsymbol{k}}) \tag{3.28}$$

as an ideal in $\mathbb{Q}[z_{i1}^{\pm 1}, \ldots, z_{in_i}^{\pm 1}]_{i\in I}$, where φ_C denotes the ring homomorphism

$$\mathbb{Q}[z_{i1}^{\pm 1}, \ldots, z_{ik_i}^{\pm 1}]_{i\in I} \xrightarrow{\varphi_C} \mathbb{Q}[z_{i1}^{\pm 1}, \ldots, z_{in_i}^{\pm 1}]_{i\in I}, \qquad z_{ia} \mapsto z_{ic_a^{(i)}} \tag{3.29}$$

Proposition 3.9. *For any $A = \{\#_{ij} \geq 0\}_{i,j\in I}$, there exist homogeneous ideals*

$$\mathcal{I}_{A,\boldsymbol{n}} \subseteq \mathbb{Q}[z_{i1}^{\pm 1}, \ldots, z_{in_i}^{\pm 1}]_{i\in I} \tag{3.30}$$

with the quotient $\mathbb{Q}[z_{i1}^{\pm 1}, \ldots, z_{in_i}^{\pm 1}]_{i\in I}/\mathcal{I}_{A,\boldsymbol{n}}$ supported on the small diagonal

$$\left\{ z_{ia} = z_{jb} \middle| \forall i,j \in I, a \in \{1,\ldots,n_i\}, b \in \{1,\ldots,n_j\} \right\} \tag{3.31}$$

and such that for all $\boldsymbol{n} \in \mathbb{N}^I$ we have

$$\mathcal{S}_{A,\pm\boldsymbol{n}} = \mathcal{V}_{A,\pm\boldsymbol{n}} \bigcap_{\mathbf{0}\leq\boldsymbol{k}\leq\boldsymbol{n}} \mathcal{I}_{A,\boldsymbol{k}}^{(\boldsymbol{n})} \tag{3.32}$$

The choice of the ideals $\mathcal{I}_{A,\boldsymbol{n}}$ is not unique, although we will give a natural construction in (3.42) below. The study of these ideals is warranted by the fact that Proposition 3.9 reduces the study of shuffle algebras to the ideals $\mathcal{I}_{A,\boldsymbol{n}}$. Moreover, as these ideals have finite codimension in any homogeneous degree,[8] then the inclusion (3.30) is cut out by finitely many linear conditions in any homogeneous degree.

PROOF OF PROPOSITION 3.9. Recall from (3.3) that $\mathcal{S}_{A,\pm\boldsymbol{n}} = \frac{\mathcal{J}_{A,\boldsymbol{n}}}{\Delta_{\boldsymbol{n}}} \cap \mathcal{V}_{A,\pm\boldsymbol{n}}$, where

$$\mathcal{J}_{A,\boldsymbol{n}} = \left(\prod_{(i,a)<(j,b)} (z_{ia} - z_{jb})^{\#_{ij}} \right)_{\text{total order } < \text{ on } \{(i,a)|i\in I, a\in\{1,\dots,n_i\}\}} \tag{3.33}$$

and $\Delta_{\boldsymbol{n}}$ is given by (3.5). It suffices to construct the ideals (3.30) so that

$$\frac{\mathcal{J}_{A,\boldsymbol{n}}}{\Delta_{\boldsymbol{n}}} = \bigcap_{\boldsymbol{0}\leq\boldsymbol{k}\leq\boldsymbol{n}} \mathcal{I}_{A,\boldsymbol{k}}^{(\boldsymbol{n})} \tag{3.34}$$

and we will do this by induction on $\boldsymbol{n}$. Let $\mathcal{I}_{A,\boldsymbol{0}} = (1)$ and assume that $\mathcal{I}_{A,\boldsymbol{k}}$ have been constructed for all $\boldsymbol{k} < \boldsymbol{n}$ (this is shorthand for $\boldsymbol{0} \leq \boldsymbol{k} \leq \boldsymbol{n}$ and $\boldsymbol{k} \neq \boldsymbol{n}$) and let us construct $\mathcal{I}_{A,\boldsymbol{n}}$. Formula (3.34) for $\boldsymbol{n}$ replaced by $\boldsymbol{k}$ implies that

$$\frac{\mathcal{J}_{A,\boldsymbol{k}}}{\Delta_{\boldsymbol{k}}} \subseteq \mathcal{I}_{A,\boldsymbol{k}} \tag{3.35}$$

However, comparing (3.33) for $\boldsymbol{n}$ and $\boldsymbol{k}$ implies that

$$\mathcal{J}_{A,\boldsymbol{n}} \subseteq \mathcal{J}_{A,\boldsymbol{k}}^{(\boldsymbol{n})} \tag{3.36}$$

(see (3.28) for the notation in the right-hand side), simply because the linear factors that generate the ideal $\mathcal{J}_{A,\boldsymbol{n}}$ are divisible by the linear factors that generate the ideal $\mathcal{J}_{A,\boldsymbol{k}}$ for any $\boldsymbol{k} < \boldsymbol{n}$. Thus, we have

$$\frac{\mathcal{J}_{A,\boldsymbol{n}}}{\Delta_{\boldsymbol{n}}} \subseteq \frac{\mathcal{J}_{A,\boldsymbol{k}}^{(\boldsymbol{n})}}{\Delta_{\boldsymbol{n}}} = \left(\frac{\mathcal{J}_{A,\boldsymbol{k}}}{\Delta_{\boldsymbol{k}}} \right)^{(\boldsymbol{n})} \tag{3.37}$$

where the equality is proved just like (3.26) (we leave the details as an exercise to the reader). Combining (3.35) and (3.37) yields

$$\frac{\mathcal{J}_{A,\boldsymbol{n}}}{\Delta_{\boldsymbol{n}}} \subseteq \bigcap_{\boldsymbol{0}\leq\boldsymbol{k}<\boldsymbol{n}} \mathcal{I}_{A,\boldsymbol{k}}^{(\boldsymbol{n})}$$

CLAIM 3.10. The quotient

$$\left(\bigcap_{\boldsymbol{0}\leq\boldsymbol{k}<\boldsymbol{n}} \mathcal{I}_{A,\boldsymbol{k}}^{(\boldsymbol{n})} \right) \Big/ \left(\frac{\mathcal{J}_{A,\boldsymbol{n}}}{\Delta_{\boldsymbol{n}}} \right) \tag{3.38}$$

is supported on the small diagonal $\{z_{ia} = z_{jb} | i, j \in I, a \in \{1, \dots, n_i\}, b \in \{1, \dots, n_j\}\}$.

Let us first conclude the proof of Proposition 3.9, and then prove Claim 3.10. Let

$$\mathcal{M} := \bigcap_{\boldsymbol{0}\leq\boldsymbol{k}<\boldsymbol{n}} \mathcal{I}_{A,\boldsymbol{k}}^{(\boldsymbol{n})} \supseteq \frac{\mathcal{J}_{A,\boldsymbol{n}}}{\Delta_{\boldsymbol{n}}} =: \mathcal{N} \tag{3.39}$$

[8]This is because the quotient $\mathbb{Q}[z_{i1}^{\pm 1}, \dots, z_{in_i}^{\pm 1}]_{i\in I}/\mathcal{I}_{A,\boldsymbol{n}}$ is a graded coherent sheaf over a punctured affine line, and thus a finite-dimensional $\mathbb{Q}$-vector space in any homogeneous degree.

and let $\mathfrak{m}$ denote the ideal $(z_{ia} - z_{jb})$ of the small diagonal $\operatorname{Spec} \mathbb{Q}[x^{\pm 1}]$. Consider the following descending sequence of ideals, as k runs over $\mathbb{N}$

$$\mathcal{M} \supseteq \cdots \supseteq \mathcal{M} \cap (\mathcal{N} + \mathfrak{m}^k) \supseteq \cdots \supseteq \mathcal{N} \tag{3.40}$$

By Claim 3.10, the successive quotients of all the inclusions above are supported on the small diagonal. But since all ideals involved are homogeneous, the fiber of (3.40) over the small diagonal is a chain of finite-dimensional $\mathbb{Q}$-vector spaces tensored with $\mathbb{Q}[x^{\pm 1}]$. By the Krull intersection theorem in the local ring at $\mathfrak{m}$, there exists d large enough such that

$$\mathcal{N} = \mathcal{M} \cap (\mathcal{N} + \mathfrak{m}^d) \tag{3.41}$$

If we let

$$\mathcal{I}_{A,\boldsymbol{n}} = \frac{\mathcal{J}_{A,\boldsymbol{n}}}{\Delta_{\boldsymbol{n}}} + \mathfrak{m}^d \tag{3.42}$$

for this large enough value of d, then (3.41) is precisely the required (3.34).

REMARK 3.11. Note that we have been quite profligate in our choice of the ideal (3.42), and other choices (such as those arising from the primary decomposition of the ideal $\frac{\mathcal{J}_{A,\boldsymbol{n}}}{\Delta_{\boldsymbol{n}}}$) might achieve (3.34) for a larger ideal $\mathcal{I}_{A,\boldsymbol{n}}$ than the one of (3.42).

Let us prove Claim 3.10. Consider any closed point $p = (p_{ia}) \in (\overline{\mathbb{Q}}^\times)^n$ with not all the p_{ia}'s equal to each other. Thus, the partition

$$\{(i,a) | i \in I, a \in \{1, \dots, n_i\}\} = C_1 \sqcup \cdots \sqcup C_k$$

where

$$p_{ia} = p_{jb} \quad \Leftrightarrow \quad (i,a),(j,b) \in C_\bullet \text{ for some } \bullet \in \{1,\dots,k\}$$

has the property that $k \geq 2$. By analogy with the proof of Claim 3.7, we have

$$\left(\frac{\mathcal{J}_{A,\boldsymbol{n}}}{\Delta_{\boldsymbol{n}}}\right)_p = \prod_{\bullet=1}^{k} \varphi_{C_\bullet}\left(\frac{\mathcal{J}_{A,\boldsymbol{n}_\bullet}}{\Delta_{\boldsymbol{n}_\bullet}}\right)_p$$

where $\boldsymbol{n}_\bullet$ is the vector counting the number of variables in the part $C_\bullet$. Just like in Lemma 3.8, one may show that

$$\prod_{\bullet=1}^{k} \varphi_{C_\bullet}\left(\frac{\mathcal{J}_{A,\boldsymbol{n}_\bullet}}{\Delta_{\boldsymbol{n}_\bullet}}\right)_p = \bigcap_{\bullet=1}^{k} \varphi_{C_\bullet}\left(\frac{\mathcal{J}_{A,\boldsymbol{n}_\bullet}}{\Delta_{\boldsymbol{n}_\bullet}}\right)_p$$

Since (3.34) is known to hold for all $\boldsymbol{n}_\bullet < \boldsymbol{n}$ by the induction hypothesis, we have

$$\bigcap_{\bullet=1}^{k} \varphi_{C_\bullet}\left(\frac{\mathcal{J}_{A,\boldsymbol{n}_\bullet}}{\Delta_{\boldsymbol{n}_\bullet}}\right)_p \supseteq \left(\bigcap_{\boldsymbol{0} \leq \boldsymbol{k} < \boldsymbol{n}} \mathcal{I}_{A,\boldsymbol{k}}^{(\boldsymbol{n})}\right)_p$$

The three equations above prove that the quotient (3.38) is 0 in the localization at p. Since this holds for all p outside of the small diagonal, we conclude Claim 3.10. □

PROOF OF THEOREM 1.4. immediately from (3.15), (3.16) and (3.32). In more detail, we can define the ideal (1.24) inductively in $\boldsymbol{n}$ by $\mathcal{I}_{p,\boldsymbol{0}} = (1)$, while for all $\boldsymbol{n} > \boldsymbol{0}$

$$\mathcal{I}_{p,\boldsymbol{n}} = \frac{\mathcal{J}_{\boldsymbol{n}}}{\Delta_{\boldsymbol{n}}} + \mathfrak{m}_p^d \tag{3.43}$$

where $\mathcal{J}_{\boldsymbol{n}}$, $\Delta_{\boldsymbol{n}}$ are defined in (3.4), (3.5) respectively, $\mathfrak{m}_p$ is the ideal of the small diagonal $(z_{ia}) \in p\mathbb{K}^\times$ and the natural number d is chosen large enough so that the following analogue of (3.41) holds

$$\left(\frac{\mathcal{J}_{\boldsymbol{n}}}{\Delta_{\boldsymbol{n}}}\right)_p = \left(\bigcap_{\boldsymbol{0}\leq \boldsymbol{k}<\boldsymbol{n}} \mathcal{I}_{p,\boldsymbol{k}}^{(\boldsymbol{n})}\right)_p \cap \left(\frac{\mathcal{J}_{\boldsymbol{n}}}{\Delta_{\boldsymbol{n}}} + \mathfrak{m}_p^d\right)_p \qquad \square$$

3.5. Let us now develop the dual (i.e. at the level of $\widetilde{\mathbf{U}}^\pm$ instead of $\mathcal{V}^\pm$, using (1.19)) treatment of the preceding Subsections. The goal is to realize the inclusion

$$\mathcal{S}^\mp \subset \mathcal{V}^\mp$$

as pairing with certain elements of $K^\pm \subset \widetilde{\mathbf{U}}^\pm$ under (2.12) and (2.13).

PROPOSITION 3.12. *For any $A = \{\#_{ij} \geq 0\}_{i,j\in I}$, there exist finite sets*

$$W_{A,\pm\boldsymbol{n},\pm d} \subset K_A^\pm = \text{Ker}\, \widetilde{\Upsilon}^\pm \subset \widetilde{\mathbf{U}}_{A,\pm\boldsymbol{n},\pm d} \tag{3.44}$$

for all $(\boldsymbol{n}, d) \in \mathbb{N}^I \times \mathbb{Z}$, such that for all $R^\pm \in \mathcal{V}_A^\pm$ we have

$$\left\langle J_A^+, R^- \right\rangle = 0 \qquad \Leftrightarrow \qquad R^- \in \mathcal{S}_A^- \tag{3.45}$$

$$\left\langle R^+, J_A^- \right\rangle = 0 \qquad \Leftrightarrow \qquad R^+ \in \mathcal{S}_A^+ \tag{3.46}$$

where $J_A^\pm = (W_{A,\pm\boldsymbol{n},\pm d})_{(\boldsymbol{n},d)\in\mathbb{N}^I\times\mathbb{Z}}$ as an ideal[9] *of $\widetilde{\mathbf{U}}_A^\pm$.*

PROOF. We will deal with the case of $\pm = +$, as the case $\pm = -$ is analogous. We assume that (3.44) have been constructed for all $\boldsymbol{0} \leq \boldsymbol{k} < \boldsymbol{n}$ such that (3.45) and (3.46) hold in all degrees $< \boldsymbol{n}$, and let us perform the construction for $\boldsymbol{n}$. Let

$$J'_{A,\boldsymbol{n}} = \bigoplus_{d\in\mathbb{Z}} J'_{A,\boldsymbol{n},d}$$

denote the degree $\boldsymbol{n}$ component of the ideal $(W_{A,\boldsymbol{k},d})_{\boldsymbol{0}\leq\boldsymbol{k}<\boldsymbol{n},d\in\mathbb{Z}} \subset \widetilde{\mathbf{U}}_A^+$.

CLAIM 3.13. For any $R^- \in \mathcal{V}_{A,-\boldsymbol{n}}$, we have

$$\left\langle J'_{A,\boldsymbol{n}}, R^- \right\rangle = 0 \qquad \Leftrightarrow \qquad R^- \in \mathcal{V}_{A,-\boldsymbol{n}} \bigcap_{\boldsymbol{0}\leq\boldsymbol{k}<\boldsymbol{n}} \mathcal{I}_{A,\boldsymbol{k}}^{(\boldsymbol{n})} \tag{3.47}$$

in the notation of (3.32).

Let us first show how Claim 3.13 allows us to complete the proof of Proposition 3.12. Let $\mathcal{M}$ and $\mathcal{N}$ be as in (3.39), and let us write $\mathcal{M}_d$ and $\mathcal{N}_d$ for their homogeneous degree d components, for any $d \in \mathbb{Z}$. The quotient $\mathcal{M}_d/\mathcal{N}_d$ is a finite dimensional $\mathbb{Q}$-vector space. Formula (3.47) tells us that if $R^- \in \mathcal{V}_{A,-\boldsymbol{n},-d}$, then

$$\left\langle J'_{A,\boldsymbol{n}}, R^- \right\rangle = 0 \qquad \Leftrightarrow \qquad R^- \in \mathcal{V}_{A,-\boldsymbol{n},-d} \cap \mathcal{M}_{-d}$$

Definition 2.4 of $\mathcal{S}^-$ reads

$$\left\langle K^+, R^- \right\rangle = 0 \qquad \Leftrightarrow \qquad R^- \in \mathcal{V}_{A,-\boldsymbol{n},-d} \cap \mathcal{N}_{-d} = \mathcal{S}_{A,-\boldsymbol{n},-d}$$

[9] In Subsection 3.7 we will prove that $J_A^\pm = K_A^\pm$.

The finite-dimensionality of $\mathcal{M}_{-d}/\mathcal{N}_{-d}$ means that we can choose finitely many elements $\phi_1, \ldots, \phi_N \in K^+$ such that

$$\left\langle J'_{A,\boldsymbol{n}} + \sum_{\bullet=1}^{N} \mathbb{Q} \cdot \phi_\bullet, R^- \right\rangle = 0 \qquad \Leftrightarrow \qquad R^- \in \mathcal{S}_{A,-\boldsymbol{n},-d}$$

In more detail, $\phi_1, \ldots, \phi_N \in K^+$ are successively defined to pair non-trivially with finitely many elements of $\mathcal{V}_{A,-\boldsymbol{n},-d} \cap (\mathcal{M}_{-d}/\mathcal{N}_{-d})$, so $\phi_1, \ldots, \phi_N$ can be constructed algorithmically as in Remark 2.19. Letting $W_{A,\boldsymbol{n},d} = \{\phi_1, \ldots, \phi_N\}$ yields (3.45) in degree $\boldsymbol{n}$, as required.

PROOF OF CLAIM 3.13. Fix any $\mathbf{0} \leq \boldsymbol{k} < \boldsymbol{n}$ and let us write $n = |\boldsymbol{n}|$ and $k = |\boldsymbol{k}|$. We will choose a relabeling $\varsigma^{i_1} + \cdots + \varsigma^{i_n} = \boldsymbol{n}$ such that $\varsigma^{i_{n-k+1}} + \cdots + \varsigma^{i_n} = \boldsymbol{k}$. Recall from (2.14) that for any $R^- \in \mathcal{V}_{-\boldsymbol{n}}$, $\phi \in J_{A,\boldsymbol{k}}$ and any $l_1, \ldots, l_{n-k} \in \mathbb{Z}$, the pairing

$$\left\langle e_{i_1,l_1} \ldots e_{i_{n-k},l_{n-k}} \phi, R^- \right\rangle \tag{3.48}$$

is computed as follows. Expand

$$\frac{R^-(z_1, \ldots, z_n)}{\prod_{\{1,\ldots,n-k\} \ni a < b \in \{1,\ldots,n\}} \zeta_{i_b i_a}\left(\frac{z_b}{z_a}\right)} \tag{3.49}$$

as a power series in $|z_1| \gg \cdots \gg |z_{n-k}|$, extract the coefficient of $z_1^{-l_1} \ldots z_{n-k}^{-l_{n-k}}$, and then pair the resulting Laurent polynomial in $z_{n-k+1}, \ldots, z_n$ with ϕ. Since l_1, $\ldots$, l_{n-k} can be arbitrary, formula (3.45) in degree $\boldsymbol{k}$ means that the vanishing of all the pairings (3.48) is equivalent to R^- lying in the ideal generated by

$$\varphi_C(\mathcal{I}_{A,\boldsymbol{k}})$$

where $\varphi_C : \mathbb{Q}[z_1^{\pm 1}, \ldots z_k^{\pm 1}] \longrightarrow \mathbb{Q}[z_1^{\pm 1}, \ldots z_n^{\pm 1}]$ is the map that sends $z_i \mapsto z_{n-k+i}$. Since R^- is symmetric, this is equivalent to R^- lying in the intersection

$$\mathcal{I}_{A,\boldsymbol{k}}^{(\boldsymbol{n})}$$

in the notation of (3.28). Finally, because $\mathbf{0} \leq \boldsymbol{k} < \boldsymbol{n}$ is arbitrary, then we obtain precisely the equivalence (3.47). □

□

REMARK 3.14. The proof given above is invariant under the shift homomorphisms (2.4), (2.6), (2.8), (2.9), in the sense that for fixed $\boldsymbol{k} \in \mathbb{Z}^I$ one can choose

$$W_{A,\pm\boldsymbol{n},\pm(d+\sum_{i\in I} n_i k_i)} = \tau_{\boldsymbol{k}}(W_{A,\pm\boldsymbol{n},\pm d}) \tag{3.50}$$

for any $(\boldsymbol{n}, d) \in \mathbb{N}^I \times \mathbb{Z}$. Thus, if $k_i > 0$ for all $i \in I$, then equality (3.50) reduces the choice of $W_{A,\pm\boldsymbol{n},\pm d}$ to finitely many choices for each $\boldsymbol{n} \in \mathbb{N}^I$.

REMARK 3.15. We have been intentionally vague concerning which kind of ideal one considers in (3.45)–(3.46). The proof given above works if $J_A^\pm$ is defined as the left ideal generated by the elements $W_{A,\pm\boldsymbol{n},\pm d}$, but a similar argument works if $J_A^\pm$ is defined as the right (or two-sided) ideal generated by the same elements.

3.6. Just like the study of shuffle algebras can be reduced to the (purely combinatorial) particular case of Example 3.1, the same can be said about the dual situation of quantum loop groups. Specifically, in the following Proposition we will show how to define the elements (1.26) for the quantum loop group associated to any choice of zeta functions (3.2), in terms of the finite sets

$$W_{A,\pm\boldsymbol{n},\pm d} \subset \widetilde{\mathbf{U}}_{A,\pm\boldsymbol{n},\pm d}$$

that we defined in (3.44), for various matrices $A = \{\#_{ij} \geq 0\}_{i,j\in I}$.

DEFINITION 3.16. For any $(\boldsymbol{n}, d) \in \mathbb{N}^I \times \mathbb{Z}$ and $p \in (\mathbb{K}^\times)^n/\mathbb{K}^\times$, assume that

$$W_{A^p,\boldsymbol{n}^p,d} = \Big\{\phi_1, \dots, \phi_N\Big\} \subset \widetilde{\mathbf{U}}_{A^p,\boldsymbol{n}^p,d}$$

where A^p and $\boldsymbol{n}^p$ are defined as in Subsection 3.2. Let us choose a representation

$$\phi_\bullet = \sum_{\substack{\{(i_1,a_1),\dots,(i_n,a_n)\}=\\ \{(i,a)|i\in I,a\in\{1,\dots,n_i\}\}}} \Big[\rho^\bullet_{(i_1,a_1),\dots,(i_n,a_n)}(z_1,\dots,z_n)e_{(i_1,a_1)}(z_1)\dots e_{(i_n,a_n)}(z_n)\Big]_{\text{ct}}$$

for each $\bullet \in \{1, \dots, N\}$, where $\rho^\bullet_{(i_1,a_1),\dots,(i_n,a_n)}$ are Laurent polynomials. Define

$$W_{p,\boldsymbol{n},d} = \Big\{\phi'_1, \dots, \phi'_N\Big\} \subset \widetilde{\mathbf{U}}_{\boldsymbol{n},d}$$

given by the following formula for each $\bullet \in \{1, \dots, N\}$

$$\phi'_\bullet = \sum_{\substack{\{(i_1,a_1),\dots,(i_n,a_n)\}=\\ \{(i,a)|i\in I,a\in\{1,\dots,n_i\}\}}} \Big[\rho^\bullet_{(i_1,a_1),\dots,(i_n,a_n)}(z_1,\dots,z_n)$$
$$\prod_{1\leq u<v\leq n} \frac{\widetilde{\zeta}_{i_v i_u}\left(\frac{z_v}{z_u}\right)\left(-\frac{z_u}{z_v}\right)^{\delta_{i_u i_v}\delta_{v\lhd u}}}{\widetilde{\zeta}^p_{(i_v,a_v)(i_u,a_u)}\left(\frac{z_v}{z_u}\right)} \cdot e_{i_1}(z_1)\dots e_{i_n}(z_n)\Big]_{\text{ct}} \tag{3.51}$$

In the right-hand side of (3.51), we fix an arbitrary total order $\lhd$ on I^p and write

$$u \lhd v \qquad \text{if} \qquad (i_u, a_u) \lhd (i_v, a_v)$$

The analogous construction with $e \leftrightarrow f$ and $\widetilde{\zeta}_{i_v i_u}\left(\frac{z_v}{z_u}\right) \leftrightarrow \widetilde{\zeta}_{i_u i_v}\left(\frac{z_u}{z_v}\right)$ defines $W_{p,-\boldsymbol{n},-d}$.

PROPOSITION 3.17. *For any $(\boldsymbol{n}, d) \in \mathbb{N}^I \times \mathbb{Z}$ and $p \in (\mathbb{K}^\times)^n/\mathbb{K}^\times$, we have $W_{p,\pm\boldsymbol{n},\pm d} \in K^\pm = \operatorname{Ker} \widetilde{\Upsilon}^\pm$. Moreover, for all $R^\pm \in \mathcal{V}^\pm$ we have*

$$\Big\langle J^+, R^-\Big\rangle = 0 \quad \Leftrightarrow \quad R^- \in \mathcal{S}^- \tag{3.52}$$

$$\Big\langle R^+, J^-\Big\rangle = 0 \quad \Leftrightarrow \quad R^+ \in \mathcal{S}^+ \tag{3.53}$$

where $J^\pm = (W_{p,\pm\boldsymbol{n},\pm d})_{p\in(K^\times)^n/\mathbb{K}^\times,\boldsymbol{n}\in\mathbb{N}^I,d\in\mathbb{Z}}$ as an ideal[10] *of $\widetilde{\mathbf{U}}^+$.*

PROOF. We will deal with the case $\pm = +$, as the case $\pm = -$ is analogous and left as an exercise to the reader. Let us show that the elements (3.51) lie in $K^+ = \operatorname{Ker} \widetilde{\Upsilon}^+$. To keep the subsequent formulas short, we will suppress the

[10] In Subsection 3.7 we will prove that $J^\pm = K^\pm$.

indexing sets of $\sum$ and ρ. As we saw in (2.23) and (2.24), $\widetilde{\Upsilon}^+$ applied to the right-hand side of (3.51) is

$$\mathrm{Sym}\left[\sum \rho(z_1,\dots,z_n) \prod_{1\le u<v\le n} \frac{\widetilde{\zeta}_{i_v i_u}\left(\frac{z_v}{z_u}\right)\left(-\frac{z_u}{z_v}\right)^{\delta_{i_u i_v}\delta_{v\lhd u}} \zeta_{i_u i_v}\left(\frac{z_u}{z_v}\right)}{\widetilde{\zeta}^p_{(i_v,a_v)(i_u,a_u)}\left(\frac{z_v}{z_u}\right)}\right]$$

where "Sym" denotes symmetrization with respect to all z_u and z_v such that $i_u = i_v$ (and one needs to perform a relabeling (2.16) in order for the Laurent polynomial above to be an element of $\mathcal{V}^+$). The expression above is equal to the symmetric rational function

$$\prod_{1\le u\neq v\le n} \zeta_{i_u i_v}\left(\frac{z_u}{z_v}\right)$$

times

$$\mathrm{Sym}\left[\sum \rho(z_1,\dots,z_n) \prod_{1\le u<v\le n} \frac{\left(1-\frac{z_v}{z_u}\right)^{\delta_{i_u i_v}\delta_{(i_u,a_u)\not\sim(i_v,a_v)}}\left(-\frac{z_u}{z_v}\right)^{\delta_{i_u i_v}\delta_{v\lhd u}}}{\zeta^p_{(i_v,a_v)(i_u,a_u)}\left(\frac{z_v}{z_u}\right)}\right]$$

where $\sim$ denotes the equivalence relation that defines I^p, see (3.12), and $\delta_{x\not\sim y}$ is equal to 1 if $x \not\sim y$ and 0 otherwise. The expression directly above is equal to

$$\prod_{\substack{1\le u\neq v\le n\\ (i_u,a_u)\lhd(i_v,a_v)}} \left(1-\frac{z_v}{z_u}\right)^{\delta_{i_u i_v}}$$

(which takes the same value for all the summands in $\sum$, after relabeling) times

$$\mathrm{Sym}\left[\sum \rho(z_1,\dots,z_n) \prod_{1\le u<v\le n} \frac{1}{\zeta^p_{(i_v,a_v)(i_u,a_u)}\left(\frac{z_v}{z_u}\right)}\right]$$

The symmetrization above is 0 because $\phi_\bullet \in K^+_{A^p}$. By the chain of equalities above, we conclude that $\phi'_\bullet \in K^+$, as required.

We have just proved that $J^+ \subseteq K^+$, and have therefore established the $\Leftarrow$ implication of (3.52). As for the $\Rightarrow$ implication, let us consider an element $R^- \in \mathcal{V}^-$ such that $\langle J^+, R^-\rangle = 0$. First of all, note that for $\phi'_\bullet$ as in (3.51), we have

$$\left\langle \phi'_\bullet, R^-\right\rangle = \sum \int_{|z_1|\gg\cdots\gg|z_n|} \rho(z_1,\dots,z_n)$$

$$\prod_{1\le u<v\le n} \frac{\widetilde{\zeta}_{i_v i_u}\left(\frac{z_v}{z_u}\right)\left(-\frac{z_u}{z_v}\right)^{\delta_{i_u i_v}\delta_{v\lhd u}}}{\widetilde{\zeta}^p_{(i_v,a_v)(i_u,a_u)}\left(\frac{z_v}{z_u}\right)} \cdot \frac{R^-(z_1,\dots,z_n)}{\prod_{1\le u<v\le n}\zeta_{i_v i_u}\left(\frac{z_v}{z_u}\right)}$$

$$= \sum \int_{|z_1|\gg\cdots\gg|z_n|} \rho(z_1,\dots,z_n) \frac{R^-(z_1,\dots,z_n)\prod_{1\le u\neq v\le n}\left(1-\frac{z_v}{z_u}\right)^{\delta_{i_u i_v}\delta_{(i_u,a_u)\lhd(i_v,a_v)}}}{\prod_{1\le u<v\le n}\zeta^p_{(i_v,a_v)(i_u,a_u)}\left(\frac{z_v}{z_u}\right)}$$

$$= \left\langle \phi_\bullet, R^-(z_1,\dots,z_n) \prod_{\substack{1\le u\neq v\le n\\ (i_u,a_u)\lhd(i_v,a_v)}} \left(1-\frac{z_v}{z_u}\right)^{\delta_{i_u i_v}}\right\rangle$$

where the pairing on the top line denotes the one between $\widetilde{\mathbf{U}}^+$ and $\mathcal{V}^-$, while the pairing on the bottom line denotes the one between $\widetilde{\mathbf{U}}^+_{A^p}$ and $\mathcal{V}^-_{A^p}$. Analogous formulas as above hold when $\phi_\bullet$ and $\phi'_\bullet$ are multiplied by various products of $e_{i,d}$'s, thus yielding general elements of $J^+_{A^p}$ and J^+, respectively. The fact that R^- pairs trivially with the whole of J^+ implies that

$$\widetilde{R}^- = R^- \prod_{\substack{1 \leq u \neq v \leq n \\ (i_u, a_u) \lhd (i_v, a_v)}} \left(1 - \frac{z_v}{z_u}\right)^{\delta_{i_u i_v}} \tag{3.54}$$

pairs trivially with $J^+_{A^p}$, which by (3.45) means that $\widetilde{R}^- \in \mathcal{S}^-_p$. By (3.32), $\mathcal{S}^-_p$ is the intersection of ideals $\mathcal{I}$ which all enjoy the following analogue of property (3.27)

$$(z_u - z_v)x \in \mathcal{I} \quad \Rightarrow \quad x \in \mathcal{I}$$

whenever $i_u = i_v$ but $(i_u, a_u) \not\sim (i_v, a_v) \Leftrightarrow p_{i_u a_u} \neq p_{i_v a_v}$ (indeed, the only non-trivial situation is when $\mathcal{I}$ is generated by polynomials which involve both z_u and z_v; but in this case, $\mathcal{I}$ is supported on the locus

$$L = \left\{ \frac{z_u}{z_v} = \frac{p_{i_u a_u}}{p_{i_v a_v}} \neq 1 \right\}$$

and $z_u - z_v$ is nowhere vanishing on L). Therefore, we conclude that $R^- \in \mathcal{S}^-_p$; since this holds for all $p \in (\mathbb{K}^\times)^n / \mathbb{K}^\times$, (3.16) implies that $R^- \in \mathcal{S}^-$, as required. $\square$

3.7. Now that we have defined the elements $W_{p, \pm \boldsymbol{n}, \pm d} \in \widetilde{\mathbf{U}}^\pm$ for any point (3.11) and any $(\boldsymbol{n}, d) \in \mathbb{N}^I \times \mathbb{Z}$, it remains to show that the ideal $J^\pm$ they generate is equal to $K^\pm$, and this will imply Theorem 1.5. Comparing (2.19) with (3.52) shows that

$$\left\langle K^+, R^- \right\rangle = 0 \quad \Leftrightarrow \quad \left\langle J^+, R^- \right\rangle = 0 \tag{3.55}$$

If $\widetilde{\mathbf{U}}^+ \otimes \mathcal{V}^- \xrightarrow{\langle \cdot, \cdot \rangle} \mathbb{K}$ were a non-degenerate pairing of finite-dimensional vector spaces (or at least finite-dimensional in any $\mathbb{Z}^I \times \mathbb{Z}$ degree), then (3.55) would imply that $J^+ = K^+$ and the proof of Theorem 1.5 would be complete. Thus, we need to adapt this line of reasoning to the setting of infinite-dimensional vector spaces.

PROOF OF THEOREM 1.5. We will prove that $J^+ = K^+$, as the situation when $+$ is replaced by $-$ is analogous. Fix $(\boldsymbol{n}, d) \in \mathbb{N}^I \times \mathbb{Z}$ and our goal will be to show that

$$J_{\boldsymbol{n},d} = J^+ \cap \widetilde{\mathbf{U}}_{\boldsymbol{n},d} \quad \text{equals} \quad K_{\boldsymbol{n},d} = K^+ \cap \widetilde{\mathbf{U}}_{\boldsymbol{n},d} \tag{3.56}$$

By construction, we have

$$J_{\boldsymbol{n},d} \subseteq K_{\boldsymbol{n},d} \tag{3.57}$$

and it remains to prove the opposite inclusion. For any finite set $T \subset \mathcal{W}_{\text{non-inc}}$ of non-increasing words of degree $(\boldsymbol{n}, d)$, define

$$\widetilde{\mathbf{U}}^{+,T} = \bigoplus_{w \in T} \mathbb{K} \cdot e_w$$

(recall that as w runs over $\mathcal{W}_{\text{non-inc}}$, the e_w's yield a basis of $\widetilde{\mathbf{U}}^+$, as shown in Proposition 2.12). Moreover, the notion of "leading word" introduced in the latter part of the proof of Proposition 2.12 allows us to define

$$\mathcal{V}^{-,T} \subset \mathcal{V}_{-\boldsymbol{n},-d} \tag{3.58}$$

to be the subspace of Laurent polynomials R^-, all of whose monomials have leading word in T. Thus, the restriction of the pairing (2.12)

$$\widetilde{\mathbf{U}}^{+,T} \otimes \mathcal{V}^{-,T} \xrightarrow{\langle \cdot,\cdot \rangle} \mathbb{K} \tag{3.59}$$

is a non-degenerate pairing of finite-dimensional vector spaces (indeed, because the two vector spaces have the same dimension, it suffices to show non-degeneracy in the second argument, which follows immediately from (2.49)).

Let us consider a finite set of non-increasing words T of degree $(\boldsymbol{n}, d)$, which will be chosen in (3.71). The only thing we postulate for the time being is that the set T can be chosen "arbitrarily large", i.e. to contain any given finite set of words. Let $\mathcal{S}^{-,T} = \mathcal{S}^- \cap \mathcal{V}^{-,T}$. Formula (3.52) implies that

$$\mathcal{S}^{-,T} \subseteq \left(J^+ \cap \widetilde{\mathbf{U}}^{+,T}\right)^\perp$$

where the orthogonal complement is defined with respect to the pairing (3.59). Our goal will be to prove the opposite inclusion, namely the following result.

CLAIM 3.18. We have:

$$\left(J^+ \cap \widetilde{\mathbf{U}}^{+,T}\right)^\perp \subseteq \mathcal{S}^{-,T} \tag{3.60}$$

Let us first use (3.60) to establish the opposite inclusion to (3.57), and thus conclude the proof of Theorem 1.5. Because (3.59) is a non-degenerate pairing of finite-dimensional vector spaces, we have

$$\mathcal{S}^{-,T} = \left(J^+ \cap \widetilde{\mathbf{U}}^{+,T}\right)^\perp \quad \Rightarrow \quad J^+ \cap \widetilde{\mathbf{U}}^{+,T} = \left(\mathcal{S}^{-,T}\right)^\perp$$

However, we have $\left(\mathcal{S}^{-,T}\right)^\perp \supseteq K^+ \cap \widetilde{\mathbf{U}}^{+,T}$ by (2.19), so the display above implies

$$J^+ \cap \widetilde{\mathbf{U}}^{+,T} \supseteq K^+ \cap \widetilde{\mathbf{U}}^{+,T}$$

As the finite set T may be chosen arbitrarily large, the formula above implies

$$J_{\boldsymbol{n},d} \supseteq K_{\boldsymbol{n},d} \tag{3.61}$$

for all $(\boldsymbol{n}, d) \in \mathbb{N}^I \times \mathbb{Z}$, which precisely provides the opposite inclusion to (3.57).

PROOF OF CLAIM 3.18. The statement we need to prove is that for any $R^- \in \mathcal{V}^{-,T}$

$$\left\langle \phi, R^- \right\rangle = 0, \ \forall \phi \in J^+ \cap \widetilde{\mathbf{U}}^{+,T} \quad \Rightarrow \quad R^- \in \mathcal{S}^{-,T} \tag{3.62}$$

Equivalently, we will prove its contrapositive statement, which states that

$$R^- \in \mathcal{V}^{-,T} \backslash \mathcal{S}^{-,T} \quad \Rightarrow \quad \exists \phi \in J^+ \cap \widetilde{\mathbf{U}}^{+,T} \text{ s.t. } \left\langle \phi, R^- \right\rangle \neq 0 \tag{3.63}$$

Given $R^- \in \mathcal{V}^{-,T}$, the fact that $R^- \notin \mathcal{S}^{-,T}$ implies via (3.52) that there exists

$$\phi = e_{i_1,l_1} \dots e_{i_{n-k},l_{n-k}} W_{p,\boldsymbol{k},d-l_1-\dots-l_{n-k}} \in J^+ \tag{3.64}$$

for some $\mathbf{0} \leq \boldsymbol{k} \leq \boldsymbol{n}$ (we will henceforth denote $n = |\boldsymbol{n}|$ and $k = |\boldsymbol{k}|$), $p \in (\mathbb{K}^\times)^n/\mathbb{K}^\times$, $i_1, \dots, i_{n-k} \in I$ and $l_1, \dots, l_{n-k} \in \mathbb{Z}$ such that

$$\left\langle \phi, R^- \right\rangle \neq 0 \tag{3.65}$$

Moreover, we may assume the word

$$v = \left[i_1^{(l_1)} \dots i_{n-k}^{(l_{n-k})}\right] \tag{3.66}$$

to be non-increasing, and maximal such that (3.65) holds. We may also write

$$W_{p,\boldsymbol{k},d-l_1-\dots-l_{n-k}} = \sum_{\substack{i_{n-k+1},\dots,i_n \in I \\ l_{n-k+1},\dots,l_n \in \mathbb{Z}}} \text{coefficient} \cdot e_{i_{n-k+1},l_{n-k+1}} \dots e_{i_n,l_n} \tag{3.67}$$

where we only require in all terms of the sum above that

$$l_{n-k+1}, \dots, l_n \text{ are within a global constant away from their average} \tag{3.68}$$

We can accomplish this because Remark 3.14 for the shift vector $(1, \dots, 1) \in \mathbb{N}^I$ allows us to obtain all the elements $W_{p,\boldsymbol{k},\bullet}$ from finitely many values of $\bullet$. Then formula (3.65) implies that there are $i_1, \dots, i_n \in I$, $l_1, \dots, l_n \in \mathbb{Z}$ as above such that

$$0 \neq \left\langle e_{i_1,l_1} \dots e_{i_n,l_n}, R^- \right\rangle$$

If we write

$$R^- = \dots + \text{coefficient} \cdot z_1^{-k_1} \dots z_n^{-k_n} + \dots \tag{3.69}$$

(with the variables relabeled as in (2.16)), then we conclude that

$$0 \neq \int_{|z_1| \gg \dots \gg |z_n|} \frac{\dots + \text{coefficient} \cdot z_1^{l_1-k_1} \dots z_n^{l_n-k_n} + \dots}{\prod_{1 \leq a < b \leq n} \zeta_{i_b i_a}\left(\frac{z_b}{z_a}\right)} \prod_{a=1}^n Dz_a$$

$$= \int_{|z_1| \gg \dots \gg |z_n|} \frac{\dots + \text{coefficient} \cdot z_1^{l_1-d_1} \dots z_n^{l_n-d_n} + \dots}{\prod_{1 \leq a < b \leq n} \left(\alpha_{i_b i_a} + \sum_{s \geq 1} \text{coefficient} \cdot \frac{z_b^s}{z_a^s}\right)} \prod_{a=1}^n Dz_a$$

where $d_1, \dots, d_n$ are related to $k_1, \dots, k_n$ via formula (2.47). The maximality of the word (3.66) implies that

$$(l_1, \dots, l_{n-k}) = (d_1, \dots, d_{n-k}) \tag{3.70}$$

for some monomial which appears with non-zero coefficient in R^- (see (3.69)).

Choice of T: Let m and M be sufficiently large natural numbers, and define

$$T = \Big\{ w = \left[i_1^{(d_1)} \dots i_n^{(d_n)}\right] \in \mathcal{W}_{\text{non-inc}} \text{ such that } \deg w = (\boldsymbol{n}, d) \text{ and}$$
$$\sum_{a \in A} d_a \geq -M|A| + m|A|^2 + \sum_{A \ni a < b \notin A} (s_{i_a i_b} + s_{i_b i_a}), \forall A \subseteq \{1, \dots, n\} \Big\} \tag{3.71}$$

(since M can be arbitrarily large, this would ensure the fact that any finite set of non-increasing words can be contained in T). The reason we add $\sum_{A \ni a < b \notin A}(s_{i_a i_b} + s_{i_b i_a})$ to the right-hand side of the inequality above is the straightforward fact (which we leave as an exercise to the interested reader) that the inequality on the second line of (3.71) holds for the leading word (2.46) of a monomial (2.45) if and only if it holds for the analogous word associated to any other order of the variables in the monomial. Then (3.70) and the fact that $R^- \in \mathcal{V}^{-,T}$ imply that

$$\sum_{a \in A} l_a \geq -M|A| + m|A|^2 + \sum_{A \ni a < b \notin A} (s_{i_a i_b} + s_{i_b i_a}) \tag{3.72}$$

where $A = B$ or $A = B \sqcup \{n-k+1, \dots, n\}$, for arbitrary $B \subseteq \{1, \dots, n-k\}$.

We will use inequality (3.72) to show that ϕ of (3.64) lies in $\widetilde{\mathbf{U}}^{+,T}$ (which would conclude the proof of (3.63), and with it Claim 3.18). Let $b \gg \max(\beta(1), \ldots, \beta(n))$, with the β's as in (2.44). For any $i_1, \ldots, i_n \in I$, $l_1, \ldots, l_n \in \mathbb{Z}$ which appear in (3.64) and (3.67), consider the largest index $x \in \{0, \ldots, n-k\}$ such that

$$l_x < l_{x+1} - b \tag{3.73}$$

(we make the convention that $l_0 = -\infty$). We may use relation (2.44) to write

$$e_{i_{x+1},l_{x+1}} \ldots e_{i_n,l_n} = \sum_{\left[j_{x+1}^{(l'_{x+1})} \ldots j_n^{(l'_n)}\right] \in \mathcal{W}_{\text{non-inc}}} \text{coefficient} \cdot e_{j_{x+1},l'_{x+1}} \ldots e_{j_n,l'_n}$$

We need to make two observations about the l''s that appear in the formula above.

- All the numbers $l'_{x+1}, \ldots, l'_n$ are within a global constant away from their average
- The large difference between l_x and l_{x+1} ensures that all concatenated words

$$\left[i_1^{(l_1)} \ldots i_x^{(l_x)} j_{x+1}^{(l'_{x+1})} \ldots j_n^{(l'_n)}\right] \tag{3.74}$$

  which arise in the procedure above are non-increasing (recall that the word (3.66) was non-increasing to begin with, and thus so are all of its prefixes)

Thus, it remains to show that the concatenated words that appear in (3.74) are in T. By (3.72), we have

$$\sum_{a\in B} l_a + \sum_{a \in C} l'_a \geq -M|A| + m|A|^2 + \sum_{A \ni a < b \notin A} (s_{i_a i_b} + s_{i_b i_a}) \tag{3.75}$$

where $A = B \sqcup C$ for arbitrary $B \subseteq \{1, \ldots, x\}$, but $C = \varnothing$ or $C = \{x+1, \ldots, n\}$. If the defining property of T were to be violated, then we would have

$$\sum_{a\in B} l_a + \sum_{a \in C} l'_a < -M|A| + m|A|^2 + \sum_{A \ni a < b \notin A} (s_{i_a i_b} + s_{i_b i_a}) \tag{3.76}$$

where $A = B \sqcup C$ for some $B \subseteq \{1, \ldots, x\}$, but C a <u>proper</u> subset of $\{x+1, \ldots, n\}$. We claim that (3.75) and (3.76) are incompatible (for m chosen large enough compared to the global constant mentioned in the first bullet above, and the s_{ij}'s). Indeed, letting μ be the average from the first bullet above, relation (3.75) implies

$$\sum_{s\in B} l_s \geq -M|B| + m|B|^2 - \varepsilon \tag{3.77}$$

$$\sum_{s\in B} l_s + \mu(n-x) \geq -M(|B|+n-x) + m(|B|+n-x)^2 - \varepsilon \tag{3.78}$$

for some sufficiently large global constant ε. Meanwhile, (3.76) implies

$$\sum_{s\in B} l_s + \mu y < -M(|B|+y) + m(|B|+y)^2 + \varepsilon \tag{3.79}$$

where $y = |C|$ lies in $\{1, \ldots, n-x-1\}$. Subtracting (3.77) from (3.79) yields

$$\mu < -M + m(y + 2|B|) + 2\varepsilon$$

and subtracting (3.79) from (3.78) yields

$$\mu > -M + m(n - x + y + 2|B|) - 2\varepsilon$$

The two inequalities above are incompatible with each other if m is chosen sufficiently larger than ε, thus yielding the desired contradiction. □

□

Acknowledgments

I would like to express my gratitude to all the mathematicians and physicists that have worked on the wonderful constructions and ideas referenced in the present paper, with special thanks to Igor Frenkel for his many beautiful and fundamental contributions.

References

[1] B. Enriquez, *On correlation functions of Drinfeld currents and shuffle algebras*, Transform. Groups **5** (2000), no. 2, 111–120, DOI 10.1007/BF01236465. MR1762114

[2] B. L. Feigin and A. V. Odesskii, *Quantized moduli spaces of the bundles on the elliptic curve and their applications*, Integrable structures of exactly solvable two-dimensional models of quantum field theory (Kiev, 2000), NATO Sci. Ser. II Math. Phys. Chem., vol. 35, Kluwer Acad. Publ., Dordrecht, 2001, pp. 123–137. MR1873568

[3] Dmitry Galakhov, Wei Li, and Masahito Yamazaki, *Toroidal and elliptic quiver BPS algebras and beyond*, J. High Energy Phys. **2** (2022), Paper No. 024, 52, DOI 10.1007/jhep02(2022)024. MR4407441

[4] James A. Green, *Quantum groups, Hall algebras and quantized shuffles*, Finite reductive groups (Luminy, 1994), Progr. Math., vol. 141, Birkhäuser Boston, Boston, MA, 1997, pp. 273–290, DOI 10.1007/s10107-012-0519-x. MR1429876

[5] Maxim Kontsevich and Yan Soibelman, *Cohomological Hall algebra, exponential Hodge structures and motivic Donaldson-Thomas invariants*, Commun. Number Theory Phys. **5** (2011), no. 2, 231–352, DOI 10.4310/CNTP.2011.v5.n2.a1. MR2851153

[6] Wei Li and Masahito Yamazaki, *Quiver Yangian from crystal melting*, J. High Energy Phys. **11** (2020), 035, 124, DOI 10.1007/jhep11(2020)035. MR4204250

[7] A. Neguţ, *Reduced quiver quantum toroidal algebras*, `arXiv:2301.00703`

[8] A. Neguţ, *Quantum loop groups for symmetric Cartan matrices*, `arXiv:2207.05504`

[9] Andrei Neguţ, *Shuffle algebras for quivers and wheel conditions*, J. Reine Angew. Math. **795** (2023), 139–182, DOI 10.1515/crelle-2022-0074. MR4542416

[10] A. Neguţ, F. Sala, and O. Schiffmann, *Shuffle algebras for quivers as quantum groups*, `arXiv:2111.00249`

[11] Go Noshita and Akimi Watanabe, *A note on quiver quantum toroidal algebra*, J. High Energy Phys. **5** (2022), Paper No. 011, 41, DOI 10.1007/jhep05(2022)011. MR4430194

[12] Tudor Pădurariu, *K-theoretic Hall algebras of quivers with potential as Hopf algebras*, Int. Math. Res. Not. IMRN **7** (2023), 5781–5815, DOI 10.1093/imrn/rnac013. MR4565701

[13] Miroslav Rapčák, Yan Soibelman, Yaping Yang, and Gufang Zhao, *Cohomological Hall algebras, vertex algebras and instantons*, Comm. Math. Phys. **376** (2020), no. 3, 1803–1873, DOI 10.1007/s00220-019-03575-5. MR4104538

[14] Marc Rosso, *Quantum groups and quantum shuffles*, Invent. Math. **133** (1998), no. 2, 399–416, DOI 10.1007/s002220050249. MR1632802

[15] Peter Schauenburg, *A characterization of the Borel-like subalgebras of quantum enveloping algebras*, Comm. Algebra **24** (1996), no. 9, 2811–2823, DOI 10.1080/00927879608825714. MR1396857

[16] Olivier Schiffmann and Eric Vasserot, *The elliptic Hall algebra and the K-theory of the Hilbert scheme of* $\mathbb{A}^2$, Duke Math. J. **162** (2013), no. 2, 279–366, DOI 10.1215/00127094-1961849. MR3018956

Department of Mathematics, MIT, Cambridge, Massachusetts 02139; and Simion Stoilow Institute of Mathematics, Bucharest 010702, Romania

Current address: École polytechnique fédérale de Lausanne (EPFL), 1015 Lausanne, Switzerland

Email address: `andrei.negut@gmail.com`

Contemporary Mathematics
Volume **817**, 2025
https://doi.org/10.1090/conm/817/16328

On the topics of my conversations with Igor Frenkel

Samson L. Shatashvili

Dedicated to Igor Frenkel on his 70th birthday

ABSTRACT. This is a summary of my lecture at Igor Frenkel's 70th birthday conference. I give a brief review of my almost forty years of scientific interactions with Igor. I focus here on three topics of joint interests: 2-cocycles, coadjoint orbits and WZW_4. At the end of each topic I comment on new developments, if any.

1. Introduction

Talks on such occasions are usually of two types: Type I - speakers say a few words about the birthday person and continue to describe their own recent work (or similar); Type II - speakers give a talk specifically designed for the birthday person covering topics of their past/present joint interests. I adopted the latter, Type II, approach in my talk [**1**].

First time Igor and I met was around Thanksgiving in 1989 during my brief visit to Yale. But I knew Igor's work much before. In 1983 L. Faddeev and I were working on quantum anomalies and group cohomology and somewhere, in one of the western countries, Faddeev met I. M. Singer who told him that together with Igor they discovered 2-cocyle in three-dimensions but did not publish because they received (I believe from R. Jackiw) our manuscript [**2**] with the same result. In fact, there is a comment in I. M. Singer's Astérisque article about this [**3**]. Later I learned from Igor how one early morning in spring/summer of 1984 Singer knocked on his door in Aspen to inform him about this "bad" news. The first part of this talk, Section 2, will be about this 2-cocycle.

Igor and I became colleagues in 1994 when I moved to Yale. He had been at Yale for over a decade and in addition to being a colleague he became a close friend who would guide me in the new environment. We had a joint seminar series between Physics and Mathematics Departments (interestingly, I. M. Singer was responsible for such a program to exist) - "Geometry, Physics, Symmetry". This was one of the most successful seminar series I have ever witnessed in my forty years in academia. It very much shaped our life at Yale. At that time I was very much interested in the work of H. Nakajima on relation between instantons on ALE spaces and Kac-Moody Algebras. I learned about Nakajima's work from C. Vafa when we were

The author's research was partly supported by the award of the Simons Foundation to the Hamilton Mathematics Institute of the Trinity College Dublin under the program "Targeted Grants to Institutes".

working on string theories on exceptional holonomy manifolds in the Spring 1994 (Cumrun was working with E. Witten on $N = 4$ SYM and S-Duality at the same time and he pointed me to [**4**] which played an important role in their work). Since instanton moduli space was one of the topics of my PhD thesis 10 years before I obviously got very interested and shared my thoughts with Igor immediately after I arrived at Yale in the Fall of 1994. Not surprisingly it turned out that Igor was thinking about exactly the same thing, completely independently and on his own. So, we invited H. Nakajima to speak at "Geometry, Physics, Symmetry" and at the same time asked I. Grojnowski (who had just arrived at Yale as a postdoc in department of mathematics) to deliver series of lectures on the topics very much related to Nakajima's work. All this led to a very successful set of papers for both Igor and me [**5**]-[**13**]. I devote the third part of this talk to some aspects of this large topic - there have been recent developments related to one of the spinoffs of it (mainly related to [**5**], [**7**], [**9**]), due to K. Costello [**14**], and I will make a few comments about it below in Section 4.

Lastly, I want to mention coadjoint orbits, one of the topics of Igor's early interests, originating in late 70s and included in his PhD thesis [**15**]. I myself became interested in this topic much after Igor, in 1988-89, when together with A. Alekseev and L. Faddeev we were looking at path integrals (quantum mechanics) over coadjoint orbits as proper "upgrade" of A. Kirillov's original approach (including Kac-Moody and Virasoro coadjoint orbits [**16**], [**17**]). I recall many discussions with Igor on this topic at Yale in the 90s. Of course, Igor had numerous ideas how to improve these methods and I hope in this talk I can present one recent result (which might, or might not, surprise Igor - as we indeed discussed something very related 30 years ago) obtained in joint work with A. Alekseev [**18**], [**19**]. This I discuss in Section 3.

2. Gauge fields and 2-cocycle

Let $M_4 = \mathbb{R}^{1,3}$ be a four-dimensional space-time with local coordinates (x_0, x_1, x_2, x_3) and Minkowski metric

$$ds^2 = \sum_{\mu,\nu} \eta_{\mu\nu} dx^\mu dx^\nu, \eta^{\mu\nu} = diag(1, -1, -1, -1).$$

Let $\mathbf{G}$ be a compact Lie group, $\mathbf{g}$ its Lie algebra and $P \to M_4$ a principal G-bundle over M_4. We write $\mathcal{A}$ for the infinite-dimensional affine space of connections in the adjoint bundle $\operatorname{ad} P = P \times_G \mathbf{g}$ and $\mathcal{G}_4 = Map(M_4, G)$ for the infinite-dimensional group of gauge transformations. [1] On the space of connections $\mathcal{A}$ gauge group acts as:

$$A \to A^g = gAg^{-1} + dgg^{-1}, \qquad A = \sum_{\mu=0}^{3} A_\mu dx^\mu \in \mathcal{A}, \quad g \in \mathcal{G}_4, \tag{1}$$

$$F \to F^g = gFg^{-1}. \tag{2}$$

[1]When gauge field interacts with 'matter' (for example fermions), one also uses the complex vector bundle $E \to M_4$ associated with a given finite-dimensional representation $\rho : G \to EndV$ of G. We will use boldface notation $\mathbf{A}$ for the connections in the infinite-dimensional affine space of connections on E (as well boldface notation for the parameters of gauge transformation) later in this text.

Curvature F of a connection A is defined as $F = (d - A)^2 = -dA + A \wedge A$. Corresponding Lie-algebraic gauge transformations are:

$$A \to A^u = du - [A, u], \quad F \to -[F, u], \quad u \in \mathfrak{g}_4, \tag{3}$$

where $\mathfrak{g}_4$ denotes the Lie algebra of $\mathcal{G}_4$.

In the Hamiltonian approach for the four-dimensional gauge theories the Gauss law plays a central role. In this approach one considers space $M_3 = \mathbb{R}^3$ (or any other three-dimensional space, but for simplicity we will treat here the case of $\mathbb{R}^3$), a time direction with coordinate $x_0 = t$ and focus is on constant time slice (for example $t = 0$). Now one has the principal G-bundle P_3 over M_3, connections (gauge fields) $A = \sum_{i=1}^3 A_i dx^i$ and time-independent gauge transformations $\mathcal{G}_3 = Map(M_3, G)$. Infinite-dimensional phase space of this Hamiltonian system has coordinates (A, E) where $E = \sum_i E_i dx^i = \sum_i F_{0i} dx^i$. Denote by t^a the generators of Lie algebra $\mathbf{g}$ ($a = 1, \ldots, n = dimG$; $\mathrm{Tr}(t^a t^b) = -2\delta^{ab}, [t^a, t^b] = \sum_c f_c^{ab} t^c$) and write $A_i(x) = \sum_a A_i^a(x) t^a, E_i(x) = \sum_a E_i^a(x) t^a$. In these coordinates the canonical symplectic form on the phase space and canonical Poisson structure are given by:

$$\Omega = \int_{M_3} Tr\delta A_i \delta E_i, \qquad \{A_i^a(x), E_j^b(y)\} = \delta^{ab}\delta_{ij}\delta^{(3)}(x - y) \tag{4}$$

and the Hamiltonian is ($B_i = \frac{1}{2}\sum_{jk} \epsilon_{ijk} F_{jk}$, ϵ_{ijk} - totally antisymmetric tensor, $\epsilon_{123} = 1$):

$$H = \int_{M_3} Tr \sum_{i=1}^{3} \frac{E_i^2 + B_i^2}{2} \tag{5}$$

Gauss Law generators are:

$$G(u) = \int_{M_3} Tr(\partial_i u - [A_i, u]) E_i \tag{6}$$

They generate the Lie-algebraic gauge transformations (3) under Poisson brackets (4):

$$\{G(u), A\} = A^u = du - [A, u], \qquad \{G(u), E\} = E^u = -[E, u] \tag{7}$$

and satisfy:

$$\{G(u), G(v)\} = G([u, v]) \tag{8}$$

If one adds the matter, for example Weyl (or Dirac) fermions ψ in the representation ρ, the phase space with its symplectic structure (Poisson brackets) and Hamiltonian will be modified appropriately, but one will always have the functional on phase space $\mathbf{G}(\mathbf{u})$ that generates gauge transformation and satisfies (8) under the corresponding Poisson brackets in the Hamiltonian approach. In fact, in the case of fermions - the modified Gauss law $\mathbf{G}(\mathbf{u}) = G(u) + \int_{M_3} Tr_\rho \mathbf{uJ}$ where $\mathbf{J}$ is a fermionic current, bilinear in ψ.

Classical "physical" phase space is given by a Hamiltonian reduction of above phase space under the contraints $\mathbf{G}(\mathbf{u}) = 0$. Upon quantization, when $\mathbf{G}(\mathbf{u})$ becomes operator in an appropriate second-quantized Hilbert space and Poisson brackets are replaced by commutators, $\{\bullet, \bullet\} \to i[\bullet, \bullet]$ the relations (8) might get modified, but $\mathbf{G}(\mathbf{u})$ is required to generate the gauge transformations (7) with Poisson bracket replaced by commutator and satisfy Jacoby identity.

The study of non-abelian quantum anomalies in [**2**], [**20**] led to a possibility of the consistent modification of (8) with a non-trivial extension given by a gauge-field dependent 2-cocycle:

$$[\mathbf{G}(\mathbf{u}), \mathbf{G}(\mathbf{v})] = \mathbf{G}([\mathbf{u}, \mathbf{v}]) + \frac{i}{12\pi^2}\int_{M_3} Tr_\rho \mathbf{A} d\mathbf{u} d\mathbf{v} \tag{9}$$

Here Tr_ρ, gauge field $\mathbf{A}$, and $\mathbf{u}$, $\mathbf{v}$ all are in the representation ρ of gauge group. This is the 2-cocycle Igor and I. Singer also found, but didn't publish, as mentioned in the introduction.

It is important to stress that the extension in (9) is not a central extension - the generators of the algebra, $\mathbf{G}(\mathbf{u})$, do not commute with it because $[\mathbf{G}(\mathbf{u}), \mathbf{A}] = d\mathbf{u} - [\mathbf{A}, \mathbf{u}]$. But direct check shows that the new commutation relations (9) satisfy Jacoby identity:

$$[\mathbf{G}(\mathbf{u_1}), [\mathbf{G}(\mathbf{u_2}), \mathbf{G}(\mathbf{u_3})]] + cyclic = 0 \tag{10}$$

and that the extension is not trivial (in the space of local functionals of gauge field).

If $\mathbf{G}(\mathbf{u})$ in quantum theory satisfies (9) instead of (8) the theory is anomalous, the operator version of classical constraints $\mathbf{G}(\mathbf{u}) = 0$ makes no sense - one needs to find a "new way" to quantize the theory.

A few words about how this 2-cocycle appeared in the theory of quantum anomalies [**21**], [**22**]. Here I will follow the paper [**2**]. One considers Yang-Mills field interacting with Weyl fermions ψ in a representation ρ of gauge group. A quantum anomaly is the assertion that the partition function $Z(\mathbf{A})$ of Weyl fermions in an external gauge field $\mathbf{A}$ (which is formally given by the determinant of Weyl operator - a "square root" of the determinant of the Dirac operator in the external gauge field $\mathbf{A}$) is not gauge invariant. One easily sees that there is a regularization (for example, using the ζ-function) under which the determinant of the Dirac operator is in fact gauge invariant. Thus, the partition function of Weyl fermions $Z(\mathbf{A})$ is invariant up to a phase:

$$Z(\mathbf{A}) \to Z(\mathbf{A}^g) = e^{-i\alpha_1(A,g)} Z(\mathbf{A}) \tag{11}$$

One can modify the action of the group:

$$Z(A) \to U(g) \cdot Z(\mathbf{A}) = e^{i\alpha_1(A,g)} Z(\mathbf{A}^g) \tag{12}$$

and under this new action $Z(\mathbf{A})$ is gauge invariant: $U(g) \cdot Z(\mathbf{A}) = \mathbf{Z}(\mathbf{A})$ as it follows from (11). Since (12) needs to be a group action, $U(g_1)U(g_2) = U(g_1 g_2)$, we conclude, when we act twice using (12), that possible $\alpha_1(A, g)$'s are restricted to those satisfying:

$$(\delta\alpha_1)(A; g_1, g_2) = \alpha_1(A^{g_1}, g_2) - \alpha_1(A, g_1 g_2) + \alpha_1(A, g_1) = 0 \qquad mod \quad 2\pi \tag{13}$$

Thus one is looking for the solution of (13) in the space of ("local") functionals of gauge field A and a group element g up to the identification $\alpha_1(A, g) \to \alpha_1(A, g) + \delta\beta, \quad \delta\beta(A) = \beta(A^g) - \beta(A)$. [2]

In general, one can introduce the "differential" δ in an abstract setup. Thus, let G be an abstract group and M a right G-module. Consider the complex $\alpha_\bullet(M, G)$

[2] Change of the regularization scheme corresponds to the redefinition of partition function $Z(\mathbf{A}) \to e^{i\beta(A)} Z(\mathbf{A})$ with some local functional $\beta(A)$ (local counterterm in physics), thus $\alpha_1(A, g)$ is defined up to the variation of this $\beta(A)$.

with cochains $\alpha_k(M,G)$, the spaces of functions on $M\times G^k$, $k=0,1,2,\dots$, and with the differential δ:

$$\delta : \alpha_k(M,G) \to \alpha_{k+1}(M,G) \tag{14}$$

defined by:

$$\begin{aligned}(\delta\alpha_k)(m;g_1,g_2,\dots,g_{k+1}) &= \alpha_{k+1}(m;g_1,g_2,\dots,g_{k+1})\\ &= \alpha_k(m\cdot g_1;g_2,\dots,g_{k+1}) - \alpha_k(m;g_1g_2,g_3,\dots,g_{k+1})\\ +\sum_{i=2}^{n}(-1)^i\alpha_k(m;g_1,\dots,g_ig_{i+1},g_{i+2},&\dots,g_{k+1}) + (-1)^{n+1}\alpha_k(m;g_1,\dots,g_k)\end{aligned} \tag{15}$$

One can check that $\delta^2=0$. This cochain was applied in [**2**] to the case when G is the gauge group and $M=\mathcal{A}$ in order to interpret the anomalous phase $\alpha_1(A,g)$ (called Wess-Zumino-Witten action in physics) in (11) as a corresponding 1-cocycle since for $n=1$ (15) gives exactly (13): $\delta\alpha_1(A,g)=0$. [**2**] suggested to use the bicomplex with operators δ and d (the exterior differentiation operator arising in the theory of Chern-Simons secondary characteristic classes) in a following way: start with manifold M^{2n+2} of dimensions $(2n+2)$ and consider the Chern polynomial

$$w_{2n+2} = \frac{1}{n!}tr_\rho F^{n+1} \tag{16}$$

Since

$$dw_{2n+2}=0, \qquad \delta w_{2n+2}=0 \tag{17}$$

one can find locally on M^{2n+2} a differential form of degree $(2n+1)$ such that

$$dw_{2n+1} = w_{2n+2} \tag{18}$$

According to Novikov, globally on M^{2n+2} it is a multivalued form:

$$w_{2n+1} = d^{-1}w_{2n+2} \tag{19}$$

but if $M^{2n+1}=\partial B^{2n+2}$ is a $2n+1$ dimensional cycle in M^{2n+2} which is a boundary of a $(2n+2)$-dimensional submanfold B^{2n+2} of M^{2n+2} one can integrate w_{2n+1} over M^{2n+1} and obtain well-defined functional called Chern-Simons action:

$$I_{CS}(A) = 2\pi\int_{B^{2n+2}} w_{2n+2}(A') = 2\pi\int_{M^{2n+1}} w_{2n+1}(A) \tag{20}$$

where one chooses any extension A' of the connection A from M^{2n+1} to B^{2n+2}. For two such extensions, the difference

$$\int_{B^{2n+2}} w_{2n+2}(A') - \int_{B^{2n+2}} w_{2n+2}(A'') \tag{21}$$

is the integral of w_{2n+2} over a top-dimensional cycle in M^{2n+2} so assuming the integrality of the cohomology class (the coefficient in the definition of w_{2n+2} was chosen appropriately)

$$[w_{2n+2}] \in H^{2n+2}(M^{2n+2},\mathbb{Z}) \tag{22}$$

one gets univalued exponent $e^{iI_{CS}(A)}$ and univaluedness of this exponent is enough (due to the path integral properties used in physics).

It is easily verifiable that the following $(2n+1)$-form is closed:

$$\delta w_{2n+1} = w_{2n+1}(A^g) - w_{2n+1}(A); \qquad d\delta w_{2n+1} = d\delta d^{-1}w_{2n+2} = \delta w_{2n+2} = 0 \tag{23}$$

It follows that, in contrast to the Chern-Simons action, its gauge variation

$$\alpha_1(A,g) = I_{CS}(A^g) - I_{CS}(A) \tag{24}$$

is well defined (modulo $\mathbb{Z}$) also in the case when M^{2n+1} is not a cycle but a chain with boundary $\partial M^{2n+1} = M^{2n}$:

$$\alpha_1(A,g) = 2\pi \int_{M^{2n+1}} \delta w_{2n+1}(a) = 2\pi \int_{M^{2n}} d^{-1}\delta w_{2n+1} = 2\pi \int_{M^{2n}} d^{-1}\delta d^{-1} w_{2n+2} \tag{25}$$

and $\delta\alpha_1(A,g) = 0$ as a consequence of $\delta^2 = 0$.

Thus (25) gives the desired expression for the phase in (12) (Wess-Zumino-Witten action) for the transformation property of the determinant of Weyl operator (11) - in four-dimensional case of Weyl fermions we just choose $n = 2$.

Finally, if one considers gauge theory with $CS(A)$ action for the $2n+1$ dimensional space-time manifold with $2n$-dimensional boundary and with anomalous chiral Weyl fermions living on this boundary - the total partition function will be gauge invariant since variation of CS action (24) will exactly cancel the anomalous phase of the chiral fermion partition function (12). This is frequently called the "anomaly inflow" mechanism [**2**], [**23**].

The "descent" procedure, described above, can be continued. At k-th step, when d^{-1} is applied k-times, one gets a closed $(2n+2-k)$-form:

$$w_{2n+2-k} = \delta d^{-1}\delta d^{-1} \ldots \delta d^{-1} w_{2n+2} \tag{26}$$

which when integrated over appropriate boundary $\partial M^{2n+2-(k-1)} = M^{2n+2-k}$ produces $(k-1)$-cocycle:

$$\alpha_{k-1}(A; g_1, \ldots, g_{k-1}) = \int_{M^{2n+2-k)}} w_{2n+2-k}, \qquad \delta\alpha_{k-1} = 0. \tag{27}$$

There is an algebraic version of this construction (equivalent to that of [**24**], [**25**] where Faddeev-Popov Ghosts were used and the BRST operator was defined instead of straightforward algebraic counterpart of δ from (15)) - k-th cochain $\mathbf{a}_k(A; u_1, \ldots, u_k)$ is a functional of gauge field and $u_1, \ldots, u_k \in \mathfrak{g}$ and δ in (15) is defined in terms of gauge transformation from Lie algebra - $A^u = du - [A, u]$, and group multiplication $g_1 g_2$ is replaced by commutator $[u_1, u_2]$. In the algebraic version multi-valuedness, discussed above, is absent.

The third step, $k = 3$, leads to a 2-cocycle:

$$\alpha_2(A; g_1, g_2) = \int_{M^{2n-1}} d^{-1}\delta d^{-1}\delta d^{-1} w_{2n+2}(A) \tag{28}$$

which may appear in the projective representation of gauge group:

$$V(g_1) \cdot V(g_2) = e^{i\alpha_2(A; g_1, g_2)} V(g_1 g_2) \tag{29}$$

and the 2-cocycle property of $\alpha_2(A; g_1, g_2), \delta\alpha_2(A; g_1, g_2) = 0$, guarantees associativity in (29).

Again, for $n = 2$ one gets the 2-cocycle in three-dimensions, since the integral is over three-manifold in (28). One can explicitly compute the algebraic form of this $\alpha_2(A; g_1, g_2)$ from the above construction (multivaluadness subtleties are absent in the algebraic setup, everything can be explicitly calculated starting descent from the Chern class $w_6 = \frac{1}{3!} tr_\rho F^3$) - the result is exactly the extension entering in the Gauss Law commutation relations given by the formula (9).

In [2] it was conjectured that in the anomalous theory of Weyl fermions interacting with gauge fields in (3+1)-dimensions the total space of quantum states is fibered over the space of all gauge fields $\mathcal{A}$ with fibers given by the fermionic Fock-space (denote the section by $\Psi(A)$) and gauge group acts as:

$$V(g)\cdot\Psi(A)=U(A,g)\Psi(A^g) \tag{30}$$

(it is useful to compare with (12)) with some operator $U(A,g)$ satisfiying:

$$U(A,g)U(A^g,h)=e^{i\alpha_2(A;g,h)}U(A,gh) \tag{31}$$

Unfortunately, for non-abelian G, such representation (in a proper Hilbert space) was not found at that time. I have not been following the topic for over 30 years. I will be more than grateful if someone could point me to a paper(s) in mathematics literature where (if) this has been done. In the path integral approach the validity of this conjecture was demonstrated in the 80s, see e.g. [26].

3. Coadjoint orbits

The first example of orbital integral of an exponential function goes back to Harish-Chandra (in modern days called Harish-Chandra-Itzykson-Zuber integral [27], [28]):

$$\int_{U(n)}du\, e^{\mathrm{Tr}(AuBu^{-1})}=C_n\frac{\sum_{\sigma\in S_n}(-1)^{|\sigma|}e^{\sum_i a_ib_{\sigma(i)}}}{\prod_{i<j}(a_i-a_j)(b_i-b_j)} \tag{32}$$

for two Hermitian $n\times n$ matrices A and B with with distinct eigenvalues $a_1,\dots,a_n$ and $b_1,\dots b_n$. Here du is the Haar measure on the group $U(n)$, $C_n=\prod_{i=1}^n i!$. The left hand side in the expression (32) can be understood, up to the inverse of Vandermond determinant $\Delta(B)=\prod_{i<j}(b_i-b_j)$ of matrix B, as an integral over the coadjoint orbit $\mathcal{O}_B=\{uBu^{-1};u\in U(n)\}$ under the action of $U(n)$ by conjugations. Individual terms in the sum on the right hand side are in one-to-one correspondence with elements of the orbit invariant under conjugations by diagonal matrices given by permutations of eigenvalues of B: $B_\sigma=\mathrm{diag}(b_{\sigma(1)},\dots,b_{\sigma(n)})$ (this sum can be rewritten as a $\det\left[e^{a_ib_j}\right]$).

Integral in (32) is an example of Duistermaat-Heckman (DH) integral [29][3]. Let M be a compact symplectic manifold of dimension $\dim_{\mathbb{R}}(M)=2d$ with symplectic form ω. We assume that it carries a Hamiltonian action of a torus T with isolated fixed points $p_1,\dots,p_m$. We denote the Lie algebra of T by $\mathfrak{t}=\mathrm{Lie}(T)$ (since fixed points are isolated weights of the T-action on tangent spaces $T_{p_i}M$, denoted by $w_{a,i}\in\mathfrak{t}^*$ for $a=1,\dots,d,i=1,\dots,m$, are non-vanishing). We denote the corresponding moment map by $\mu:M\to\mathfrak{t}^*$. The Duistermaat-Heckman formula computes the following oscillating integral

$$I_M(\xi)=\int_M\frac{\omega^d}{d!}e^{i\langle\mu,\xi\rangle}, \tag{33}$$

where $\xi\in\mathfrak{t}$. The result is given by the localization formula:

$$I_M(\xi)=\left(\frac{2\pi}{\sqrt{-1}}\right)^d\sum_{i=1}^m\frac{e^{i\langle\mu(p_i),\xi\rangle}}{\prod_{a=1}^d\langle w_{a,i},\xi\rangle} \tag{34}$$

[3] One of the first papers expressing the Kirillov's integral over coadjoint orbits of compact Lie groups [30] as a sum (32) was [31].

which is known to be a scaling limit of localization formula for the T-equivariant index of a symplectic Dirac operator twisted by the pre-quantum line bundle.

For coadjoint orbits corresponding to compact Lie groups (orbit $\mathcal{O}_B$ in case of (32)) the DH integral coincides with (32). According to Kirillov, who originally introduced the orbit method in 1961 [**30**], the character of the irreducible representation of the compact Lie group corresponding to the orbit ($\mathcal{O}_B$ in (32)) is given by (33) (where M is a coadjoint orbit) but up to a factor $p(\xi)$ which is universal and doesn't depend on orbit (and is identity at $\xi = 0$) - see formula Φ in [**30**]. Kirillov also mentioned in [**30**] that proper integral formula for the character, which automatically gives this pre-factor $p(\xi)$, should be the functional integral (Feynman integral, which explains why the main formula was denoted as Φ) over coadjoint orbit and not an ordinary integral like in (33) (thus he thought character should be obtained by quantum mechanics on coadjoint orbit), but such path integral was not known at that time; corresponding path integral for compact Lie groups $SU(N)$ and $SO(N)$ was obtained in 1980's, see [**32**].

Igor pioneered the topic of orbital integrals for the infinite-dimensional case. Basically he introduced the infinite-dimensional version of (33) for the coadjoint orbits of loop group $\widehat{LG}$ [**15**] suggesting the Wiener type measure (his integral probably best can be thought of in the spirit of the integration over the group manifold as in (32)). Calculating the integral he showed that answer can be written in a very similar form to (34), see formula (53) below. After Igor's work the orbit method for loop groups was developed in many papers, both in mathematics and physics literature. Here I list a few, in addition to those already mentioned. In physics paper [**33**] was very similar to Igor's approach. Related, and very recently, a new supersymmetric localization principle for the path integral over the tangent bundle to LG was formulated in [**34**], applied to derive an exact formula for the propagator on a group manifold as well as the Selberg trace formula [**35**]. Symplectic geometry approach to formal DH integral on $\widehat{LG}$ coadjoint orbits has been developed in [**36**], and more recently addressed by [**37**]. The probabilistic approach to $\widehat{LG}$ orbital integrals has been further developed in [**38**], [**39**]. Rigorous mathematical derivation of DH integral for infinite-dimensional case of $\widehat{LG}$ coadjoint orbits is contained in the approach of [**40**], [**41**] using hypoelliptic Laplacian on $G \times \mathbf{g}$.

The orbit method appeared recently in a wide variety of topics in modern theoretical physics. To list just a few interesting connections, we mention the SYK model [**42**], [**43**] as studied in [**44**]-[**52**], and the vacuum decay in CFT of [**53**]. In the SYK model, and related generalizations, one is interested in $\widehat{\mathrm{Diff}}(S^1)$ coadjoint orbits. For example the infinite-dimensional orbital integrals of the type

$$I(\tau) = \int_{\mathrm{Diff}(S^1)} \mathcal{D}f \exp\left(\tau \int_{S^1} S(f(s))ds\right) \tag{35}$$

were recently studied in [**52**], [**54**] (here $f(s)$ is a diffeomorphism $f(s) \in \widehat{\mathrm{Diff}}(S^1)$ and $S(f)$ is a Schwarzian derivative as a Hamiltonian on coadjoint orbit, see next paragraph) in relation to SYK.

As mentioned above - already Kirilliov in the 60s stressed that "correct" approach to the relation between characters and orbital integrals should be in terms of path integral. In quantum mechanics, for the phase space with local coordinates $(p, q) \equiv (p_i, , , p_n; q_i, \dots, q_n)$, Hamiltonian function $H(p, q)$ and Planck constant $\hbar$ the partition function is given by path integral (integral over closed paths,

parametrized by t, of length T, on the phase space $(p(t), q(t))$:
(36)

$$Z_\hbar = \int \prod_t \prod_i dp_i dq_i e^{\frac{i}{\hbar}\int(\sum_i p_i \dot{q}_i - H(p,q))dt} = \int \prod_t \prod_i dp_i dq_i e^{\frac{i}{\hbar}\int(d^{-1}\omega - H(p,q)dt)}$$

In the limit $\hbar \to 0, T \to 0, T/\hbar = \tau = fixed$ (classical limit) this integral becomes an ordinary integral over phase space as in (33):[4]

$$Z_{\hbar\to 0} = \int \prod_i dp_i dq_i e^{-iH(p,q))\tau} \tag{37}$$

thus assuming that character of irreducible representation, when phase space is given by coadjoint orbit, coincides with the partition function (36), one expects [**32**], [**44**], [**19**], [**54**] that this classical limit will be given by corresponding DH formula.

This idea can be verified for the cases when both character formula and DH formula are known independently. It can be considered as a conjecture when the answer for DH integral is not known but character is known, e.g. from algebraic methods. One of the infinite-dimensional examples, when both the character and DH integral formula are known, is precisely the case studied by Igor [**15**] which we will consider in a moment.

Before considering the scaling for the infinite-dimensional case and Igor's formula we will illustrate this scaling property of the character formula on the examples of coadjoint orbits for the finite-dimensional compact Lie groups. Let G be a compact Lie group. Consider a character $\chi_\lambda(g)$ of an irreducible representation of highest weight λ. There is a coadjoint orbit $\mathcal{O}_\lambda$ associated to this representation. We denote its dimension by $\dim(\mathcal{O}_\lambda) = 2d$. Then, the character admits the following interesting asymptotic expansion:

$$\chi_{k\lambda}\left(e^{\frac{h}{k}}\right) = \left(\frac{k}{2\pi}\right)^d I^{\mathrm{DH}}_{\mathcal{O}_\lambda}(h) + \dots. \tag{38}$$

Here $k \in \mathbb{Z}$ is a large parameter, $h \in \mathfrak{t} = \mathrm{Lie}(T)$ is in the Lie algebra of the maximal torus $T \subset G$, $I^{\mathrm{DH}}_{\mathcal{O}_\lambda}(h)$ is a DH integral formula (33), (34) for $M = \mathcal{O}_\lambda$ (h plays the role of ξ), and $\dots$ stands for subleading terms in k. Note that in (38) we scale the highest weight λ with a large factor k and that at the same time we scale down the group element $\exp(h/k)$ so as it gets closer and closer to the group unit.

The first example we consider, in order to demonstrate the validity of (38), is the case of the character of the compact group $SU(2)$ and corresponding DH integral (coadjoint orbit is a sphere S^2_s of radius s - radius corresponds to spin):

$$\chi_s(\theta) = \frac{sin(s+\frac{1}{2})\theta}{sin\frac{\theta}{2}} \tag{39}$$

and scaling (38) gives:

$$I_{S^2}(\theta) = \lim_{k\to\infty}\frac{2\pi}{k}\frac{\sin\left(\left(ks+\frac{1}{2}\right)\frac{\theta}{k}\right)}{\sin\left(\frac{\theta}{2k}\right)} = 2\pi\frac{e^{is\theta} - e^{-is\theta}}{i\theta} = \int_{S^2_s} e^{iz\theta} dz \wedge d\phi \tag{40}$$

which is exactly the DH integral for the sphere S^2_s expressed in cylindrical coordinates.

[4]This argument doesn't take into account certain normalization factors, but scaling property is captured properly, as shown by examples below.

Similarly, for a coadjoint orbit O_λ of a compact connected Lie group G we obtain the following scaling

(41)
$$\lim_{k\to\infty}\left(\frac{2\pi}{k}\right)^d\sum_{w\in W}\frac{e^{ik\langle w(\lambda),\xi/k\rangle}}{\prod_{\alpha\in\Delta_+}(1-e^{-i\langle w(\alpha),\xi/k\rangle})}=\left(\frac{2\pi}{\sqrt{-1}}\right)^d\sum_{w\in W}\frac{e^{i\langle w(\lambda),\xi\rangle}}{\prod_{\alpha\in\Delta_+}\langle w(\alpha),\xi\rangle}.$$

The left hand side is scaled character and the right hand side of this equation is the Duistermaat-Heckman localization formula for the orbital integral (34). Thus the heuristic argument with path integrals in (36), (37) turns out to be precisely correct with appropriate normalization factor as in (38).

Now we move on to the infinite-dimensional situation, the case of loop groups (we focus on the case of $\widehat{\mathrm{su}(2)}_l$ for simplicity). For $\mathfrak{g}=\mathrm{su}(2)$ and l a positive integer, the Kac-Moody algebra $\widehat{\mathrm{su}(2)}_l$ admits $l+1$ integrable irreducible representations labeled by spins $j=0,1/2,\ldots,l/2$. The Kac-Weyl character formula gives characters of these representations:

$$\chi_{l,j}(\tau,\gamma)=\mathrm{Tr}_{V_{l,j}}\left(e^{2\pi i\tau L_0+i\gamma J_0^z}\right)=\frac{\Delta_{l,j}(\tau,\gamma)}{D(\tau,\gamma)},\tag{42}$$

(43)
$$\begin{aligned}&\Delta_{l,j}(\tau,\gamma)\\&\quad=e^{2\pi i\tau j(j+1)/(l+2)}\sum_{n\in\mathbb{Z}}e^{2\pi i\tau((l+2)n^2+n(2j+1))}(e^{i(j+(l+2)n)\gamma}-e^{-i(j+1+(l+2)n)\gamma}),\end{aligned}$$

and the denominator does not depend on j:

$$D(\tau,\gamma)=(1-e^{-i\gamma})\prod_{r=1}^{\infty}(1-e^{2\pi ir\tau})(1-e^{2\pi ir\tau}e^{i\gamma})(1-e^{2\pi ir\tau}e^{-i\gamma}).\tag{44}$$

Inspired by the finite dimensional case, we introduce the following scaling:

$$l\mapsto kl,j\mapsto kj,\tau\mapsto\frac{\tau}{k},\gamma\mapsto\frac{\gamma}{k}.\tag{45}$$

Under scaling transformations, the numerator of the character formula behaves as follows:

$$\Delta_{kl,kj}\left(\frac{\tau}{k},\frac{\gamma}{k}\right)\to_{k\to\infty}e^{2\pi i\tau j^2/l}\sum_{n\in\mathbb{Z}}e^{2\pi i\tau((ln^2+2jn)}(e^{i(j+ln)\gamma}-e^{-i(j+ln)\gamma}).\tag{46}$$

Note that it has a well defined limit, and that all the terms of the Kac-Weyl formula are still present with slightly simplified exponents.

Now we turn to the large k behaviour of the denominator of the character formula. Recall that the Dedekind η-function $\eta(\tau)=q^{1/24}\prod_{n=1}^{\infty}(1-q^n)$, where $q=\exp(2\pi i\tau)$ and $\mathrm{Im}\,\tau>0$ verifies the modular property

$$\eta(-1/\tau)=\sqrt{-i\tau}\eta(\tau).\tag{47}$$

This implies

$$\frac{1}{\prod_{n=1}^{\infty}(1-q^n)}=\frac{1}{\prod_{n=1}^{\infty}(1-e^{2\pi in\tau})}=\frac{\sqrt{-i\tau}e^{\pi i(\tau+\tau^{-1})/12}}{\prod_{n=1}^{\infty}(1-e^{-2\pi in/\tau})}.\tag{48}$$

Furthermore, recall the following identity:

$$(1-e^{-i\gamma})\prod_{l=1}^{\infty}(1-e^{2\pi il\tau}e^{i\gamma})(1-e^{2\pi il\tau}e^{-i\gamma})=(q^{x},q)_\infty(q^{1-x},q)_\infty, \tag{49}$$

where $(a,q)_\infty = \prod_{k=0}^{\infty}(1-aq^k)$ is the Pochhammer symbol and $x = -\gamma/2\pi\tau$. The combination of Pochhammer symbols on the right hand side of (49) has the following asymptotic behaviour (see Corollary 3.3 in **[55]** and Corollary 1.3 in **[56]**):

$$(q^{\tau},q)_\infty(q^{1-\tau},q)_\infty \sim_{q\to 1} 2\sin(\pi x)e^{\pi^2/3\ln(q)}q^{-1/2(1/6-x+x^2)}. \tag{50}$$

Applying rescaling (45), we obtain the asymptotic behaviour of the inverse denominator for $k\to\infty$:

$$\begin{aligned} D\left(\frac{\tau}{k},\frac{\gamma}{k}\right)^{-1} &\sim_{k\to\infty} \frac{1}{2\sin(\gamma/2\tau)}\left(\frac{-i\tau}{k}\right)^{1/2} e^{i\pi k/4\tau} \\ &= \left(\frac{2\pi}{k}\right)^{1/2} e^{i\pi k/4\tau}\cdot\frac{(-i\tau)^{1/2}}{2\sqrt{2\pi}\sin(\gamma/2\tau)}. \end{aligned} \tag{51}$$

The divergent factor is of the form $(k/2\pi)^d\exp(i\pi\alpha k/12\tau)$ with $d=-1/2$ and $\alpha=3$. The factor $(k/2\pi)^d$ is similar to the finite dimensional case even though the "dimension" d is not an integer and is negative. The factor $\exp(i\pi\alpha k/12\tau)$ has an essential singularity at $k=\infty$, and this singularity depends on the Lie algebra parameter τ.[5]

In conclusion, the behavior of the character under the scaling transformation is given by the formula

$$\chi_{kl,kj}\left(\frac{\tau}{k},\frac{\gamma}{k}\right) \sim_{k\to\infty} \left(\frac{2\pi}{k}\right)^{1/2} e^{i\pi k/4\tau}\cdot I_{l,j}(\tau,\gamma), \tag{52}$$

where

$$I_{l,j}(\tau,\gamma)=\frac{(-i\tau)^{1/2}}{2\sqrt{2\pi}\sin(\gamma/2\tau)}e^{2\pi i\tau j^2/l}\sum_{n\in\mathbb{Z}}e^{2\pi i\tau((ln^2+2jn)}(e^{i(j+ln)\gamma}-e^{-i(j+ln)\gamma}). \tag{53}$$

The expression (53), appearing in the right hand side of scaling property (52), is exactly the formula derived by Igor **[15]** as a loop group generalization of Kirillov's orbit integral (with some simple identification of parameters τ,γ with those used in **[15]**, see **[19]** for details). Exponential pre-factor, similar to one in (52), also appears in the scaling relation between Virasoro character and DH integral over $\widehat{\mathrm{Diff}}(S^1)$ coadjoint orbit **[52]** as shown in **[19]**.

4. WZW_4 and holomorphic CS

WZW_4 theory is a four-dimensional avatar of the WZW_2 model.[6] Let X_4 be a four-manifold equipped with a metric $h_{\mu\nu}$ and a closed 2-form $\omega\in\Omega^2(X;\mathbb{R})$. Let

---

[5] A possible CFT interpretation of the exponent α is the number of free fields in the theory. Indeed, irreducible representations of $\widehat{LSU(2)}$ can be resolved in terms of free field Wakimoto modules with 3 free fields. Thus, one might conjecture that $\alpha = dimG$ for higher rank groups.

[6] During my lecture at Igor's 70th birthday conference due to the time limit I was not able to cover this part, I talked about it in **[57]** instead.

$g \in Map(X_4, G)$ for a Lie group G. Fix a reference field configuration $g_0(x)$. Then, for any $g(x)$ in the same homotopy class as $g_0(x)$ we may define the Lagrangian:

$$S_\omega[g; g_0] = \frac{f_\pi^2}{8\pi} \int_{X_4} Trg^{-1}dg \wedge *(g^{-1}dg) + \frac{i}{12\pi} \int_{X_5} \omega \wedge Tr(g^{-1}dg)^3. \tag{54}$$

Here f_π is a dimensionful parameter. Tr is in adjoint representation. In the integral over $X_5 = X_4 \times I$ in (54) ω is independent of the fifth coordinate; moreover, we use a homotopy of g to g_0. The action is independent of the choice of homotopy up to a multiple of the periods of ω. The classical equations of motion, following from (54), are:

$$d * g^{-1}dg + \omega \wedge g^{-1}dg \wedge g^{-1}dg = 0 \tag{55}$$

When X_4 is a complex four-manifold with Kähler metric $\omega = \frac{i}{2} f_\pi^2 h_{i\bar{j}} dz^i \wedge dz^{\bar{j}}$ one gets the point in the space of Lagrangians, the "Kähler point." Now one can rewrite the action as:

$$S_\omega[g] = -\frac{i}{4\pi} \int_{X_4} \omega \wedge Tr(g^{-1}\partial g \wedge g^{-1}\bar{\partial} g) + \frac{i}{12\pi} \int_{X_5} \omega \wedge Tr(g^{-1}dg)^3. \tag{56}$$

The equations of motion following from (56) are:

$$\omega \wedge \bar{\partial}(g^{-1}\partial g) = 0 \tag{57}$$

which means that (2,1)-form current $J^{(2,1)} = \omega \wedge g^{-1}\partial g$ is $\bar{\partial}$ closed. At this special Kähler point WZW_4 theory is equivalent to the classical field theory of the N $= 2$ open strings for any group G [**58**], [**59**], [**7**], [**9**]. [7]

From now on we will only focus on Kähler point. The coefficient of the WZ term is quantized. Two different homotopies to g_0 define a map $g : S^1 \times X_4 \to G$. Thus, if the group G is non-abelian, the measure $\exp iS$ in the path integral is only well-defined if

$$\omega \wedge \frac{1}{12\pi} Tr(g^{-1}dg)^3 \in H^5(S^1 \times X_4; 2\pi\mathbb{Z}) \tag{58}$$

which forces the cohomology class $[\omega]$ to lie in the lattice:

$$[\omega] \in H^2(X_4; \mathbb{Z}). \tag{59}$$

The class $[\omega]$ is the four-dimensional analog of level k in WZW_2. Note that although $[\omega]$ is quantized the Lagrangian depends on the representative of the class. Since ω is of type $(1,1)$ condition (59) implies $[\omega] \in H^2(X; \mathbb{Z}) \cap H^{1,1}(X; \mathbb{R})$ so the metric is Hodge, and, by the Kodaira embedding theorem, if X_4 is compact, it must be algebraic.

The important property of WZW_4 theory is that at the Kähler point quantum theory is finite [**7**] (this is a very special point - quantum theory can not be deformed, away this point it becomes non renormalizable). The argument is very similar to that of in WZW_2 - first one demonstrates it is finite in one-loop by calculating one loop determinant in the background field method, after which one uses the holomorphicity property of $(2,1)$ current $J^{(2,1)}$ in (57) to claim it must be finite to all loop orders (like in WZW_2 this holomorphicity will be violated if there are divergences at higher loops). This all loop finiteness statement is also well supported

[7]String investigations of this theory have focused on the S-matrix for π, defined as $g = e^{i\pi}$, hence one may expect the connections with the nonperturbative N $= 2$ string theory.

by studies of perturbative amplitudes in open $N = 2$ strings (self-dual YM) which show no loop divergences.

Quantum WZW_4 theory has infinite-dimensional global, two-loop, symmetry $g(z, \bar{z}) \to h_L(\bar{z})g(z, \bar{z})h_R(z)$ (not to be confused with gauge symmetry) and this allows to calculate many correlation functions (of surface operators); 1-cocycle properties (from Section 2) of the WZW_4 action play a key role in the quantum theory. Conservation law, holomorphicity of $(2, 1)$ current $J^{(2,1)}$ in (57), again leads to useful Ward identities. The five-dimensional viewpoint ("Kähler-Chern-Simons" theory), similar to 3d CS/WZW_2 relation, allows to define and calculate "four-dimensional Verlinde formula" as well as establish the connection to Donaldson theory, etc. All these can be found in [**7**].

One needs to emphasize again that WZW_4 theory is not a gauge theory. Rather it is a very special sigma-model, and its infinite-dimensional symmetries are not gauge symmetries - they are global symmetries of the theory (similar to the infinite-dimensional symmetries of WZW_2 which are global symmetries there too). Coupling of WZW_4 to gravity is also a well-defined quantum theory for any group and is equivalent to open/closed $N = 2$ strings [**58**], [**9**] - WZW_4 is a well-defined theory, with or without coupling to the dynamical gravity and for any group G.

A bit of history - the Lagrangian (56) was first written in [**60**], the equations of motion (57) appeared even before in [**61**] as being equivalent to self-dual Yang-Milles equations (for Hermitian g and after solving part of the self-dual YM equations) and are often called Yang equations. It was studied in [**62**], [**63**] as a natural generalization of the $2d$ CFT/ 3d CS correspondence.

The equations of motion (57) are integrable and can be solved using the twistor transform (see brief description in [**7**], page 3). Actually, as explained in [**64**], [**65**], instantons on $\mathbb{S}^4$ correspond to holomorphic bundles $F^{0,2}(A) = \bar{\partial}\bar{A} + \bar{A}^2 = 0$ on $\mathbb{CP}^3$ (non-compact version relates equations on $\mathbb{R}^4$ and twistor space $\mathbb{PT}$):

$$(60) \quad \hat{x} = q_2^{-1}q_1, \quad \hat{x} = x^0 + i\sum_{i=1}^{3} x^i\sigma^i, \quad q_1 = \begin{pmatrix} z_1 & z_2 \\ -\bar{z}_2 & \bar{z}_1 \end{pmatrix}, \quad q_2 = \begin{pmatrix} z_3 & z_4 \\ -\bar{z}_4 & \bar{z}_3 \end{pmatrix}$$

where σ^i are Pauli matrices, $x \in \mathbb{R}^4$, $z \in \mathbb{C}^4$ and q_i are quaternions. Now one can check that if gauge field $a = \sum_\mu a_\mu dx^\mu$ is instanton, satisfies self-duality condition $F(a) = \star F(a)$, then $A_i(z, \bar{z}) = \sum_\mu a_\mu(x)\frac{dx^\mu}{dz^i}$ satisfies $F^{0,2}(A) = 0$ (in fact - this is an equation in $\mathbb{CP}^3$ and not $\mathbb{C}^4$ due to the scale invariance of (60) under $q_i \to \lambda q_i$). The reverse, exactly how to obtain the instanton on $\mathbb{S}^4$ using this equivalence, is done through $ADHM$ construction [**66**].

Let us stress two moments here: a) solutions to self-dual YM equations on $\mathbb{S}^4$ modulo its gauge transformation have been mapped to the solutions of $F^{0,2}(A) = 0$ on $\mathbb{CP}^3$ modulo its gauge transformations, b) Yang equation (57) is a result of solving the parts of the self-duality equations ($F^{2,0}(a) = 0$ and $F^{0,2}(a) = 0$ parts) and consecutively the final equation (57) has no remaining gauge symmetry.

Now let me turn to more recent developments. The relation between equation (57) on $\mathbb{R}^4$ and $F^{0,2}(A) = 0$ on twistor space $\mathbb{PT}$, which is true for any group G, can be written as a classical equivalence between action (56) and holomorphic CS action on $\mathbb{PT}$:

$$(61) \qquad S_{HCS} = \int_X \Omega \wedge Tr(\bar{A} \wedge \bar{\partial}\bar{A} + \frac{2}{3}\bar{A} \wedge \bar{A} \wedge \bar{A})$$

$X = \mathbb{PT}$ and Tr is again in the adjoint representation. One can consult [**14**], [**67**], [**68**], [**69**] for detailed definition of (61) on $\mathbb{PT}$ (including that of holomorphic $(3,0)$-form Ω on $\mathbb{PT}$ which is used here - actually meromorphic form is used, poles are allowed and their boundary contributions are eliminated by imposing appropriate conditions on the gauge field so equations of motion derived from (61) are $F^{0,2} = 0$, and the action (61) is well-defined as integral converges over $X = \mathbb{CP}^3$). In these papers one can also find the explicit formulae relating g in (56) and $\bar{A}$ in (61) such that the actions (61) and (56) become identical after solving the part of the six-dimensional equations for unphysical modes.

Here we need one important element of the above classical "off-shell" equivalence between WZW_4 in 4d and holomorphic CS in 6d. In establishing this equivalence the twistor space $\mathbb{PT}$ is viewed as a $\mathbb{CP}^1$ bundle over $\mathbb{R}^4$ and the six-dimensional theory is treated as a four-dimensional theory on $\mathbb{R}^4$ with infinitely many fields with only zero-modes being physical - other modes turn out to be unphysical, either auxiliary fields not containing space-time derivatives along $\mathbb{R}^4$ or "non-propagating" degrees which are guage transformations (see [**68**]). All these unphysical degrees of freedom are classically "integrated out" (by solving the equations of motion for these fields and substitutes back into the action) or eliminated by the gauge transformations, after which the actions (61) and (56) become identical.

Let's stress once more - in this "off-shell" equivalence it is very important that some fields (auxiliary/unphysical from 4d point of view) are classically consistently eliminated and in that the gauge invariance of classical action (equations of motion) is used.

Thus, if we collect the information presented so far, the conclusion is that there are several classical equivalences (with non-dynamical gravity, in the fixed gravitational background) for any group G, on the level of classical actions together with the equations of motion, and two of them involve WZW_4 (the equivalence which doesn't involve WZW_4, 2nd one below, is important here since it is the one which leads to a puzzle we would like to discuss). These equivalences are the main reason of current presentation:

I. between WZW_4 and $N = 2$ open strings (self-dual Yang-Mills as mentioned before),

II. between holomorphic CS theory on (compact) Calabi-Yau manifold X in (61) and open topological strings [**70**],

III. between holomorphic CS theory on twistor space $\mathbb{PT}$ and WZW_4 on $\mathbb{R}^4$ (or in compactified version - between holomorphic CS on $\mathbb{CP}^3$ and WZW_4 on $\mathbb{S}^4$).

All three of these equivalences have a version with dynamical gravity since all of them have the interacting open/closed string theory counterparts.

At the quantum level the situation with these equivalences is more complicated. I was discussing the corresponding issues, quantum equivalences both without and with dynamical gravity, in the summer 2022, around the time of Igor's '70th conference, with F. Nieri, J. Manschot, C. Vafa and J. Walcher [**71**]. The conclusion is that: **Equiv I** holds on quantum level (both without and with dynamical gravity), this is an old and established fact [**58**], **Equiv II** holds on quantum level only when coupled to the dynamical gravity and for a very special gauge groups [**72**], [**73**], [**14**], [**67**], [**68**], with **Equiv III** there are some serious difficulties on quantum level. Below I comment on **Equiv II** and **Equiv III** based on these conversations and my current understanding.

Equivalence II. In topological string theory target space is Calabi-Yau manifold, so $X = CY$, thus integration over $X = \mathbb{PT}$ $(\mathbb{CP}^3)$ in (61) needs to be replaced by the integration over CY and Ω now is CY holomorphic 3-form in some fixed complex structure. In order to properly couple it with background gravitational fields one can deform this complex structure replacing $\bar\partial \to \bar\nabla = \bar\partial - \mu^i \partial_i$ where μ^i is Beltrami differential: $\mu^i\partial_i = \mu^i_{\bar j} dz^{\bar j}\partial_i$. After this (61) will depend on two classical closed string backgrounds Ω and μ which must satisfy (on-shell) equations of motion of closed topological strings: $\bar\nabla\Omega + \partial_i\mu^i\Omega = 0$ and $\bar\partial\mu^i - \mu^j\partial_j\mu^i = 0$, meaning that Ω is holomorphic in the complex structure associated with μ^i, and this μ^i defines the integrable complex structure satisfying the Kodaira-Spencer equation. One can write the action (61) on CY manifold in an arbitrary closed string background (off-shell) by introducing many auxiliary fields [**74**]. And, these background fields can be made dynamical - in this case one gets the open/closed topological strings (B-model), holomorphic CS coupled to Kodaira-Spencer gravity.

It was pointed out already in [**72**], from the world sheet considerations, that open topological string at the quantum level has anomalous behavior that needs to be taken care of. This is an internal anomaly and such anomalous theories are usually not consistent quantum-mechanically. Solution proposed in [**72**] involves coupling to closed strings (gravity) after the orientifold projection and cancelling the orientifold charge by D-branes.[8]

This topic was long forgotten until the space-time theory was studied in [**14**], [**67**], [**68**] where the proper and careful treatment of the space-time theory was given. Because of the chiral origin of the open string sector of the space-time theory ((61), which depends only on $\bar A$), the proper definition of quantum theory suffers anomalies, similar to those described in Section 2 for chiral fermions. Again, here for the properly defined path integral over the open string modes ($\bar A$ and all other auxiliary fields mentioned above) in a fixed gravitational background (Ω, μ and corresponding gravitational auxiliary fields) at the one loop one finds the anomaly [**14**], [**67**], [**68**]. There is an important difference between the chiral fermions in the background gauge field (where anomaly is "harmless", unless these background gauge fields are replaced by the dynamical gauge fields) and the holomorphic CS in a fixed gravitational background. Latter is a non-linear self-interacting theory and such anomaly would be again (as in the world-sheet approach mentioned above) an "internal anomaly" leading to the inconsistent quantum theory at the higher loops. The explicit formula for this chiral anomaly, as always, can be obtained by the same cohomological descent procedure as in the Section 2 giving now the 1-cocycle with the values in the space of functionals of $\bar A$, $\alpha_1(\bar A, u)$ ($\mathbf{a}(\bar A, u)$ for algebraic 1-cocycle/anomaly). Since now we are in six-dimensions we need to start the descent from $w_8 = TrF^4$ (Tr in the adjoint representation) and look for the holomorphic representatives, $\alpha_1(\bar A, g)$ holomorphic in gauge field (in some fixed complex structure, or holomorphic both in $\bar A$ and μ if one varies the complex structure). Actually, in order to see the anomaly, it is enough to consider the algebraic descent which leads (for a fixed complex structure) to $\mathbf{a}_1(\bar A, u) = \int_X Tr(u(\partial \bar A)^3)$. This doesn't

[8]As explained by [**75**] , and used in [**72**], orientifold plane carries a D-brane charge which needs to be cancelled by D-branes to get the anomaly free theory. The focus of [**72**] was the A-model, but when translated to the B-model by mirror symmetry/T-duality, the [**75**] charge that needs to be cancelled is $2^{d/2}$ (d is the real critical dimension of the target space) and this implies that non-anomalous gauge group for 6-dimensional CY must be $SO(8)$, although this was not written explicitly in [**72**].

vanish. So, holomorphic CS theory in non-dynamical gravitation background is anomalous and inconsistent.

As in the world-sheet approach to topological strings, described above, the space-time theory can be made consistent quantum theory if coupled to the dynamical gravity (closed strings). Coupling to the dynamical gravity opens the opportunity to get the anomaly free theory if one can show that one loop quantum anomaly from the holomorphic CS can be cancelled by the classical transformations from the gravity sector. Such mechanism was found in [**14**], [**67**], [**68**] where it was shown explicitly that if the gravitational fields are made dynamical (thus properly coupling the holomorphic CS to the appropriate Kodaira-Spencer theory) this is indeed the case - anomaly is cancelled from the contributions arising due to the transformations of gravitational fields via Green-Schwarz type mechanism, but only for the choice of the gauge group $G = SO(8)$ in (61).

Thus, following [**14**], [**67**], [**68**], we conclude that the only consistent, anomaly free, six-dimensional quantum theory is the holomorphic CS theory coupled to the dynamical gravity (open/closed topological string theory) with the gauge group $G = SO(8)$. Purely holomorphic CS with non-dynamical gravitational background is inconsistent on quantum level. Simply stated - **Equiv II** can be extended to the quantum level only when a) properly coupled to the dynamical gravity and b) only for the gauge group $SO(8)$.

One may ask - does this conclusion lead to the consequences for the **Equiv III**? Does this mean that quantum WZW_4 is also inconsistent unless it is coupled to the dynamical gravity (coupled open/closed $N = 2$ string theory) and only for the gauge group $SO(8)$? If so - why such restriction was not found in 90s when studying $N = 2$ strings, WZW_4, etc?

Equivalence III. In order to address this relation one needs to replace $X = CY$, discussed in **Equiv II**, by non-Calabi-Yau $X = \mathbb{PT}$ (or $\mathbb{CP}^3$). In the six-dimensional quantum theory, defined by the action functional (61), in the non-dynamical gravitational background such replacement will not affect the anomaly, so quantum theory again will be consistent only for $G = SO(8)$ and only after coupling to the dynamical gravity.

Thus, assuming **Equiv III** is true on quantum level, one may conclude that the same restrictions should apply to quantum WZW_4: this 4d theory is consistent quantum field theory only when properly coupled to the dynamical gravity and only for $G = SO(8)$.

This conclusion creates a puzzle. Classical theory on $\mathbb{PT}$ ($\mathbb{CP}^3$) was equivalent to four-dimensional WZW_4 theory on $\mathbb{R}^4$ ($\mathbb{S}^4$) and this four-dimensional theory, as we stressed above, doesn't require any kind of restrictions from the previous paragraph on quantum level.

What is the resolution of the puzzle? Most likely resolution is that at the quantum level the **Equiv III** fails. Correct logic should be - instead of claiming that the restrictions from **Equiv II** imply that 4d theory should be also restricted (by the coupling to gravity and $G = SO(8)$), the logical conclusion should be that there is no quantum **Equiv III** even if we couple 4d WZW_4 theory to dynamical gravity (except for the coupled theory and group $G = SO(8)$ when one might certainly expect such equivalence to exist).

This conclusion is hidden in the details of exactly how the classical equivalence was established. Mainly, from thinking about the classical theory on $\mathbb{PT}$ as a four-dimensional theory on $\mathbb{R}^4$ with the infinitely many unphysical degrees of freedom associated to $\mathbb{CP}^1$ (which were integrated out and consistently eliminated, classically). For generic group G we know that six-dimensional theory is anomalous - not gauge-invariant already at one loop. This in particular means that in the six-dimensional quantum theory above unphysical degrees of freedom ("longitudinal", gauge transformation, degrees of freedom) quantum-mechanically are not unphysical anymore - they are propagating degrees of freedom in quantum theory (as is always the case in theories with the internal quantum anomalies) and can not be consistently integrated out/eliminated in the same simple way as it was done classically. So, we can not establish quantum equivalence anymore in the same fashion as was done classically - when describing the classical **Equiv III** we stressed the importance of the gauge symmetry for such equivalence to exist.

To reiterate - for the classical **Equiv III** it was crucial that we solved the part of the classical equations of motions in six-dimensions eliminating auxiliary and unphysical degrees of freedom; thus, since six-dimensional theory is anomaly free only when coupled to dynamical gravity and only for $SO(8)$ gauge group, integrating out the same degrees of freedom (as in classical theory) in quantum theory is a harmless operation only for the holomorphic CS theory coupled to dynamical gravity and only for the $SO(8)$ gauge group - **Equiv III** holds on quantum level only for the coupled theory and $G = SO(8)$. For other groups, coupled to dynamical gravity or not, one can not eliminate consistently these degrees of freedom in quantum theory, so for other groups **Equiv III** fails. Important conclusion - WZW_4 does not suffer any problems if G is not $SO(8)$. At least these problems can not arise from those in $6d$ for such groups, since 4d theory is no longer equivalent to 6d theory, the restrictions in 6d theory have no consequences for the existence (or non-existence) of quantum theory in 4d.

For Igor and I, as I already mentioned in the introduction, the study of WZW_4 theory was a spinoff of our attempts to find the quantum field theory interpretation of Nakajima's results. We have had many other spinoffs that led to many interesting developments - I only covered WZW_4 here since this topic has recently seen new important developments.

Acknowledgments

I thank A. Alekseev for the long-standing collaboration and for the discussions during the preparation of this text. I thank J. Manschot, F. Nieri, C. Vafa and J. Walcher for many enlightening discussions that led to my understanding presented here in Section 4. I thank M. Gualtieri and Z. Komargodski for the discussions during the preparation of this manuscript and A. Gerasimov for the discussions of closely related topics over the years.

References

[1] S. Shatashvili, `https://scgp.stonybrook.edu/video/video.php?id=5229`

[2] L. D. Faddeev and S. L. Shatashvili, *Algebraic and Hamiltonian Methods in the Theory of Nonabelian Anomalies*, Teor. Mat. Fiz. 60 (1984) 206–217, Theor. Math. Phys. 60 (1985) 770–778.

[3] I. M. Singer, *Families of Dirac operators with applications to physics*, Astérisque **Numéro Hors Série** (1985), 323–340. The mathematical heritage of Élie Cartan (Lyon, 1984). MR837207

[4] Hiraku Nakajima, *Gauge theory on resolutions of simple singularities and simple Lie algebras*, Internat. Math. Res. Notices **2** (1994), 61–74, DOI 10.1155/S1073792894000085. MR1264929

[5] Igor B. Frenkel and Boris A. Khesin, *Four-dimensional realization of two-dimensional current groups*, Comm. Math. Phys. **178** (1996), no. 3, 541–562. MR1395204

[6] Igor B. Frenkel and Andrey N. Todorov, *Complex counterpart of Chern-Simons-Witten theory and holomorphic linking*, Adv. Theor. Math. Phys. **11** (2007), no. 4, 531–590. MR2354075

[7] Andrei Losev, Gregory Moore, Nikita Nekrasov, and Samson Shatashvili, *Four-dimensional avatars of two-dimensional RCFT*, Nuclear Phys. B Proc. Suppl. **46** (1996), 130–145, DOI 10.1016/0920-5632(96)00015-1. S-duality and mirror symmetry (Trieste, 1995). MR1411466

[8] Andrei Losev, Gregory Moore, Nikita Nekrasov, and Samson Shatashvili, *Central extensions of gauge groups revisited*, Selecta Math. (N.S.) **4** (1998), no. 1, 117–123, DOI 10.1007/s000290050026. MR1623710

[9] Andrei Losev, Gregory Moore, Nikita Nekrasov, and Samson Shatashvili, *Chiral Lagrangians, anomalies, supersymmetry, and holomorphy*, Nuclear Phys. B **484** (1997), no. 1-2, 196–222, DOI 10.1016/S0550-3213(96)00612-8. MR1431695

[10] Gregory Moore, Nikita Nekrasov, and Samson Shatashvili, *Integrating over Higgs branches*, Comm. Math. Phys. **209** (2000), no. 1, 97–121, DOI 10.1007/s002200050016. MR1736944

[11] Gregory Moore, Nikita Nekrasov, and Samson Shatashvili, *D-particle bound states and generalized instantons*, Comm. Math. Phys. **209** (2000), no. 1, 77–95, DOI 10.1007/s002200050016. MR1736943

[12] A. Losev, N. Nekrassov, and S. Shatashvili, *Testing Seiberg-Witten solution*, Strings, branes and dualities (Cargèse, 1997), NATO Adv. Sci. Inst. Ser. C: Math. Phys. Sci., vol. 520, Kluwer Acad. Publ., Dordrecht, 1999, pp. 359–372, DOI 10.1007/978-94-011-4730-9_13. MR1688304

[13] A. Losev, N. Nekrasov, and S. Shatashvili, *Issues in topological gauge theory*, Nuclear Phys. B **534** (1998), no. 3, 549–611, DOI 10.1016/S0550-3213(98)00628-2. MR1663467

[14] K. Costello, *Topological strings, twistors and Skyrmions*, `http://web.math.ucsb.edu/~drm/WHCGP/#past20`

[15] I. B. Frenkel, *Orbital theory for affine Lie algebras*, Invent. Math. **77** (1984), no. 2, 301–352, DOI 10.1007/BF01388449. MR752823

[16] A. Alekseev and S. Shatashvili, *Path integral quantization of the coadjoint orbits of the Virasoro group and 2-d gravity*, Nuclear Phys. B **323** (1989), no. 3, 719–733, DOI 10.1016/0550-3213(89)90130-2. MR1014632

[17] A. Alekseev and S. Shatashvili, *From geometric quantization to conformal field theory*, Comm. Math. Phys. **128** (1990), no. 1, 197–212. MR1042451

[18] Anton Alekseev and Samson L. Shatashvili, *Coadjoint orbits, cocycles and gravitational Wess-Zumino*, Rev. Math. Phys. **30** (2018), no. 6, 1840001, 15, DOI 10.1142/S0129055X18400019. MR3820680

[19] Anton Alekseev and Samson L. Shatashvili, *Characters, coadjoint orbits and Duistermaat-Heckman integrals*, J. Geom. Phys. **170** (2021), Paper No. 104386, 20, DOI 10.1016/j.geomphys.2021.104386. MR4321656

[20] Jouko Mickelsson, *Chiral anomalies in even and odd dimensions*, Comm. Math. Phys. **97** (1985), no. 3, 361–370. MR778620

[21] S. L. Adler, *Axial-Vector Vertex in Spinor Electrodynamics*, Phys. Rev. 177: 2426- 2438 (1969).

[22] J. S. Bell and R. Jackiw, *A PCAC Puzzle: $\pi_0 \to \gamma\gamma$ in the σ-Model,* Nuovo Cimento A 60: 47-61 (1969).

[23] C. G. Callan Jr. and J. A. Harvey, *Anomalies and fermion zero modes on strings and domain walls*, Nuclear Phys. B **250** (1985), no. 3, 427–436, DOI 10.1016/0550-3213(85)90489-4. MR779916

[24] Bruno Zumino, *Chiral anomalies and differential geometry*, Relativity, groups and topology, II (Les Houches, 1983), North-Holland, Amsterdam, 1984, pp. 1291–1322. Notes prepared by K. Sibold. MR830254

[25] Raymond Stora, *Algebraic structure and topological origin of anomalies*, Progress in gauge field theory (Cargèse, 1983), NATO Adv. Sci. Inst. Ser. B: Phys., vol. 115, Plenum, New York, 1984, pp. 543–562. MR782343

[26] A. Yu. Alekseev, Ya. Madaĭchik, L. D. Faddeev, and S. L. Shatashvili, *Derivation of an anomalous commutator in the formalism of a functional integral* (Russian, with English summary), Teoret. Mat. Fiz. **73** (1987), no. 2, 187–190. MR933545

[27] Harish-Chandra, *Differential operators on a semisimple Lie algebra*, Amer. J. Math. **79** (1957), 87–120, DOI 10.2307/2372387. MR84104

[28] C. Itzykson and J. B. Zuber, *The planar approximation. II*, J. Math. Phys. **21** (1980), no. 3, 411–421, DOI 10.1063/1.524438. MR562985

[29] J. J. Duistermaat and G. J. Heckman, *On the variation in the cohomology of the symplectic form of the reduced phase space*, Invent. Math. **69** (1982), no. 2, 259–268, DOI 10.1007/BF01399506. MR674406

[30] A. A. Kirillov, *Elements of the theory of representations*, Grundlehren der Mathematischen Wissenschaften, Band 220, Springer-Verlag, Berlin-New York, 1976. Translated from the Russian by Edwin Hewitt. MR412321

[31] M. A. Semenov-Tjan-Šanskiĭ, *A certain property of the Kirillov integral* (Russian), Zap. Naučn. Sem. Leningrad. Otdel. Mat. Inst. Steklov. (LOMI) **37** (1973), 53–65. Differential geometry, Lie groups and mechanics. MR376968

[32] A. Alekseev, L. Faddeev, and S. Shatashvili, *Quantization of symplectic orbits of compact Lie groups by means of the functional integral*, J. Geom. Phys. **5** (1988), no. 3, 391–406, DOI 10.1016/0393-0440(88)90031-9. MR1048508

[33] R. F. Picken, *The propagator for quantum mechanics on a group manifold from an infinite-dimensional analogue of the Duistermaat-Heckman integration formula*, J. Phys. A **22** (1989), no. 13, 2285–2297. MR1003731

[34] Changha Choi and Leon A. Takhtajan, *Supersymmetry and trace formulas. Part I. Compact Lie groups*, J. High Energy Phys. **6** (2024), Paper No. 26, DOI 10.1007/jhep06(2024)026. MR4763003

[35] Changha Choi and L. Takhtajan, *Supersymmetry and trace formulas II. Selberg trace formula,* `arXiv:2306.13636`.

[36] Robert Wendt, *A symplectic approach to certain functional integrals and partition functions*, J. Geom. Phys. **40** (2001), no. 1, 65–99, DOI 10.1016/S0393-0440(01)00030-4. MR1854316

[37] Lisa C. Jeffrey and James A. Mracek, *Hyperfunctions, the Duistermaat-Heckman theorem and loop groups*, Geometry and physics. Vol. I, Oxford Univ. Press, Oxford, 2018, pp. 319–346. MR3932265

[38] Philippe Bougerol and Manon Defosseux, *Pitman transforms and Brownian motion in the interval viewed as an affine alcove* (English, with English and French summaries), Ann. Sci. Éc. Norm. Supér. (4) **55** (2022), no. 2, 429–472, DOI 10.24033/asens.2499. MR4411878

[39] Manon Defosseux, *Kirillov-Frenkel character formula for loop groups, radial part and Brownian sheet*, Ann. Probab. **47** (2019), no. 2, 1036–1055, DOI 10.1214/18-AOP1278. MR3916941

[40] Jean-Michel Bismut, *The hypoelliptic Laplacian on a compact Lie group*, J. Funct. Anal. **255** (2008), no. 9, 2190–2232, DOI 10.1016/j.jfa.2008.07.017. MR2473254

[41] Jean-Michel Bismut, *Hypoelliptic Laplacian and orbital integrals*, Annals of Mathematics Studies, vol. 177, Princeton University Press, Princeton, NJ, 2011, DOI 10.1515/9781400840571. MR2828080

[42] S. Sachdev, J.-W. Ye, *Gapless spin fluid ground state in a random, quantum Heisenberg magnet,* Phys. Rev. Lett. 70 (1993) 3339, `arXiv:cond-mat/9212030`.

[43] A. Kitaev, Talk given at the Fundamental Physics Prize Symposium, Nov. 10, 2014; A. Kitaev, KITP seminar, Feb. 12, 2015; ÒA simple model of quantum holography,Ó talks at KITP, April 7, 2015 and May 27, 2015.

[44] Gustavo J. Turiaci and Herman Verlinde, *On CFT and quantum chaos*, J. High Energy Phys. **12** (2016), 110, front matter+21, DOI 10.1007/JHEP12(2016)110. MR3602310

[45] Juan Maldacena and Douglas Stanford, *Remarks on the Sachdev-Ye-Kitaev model*, Phys. Rev. D **94** (2016), no. 10, 106002, 43, DOI 10.1103/physrevd.94.106002. MR3751002

[46] K. Jensen, *Chaos in AdS_2 holography,* Phys. Rev. Lett. 117, 111601 (2016) `arXiv:1605.06098`.

[47] J. Maldacena, *The SYK Model and Near Extremal Black Holes,* `http://scgp.stonybrook.edu/video/video.php?id=2733`.

[48] Juan Maldacena, Douglas Stanford, and Zhenbin Yang, *Conformal symmetry and its breaking in two-dimensional nearly anti-de Sitter space*, PTEP. Prog. Theor. Exp. Phys. **12** (2016), 12C104, 26, DOI 10.1093/ptep/ptw124. MR3628687

[49] Julius Engelsöy, Thomas G. Mertens, and Herman Verlinde, *An investigation of* AdS_2 *backreaction and holography*, J. High Energy Phys. **7** (2016), 139, front matter+29, DOI 10.1007/JHEP07(2016)139. MR3550234

[50] Gautam Mandal, Pranjal Nayak, and Spenta R. Wadia, *Coadjoint orbit action of Virasoro group and two-dimensional quantum gravity dual to SYK/tensor models*, J. High Energy Phys. **11** (2017), 046, front matter+40, DOI 10.1007/jhep11(2017)046. MR3747261

[51] Davide Gaiotto and Miroslav Rapčák, *Vertex algebras at the corner*, J. High Energy Phys. **1** (2019), 160, front matter+85, DOI 10.1007/jhep01(2019)160. MR3919335

[52] Douglas Stanford and Edward Witten, *Fermionic localization of the schwarzian theory*, J. High Energy Phys. **10** (2017), 008, front matter+27, DOI 10.1007/jhep10(2017)008. MR3731878

[53] Guilherme L. Pimentel, Alexander M. Polyakov, and Grigory M. Tarnopolsky, *Vacuum decay in CFT and the Riemann-Hilbert problem*, Nuclear Phys. B **907** (2016), 617–632, DOI 10.1016/j.nuclphysb.2016.03.039. MR3498446

[54] Gustavo J. Turiaci and Edward Witten, $\mathcal{N}=2$ *JT supergravity and matrix models*, J. High Energy Phys. **12** (2023), Paper No. 3, 121, DOI 10.1007/jhep12(2023)003. MR4675222

[55] Shubho Banerjee and Blake Wilkerson, *Asymptotic expansions of Lambert series and related q-series*, Int. J. Number Theory **13** (2017), no. 8, 2097–2113, DOI 10.1142/S1793042117501135. MR3681692

[56] Masanori Katsurada, *Asymptotic expansions of certain q-series and a formula of Ramanujan for specific values of the Riemann zeta function*, Acta Arith. **107** (2003), no. 3, 269–298, DOI 10.4064/aa107-3-5. MR1970926

[57] S. Shatashvili, `https://scgp.stonybrook.edu/video/video.php?id=5365`

[58] Hirosi Ooguri and Cumrun Vafa, *Self-duality and* $N=2$ *string magic*, Modern Phys. Lett. A **5** (1990), no. 18, 1389–1398, DOI 10.1142/S021773239000158X. MR1063040

[59] N. Berkovits, *Super-Poincare Invariant Superstring Field Theory,* `arXiv:hep-th/9503099`

[60] S. K. Donaldson, *Anti self-dual Yang-Mills connections over complex algebraic surfaces and stable vector bundles*, Proc. London Math. Soc. (3) **50** (1985), no. 1, 1–26, DOI 10.1112/plms/s3-50.1.1. MR765366

[61] Chen Ning Yang, *Condition of self-duality for* su(2) *gauge fields on Euclidean four-dimensional space*, Phys. Rev. Lett. **38** (1977), no. 24, 1377–1379, DOI 10.1103/PhysRevLett.38.1377. MR475357

[62] V.P. Nair and Jeremy Schiff, *Kähler Chern Simons theory and symmetries of antiselfdual equaons,* Nucl. Phys. B371:329-352,1992; *A Kähler Chern-Simons theory and quantization of the moduli of antiselfdual instantons*, Phys. Lett. B246:423-429, 1990; *Topological gauge theory and twistors*, Phys. Lett. B233:343, 1989.

[63] V.P.Nair, *Kähler-Chern-Simons Theory,* `arXiv:hep-th/9110042`, 1991

[64] R. S. Ward, *On self-dual gauge fields*, Phys. Lett. A **61** (1977), no. 2, 81–82, DOI 10.1016/0375-9601(77)90842-8. MR443823

[65] M. F. Atiyah and R. S. Ward, *Instantons and algebraic geometry*, Comm. Math. Phys. **55** (1977), no. 2, 117–124. MR494098

[66] M. F. Atiyah, N. J. Hitchin, V. G. Drinfel′d, and Yu. I. Manin, *Construction of instantons*, Phys. Lett. A **65** (1978), no. 3, 185–187, DOI 10.1016/0375-9601(78)90141-X. MR598562

[67] Kevin Costello and Si Li, *Anomaly cancellation in the topological string*, Adv. Theor. Math. Phys. **24** (2020), no. 7, 1723–1771, DOI 10.4310/ATMP.2020.v24.n7.a2. MR4313234

[68] K. Costello, *Quantizing holomorphic field theories on twistor space*, `arXiv:2111.08879v1`.

[69] Roland Bittleston and David Skinner, *Twistors, the ASD Yang-Mills equations and 4d Chern-Simons theory*, J. High Energy Phys. **2** (2023), Paper No. 227, 49, DOI 10.1007/jhep02(2023)227. MR4552493

[70] E. Witten, *Chern-Simons gauge theory as a string theory*, The Floer memorial volume, Progr. Math., vol. 133, Birkhäuser, Basel, 1995, pp. 637–678. MR1362846

[71] Discussions with J. Manschot, F. Nieri, C. Vafa and J. Walcher, 2022.

[72] Johannes Walcher, *Evidence for tadpole cancellation in the topological string*, Commun. Number Theory Phys. **3** (2009), no. 1, 111–172, DOI 10.4310/CNTP.2009.v3.n1.a3. MR2504755

[73] Paul L. H. Cook, Hirosi Ooguri, Jie Yang, *New Anomalies in Topological String Theory,* Prog. Theor. Phys. Suppl. 177 (2009) 120-127, `arXiv:0804.1120`

[74] Stefano Giusto, Camillo Imbimbo, and Dario Rosa, *Holomorphic Chern-Simons theory coupled to off-shell Kodaira-Spencer gravity*, J. High Energy Phys. **10** (2012), 192, front matter + 21, DOI 10.1007/JHEP10(2012)192. MR3033807

[75] S. Sinha and C. Vafa, *SO and Sp Chern-Simons at large N*, `arXiv:hep-th/0012136` (2000).

The Hamilton Mathematics Institute, Trinity College Dublin, Ireland; The School of Mathematics, Trinity College Dublin, Ireland; and Simons Center for Geometry and Physics, Stony Brook University, Stony Brook, New York 11790

Email address: `samson@math.tcd.ie; sshatashvili@scgp.stonybrook.edu`

Contemporary Mathematics
Volume **817**, 2025
https://doi.org/10.1090/conm/817/16324

Superopers revisited

Anton M. Zeitlin

ABSTRACT. The relation between special connections on the projective line, called Miura opers, and the spectra of integrable models of Gaudin type provides an important example of the geometric Langlands correspondence. The possible generalization of that correspondence to simple Lie superalgebras is much less studied. Recently some progress has been made in understanding the spectra of Gaudin models and the corresponding Bethe ansatz equations for some simple Lie superalgebras. At the same time, the original example was reformulated in terms of an intermediate object: Miura-Plücker oper. It has a direct relation to the so-called qq-systems, the functional form of Bethe ansatz, which, in particular, allows q-deformation. In this note, we discuss the notion of superoper and relate it to the examples of qq-systems for Lie superalgebras, which were recently studied in the context of Bethe ansatz equations. We also briefly discuss the q-deformation of these constructions.

CONTENTS

1. Introduction

One of the most well-understood examples of geometric Langlands correspondence, studied by E. Frenkel and collaborators [**FFR1**], [**F1**], [**F2**],[**F3**], [**FFTL**], [**FFR2**], is the relation between the spectrum of Gaudin models for simple Lie algebra $\mathfrak{g}$ and Miura ${}^LG$-opers, namely certain meromorphic connections for the principal ${}^LG$-bundles over $\mathbb{P}^1$. Here ${}^LG$ is a simple Lie group of adjoint type with a Lalnglands dual Lie algebra ${}^L\mathfrak{g}$. In particular, the set of algebraic equations, known as Bethe equations, which describe the spectrum of the Gaudin model, provides the relation between the "moduli" parameters determining the corresponding oper connections.

Recently a q-deformation of this correspondence was studied [**KSZ**], [**FKSZ**], which led to various generalizations [**KZ1**], [**KZ4**] and applications [**KZ3**], [**KZ2**].

The author was partially supported by Simons Collaboration Grant 578501 and NSF grant DMS-2526435 (formerly DMS-2203823).

In particular, the concepts of Z-twisted Miura oper and Z-twisted Miura-Plücker oper were introduced, which have analogues in the differential case [**BSZ**], [**Z1**]. The first notion means that a meromorphic gauge transformation produces a constant connection Z out of a given Miura oper. The second, Z-twisted Miura-Plücker oper, is a less restrictive notion, which one can view as a first approximation to a Z-twisted condition. Namely, one applies Z-twisted condition only to the induced $GL(2)$-oper connections in the 2-dimensional subbundle of the associated bundle for each fundamental representation, corresponding to the vectors of fundamental weight and its elementary Weyl reflection. As a result, one obtains one-to-one correspondence between the data of Z-twisted Miura-Plücker opers and the functional system of equations called qq-system (QQ-system in q-deformed case). Such functional relations previously emerged in the study of the Gaudin model and other spin chain models. In that case, parameter Z is a twist parameter for the boundary condition. Under certain nondegeneracy conditions, these functional relations lead to the Bethe equations describing the spectrum of the related integrable model. It was also shown in [**BSZ**] that upon these nondegeneracy conditions, Z-twisted Miura-Plücker opers turn into Z-twisted Miura opers, where Z is an element of Cartan. An important notion in proving that is the notion of Bäcklund transformations, the gauge transformations, which produce $w(Z)$-twisted Miura oper from a given one, where w is an element of the Weyl group of ${}^LG$. Miura oper connections are specializations of more general oper connections: they are required to preserve specific reduction to the Borel subgroup of the principal bundle to the Borel subgroup. In other words, the Bäcklund transformations allow to "travel" between various Miura oper connections corresponding to one single oper connection. In the context of Bethe ansatz on the level of qq-systems, these transformations are known as *reproduction procedure* and *populations*, and were tied to the transformations of scalar differential operators, see, e.g., [**MV3**], [**MV2**].

One would expect to repeat that story in the context of simple superalgebras and supergroups: that, in particular, would give an example of Langlands duality in the super-context. The quest for finding a notion of oper in the context of simple supergroups started in [**Z2**], where the notion of oper on a super Riemann surface, called *superoper*, was introduced, which uses the concept of flat superholomorphic connection. We considered the case of supergroups that allow a purely fermionic system of simple roots. In that case, there is an explicit representation of the space of opers in terms of sections of super projective connections and superconformal vector bundles, known as superconformal fields. Also, as a particular example, we have shown in [**Z2**] the correspondence between the $OSP(1|2)$-superoper on a super Riemann sphere and $\mathfrak{osp}(1|2)$-Gaudin model studied in [**KM1**], [**KM2**].

Here we consider the superopers for general simple simply-connected Lie supergroups on a super Riemann sphere. We associate with them a purely even connection on a projective line, which is an oper connection for the reductive purely even subgroup. We define the notion of Z-twisted Miura-Plücker oper in this case and formulate more general notions of qp-system and qq-systems describing those and their specializations. The difference between purely even cases and supergroup cases is that there are several Dynkin diagrams for simple superalgebras, each generally producing a gauge inequivalent oper connection. At the same time, one can "travel" between various Dynkin diagrams for a given Lie supergroup if one adds odd Weyl reflections generated by odd roots, which are not automorphisms of Lie superalgebra and certainly do not lift to the Lie supergroup.

In this note, we propose to unify the Miura oper connections corresponding to all Dynkin diagrams by introducing formal Bäcklund transformations, which are no longer produced by the gauge transformations of a given connection, which follow the generalized Weyl transformation property on the level of qq-systems. We conjecture that satisfying these Bäcklund transformations under certain non-degeneracy conditions will put enough constraints on the qq-system to be in 1-to-1 correspondence with Bethe equations for the corresponding Gaudin model.

Recently, in [**HMVY**], the authors looked at the qq-system, which describes the Bethe ansatz equations for $\mathfrak{gl}(n|m)$ Gaudin model, and studied their populations along the lines of [**MV2**]. That work was continued in paper [**LM**], where the authors aimed to do the same for $\mathfrak{osp}(n|2m)$ Gaudin Bethe ansatz using the same technique of the transformations of the factorizations of pseudo-differential operators. The resulting transformations between data of the populations of these qq-systems bring remarkable transformations for the data which characterize the Miura-Plücker oper. We hope this more geometric approach can illuminate these reproduction procedures.

The structure of the paper is as follows. In Section 2 we discuss, following [**Z2**], the notion of Miura superoper, which a superholomorphic connection [**RT**] on a super Riemann surface [**BS**], [**F**]. We also define its reduction to the oper connection on a Riemann surface. In Section 3, we introduce the updated notion of Miura-Plücker opers and the qp-system, which generalizes qq-system for opers with regular singularities in pure even case. Section 4 is mainly devoted to the discussion of Bäcklund transformations for Miura opers related to the extended Weyl group and relation to the results of [**HMVY**], [**LM**]. In the end, we also speculate about the q-deformation of these constructions, which should be related to the recent work [**HLM**].

2. Superopers on supercurves and their reduction

2.1. A reminder on super Riemann surfaces and super holomorphic connections. For the general information on supermanifolds and superschemes one can consult [**M**], [**B**], [**L**]. For supercurves and Riemann surfaces one could follow [**BR**], [**W**]; for connections for vector bundles over super-Riemann surfaces we refer to [**RT**].

A *supercurve* of dimension $(1|1)$ over some fixed Grassmann algebra S (which is fixed throughout this paper) is a pair $(X, \mathcal{O}_X)$, where X is a topological space and $\mathcal{O}_X$ is a sheaf of supercommutative S-algebras over X such that $(X, \mathcal{O}_X^{\text{red}})$ is an algebraic curve: $\mathcal{O}_X^{\text{red}}$ is obtained from $\mathcal{O}_X$ by getting rid of nilpotents. Locally, for some open sets $U_\alpha \subset X$ and some linearly independent elements $\{\theta_i\}$ we have $\mathcal{O}_{U_i} = \mathcal{O}_{U_i}^{\text{red}} \otimes S[\theta_i]$.

Such collection of open sets $\{U_i\}$ serve as coordinate neighborhoods for supercurves with coordinates (z_i, θ_i). The coordinate transformations on the overlaps $U_i \cup U_j$ are given by the following formulas: $z_i = F_{ij}(z_j, \theta_j)$, $\theta_i = \Phi_{ij}(z_j, \theta_j)$, where $\{F_{ij}\}$, $\{\Phi_{ij}\}$ are even and odd functions correspondingly.

A super Riemann surface Σ over some Grassmann algebra S is a supercurve of dimension $1|1$ over S, with one more extra structure: there is a subbundle $\mathcal{D}$ of $T\Sigma$ of dimension $0|1$, such that for any nonzero section D of $\mathcal{D}$ on an open subset U of

Σ, D^2 is nowhere proportional to D, i.e. one obtains the exact sequence:

$$0 \to \mathcal{D} \to T\Sigma \to \mathcal{D}^2 \to 0. \tag{2.1}$$

One can pick the holomorphic local coordinates in such a way that this odd vector field will have the form $f(z,\theta)D_\theta$ for non-vanishing function $f(z,\theta)$ and

$$D_\theta = \partial_\theta + \theta\partial_z, \quad D_\theta^2 = \partial_z. \tag{2.2}$$

Such coordinates are called *superconformal.* The transformation between two superconformal coordinate systems (z,θ), (z',θ') is determined by the condition that $\mathcal{D}$ should be preserved:

$$D_\theta = (D_\theta\theta')D_{\theta'}, \tag{2.3}$$

so that the constraint on the transformation coming from the local change of coordinates is $D_\theta z' - \theta' D_\theta \theta' = 0$. An important example of a super Riemann surface is the super Riemann sphere SC^*: there are two charts (z,θ), (z,θ') so that

$$z' = -\frac{1}{z}, \quad \theta' = \frac{\theta}{z}.$$

We call the sections of $\mathcal{D}^n$ the *superconformal fields* of dimension $-n/2$ following [**W**]. In particular, taking the dual of the exact sequence (2.1), we find that a bundle of superconformal fields of dimension 1, namely $\mathcal{D}^{-2}$, is a subbundle in $T^*\Sigma$. Considering the superconformal coordinate system, a nonzero section of this bundle is generated by $\eta = dz - \theta d\theta$, which is orthogonal to D_θ under the standard pairing.

Let us consider a principal bundle $\mathcal{F}_G$ over the super Riemann surface with the Lie supergroup G over Grassmann algebra S. As usual, locally one can associate to the connection a differential operator, so that in the chart (z,θ) the connection has the following form:

$$\begin{aligned} &d_A = d + A = d + (\eta A_z + d\theta A_\theta) + (\bar\eta A_{\bar z} + d\bar\theta A_{\bar\theta}) = \\ &(\partial + \eta A_z + d\theta A_\theta) + (\bar\partial + \bar\eta A_{\bar z} + d\bar\theta A_{\bar\theta}) = (\eta D_z^A + d\theta D_\theta^A) + (\bar\eta D_{\bar z}^A + d\bar\theta D_{\bar\theta}^A). \end{aligned} \tag{2.4}$$

Here A takes values in $\mathfrak{g}_S$, the Lie algebra of G. We note, that we used here the fact that $d = \partial + \bar\partial$ and $\partial = \eta\partial_z + d\theta D_\theta$. The expression for the curvature is:

$$\begin{aligned} F = d_A^2 = d\theta d\theta F_{\theta\theta} + \eta d\theta F_{z\theta} + d\bar\theta d\bar\theta F_{\bar\theta\bar\theta} + \bar\eta d\bar\theta F_{\bar z\bar\theta} + \eta\bar\eta F_{z\bar z} + \\ \eta d\bar\theta F_{z\bar\theta} + \bar\eta d\theta F_{\bar z\theta} + d\theta d\bar\theta F_{\theta\bar\theta}, \end{aligned}$$

where

$$\begin{aligned} F_{\theta\theta} = -{D_\theta^A}^2 + D_z^A, F_{z\theta} = [D_z^A, D_\theta^A], F_{z,\bar z} = [D_z^A, D_{\bar z}^A], F_{z\bar\theta} \\ = [D_z^A, D_{\bar\theta}^A], F_{\theta\bar\theta} = -[D_\theta^A, D_{\bar\theta}^A]. \end{aligned}$$

It appears that if the connection d_A offers partial flatness, which implies $F_{\theta\theta} = F_{z\theta} = F_{\bar\theta\bar\theta} = F_{\bar z\bar\theta} = 0$, then there is a superholomorphic structure on any associated bundle (i.e. transition functions of the bundle can be made superholomorphic) [**RT**]. We are interested in the flat superholomorphic connections. In this case, since $F_{\theta\theta} = 0$, the connection is fully determined by the D_θ^A locally. In other words it is determined by the following odd differential operator, which from now on will denote $\hat\nabla$:

$$\hat\nabla = D_\theta + A_\theta(z,\theta), \tag{2.5}$$

so that the gauge transformation properties for A_θ are: $A_\theta \to gA_\theta g^{-1} - D_\theta g g^{-1}$, where g is a superholomorphic function providing change of trivialization.

2.2. Miura Superopers.

2.2.1. *Notations.* We refer to [**K2**], [**K1**], [**B**], [**CW**], [**FSS**] for further information regarding simple Lie supergroups, superalgebras, and their representations. Let G is a simple simply connected Lie supergroup of rank r over some Grassmann algebra S, B_- is its fixed Borel subgroup associated to a given Dynkin diagram with unipotent radical $N_- = [B_-, B_-]$. Let B_+ be the opposite Borel subgroup containing H and $N_+ = [B_+, B_+]$. Note that the Lie algebra $\mathfrak{g}_S$ of G is a module over S, namely $\mathfrak{g}_S = S \otimes \mathfrak{g}$, where $\mathfrak{g}$ is a simple Lie superalgebra over $\mathbb{C}$. Let $\{\alpha_1, \dots, \alpha_r\}$ be the set of positive simple roots for the pair $H \subset B_+$. For a given Dynkin diagram, we divide the index set of simple roots $I = \{1, \dots, r\}$ into union $I = I_w \sqcup I_g \sqcup I_b$ corresponding to the index set of white, grey, and black roots. W stands for the Weyl group generated the Weyl reflections corresponding to $\{\alpha_i\}_{i \in I_w}$.

Let $\{e_i, f_i, \check{\alpha}_i\}_{i=1,\dots,r}$ be the Chevalley generators of $\mathfrak{g}$, $a_{ji} = \langle \check{\alpha}_j, \alpha_i \rangle$ is the Cartan matrix. We note also that for grey roots $a_{ii} = 0$ for $i \in I_g$, and, in additional to standard Serre relations for $\{e_i\}$, $\{f_i\}$ there are extra Serre relations related to the grey root generators.

The Lie superalgebra $\mathfrak{b}_{-,S} = \mathfrak{b}_- \otimes S = \mathrm{Lie}(B_-)$ is generated by the f_i's and the $\check{\alpha}_i$'s, and $\mathfrak{b}_{+,S} = \mathfrak{b}_+ \otimes S = \mathrm{Lie}(B_+)$ is generated by the e_i's and the $\check{\alpha}_i$'s.

Let $\pi : S \to \mathbb{C}$ is the natural projection. That can be extended to $\pi : G \to \bar{G}$, where $\bar{G}$ is the underlying reductive simply connected group over $\mathbb{C}$. We denote by $\bar{\mathfrak{g}}$, $\bar{\mathfrak{b}}_\pm$, $\bar{\mathfrak{h}}$, $\bar{\mathfrak{n}}_\pm$ the corresponding pure even reductive Lie algebra, and the pure even versions of its Borel, Cartan and the maximal nilpotent subalgebras, while $\bar{G}$, $\bar{B}_\pm$, $\bar{H}$, $\bar{N}_\pm$ the corresponding subgroups.

2.2.2. *Definition of superopers.* Now we are ready to define the notion of superoper following a similar definition in the pure even case [**BD**], [**F2**] and inspired by the study of integrable hierarchies of Drinfeld-Sokolov type and related integrable models [**O**], [**IK**], [**GPZ**], [**DG**].

Let us consider a principal G-bundle $\mathcal{F}_G$ over a super Riemann surface X and its reduction $\mathcal{F}_{B_-}$ to the Borel subgroup B_-. We assume that it has a flat superholomorphic connection determined by $\hat{\nabla}$. suppose $\hat{\nabla}'$ is another superholomorphic connection, which preserves $\mathcal{F}_{B_-}$. Then the difference $\hat{\nabla}' - \hat{\nabla}$ has a structure of superconformal field of dimension 1/2 with values in the associated bundle $\mathfrak{g}_{\mathcal{F}_{B_-}}$.

Following the purely even case we define an open B_--orbit

$$\mathbf{O}_S \subset [\mathfrak{n}_{-_S}, \mathfrak{n}_{-,S}]^\perp / \mathfrak{b}_{-,S},$$

consisting of vectors, stabilized by N_- and such that all the simple root components of these vectors with respect to the adjoint action of H are non-zero. Here the orthogonal component is taken with respect to the nondenerate form for a given simple Lie superalgebra.

Definition 2.1 ([**Z2**]). A *G-superoper* on a super Riemann surface Σ is the triple $(\mathcal{F}, \mathcal{F}_{B_-}, \nabla)$, where $\mathcal{F}$ is a principle G-bundle, $\mathcal{F}_{B_-}$ is its B_--reduction and ∇ is a long superderivative on $\mathcal{F}$, such that $\nabla / \mathcal{F}_{B_-}$ takes values in $\mathbf{O}_{\mathcal{F}_B}$.

Therefore, locally on the open subset U, with coordinates (z,θ), with respect to the trivialization of $\mathcal{F}_B$, the structure of the superholomorphic connection is:

$$\hat{\nabla} = D_\theta + \sum_{i=1}^{r} a_i(z,\theta)e_i + b(z,\theta), \tag{2.6}$$

where each $a_i(z,\theta)$ is a nonzero function of opposite parity to e_i and $b(z,\theta)$ is an odd $\mathfrak{b}_{-,S}$-valued function.

2.2.3. *Miura superopers and their pure even counterparts.* Let us start from a definition of Miura superoper.

DEFINITION 2.2. A *Miura G-superoper* on X is a quadruple $(\mathcal{F}_G, \nabla, \mathcal{F}_{B_-}, \mathcal{F}_{B_+})$, where $(\mathcal{F}_G, \hat{\nabla}, \mathcal{F}_{B_-})$ is a meromorphic G-oper on SC^* and $\mathcal{F}_{B_+}$ is a reduction of the G-bundle $\mathcal{F}_G$ to B_+ that is preserved by the connection $\hat{\nabla}$.

From now on we set $\Sigma = SC^*$. Then one can put Miura superoper in the following canonical form.

PROPOSITION 2.3. *For any Miura G-oper on SC^*, there exists a trivialization of the underlying G-bundle $\mathcal{F}_G$ on an open dense subset of SC^* for which the superoper connection has the form*

$$\hat{\nabla} = D_\theta + \sum_{i=1}^{r} g_i(z,\theta)\check{\alpha}_i + \sum_{i=1}^{r} a_i(z,\theta)e_i, \tag{2.7}$$

where $g_i(z,\theta), a_i(z) \in S(z)[\theta]$, so that $g_i(z,\theta)$ are all even and $a_i(z,\theta)$ are opposite in parity to e_i.

The proof of this proposition is similar to the one in [**BSZ**] with the use of the cell partition via super extension of Weyl group in [**VM1, VM2**], see also [**P**].

From now on we assume that $\{D_\theta a_i(z,\theta)\}_{i\in I_w}$, as well as $\{a_i(z,\theta)\}_{i\in I_g\cup I_b}$ are invertible.

DEFINITION 2.4. A *$\hat{Z}$-twisted G-superoper* on SC^* is a G-superoper that is equivalent to the constant element $A(\theta,z) = \theta\hat{Z}$, where $\hat{Z} \in \mathfrak{g}_S \subset \mathfrak{g}(z,\theta)$ under the gauge action of $G(z,\theta)$.

For simplicity from now on we will assume that $\hat{Z} \in \mathfrak{h}_S$ is regular semisimple. One can generalize most of the results to $Z \in \mathfrak{b}_{+,S}$ as it was done in [**BSZ**].

REMARK 2.5. Note, that instead one could consider any element $Z' = \xi + \theta Z$, where ξ i an odd element of $\mathfrak{g}_S$, instead of θZ in the above definition, but one can remove ξ by gauge transformation given by $\exp(\theta\xi) \in G(\theta)$.

Now we want to get rid of extra odd variables to see the relation of superopers to opers on $\mathbb{P}^1$ for a certain reductive group.

First, we will get rid of the θ variable. Let us represent the superoper connection as

$$\hat{\nabla} = D_\theta + \theta M(z) + N(z) \tag{2.8}$$

making the dependence on θ explicit. The gauge transformations can be factorized this way:

$$g(z,\theta) = (1 - \theta R(z))U(z),$$

where $R(z) \in \mathfrak{g}(z)$, $U(z) \in G(z)$. There is a unique $R(z)$, namely $R(z) = N(z)$, such that

$$g(z,\theta)^{-1}\hat{\nabla}g(z,\theta) = \partial_\theta + \theta U^{-1}(z)\tilde{\nabla}U(z), \tag{2.9}$$

where the meromorphic connection $\tilde{\nabla}$ is given locally by a differential operator

$$\tilde{\nabla} = \partial_z + \frac{1}{2}[N(z), N(z)] + M(z). \tag{2.10}$$

Thus we obtain the following Corollary.

COROLLARY 2.6. *Z-twisted condition for Miura G-superoper implies that connection $\tilde{\nabla}$ on $\mathbb{P}^1$ is gauge equivalent to a constant connection $\hat{Z} \in \mathfrak{g}$.*

Now let us proceed to Miura G-superoper, so that the resulting connection is as in (2.7) and let's construct such connection $\tilde{\nabla}$. This way, we obtain G-connection on $\mathbb{P}^1$. Now, let us apply the map $\pi : S \to \mathbb{C}$, which strips dependence on all the odd parameters. This way, we obtain the connection $\nabla = \pi(\tilde{\nabla})$ on a principal $\bar{G}$-bundle, which have locally the following form:

$$\nabla \equiv \pi(\tilde{\nabla}) = \partial_z + u(z) + \sum_{i\in I_w} L_i(z)e_i + \frac{1}{2}\sum_{i,j\in I_b\cup I_g} L_i(z)L_j(z)[e_i, e_j], \tag{2.11}$$

where $\{L_i(z)\}_{i=1,\dots,r}$ are nonzero rational functions, so that in the original connection (2.7):

$$\pi(a_i(z,\theta)) = \begin{cases} L_i(z) & \text{if } i \in I_w, \\ \theta L_i(z) & \text{if } i \in I_b \cup I_g. \end{cases} \tag{2.12}$$

That gives rise to the following definition.

DEFINITION 2.7. We say that the quadruple $(\mathcal{F}_{\bar{G}}, \mathcal{F}_{\bar{B}_+}, \mathcal{F}_{\bar{B}_-}, \nabla)$, where ∇ is a connection on a principal bundle $\mathcal{F}_{\bar{G}}$ over $\mathbb{P}^1$ together with reductions $\mathcal{F}_{\bar{B}_\pm}$ to Borel subgroups $\bar{B}_\pm$ is a Miura oper associated to Miura G-superoper $(\mathcal{F}_G, \mathcal{F}_{B_+}, \mathcal{F}_{B_-}, \hat{\nabla})$.

In particular, we notice that the differential operator which isolates the Cartan part of (2.11):

$$\nabla^H = \partial_z + u(z) \tag{2.13}$$

defines an H-connection on $\mathbb{P}^1$, which we call a *Cartan connection* ∇^H associated to Miura oper ∇. We say that H-connection is Z-twisted if it is gauge equivalent to the constant connection $\partial_z + Z$, where $Z \in \mathfrak{h}$.

REMARK 2.8.

(i) We note here that in general, even for distinguished Dynkin diagrams the collection $\{\check{\alpha}_i\}_{i=1,\dots,r}$, $[e_i, e_j]$ for all $i, j \in I_g \cup I_b$ and e_i for all $i \in I_w$ do not give Chevalley generators producing $\bar{B}_+$, but instead a Borel subgroup of a smaller reductive group.
(ii) If the superoper was Z-twisted, the resulting connection ∇ is gauge equivalent to $\pi(Z)$.

3. Miura-Plücker opers, qp-systems, and qq-systems

3.1. Z-twisted Cartan connections. Let us fix the component notation for the Cartan connection, associated to a given superoper and the corresponding twist element Z:

$$u(z) = \sum_{i\in I_w\cup I_g} u^i(z)\check{\alpha}_i + \sum_{i\in I_b} u^i(z)\frac{\check{\alpha}_i}{2}; \quad Z = \sum_{i\in I_w\cup I_g} \zeta_i\check{\alpha}_i + \sum_{i\in I_b} \zeta_i\frac{\check{\alpha}_i}{2} \tag{3.1}$$

This leads to the following proposition.

PROPOSITION 3.1. *If the Cartan connection parametrized by $u(z)$ in (2.11) is Z-twisted, then there exist rational functions $\{p_i(z)\}_{i=1,\dots,r}$ such that*

$$u_i(z) = \zeta_i + \ln'\big[p_i(z)\big], \quad 1,\dots,r. \tag{3.2}$$

One can view it as a first approximation for Z-twisted condition for the Miura $\bar{G}$-oper connection (2.11).

Now let us consider in detail the rank $r=1$ examples of Z-twisted Miura opers.

3.2. Low rank cases.

3.2.1. *Z-twisted Miura $SL(2)$-opers.* In this case we are dealing with the meromorphic $SL(2)$-connection for $SL(2)$ bundle which has the following form:

$$\nabla = \partial_z + u(z)\check{\alpha} + L(z)e \tag{3.3}$$

where $e, f, \check{\alpha}$ are the Chevalley generators of $\mathfrak{sl}(2)$. The Z-twisted condition states that there exists $U(z)\in B_+(z)$, such that

$$U(z)\nabla U^{-1}(z) = \partial_z + Z. \tag{3.4}$$

where $Z = \zeta\check{\alpha}$. We can represent the resulting group element as

$$U(z) = e^{q(z)e}p(z)^{\check{\alpha}},$$

where $p(z), q(z)\in\mathbb{C}(z)$. Looking at the coefficient of $\check{\alpha}$ in (3.4), we obtain

$$u(z) = \zeta + \ln'\big[p(z)\big] \tag{3.5}$$

The e-coefficient gives the equation:

$$q'(z) + 2\zeta q(z) = p^2(z)L(z).$$

Representing $q(z) = \frac{q_-(z)}{q_+(z)}$, so that $q_\pm(z)\in\mathbb{C}[z]$, $q_+(z)$ is monic, we obtain the following equations:

$$W(q_+, q_-)(z) + 2\zeta q_-(z)q_+(z) = \Lambda(z), \tag{3.6}$$

where $W(f,g)(z) = f(z)g'(z) - f'(z)g(z)$, and

$$\Lambda(z) = q_+^2(z)p^2(z)L(z) \tag{3.7}$$

is a polynomial. Expanding $u(z) = \frac{\tilde{u}(z)}{\tilde{\Lambda}(z)}$ and $p(z) = \frac{p_-(z)}{p_+(z)}$, where $p_\pm(z)$ do not have common roots and $p_-(z)$ is chosen to be monic, we obtain:

$$\tilde{\Lambda}(z) = p_+(z)p_-(z). \tag{3.8}$$

Note, that given the factorization of denominator $\tilde{\Lambda}$, the numerator is determined uniquely:

$$\tilde{u}(z) = W(p_+, p_-)(z) + \zeta p_-(z)p_+(z).$$

Thus let us call the following system of equations:

$$W(q_+,q_-)(z)+2\zeta q_-(z)q_+(z)=\frac{q_+^2(z)p_-^2(z)L(z)}{p_+(z)^2},\qquad (3.9)$$
$$p_+(z)p_-(z)=\tilde{\Lambda}(z)$$

the *pq-system* for $\mathfrak{sl}(2)$.

PROPOSITION 3.2. *There is one-to-one correspondence between Z-twisted Miura $SL(2)$-opers with the connection* (3.3) *and solutions of the pq-system* (3.9), *where $p(z)=\frac{p_-(z)}{p_+(z)}$ as well as $q(z)=\frac{q_-(z)}{q_+(z)}$ are irreducible fractions, so that $\tilde{u}(z)=W(p_+,p_-)(z)+\zeta p_-(z)p_+(z)$.*

A simplification of the $\mathfrak{sl}(2)$ pq-system is given by the following identification

$$\tilde{\Lambda}(z)=p_+(z)=q_+(z),\qquad (3.10)$$

which leaves just one equation:

$$W(q_+,q_-)(z)+2\zeta q_-(z)q_+(z)=\Lambda(z),\qquad (3.11)$$

so that $u(z)=\zeta-\ln'(q_+(z))$ and $L(z)=\Lambda(z)$ is a polynomial. That is known as $\mathfrak{sl}(2)$ qq-system, which is in one-to one correspondence with the $SL(2)$-opers with regular singularities: the positions of the singularities on $\mathbb{P}^1$ are given by the roots polynomial $\Lambda(z)$. Under nondegeneracy conditions, namely $q_+(z)$ have distinct roots and have no common roots with $\Lambda(z)$, a simple calculation shows that there is a bijection between such nondegenerate solutions of $\mathfrak{sl}(2)$ qq-system and the the Bethe equations of $\mathfrak{sl}(2)$ Gaudin model:

$$2\zeta+\partial_z\log\Big[\Lambda(z)(z-w_\ell)^2\Big]\Big|_{z=w^i_\ell}=0,\qquad (3.12)$$
$$i=1,\dots,r;\quad \ell=1,\dots,\deg(q_+(z)).$$

3.2.2. *Z-twisted Miura $SL(1|1)$ opers and abelian connections.* Although it is not a simple superalgebra because of nontrivial center, one can apply the notion of oper to $SL(1|1)$-group, which technically corresponds to the grey node of Dynkin diagram, and it is still useful to consider it. We see that in this case the connection $\bar{\nabla}$ and ∇ are reduced to the abelian connection corresponding to the central element $\check{\alpha}$ of $SL(1|1)$:

$$\nabla=\partial_z+u(z)\check{\alpha}.\qquad (3.13)$$

The Z-twisted condition leads to the equation (3.5). Expressing $u(z)=\frac{\tilde{u}(z)}{\tilde{\Lambda}(z)}$, one can say that this is a particular case of the first example when $L(z)=0$, so one can call the equation

$$p_+(z)p_-(z)=\tilde{\Lambda}(z)\qquad (3.14)$$

the $\mathfrak{sl}(1|1)$ pq-system, although q-part is absent here.

PROPOSITION 3.3. *There is one-to-one correspondence between $SL(1|1)$ -opers* (3.13) *and the solutions to the equation* (3.14), *so that $p_\pm(z)$ have no common roots and*

$$u(z)=\tilde{\Lambda}(z)^{-1}\Big[W(p_+,p_-)(z)+\zeta p_+(z)p_-(z)\Big].$$

The equation (3.14) actually gives the Bethe ansatz solutions for $\mathfrak{gl}(1|1)$ Gaudin model. Namely, let us suppose we can re-express

$$\tilde{\Lambda}(z) = \ln'(\Lambda(z))\pi(z), \tag{3.15}$$

where $\pi(z) = \prod_{k=1}^{n}(z - z_k)$ where z_k are all distinct roots of $\Lambda(z) = \prod_{k=1}^{n}(z - z_k)^{d_k}$. In this case the Λ determines the weights for the appropriate $\mathfrak{gl}(1|1)$ representation and the equation (3.14) is equivalent to the Bethe ansatz equations for $\mathfrak{gl}(1|1)$ Gaudin model [**MVY**], [**HMVY**]:

$$\sum_{k=1}^{n} \frac{d_k}{w_j - z_k} = 0, \quad j = 1, \dots \deg(p_+(z)), \tag{3.16}$$

where $\{w_j\}$ are the roots of $p_+(z)$.

3.2.3. *Z-twisted Miura $OSP(1|2)$-opers.* Consider an $\mathfrak{osp}(1|2)$-triple: $e, f, \check{\alpha}$, where e, f are odd Chevalley generators. The corresponding $OSP(1|2)$-Miura oper connection is:

$$\nabla = \partial_z + \frac{\check{\alpha}}{2}u(z) + \frac{1}{2}L^2(z)[e, e]. \tag{3.17}$$

Notice that this is exactly the Miura $SL(2)$-oper we considered above, since $\frac{\check{\alpha}}{2}$, $[e, e]$ are Chevalley generators of $\mathfrak{sl}(2)$ subalgebra. The only difference is that we have a square as a coefficient of e^2.

Thus we have the following Proposition.

PROPOSITION 3.4. *There is a one-to-one correspondence between Z-twisted $OSP(1|2)$-opers and Z-twisted Miura $SL(2)$ opers where $\Lambda(z) = L^2(z)$, $L(z) \in \mathbb{C}[z]$.*

That correspondence was first noted in [**Z2**]: in particular, it was shown that the resulting Bethe ansatz equations describing coincide with the Bethe ansatz equations for $\mathfrak{osp}(1|2)$ Gaudin model [**KM1**], [**KM2**].

3.3. Miura-Plücker opers. Z-twisted condition is a complicated one to solve. Instead one can look at the intermediate object. We already have seen the first approximation to that condition given by the Proposition 3.1. Now, we introduce a useful object, known as Z-twisted Miura-Plücker $\bar{G}$-oper, which is a next iteration approximating Z-twisted condition. In pure even case, for Miura G-opers with regular singularities, it was introduced in [**BSZ**] following the q-deformed version in [**FKSZ**].

Let us consider the induced Miura $\bar{G}$-oper $\bar{B}_+$-bundle connections on $\mathcal{V}_i$: the associated bundles, corresponding to highest weight irreducible modules of $\bar{\mathfrak{g}}$: *i)* V_{ω_i} if $i \in I_w \cup I_g$; *ii)* $V_{2\omega_i}$ when $i \in I_b$.

Let us define a B_+-subbundle $\mathcal{W}_i$, the rank of which will depend on i.

- If $i \in I_w$, W_i is spanned by the line subbundles $\mathcal{L}_i, \tilde{\mathcal{L}}_i$, which correspond to the vectors of weight ω_i, $\omega_i - \alpha_i$.
- If $i \in I_b$, let $\mathcal{W}_i$ be spanned by $\mathcal{L}_i$, $\tilde{\mathcal{L}}_i$ corresponding to the vectors of weight $2\omega_i$, $2\omega_i - 2\alpha_i$.
- If $i \in I_g$ we take W_i to be a line bundle which correspond to the vector of highest weight ω_i.

Let ∇_i be the induced connection on W_i.

DEFINITION 3.5. We say that Miura $\bar{G}$-oper is Z-twisted Miura-Plücker if there exists $v(z) \in B_+(z)$ such that

$$\nabla_i = v(z)(\partial_z + Z)v(z)^{-1}|_{W_i} = v_i(z)(\partial_z + Z_i)v_i(z)^{-1}, \tag{3.18}$$

where $v_i(z) = v(z)|_{W_i}$ and $Z_i = Z|_{W_i}$.

The element $v(z)$ is not uniquely determined by Miura-Plücker oper. Let $\tilde{N}_+(z)$ be the subgroup, generated by all even commutators of $[e_i, e_j]$, $i \neq j$. We have the following proposition, which gives equivalence classes of such $v(z)$.

PROPOSITION 3.6. *For a given $v(z)$ from the Definition 3.5 any element of coset*

$$v(z)H\tilde{N}_+(z)$$

also satisfies (3.18).

Following [**BSZ**], we call such a coset a *framing* of Miura-Plücker oper.

DEFINITION 3.7. The *Miura-Plücker datum* is a pair $(\nabla, v(z)\tilde{N}_+(z))$ consisting of Miura-Plücker oper and related framing.

One can fix the corresponding representative in the coset as follows:

$$v(z) = \prod_{i\in I_w\cup I_g} p^i(z)^{-\check{\alpha}_i} \prod_{j\in I_b} p^j(z)^{-\frac{\check{\alpha}_j}{2}} \prod_{i\in I_w} e^{-q^i(z)e_i} \prod_{j\in I_b} e^{-\frac{1}{2}q^j(z)[e_j,e_j]}. \tag{3.19}$$

Now we will explore the condition (3.18). For the Cartan part of ∇ we obtain the following equations:

$$u^i(z) = \zeta_i + \ln'(p^i(z)). \tag{3.20}$$

Now we will show how it works off-diagonal in each of the cases:

(1) Let $i \in I_w$. Then we have the following condition:

We first compute the matrix of $v(z)$ and Z acting on the two-dimensional subspace W_i. The following calculation gives

(3.21)

$$v(z)|_{W^i} = \begin{pmatrix} p^i(z)^{-1} & 0 \\ 0 & p^i(z)\prod_{j\neq i, j\in I_w\cup I_g} p^j(z)^{a_{ji}} \prod_{j\neq i, j\in I_b} p^j(z)^{a_{ji}/2} \end{pmatrix} \begin{pmatrix} 1 & -\frac{q^i_-(z)}{q^i_+(z)} \\ 0 & 1 \end{pmatrix}$$

and

$$Z^H|_{W_i} = \begin{pmatrix} \zeta_i & 0 \\ 0 & -\zeta_i - \sum_{j\neq i} a_{ji}\zeta_j \end{pmatrix}. \tag{3.22}$$

That implies the following equation from the top right corner of 2×2 block:

$$\partial_z q^i(z) + \langle Z, \alpha_i\rangle q^i(z) = L_i(z)\Big[p^i(z)\Big]^2 \prod_{j\neq i, j\in I_w\cup I_g} p^j(z)^{a_{ji}} \prod_{j\neq i, j\in I_b} p^j(z)^{a_{ji}/2}. \tag{3.23}$$

(2) If $i \in I_g$, we do not have any extra equations in addition to (3.20).

(3) If $i \in I_b$, we are dealing with the same 2×2 block as in $i \in I_w$. Thus we have the following equation:

$$\partial_z q^i(z) + \langle Z, \alpha_i\rangle q^i(z) = \Big[L_i(z)p^i(z)\Big]^2 \prod_{j\neq i, j\in I_w\cup I_g} p^j(z)^{2a_{ji}} \prod_{j\neq i, j\in I_b} p^j(z)^{a_{ji}}. \tag{3.24}$$

3.4. qp-and qq-system for Lie superalgebra $\mathfrak{g}$.

3.4.1. *Definition of the pq-system and relation to Z-twisted Miura-Plücker opers.* Let us choose the simple root system of Lie superalgebra $\mathfrak{g}$ with the index set $I = I_w \cup I_b \cup I_g$ and the datum of rational functions $\{L_i\}_{i\in I}(z)$, polynomial functions $\{\tilde{\Lambda}_i\}_{i\in I}$ and rational functions $\{p^i(z), q^i(z)\}_{i\in I}$ be the irreducible fractions:

$$p^i(z) = \frac{p^i_-(z)}{p^i_+(z)}, \quad q^i(z) = \frac{q^i_-(z)}{q^i_+(z)}, \quad i \in I. \tag{3.25}$$

We call the following system of equations:

$$\partial_z q^i(z) + \langle Z, \alpha_i\rangle q^i(z) = F_i(z), i \in I_w \cup I_b, \tag{3.26}$$

$$F_i(z) = \begin{cases} L_i(z)\left[p^i(z)\right]^2 \prod_{j\neq i, j\in I_w\cup I_g} p^j(z)^{a_{ji}} \prod_{j\in I_b} p^j(z)^{a_{ji}/2} & \text{if } i \in I_w \\ \left[L_i(z)p^i(z)\right]^2 \prod_{j\in I_w\cup I_g} p^j(z)^{2a_{ji}} \prod_{j\neq i, j\in I_b} p^j(z)^{a_{ji}} & \text{if } i \in I_b \end{cases}$$

$$p^i_+(z)p^i_-(z) = \tilde{\Lambda}_i(z), \quad i = 1, \ldots, r,$$

the *pq-system* associated to superalgebra $\mathfrak{g}$.

Using this definition we can restate the result of the previous section as follows.

THEOREM 3.8. *There is a bijection between the set of data of Z-twisted Miura-Plücker opers and solutions of the generalized pq-systems, so that $q^i_+(z), q^i_-(z)$ as well as $p^i_+(z), p^i_-(z)$ for all $i = 1, \ldots, r$ have no common roots and*

$$u_i(z) = \frac{\tilde{u}_i(z)}{\tilde{\Lambda}_i(z)} = \zeta_i + \ln'(p_i)(z). \tag{3.27}$$

REMARK 3.9. We remark here, that we do not make any assumptions/nondegeneracy conditions on far on the roots/poles of the data $\{\tilde{\Lambda}_i\}$, $\{L_i(z)\}$. In the following we will see examples when $\{\tilde{\Lambda}_i(z)\}_{i\in I_g}$ depends on $q^i(z)$. Notice also that there are no equations on $\{L_i(z)\}_{i\in I_g}$. Those will be taken into account when the full Z-twisted condition is implemented.

3.4.2. *Reduction to the qq-system, nondegeneracy conditions and Bethe equations.* To get in touch with the Bethe ansatz equations for Gaudin model, we will impose the following condition on the *qp*-system: we require that the reduction of the *pq*-system to the simple even subgroups of $\bar{\mathfrak{g}}$ reproduces the *qq*-system for Miura opers with regular singularities [**BSZ**]. Namely, we assume:

$$\tilde{\Lambda}_i(z) = p^i_+(z) = q^i_+(z) \text{ for } i \in I_w \cup I_b. \tag{3.28}$$

Also, we redefine:

$$p^i_\pm(z) \equiv q^i_\pm(z), \quad \text{for} \quad i \in I_g. \tag{3.29}$$

We call the resulting system of equations on the *pq*-system data, the *qq*-system associated to $\bar{\mathfrak{g}}$.

$$W(q^i_+, q^i_-)(z) + \langle Z, \alpha_i\rangle q^i_+(z)q^i_-(z) = F_i(z), \quad i \in I_w \cup I_b, \tag{3.30}$$

$$F_i(z) = \begin{cases} L_i(z)\prod_{j\neq i, j\in I_w} q^j_+(z)^{-a_{ji}} \prod_{j\in I_b} q^j_+(z)^{-a_{ji}/2} \prod_{j\in I_g} q^j(z)^{a_{ji}} & \text{if } i \in I_w \\ \left[L_i(z)\right]^2 \prod_{j\in I_w} q^j_+(z)^{-2a_{ji}} \prod_{j\neq i, j\in I_b} q^j_+(z)^{-a_{ji}} \prod_{j\in I_g} q^j(z)^{2a_{ji}} & \text{if } i \in I_b \end{cases}$$

$$q^i_+(z)q^i_-(z) = \tilde{\Lambda}_i(z), \quad i \in I_g. \tag{3.31}$$

This way we reduced the number of independent functions to $\{L_i(z)\}_{i\in I_w\cup I_b}$ and $\{\tilde{\Lambda}_i(z)\}_{i\in I_g}$. Under certain conditions, one can write solutions to the (3.30) in terms of algebraic equations. First, we introduce the nondegeneracy conditions:

DEFINITION 3.10. The solutions to the qq-system is called *nondegenerate*, if $F_i(z)$ has no common roots with $q^i_+(z)$ and all the roots of $q^i_+(z)$ are distinct for all i. Moreover $p^i_+(z)$ and $p^i_-(z)$ have no common roots.

Then Proposition 3.11 is true.

PROPOSITION 3.11. *If Z is regular semisimple, there is a bijection between nondegenerate solutions to* (3.30) *and the following algebraic equations for the roots* $\{w^i_\ell\}_{\ell=1,\dots,\deg(q^i_+(z))}$ *of* $q^i_+(z)$*:*

$$\langle\alpha_i, Z\rangle + \partial_z \log\left[F_i(z)(z-w^i_\ell)^2\right]\Bigg|_{z=w^i_\ell} = 0, \tag{3.32}$$
$$i=1,\dots,r;\quad \ell=1,\dots,\deg(q^i_+(z)).$$

The equations (3.32) are known in particular cases as Bethe equations for Gaudin model associated with simple Lie algebras, we will refer to them as Bethe equations of *even type.* The algebraic equations emerging from (3.31), which are solved just by division are Bethe equations of *odd type.* Of course, in the case, when $\mathfrak{g}$ is a simple algebra, one has only equations of even type. The following proposition follows.

In the case when $\mathfrak{g}$ is a simple Lie algebra, i.e. $I = I_w$ the qq-system (3.30),(3.31) reduces to well-known qq-system from [**MV3**], [**BSZ**], imposing the condition that $L_i(z) = \Lambda_i(z)$ is a polynomial:

$$W(q^i_+,q^i_-)(z) + \langle Z,\alpha_i\rangle q^i_+(z)q^i_-(z) = \Lambda_i(z)\prod_{j\neq i} q^j_+(z)^{-a_{ji}},\quad i=1,\dots,r, \tag{3.33}$$

where a_{ji} is a Cartan matrix of $\mathfrak{g}$.

The corresponding G-oper connections are called Miura G-opers with regular singularities, where the position of singularities are regulated by polynomials $\{\Lambda_i(z)\}_{i=1,\dots,r}$ [**BSZ**]. Locally the connection has the form

$$\nabla = \partial_z + Z - \sum_{i=1}^{r} \ln'\left[q^i_+(z)\right]\check{\alpha}_i + \sum_{i=1}^{r}\Lambda_i(z)e_i. \tag{3.34}$$

4. Z-twisted Miura opers and the extended Weyl group

4.1. Overview of the pure even case. In [**BSZ**], in the case of simple Lie algebras for opers with regular singularities, we have shown that under the nondegeneracy conditions Z-twisted Miura-Plücker opers turn out to be Z-twisted Miura opers, i.e., the solutions of the qq-system completely determine Z-twisted Miura opers.

Let us look at the related Z-twisted oper for regular semisimple Z. We find that there are precisely $|W|$ (W is a Weyl group of $\mathfrak{g}$) related Z-twisted Miura opers, when Z is regular semisimple, each described by the solutions of the corresponding qq-systems. The action of the Weyl group on the space of solutions of such qq-systems is given using the following transformations, corresponding to elementary

Weyl reflections w_i:

$$Z \to w_i(Z), \quad q^j_\pm(z) \to \begin{cases} q^i_\mp(z) & \text{if } j = i \\ q^j_\pm(z) & \text{if } j \neq i \end{cases}. \tag{4.1}$$

On the level of Miura G-oper connections that can be achieved by special gauge transformations from $B_-(z)$:

$$\nabla \to e^{\mu_i(z) f_i} \nabla e^{-\mu_i(z) f_i}, \quad \mu_i(z) = \Lambda_i(z)^{-1}\left[\partial_z \log\left(\frac{q^i_-(z)}{q^i_+(z)}\right) + \langle \alpha_i, Z\rangle\right], \tag{4.2}$$

which we called *Bäcklund transformations* in [**FKSZ**], [**BSZ**]. These transformations were previously discussed in the context of Bethe ansatz equations [**MV3**], [**MV1**] leading to the so-called "populations" of Bethe ansatz equations.

4.2. Conjectures for simple superalgebras. In the case of superalgebra $\mathfrak{g}$, Weyl reflections of even roots generate the Weyl group W. One can construct a larger group $\tilde{W}$ by adding the reflections, corresponding to the odd roots, which changes the Dynkin diagram for $\mathfrak{g}$ [**LSS**], [**FSS**]. In particular, applying such reflections to a given system of simple roots, one can generate all systems of simple roots for a given superalgebra $\mathfrak{g}$. However, these are not automorphisms of $\mathfrak{g}$; of course, one cannot lift these transformations to G.

The relation between Z-twisted Miura opers corresponding to a given oper in the pure even case, as discussed in the previous subsection, motivates introducing the following notion, generalizing the one from [**BSZ**].

DEFINITION 4.1. Consider two qq-systems based on Dynkin diagrams based on simple root systems related by a simple reflection $s_i \in \tilde{W}$. We say that two solutions of such qq-systems are *i-composable* if they are obtained from each other by the transformation (4.1) accompanied with certain transformations $\{L_i\}, \{\tilde{\Lambda}_i\} \to \{L_i^{w_i}\}, \{\tilde{\Lambda}_i^{w_i}\}$. We call two solutions of qq-systems *w-composable*, where $w \in \tilde{W}$ if a sequence of such transformations relates them. We call the related Z-twisted Miura-Plücker opers *w-composable* if their datum expressed via the solution of the qq-system is w-composable.

Notice, that we still did not specify the transformations of $\{L_i(z)\}_{i \in I_b \cup I_w}$, $\{\tilde{\Lambda}_i(z)\}_{i=1,\dots,r}$, for the w-composable qq-systems because they may vary from one qq-system to another, unlike the pure even case. Also, those may depend on $\{q^j_\pm(z)\}_{i=1,\dots,r}$ if j and i are adjacent on the Dynkin diagram. In fact, we will see below such an example of w-composable family of qq-systems associated with $\mathfrak{sl}(n|m)$.

Let us formulate the following conjecture, which is the analogue of the main results of [**BSZ**].

CONJECTURE 4.2.

(i) *Under certain nondegeneracy conditions, Z-twisted Miura-Plücker opers are Z-twisted Miura opers for certain choices of $\{L_i(z)\}_{i \in I_g}$.*

(ii) *There exists such data of $\{L_i^w(z)\}_{i \in I}$ and $\{\tilde{\Lambda}_i^w(z)\}_{i \in I}$ for the qq-systems associated to $\mathfrak{g}$ such that there exists a family of w-composable Z-twisted Miura opers, which are described by the Bethe equations of Gaudin model for a certain simple superalgebra ${}^L\mathfrak{g}$.*

Indeed, this way, we relate various Z-twisted opers for different Dynkin diagrams making an object, which one may call a *generalized* Z-twisted superoper, which unifies all w-composable Z-twisted Miura opers. We do not know what ${}^L\mathfrak{g}$ could be for the cases beyond $\mathfrak{g}=\mathfrak{sl}(m|n)$: in the next subsection we will discuss the discovered class of such w-composable qq-systems.

4.3. What is known: qq-systems for $\mathfrak{sl}(n|m)$ and Gaudin models. In [**HLM**] a certain version of qq-system was considered in the case of $\mathfrak{sl}(n|m)$ and $Z=0$. For a given Dynkin diagram for $\mathfrak{g}$ one has

$$
\begin{aligned}
W(q_+^i, q_-^i)(z) &= \Lambda_i(z) q_+^{i+1}(z) q_+^{i-1}(z) \text{ if } i \in I_w, \\
q_+^i(z) q_-^i(z) &= \tilde{\Lambda}_i(z), \text{ if } i \in I_g.
\end{aligned}
\tag{4.3}
$$

Here

$$
\Lambda_i(z) = \frac{T_i(z)}{T_{i+1}(z)}; \quad \tilde{\Lambda}_i(z) = \ln'\left(\frac{T_i(z)T_{i+1}(z)q_+^{i-1}(z)}{q_+^{i+1}(z)}\right)\pi_i(z) q_+^{i+1}(z) q_+^{i-1}(z), \tag{4.4}
$$

so that $\{\Lambda_i(z)\}$, $\{T_i(z)\}$ are polynomials and $\pi_i(z)=\prod_k(z-z_k)$, where z_k are distinct roots of $T_i(z)T_{i+1}(z)$. The solution of this qq-system under nondegeneracy condition is in 1-to-1 correspondence with Bethe ansatz equations for $\mathfrak{gl}(n|m)$ Gaudin model. The authors of [**HMVY**] produced the w_i-composable solutions, which they call populations as in original papers [**MV2**]. We note, that in this case the extended Weyl group $\tilde{W}$ can be identified with S_{n+m} and in defining representation it can be realized in the standard way using permutation matrices. The architecture of the reproduction procedure is linked to pseudo-differential operator, which for distinguished Dynkin daigram, where one grey root separates Dynkin diagrams for $\mathfrak{sl}(n)$ and $\mathfrak{sl}(m)$) looks as follows:

$$
R(z) = \prod_{i=1}^{n}\left(\partial_z - \log'\left[T_i(z)\frac{q_+^{i-1}(z)}{q_+^i(z)}\right]\right)\prod_{n+1}^{m+n}\left(\partial_z + \log'\left[T_i(z)\frac{q_+^{i-1}(z)}{q_+^i(z)}\right]\right)^{-1}, \tag{4.5}
$$

where $q_+^0(z) = q_+^{m+n}(z) = 1$.

$$
R^w(z) = \prod_{i=1}^{n+m}\left(\partial_z - s_i(w)\log'\left[T_i^w(z)\frac{q_+^{w,i-1}(z)}{q_+^{w,i}(z)}\right]\right)^{s_i(w)}. \tag{4.6}
$$

Here $s_i(w)=\pm 1$ and correspond to original permutation w of original s_i in $R(z)$ for distinguished Dynkin diagram. By extra index w we denoted the qq-system corresponding to w-transformed Dynking diagram. It turns out that w-composability of the related solutions implies identification of R^w and R [**HMVY**], using the following transformations corresponding to elementary Weyl reflections w_i (this also implies the proper formulas for the $T_i^w(z)$):

$$
\begin{aligned}
&\left(\partial_z - s_i(w)\log'\left[T_i^w(z)\frac{q_+^{w,i-1}(z)}{q_+^{w,i}(z)}\right]\right)^{s_i(w)}\left(\partial_z - s_{i+1}(w)\log'\left[T_{i+1}^w(z)\frac{q_+^{w,i}(z)}{q_+^{w,i+1}(z)}\right]\right)^{s_{i+1}(w)} \\
&\qquad = \left(\partial_z - s_{i+1}(w)\log'\left[T_i^{s_i w}(z)\frac{q_+^{w,i-1}(z)}{q_-^{w,i}(z)}\right]\right)^{s_{i+1}(w)} \\
&\qquad \times \left(\partial_z - s_i(w)\log'\left[T_{i+1}^{s_i w}(z)\frac{q_-^{w,i}(z)}{q_+^{w,i+1}(z)}\right]\right)^{s_i(w)}.
\end{aligned}
$$

This is generalization of original work from [**MV2**], where it was done for the case of $\mathfrak{sl}(n)$ and differential operators. At the same time, in that pure even case the shortcut (see [**MTV**], [**CT**], [**TC**]) between the Gaudin model and opers was achieved by taking a certain analogue of determinant of KZ connection:

$$D^{KZ}_{k,l} = \delta_{k,l}\partial_z - \sum_{\nu=1,\dots,N} \frac{\Phi^{\nu}_{k,l}}{z-z_\nu}, \quad k,l=1,\dots,n, \tag{4.7}$$

where $\Phi^{\nu}_{k,l} \in U(\mathfrak{gl}(n))^{\otimes N}$ is the operator acting as the $\mathfrak{gl}(n)$ generator $e_{k,l}$ in the ν-th place in that tensor product. This determinant is understood in a formal way using permutations description. The coefficients in the expansion of the resulting differential operator are the Gaudin Hamiltonians. The factorization of that differential operator of type $R(z)$ corresponds to bringing that operator to the normal form in the sense of [**CT**], [**TC**], which coincides with the operator-valued Miura oper connection; in this way the determinant, when applied to a Bethe vector gives exactly the factorization of $R(z)$.

It is natural to expect that a similar formula should exist in the $\mathfrak{gl}(m|n)$ case, since (4.5) is very reminiscent of a Berezinian of the corrresponding differential operator. To obtain R^w factorization one can apply the Weyl transformation $w \in \tilde{W}$ represented as a permutation $\tilde{w}$ on $(m|n)$ superspace:

$$Ber(\tilde{w}[D^{KZ}_{k,l}]\tilde{w}^{-1}).$$

One has to emphasize that w here is not an element of supergroup $SL(m|n)$, but only a formal permutation. Bringing the D^{KZ}-operator to various versions of analogue of normal form for $(m|n)\times(m|n)$ supermatrix should produce an oper connection, which we discussed: after applying formal Berezinian one obtains the $R^w(z)$ operator. We mention here a related paper [**MR**] studying center of affine superalgebra $\mathfrak{gl}(n|m)$, where Berezinian formulas naturally emerged; the relation between the center of affine algebras and opers in the pure even case was thorougly investigated (see e.g. [**F4**] for review).

We note, that the generalization of such qq-system to the case of $\mathfrak{g} = \mathfrak{osp}(m|2n)$ was introduced in [**LM**], based on appropriate Bethe equations:

$$\begin{aligned} &W(q^i_-, q^i_+)(z) = \Lambda_i(z)\prod_i [q^i_+(z)]^{-c_{ij}} \text{ if } i \in I_w \cup I_b \\ &q^i_-(z) q^i_+(z) = \tilde{\Lambda}_i(z), \text{ if } i \in I_g. \end{aligned} \tag{4.8}$$

Here

$$\tilde{\Lambda}_i(z) = \log'\Big[\Lambda_i(z)\prod_j q^j_+(z)^{-c_{ij}}\Big]\pi_i(z) \prod_{j, c_{ij}\neq 0} q^j_+(z), \tag{4.9}$$

where $\{c_{ij}\}$ is the corresponding Cartan matrix, $\Lambda_i(z) \in \mathbb{C}[z]$ for all $i = 1, \dots, r$, so that π_i is the denominator of the fraction $\log'(\Lambda_i(z))$ of minimal possible degree.

The construction of the w-composability (or reproduction procedure) in the language of some pseudo-differential operators similar to $\mathfrak{sl}(n|m)$ case, but much more involved.

4.4. (G,q)-opers for supergroups. There is a generalization of the notion of oper for simple simply-connected Lie groups to (G,q)-oper: the difference analogue of G-oper connection, which uses natural multiplicative action of $\mathbb{C}_q^\times$ on $\mathbb{P}^1$, i.e. $z \to qz$.[1]

That(G,q)-oper is again a triple $(\mathcal{F}_G, \mathcal{F}_{B_-}, A)$, where $\mathcal{F}_G, \mathcal{F}_{B_-}$ are the principal G-bundle on $\mathbb{P}^1$ and its B_--reduction correspondingly. The q-oper connection A is an element of $Hom(\mathcal{F}_G, \mathcal{F}_G^q)$, where $\mathcal{F}_G^q$ is a pull-back bundle with respect to $\mathbb{C}_q^\times$-action. Here A belongs to the Coxeter cell: $B_-(z)c(z)B_-(z)$ in the Bruhat decomposition of $G(z)$, where $c(z)$ is a lift of Coxeter element to $G(z)$, which as usual is a product of the lifts of simple Weyl reflections to $G(z)$. The Miura condition adds the requirement that there exists another reduction $\mathcal{F}_{B_+}$ to the Borel subgroup which A preserves. The Z-twist condition means, as in differential case, gauge euivalence of the q-connection A to the constant q-connection $Z \in H$.

In [**FKSZ**] we used the notion of Z-twisted Miura-Plücker (G,q)-oper to find correspondence with q-defomation of the qq-system, known as QQ-system. Following that path we constructed the explicit correspondence between Z-twisted (G,q)-opers and the XXZ spin chain models XXX spin chains if we use additive action), which are deformations of the Gaudin model, constructing an example of q-Langlands correspondence [**KSZ**], [**AFO**].

Following same principles, one can build a q-oper analogue of $\bar{\nabla}$, ∇ for a given simple supergroup and a chosen Dynkin diagram, as an element of:

$$B_-(z)\Big[\prod_{i\in I_w}\tilde{w}_i(z)\prod_{j\le k,j,k\in I_g\cup I_b}\tilde{w}_{jk}(z)\Big]B_-(z), \tag{4.10}$$

where B_- stands for the Borel subgroup for a chosen Dynkin diagram either for supergroup G or the reductive group $\bar{G}$, and $\tilde{w}_i(z)$, $\tilde{w}_{jk}(z)$ stand for the lifts of Weyl reflections corresponding to the roots α_i, $\alpha_j + \alpha_k$ correspondingly.

The explicit expression for Miura(G,q)-oper may be quite complicated because of the absence of usual Bruhat decomposition for simple supergroups. At the same time, following the results of [**FKSZ**] one can obtain that the corresponding $(\bar{G},q)$-oper connection has the following form:

$$\prod_{i=1}^{r}[r_i(z)]^{\check{\alpha}_i}\prod_{i\in I_w}e^{R_i(z)e_i}\prod_{j,k\in I_g\cup I_b}e^{R_{ij}(z)[e_j,e_k]}, \tag{4.11}$$

where $r_i(z)$, $R_{jk}(z)$ are rational functions. Using the expression (4.11) one can relate the notion of Z-twisted Miura-Plücker $(\bar{G},q)$-oper and the QQ-systems for simple superalgebras as it is done in the differential case. One may expect that it should lead, using the analogue of the notion of w-composable QQ-systems to Bethe ansatz equations for XXX, XXZ models related to simple superalgebras. In the case of $\mathfrak{gl}(n|m)$ such examples are also known [**HLM**], which are based on the study of rational difference operators: a deformation of the constructions of [**HMVY**].

References

[AFO] M. Aganagic, E. Frenkel, and A. Okounkov, *Quantum q-Langlands correspondence*, Trans. Moscow Math. Soc. **79** (2018), 1–83, DOI 10.1090/mosc/278. MR3881458

[1] Such difference connection could be defined for the additive action as well.

[B] F. A. Berezin, *Introduction to superanalysis*, Mathematical Physics and Applied Mathematics, vol. 9, D. Reidel Publishing Co., Dordrecht, 1987. Edited and with a foreword by A. A. Kirillov; With an appendix by V. I. Ogievetsky; Translated from the Russian by J. Niederle and R. Kotecký; Translation edited by Dimitri Leĭtes, DOI 10.1007/978-94-017-1963-6. MR914369

[BD] A. Beilinson and V. Drinfeld, *Opers*, Preprint, `arXiv:math/0501398`, 2005.

[BR] M. J. Bergvelt and J. M. Rabin, *Supercurves, their Jacobians, and super KP equations*, Duke Math. J. **98** (1999), no. 1, 1–57, DOI 10.1215/S0012-7094-99-09801-0. MR1683200

[BS] M. A. Baranov and A. S. Shvarts, *Multiloop contribution to string theory* (Russian), Pis′ma Zh. Èksper. Teoret. Fiz. **42** (1985), no. 8, 340–342; English transl., JETP Lett. **42** (1985), no. 8, 419–421. MR875755

[BSZ] T. J. Brinson, D. S. Sage, and A. M. Zeitlin, *Opers on the projective line, Wronskian relations, and the Bethe Ansatz*, J. Geom. Phys. **202** (2024), Paper No. 105222, 25, DOI 10.1016/j.geomphys.2024.105222. MR4746982

[CT] A. Chervov and D. Talalaev, *Quantum spectral curves, quantum integrable systems and the geometric Langlands correspondence*, Preprint, `arXiv:hep-th/0604128`, 2006.

[TC] D. Talalaev and A. Chervov, *The KZ equation, G-opers, the quantum Drinfel′d-Sokolov reduction, and the quantum Cayley-Hamilton identity* (Russian, with English and Russian summaries), Zap. Nauchn. Sem. S.-Peterburg. Otdel. Mat. Inst. Steklov. (POMI) **360** (2008), no. Teoriya Predstavleniĭ, Dinamicheskie Sitemy, Kombinatornye Metody. XVI, 246–259, 299–300, DOI 10.1007/s10958-009-9415-1; English transl., J. Math. Sci. (N.Y.) **158** (2009), no. 6, 904–911. MR2759749

[CW] S.-J. Cheng and W. Wang, *Dualities and representations of Lie superalgebras*, Graduate Studies in Mathematics, vol. 144, American Mathematical Society, Providence, RI, 2012, DOI 10.1090/gsm/144. MR3012224

[DG] F. Delduc and L. Gallot, *Supersymmetric Drinfeld-Sokolov reduction*, J. Math. Phys. **39** (1998), no. 9, 4729–4745, DOI 10.1063/1.532532. MR1643329

[F1] E. Frenkel, *Affine algebras, Langlands duality and Bethe ansatz*, XIth International Congress of Mathematical Physics (Paris, 1994), Int. Press, Cambridge, MA, 1995, pp. 606–642. MR1370720

[F2] E. Frenkel, *Opers on the projective line, flag manifolds and Bethe ansatz* (English, with English and Russian summaries), Mosc. Math. J. **4** (2004), no. 3, 655–705, 783, DOI 10.17323/1609-4514-2004-4-3-655-705. MR2119144

[F3] E. Frenkel, *Gaudin model and opers*, Infinite dimensional algebras and quantum integrable systems, Progr. Math., vol. 237, Birkhäuser, Basel, 2005, pp. 1–58, DOI 10.1007/3-7643-7341-5_1. MR2160841

[F4] E. Frenkel, *Langlands correspondence for loop groups*, Cambridge Studies in Advanced Mathematics, vol. 103, Cambridge University Press, Cambridge, 2007. MR2332156

[FFR1] B. Feigin, E. Frenkel, and N. Reshetikhin, *Gaudin model, Bethe ansatz and critical level*, Comm. Math. Phys. **166** (1994), no. 1, 27–62. MR1309540

[FFR2] B. Feigin, E. Frenkel, and L. Rybnikov, *Opers with irregular singularity and spectra of the shift of argument subalgebra*, Duke Math. J. **155** (2010), no. 2, 337–363, DOI 10.1215/00127094-2010-057. MR2736168

[F] D. Friedan, *Notes on string theory and two-dimensional conformal field theory*, Workshop on unified string theories (Santa Barbara, Calif., 1985), World Sci. Publishing, Singapore, 1986, pp. 162–213. MR849106

[FFTL] B. Feigin, E. Frenkel, and V. Toledano Laredo, *Gaudin models with irregular singularities*, Adv. Math. **223** (2010), no. 3, 873–948, DOI 10.1016/j.aim.2009.09.007. MR2565552

[FKSZ] E. Frenkel, P. Koroteev, D. S. Sage, and A. M. Zeitlin, *q-opers, QQ-systems, and Bethe ansatz*, J. Eur. Math. Soc. (JEMS) **26** (2024), no. 1, 355–405, DOI 10.4171/jems/1268. MR4705654

[FSS] L. Frappat, A. Sciarrino, and P. Sorba, *Dictionary on Lie algebras and superalgebras*, Academic Press, Inc., San Diego, CA, 2000. With 1 CD-ROM (Windows, Macintosh and UNIX). MR1773773

[GPZ] A. Gualzetti, S. Penati, and D. Zanon, *Quantum conserved currents in supersymmetric Toda theories*, Nuclear Phys. B **398** (1993), no. 3, 622–658, DOI 10.1016/0550-3213(93)90607-Q. MR1223499

[HLM] C. Huang, K. Lu, and E. Mukhin, *Solutions of* $\mathfrak{gl}_{m|n}$ *XXX Bethe ansatz equation and rational difference operators*, J. Phys. A **52** (2019), no. 37, 375204, 31, DOI 10.1088/1751-8121/ab1960. MR4002511

[HMVY] C. Huang, E. Mukhin, B. Vicedo, and C. Young, *The solutions of* $\mathfrak{gl}_{M|N}$ *Bethe ansatz equation and rational pseudodifferential operators*, Selecta Math. (N.S.) **25** (2019), no. 4, Paper No. 52, 34, DOI 10.1007/s00029-019-0498-3. MR3997135

[IK] T. Inami and H. Kanno, *Lie superalgebraic approach to super Toda lattice and generalized super KdV equations*, Comm. Math. Phys. **136** (1991), no. 3, 519–542. MR1099694

[K1] V. G. Kac, *Representations of classical Lie superalgebras*, Differential Geometrical Methods in Mathematical Physics II. Proceedings, University of Bonn, July 13–16, 1977, pp. 597–626.

[K2] V. G. Kac, *Lie superalgebras*, Adv. Math. **26** (1977), no. 1, 8–96.

[KM1] P. P. Kulish and N. Manojlović, *Bethe vectors of the* osp(1|2) *Gaudin model*, Lett. Math. Phys. **55** (2001), no. 1, 77–95, DOI 10.1023/A:1010950003268. MR1845802

[KM2] P. P. Kulish and N. Manojlović, *Creation operators and Bethe vectors of the* osp(1|2) *Gaudin model*, J. Math. Phys. **42** (2001), no. 10, 4757–4778, DOI 10.1063/1.1398584. MR1855095

[KSZ] P. Koroteev, D. S. Sage, and A. M. Zeitlin, $(\mathrm{SL}(N), q)$*-Opers, the* q*-Langlands correspondence, and quantum/classical duality*, Comm. Math. Phys. **381** (2021), no. 2, 641–672, DOI 10.1007/s00220-020-03891-1. MR4207453

[KZ1] P. Koroteev and A. M. Zeitlin, q*-opers, QQ-systems, and Bethe Ansatz II: generalized minors*, J. Reine Angew. Math. **795** (2023), 271–296, DOI 10.1515/crelle-2022-0084. MR4542420

[KZ2] P. Koroteev and A. M. Zeitlin, *The zoo of opers and dualities*, Int. Math. Res. Not. IMRN **8** (2024), 6850–6878, DOI 10.1093/imrn/rnad270. MR4735647

[KZ3] P. Koroteev and A. M. Zeitlin, *3D mirror symmetry for instanton moduli spaces*, Comm. Math. Phys. **403** (2023), no. 2, 1005–1068, DOI 10.1007/s00220-023-04831-5. MR4645733

[KZ4] P. Koroteev and A. M. Zeitlin, *Toroidal* q*-opers*, J. Inst. Math. Jussieu **22** (2023), no. 2, 581–642, DOI 10.1017/S1474748021000220. MR4557903

[L] D. A. Leites, *Theory of supermanifolds*, KF Akad. Nauk SSSR, Petrozavodsk, 1983.

[LM] K. Lu and E. Mukhin, *Bethe Ansatz equations for orthosymplectic Lie superalgebras and self-dual superspaces*, Ann. Henri Poincaré **22** (2021), no. 12, 4087–4130, DOI 10.1007/s00023-021-01091-8. MR4339086

[LSS] D. A. Leĭtes, M. V. Saveliev, and V. V. Serganova, *Embeddings of* osp$(N/2)$ *and the associated nonlinear supersymmetric equations*, Group theoretical methods in physics, Vol. I (Yurmala, 1985), VNU Sci. Press, Utrecht, 1986, pp. 255–297. MR919750

[M] Y. I. Manin, *Gauge field theory and complex geometry*, Grundlehren der mathematischen Wissenschaften [Fundamental Principles of Mathematical Sciences], vol. 289, Springer-Verlag, Berlin, 1988. Translated from the Russian by N. Koblitz and J. R. King. MR954833

[MR] A. I. Molev and E. Ragoucy, *The MacMahon master theorem for right quantum superalgebras and higher Sugawara operators for* $\widehat{\mathfrak{gl}}_{m|n}$ (English, with English and Russian summaries), Mosc. Math. J. **14** (2014), no. 1, 83–119, 171, DOI 10.17323/1609-4514-2014-14-1-83-119. MR3221948

[MTV] E. Mukhin, V. Tarasov, and A. Varchenko, *The B. and M. Shapiro conjecture in real algebraic geometry and the Bethe ansatz*, Ann. of Math. (2) **170** (2009), no. 2, 863–881, DOI 10.4007/annals.2009.170.863. MR2552110

[MV1] E. Mukhin and A. Varchenko, *Miura opers and critical points of master functions*, Cent. Eur. J. Math. **3** (2005), no. 2, 155–182, DOI 10.2478/BF02479193. MR2129922

[MV2] E. Mukhin and A. Varchenko, *Critical points of master functions and flag varieties*, Commun. Contemp. Math. **6** (2004), no. 1, 111–163, DOI 10.1142/S0219199704001288. MR2048778

[MV3] E. Mukhin and A. Varchenko, *Quasi-polynomials and the Bethe ansatz*, Groups, homotopy and configuration spaces, Geom. Topol. Monogr., vol. 13, Geom. Topol. Publ., Coventry, 2008, pp. 385–420, DOI 10.2140/gtm.2008.13.385. MR2508216

[MVY] E. Mukhin, B. Vicedo, and C. Young, *Gaudin models for* $\mathfrak{gl}(m|n)$, J. Math. Phys. **56** (2015), no. 5, 051704, 30, DOI 10.1063/1.4919652. MR3390984

[O] M. A. Olshanetsky, *Supersymmetric two-dimensional Toda lattice*, Comm. Math. Phys. **88** (1983), no. 1, 63–76. MR691198

[P] I. B. Penkov, *Borel-Weil-Bott theory for classical Lie supergroups* (Russian), Current problems in mathematics. Newest results, Vol. 32, Itogi Nauki i Tekhniki, Akad. Nauk SSSR, Vsesoyuz. Inst. Nauchn. i Tekhn. Inform., Moscow, 1988, pp. 71–124. Translated in J. Soviet Math. **51** (1990), no. 1, 2108–2140. MR957752

[RT] M. Rakowski and G. Thompson, *Connections on vector bundles over super Riemann surfaces*, Phys. Lett. B **220** (1989), no. 4, 557–561, DOI 10.1016/0370-2693(89)90786-7. MR993273

[VM1] A. A. Voronov and Yu. I. Manin, *Supercellular partitions of flag superspaces* (Russian), Current problems in mathematics. Newest results, Vol. 32, Itogi Nauki i Tekhniki, Akad. Nauk SSSR, Vsesoyuz. Inst. Nauchn. i Tekhn. Inform., Moscow, 1988, pp. 27–70. Translated in J. Soviet Math. **51** (1990), no. 1, 2083–2108. MR957751

[VM2] A. A. Voronov and Yu. I. Manin, *Schubert supercells* (Russian), Funktsional. Anal. i Prilozhen. **18** (1984), no. 4, 75–76. Translated in J. Soviet Math. **18** (1984), no. 4, 329–330. MR775936

[W] E. Witten, *Notes on super Riemann surfaces and their moduli*, Pure Appl. Math. Q. **15** (2019), no. 1, 57–211, DOI 10.4310/PAMQ.2019.v15.n1.a2. MR3946083

[Z1] A. M. Zeitlin, *On Wronskians and qq-systems*, Algebraic and topological aspects of representation theory, Contemp. Math., vol. 791, Amer. Math. Soc., [Providence], RI, [2024] ©2024, pp. 91–103, DOI 10.1090/conm/791/15873. MR4694169

[Z2] A. M. Zeitlin, *Superopers on supercurves*, Lett. Math. Phys. **105** (2015), no. 2, 149–167, DOI 10.1007/s11005-014-0735-9. MR3302224

School of Mathematics, Georgia Institute of Technology, 686 Cherry St NW, Atlanta, Georgia 30332

Email address: `zeitlin@gatech.edu`

URL: `https://zeitlin.math.gatech.edu`

Contemporary Mathematics
Volume **817**, 2025
https://doi.org/10.1090/conm/817/16329

Variation diminishing operators and unitary representations SL_2 over semifields

Yongchang Zhu

1. Introduction

A real $m \times n$ matrix A is said to be totally positive if all its minors, of any order, are non-negative. The study of such matrices was originated in the 1930's by Schoenberg [**Sch1**], and by Gantmacher and Krein [**GK**]. In [**Sch1**], it was shown that if A is totally positive, then the linear transformation $\mathbb{R}^n \to \mathbb{R}^m, x \mapsto Ax$ is variation-diminishing. This means that if $v(x)$ represents the number of sign changes in the sequence $x = (x_1, \ldots, x_n)$, and $v(Ax)$ represents the number of sign changes in the sequence for Ax, then $v(Ax) \leq v(x)$ for all $x \in \mathbb{R}^n$. The converse also holds to some extent, as explained in [**Sch1**] [**Mo**]. In [**Sch2**], Schoenberg examined variation diminishing operators on functions over $\mathbb{R}$. He specifically characterized operators of convolution type. Hirschman later generalized Schoenberg's work and provided a characterization of Hankel convolution operators with the variation diminishing property [**Hir**]. The variation diminishing linear operators on functions on $\mathbb{R}$ that preserves nonnegative functions can be considered as infinite dimensional generalization of totally positive matrices.

In [**Lus1**], Lusztig introduced the total positivity for arbitrary split reductive group G. The totally nonnegative part $G_{\geq 0}$ of $G(\mathbb{R})$ forms a moniod, in the special case $G = GL_n$, $G_{\geq 0}$ is the invertible $n \times n$ totally positive matrices. He also introduced $G(K)$ for a Kac-Moody group G over a semifield [**Lus2**]. Therefore the total positivity can be studied in the framework of Lie group.

In this work, we introduce unitary representations of SL_2 over semifields $\mathbb{R}_{>0}$ and $\mathbb{R}_{>0}^{\text{trop}}$. These representations are defined based on a $*$-structure on the monoid $SL_2(K)$. We classify irreducible unitary representations and provide explicit models of these representations on function spaces over $\mathbb{R}_{>0}$. Additionally, we demonstrate that all elements in $SL_2(\mathbb{R}_{>0})$ or $SL_2(\mathbb{R}_{>0}^{\text{trop}})$ act as variation diminishing operators and preserve nonnegative functions. This implies that the "infinite matrices" for these operators are totally positive. The proof utilizes the results of Schoenberg and Hirschman mentioned earlier.

We follow [**FZ**] to define a semifield as a commutative multiplicative group $(K, \cdot)$ with an addition $+$ such that $+$ is commutative and associative, and distributive law holds

$$a \cdot (b + c) = a \cdot b + a \cdot c.$$

The research was supported by Hong Kong RGC grant 16302521.

In this work, we only consider the following two semifields: $\mathbb{R}_{>0}$, the multiplicative group of positive real numbers with usual addition; the tropical semifield $\mathbb{R}_{>0}^{\mathrm{trop}}$, the multiplicative group $\mathbb{R}_{>0}$ with the addition given by

$$\max(a, b).$$

The semifield $\mathbb{R}_{>0}^{\mathrm{trop}}$ is isomorphic to the additive semifield $(\mathbb{R}, +, \min)$ by the map $a \mapsto -\log a$.

It is natural to consider whether the results obtained in this work can be extended to higher rank semisimple group G. One needs to first classify $*$-structures on $G(K)$, this does not seem to be a difficult question. A more challenging question is to generalize the notion of variation diminishing operators on function spaces over spaces of dimension larger than 1.

All the representations of $SL_2(\mathbb{R}_{>0}^{\mathrm{trop}})$ have the structure of an abstract scattering system, which is the simplest kind of system according to [**LP**]. In Section 4, we introduce a monoid $S(\mathbb{R}_{>0}^{\mathrm{trop}})$ that is a cover of $SL_2(\mathbb{R}_{>0}^{\mathrm{trop}})$ and show that its unitary representations are in one-to-one correspondence with abstract casual scattering systems in the sense of [**LP**]. This raises the question of characterizing the unitary representations of $S(\mathbb{R}_{>0}^{\mathrm{trop}})$ in which all elements in $S(\mathbb{R}_{>0}^{\mathrm{trop}})$ act as variation diminishing operators and preserve non-negative functions. Furthermore, we are interested in exploring the implications of total positivity on the corresponding scattering system. We plan to address these questions in future research.

This paper is organized as follows. In Section 2, we recall the definition of $SL_2(K)$ for a semifield K as given in [**Lus2**]. We then define a $*$-structure on $SL_2(K)$ and use it to define unitary representations of $SL_2(K)$ for $K = \mathbb{R}_{>0}$ or $K = \mathbb{R}_{>0}^{\mathrm{trop}}$. We give a list of irreducible unitary representations. Theorem 2.2 claims that the list is complete for $\mathbb{R}_{>0}$, we sketch the proof that uses the classification of irreducible unitary representations of the universal cover of $SL_2(\mathbb{R})$. Theorem 2.3 claims the list is complete for $K = \mathbb{R}_{>0}^{\mathrm{trop}}$, we will give a proof of this result in a later publication. In Section 3, we prove that the model of all irreducible unitary representations in Section 2 are positive in the sense that elements in $SL_2(K)$ act as variation diminishing operators and preserve non-negative functions (Theorem 3.1). In Section 4, we show that each irreducible unitary representation of $SL_2(\mathbb{R}_{>0}^{\mathrm{trop}})$ gives arise an abstract scattering systems as defined [**LP**], conversely for an abstract scattering systems, we can construct operators $h(a), x_+(a), x_-(a)$ satisfying all the defining relations for $SL_2(\mathbb{R}_{>0}^{\mathrm{trop}})$ except the relation (2.5) for $ab > 1$. This leads us to introduce the monoid $S(\mathbb{R}_{>0}^{\mathrm{trop}})$ and show that the unitary representations of $S(\mathbb{R}_{>0}^{\mathrm{trop}})$ are in one-to-one correspondence with abstract causal scattering systems.

2. Unitary representations of $SL_2(\mathbb{R}_{>0})$ and $SL_2(\mathbb{R}_{>0}^{\mathrm{trop}})$ and classifications

2.1. $SL_2(K)$ **over a semifield and unitary representations.** Let K be a semifield with the multiplicative identity 1, $SL_2(K)$ is the monoid generated by symbols

$$x_+(a), \quad x_-(a), \quad h(a), \quad a \in K$$

with relations

$$h(1) = 1, \quad h(a)h(b) = h(ab) \tag{2.1}$$

$$x_\pm(b_1)x_\pm(b_2) = x_\pm(b_1 + b_2) \tag{2.2}$$

$$h(a)x_+(b) = x_+(a^2b)h(a) \tag{2.3}$$

$$h(a)x_-(b) = x_-(a^{-2}b)h(a) \tag{2.4}$$

$$x_+(a)x_-(b) = x_-(\frac{b}{1+ab})h(1+ab)x_+(\frac{a}{1+ab}) \tag{2.5}$$

We can prove the following relation

$$x_-(a)x_+(b) = x_+(\frac{b}{1+ab})h((1+ab)^{-1})x_-(\frac{a}{1+ab}). \tag{2.6}$$

We define an order 2 automorphism $S : SL_2(K) \to SL_2(K)$ by

$$Sx_+(a) = x_-(a), \quad Sx_-(a) = x_+(a), \quad Sh(a) = h(a^{-1}). \tag{2.7}$$

We introduce the anti-involution $*$ which will be used to define the unitary representations:

$$x_+(a)^* = x_+(a), \quad x_-(a)^* = x_-(a), \quad h(a)^* = h(a^{-1}) \tag{2.8}$$

The monoid $SL_2(\mathbb{R}_{>0})$ can be embedded into $SL_2(\mathbb{R})$ by

$$x_+(a) \mapsto \begin{pmatrix} 1 & a \\ 0 & 1 \end{pmatrix}, \quad x_-(a) \mapsto \begin{pmatrix} 1 & 0 \\ a & 1 \end{pmatrix}, \quad h(a) \mapsto \begin{pmatrix} a & 0 \\ 0 & a^{-1} \end{pmatrix}$$

DEFINITION. For $K = \mathbb{R}_{>0}$ or $\mathbb{R}^{\text{trop}}_{>0}$, a unitary representation of $SL_2(K)$ on a Hilbert space H is a monoid homomorphism $\pi : SL_2(K) \to B(H)$ such that

$$\pi(g)^* = \pi(g^*) \tag{2.9}$$

where $B(H)$ is the algebra of bounded linear maps $H \to H$, $*$ in the left indicates the dual operator, the $*$ in the right is the anti-involution in (2.8). Moreover we require that for all $v \in H$, the maps $\mathbb{R}_{>0} \to H$ given by

$$a \mapsto \pi(x_+(a))v, \quad a \mapsto \pi(x_-(a))v, \quad a \mapsto \pi(h(a))v$$

are continuous.

By definition, the operators $x_\pm(a)$ are self-adjoint and $h(a)$'s are unitary. For the case $K = \mathbb{R}_{>0}$, $SL_2(\mathbb{R}_{>0})$ is a submonoid of $SL_2(\mathbb{R})$. However, the restriction of a unitary representation of $SL_2(\mathbb{R})$ to $SL_2(\mathbb{R}_{>0})$ is *not* unitary, as (2.9) is not satisfied.

A unitary representation is called irreducible if has no proper nonzero closed subspaces invariant under the action of $SL_2(K)$.

2.2. Classifications of irreducible unitary representations of $SL_2(\mathbb{R}_{>0})$. We first give a list of irreducible unitary representations H_w parameterized $w \in \mathbb{R}_{>0}$ of $SL_2(\mathbb{R}_{>0})$ and then we prove the list is complete. For $w > 0$, the representation H_w, as a Hilbert space, is

$$H_w = L^2(\mathbb{R}_{>0}, x^{w-1}dx)$$

where dx is the usual Lebesgue measure restricted on $\mathbb{R}_{>0}$. The operator $x_+(b)$, $b > 0$, is

$$(x_+(b)f)(x) = e^{-\pi bx^2} f(x). \tag{2.10}$$

The operator $h(a)$ is

$$(h(a)f)(x) = a^{\frac{w}{2}} f(ax). \tag{2.11}$$

The action of $x_-(b)$, $b > 0$, will be defined as a Hankel convolution operator. We first recall the following variant of the Hankel transform:

$$F_w : L^2(\mathbb{R}_{>0}, x^{w-1}dx) \to L^2(\mathbb{R}_{>0}, x^{w-1}dx)$$

$$F_w f(x) = 2\pi \int_0^\infty (xy)^{-\frac{w-2}{2}} J_{\frac{w-2}{2}}(2\pi xy) f(y) y^{w-1} dy \tag{2.12}$$

where $J_\alpha(x)$ is the Bessel function given by

$$J_\alpha(x) = \sum_{k=0}^{\infty} \frac{(-1)^k (x/2)^{2k+\alpha}}{k!\Gamma(k+\alpha+1)}$$

Using the standard formula for Hankel transforms, e.g., [**AAR**] p.222, [**Bateman**], one can prove that

$$F_w e^{-\pi t x^2} = t^{-\frac{w}{2}} e^{-\pi t^{-1} x^2}. \tag{2.13}$$

From this one checks directly that

$$\int_0^\infty F_w e^{-\pi t_1 x^2} F_w e^{-\pi t_2 x^2} x^{w-1} dx = \int_0^\infty e^{-\pi t_1 x^2} e^{-\pi t_2 x^2} x^{w-1} dx,$$

i.e., F_w preserves the inner product in $\mathrm{Span}(e^{-\pi t x^2}, t > 0)$. Since the later space is dense in $L^2(\mathbb{R}_{>0}, x^{w-1}dx)$, so F_w is an isometry. Since $F_w^2 e^{-\pi t x^2} = e^{-\pi t x^2}$ for all $t > 0$, $F_w^2 = 1$. We define

$$x_-(b) = F_w x_+(b) F_w^{-1}. \tag{2.14}$$

which is the Hankel convolution operator with kernel $b^{-\frac{w}{2}} e^{-\pi b^{-1} x^2}$.

In the case $w = n$ is a positive integer, the space H_n is isomorphic to the space of the radial L^2-functions on $\mathbb{R}^n$, $L^2(\mathbb{R}^n)^{O_n}$. The transform F_n corresponds to the usual Fourier transform on $L^2(\mathbb{R}^n)$.

PROPOSITION 2.1. *H_w with the given action is an irreducible unitary representation of $SL_2(\mathbb{R}_{>0})$.*

PROOF. It is clear that $h(a), S = F_w$ are unitary operators and $x_\pm(a)$ are bounded, and the relation $x_\pm(a)x_\pm(a') = x_\pm(a+a')$ is satisfied. It is easy to check

$$h(a)x_+(b) = x_+(a^2 b)h(a), \quad h(a)F_w = F_w h(a^{-1})$$

This implies that

$$h(a)x_-(b) = x_-(a^{-2}b)h(x)$$

It remains to check the relation (2.5). We test it on the functions $e^{-\pi t x^2}$:

$$\begin{aligned} x_-(b)e^{-\pi t x^2} &= t^{-\frac{w}{2}} F_w x_+(b) e^{-\pi t^{-1} x^2} \\ &= t^{-\frac{w}{2}} F_w e^{-\pi (b+t^{-1}) x^2} \\ &= t^{-\frac{w}{2}} (b+t^{-1})^{-\frac{w}{2}} e^{-\pi (b+t^{-1})^{-1} x^2} \\ &= (1+tb)^{-\frac{w}{2}} e^{-\pi \frac{t}{1+bt} x^2} \end{aligned}$$

Let $B(t)$ denote the function

$$B(t)(x) = e^{-\pi t x^2}.$$

Then we have

$$\begin{aligned} x_+(a)B(t) &= B(a+t) \\ h(a)B(t) &= a^{\frac{w}{2}} B(a^2 t) \\ x_-(a)B(t) &= (1+at)^{-\frac{w}{2}} B(\frac{t}{1+at}) \\ F_w B(t) &= t^{-\frac{w}{2}} B(t^{-1}) \end{aligned}$$

This can be used to prove (2.5) holds on functions $B(t)$. Since $\mathrm{span}\{B(t), t > 0\}$ is dense in H_w, therefore (2.5) holds on all the functions. The irreducibility is clear. □

THEOREM 2.2. *Every non-trivial irreducible unitary representation of $SL_2(\mathbb{R}_{>0})$ is isomorphic to H_w for unique $w > 0$.*

SKETCH OF PROOF. Let $\mathcal{H}$ be an irreducible unitary representation. By the relation $h(a)x_+(b)h(a)^{-1} = x_+(a^2 b)$, $h(a)$ is unitary, we see that all $x_+(b)$ have equal operator norm. Since $x_+(b)^2 = x_+(2b)$, see that $x_+(b)$ is positive definite and has operator norm 1. The operators $x_-(b)$ have similar properties. By [**HP**] Theorem 22.3.1 about semigroup of bounded self-adjoint operators, there exists a projection operator J, $\lim_{b\to+0} x_+(b)$ converges strongly to J, i.e., $\lim_{b\to+0} x_+(b)v = Jv$ for all $v \in H$. We have $Jx_+(b) = x_+(b)$ for all $b > 0$. One can prove that $J\mathcal{H}$ is a closed subrepresentation, since $\mathcal{H}$ is irreducible, so $J\mathcal{H} = \mathcal{H}$, so J is the identity operator 1. Similarly, one can prove $\lim_{b\to+0} x_-(b)$ converges to 1 strongly. Let $X_+(X_-)$ be the infinitesimal generator of semigroup $\{x_+(b)\}(\{x_-(b)\})$, H be the infinitesimal generator of one-parameter subgroup $\{h(e^t)\}$. The operators X_+, X_-, H preserve certain dense subspace C of $\mathcal{H}$ and satisfies the standard sl_2 relations. We see that $iX_+, -iX_-, H$ are skew hermitian operators on V, they also satisfies the standard sl_2 relation, therefore they integrate to obtain an irreducible unitary representation of the universal cover $\widetilde{SL_2(\mathbb{R})}$ of $SL_2(\mathbb{R})$ on $\mathcal{H}$. Using the classification of irreducible unitary representation of $\widetilde{SL_2(\mathbb{R})}$ [**Pu**] and the fact that $e^{bX_+} = x_+(b), e^{bX_-} = x_-(b)$ are bounded operators for all $b > 0$, we see that $\mathcal{H}$, as a unitary representation of $\widetilde{SL_2(\mathbb{R})}$, is a discrete series representation, which is a highest weight module with negative highest weight (not necessary an integer). This implies $\mathcal{H}$ is isomorphic to H_w as a representation of $SL_2(\mathbb{R}_{>0})$. □

2.3. Classifications of irreducible unitary representations of $SL_2(\mathbb{R}^{\mathrm{trop}}_{>0})$.

2.3.1. *Representations H_c.* For each $0 < c \leq 1$, we construct a representation H_c. As a Hilbert space,

$$H_c = L^2(\mathbb{R}_{>0}, x^{-1}dx).$$

We define the operators $x_+(b), S, h(a), x_-(b)$ by

$$(x_+(b)f)(x) = 1_{[b,\infty)}(x)f(x) \tag{2.15}$$

$$(Sf)(x) = f(cx^{-1}) \tag{2.16}$$

$$(h(a)f)(x) = f(a^{-2}x) \tag{2.17}$$

$$x_-(b) = Sx_+(b)S^{-1} \tag{2.18}$$

where $1_{[b,\infty)}(x)$ is the indicator function of the interval $b \leq x < \infty$. One can check that

$$(x_-(b)f)(x) = 1_{(0,b^{-1}c]}f(x). \tag{2.19}$$

It is easy to check the relations (2.1)–(2.4) and

$$S^2 = 1, \quad Sh(a)S^{-1} = h(a^{-1}), \quad Sx_\pm(a)S^{-1} = x_\mp(a).$$

For the semifield $\mathbb{R}_{>0}^{\text{trop}}$, the relation (2.5) reads as

$$x_+(a)x_-(b) = x_-(b\max(1,ab)^{-1})h(\max(1,ab))x_+(a\max(1,ab)^{-1}). \tag{2.20}$$

The left hand side is the multiplication operator $1_{[a,\infty)}(x)1_{(0,b^{-1}c]}(x)$. For $ab \geq 1$, the measure $[a,\infty) \cap (0, b^{-1}c]$ is 0, so the left hand side is 0. The right hand side is

$$x_-(a^{-1})h(ab)x_+(b^{-1}) = x_-(a^{-1})x_+(a^2b)h(ab),$$

we see

$$x_-(a^{-1})x_+(a^2b) = 1_{(0,ac]}(x)1_{[a^2b,\infty)}(x) = 0,$$

thanks to $ab \geq 1, c \leq 1$. So both sides of (2.20) are 0 for $ab \geq 1$. For $ab < 1$, it is obvious that the two sides of (2.20) are equal. This proves H_c is a unitary representation of $SL_2(\mathbb{R}_{>0}^{\text{trop}})$. The irreducibility is clear.

2.3.2. *Representations D_λ.* In this subsection, we construct for each $\lambda > 0$, an irreducible unitary representation D_λ for $SL_2(\mathbb{R}_{>0}^{\text{trop}})$. As a Hilbert space, $D_\lambda = L^2(\mathbb{R}_{\geq 0}, x^{-\lambda-1}dx)$. The operator $x_+(a)$ is

$$(x_+(a)f)(x) = 1_{[a,\infty)}(x)f(x).$$

We define the operator S by

$$(Sf)(x) = \lambda \int_{x^{-1}}^{\infty} f(y)y^{-\lambda-1}dy - x^\lambda f(x^{-1}) \tag{2.21}$$

and define $h(a)$ by

$$(h(a)f)(x) = a^\lambda f(a^{-2}x)$$

and $x_-(a) = Sx_+(a)S$.

We first check S is a unitary operator and $S^2 = 1$. For each $b > 0$, we introduce function $A(b)(x)$ on $\mathbb{R}_{>0}$ by

$$A(b)(x) = \begin{cases} 1 & \text{for } x \geq b, \\ 0 & \text{for } x < b. \end{cases}$$

Since $\lambda > 0$, $A(b) \in L^2(\mathbb{R}_{\geq 0}, x^{-\lambda-1}dx)$. We have, for $a' \leq a$,

$$(A(a'), A(a)) = \int_0^\infty A(a')(x)A(a)(x)x^{-\lambda-1}dx = \int_a^\infty x^{-\lambda-1}dx = \lambda^{-1}a^{-\lambda}$$

We compute $SA(a)$:

$$\begin{aligned}(SA(a))(x) &= \lambda \int_{x^{-1}}^{\infty} A(a)(y) y^{-\lambda-1} dy - x^{\lambda} A(a)(x^{-1}) \\ &= \lambda \int_{\max(a,x^{-1})}^{\infty} y^{-\lambda-1} dy - x^{\lambda} 1_{(0,a^{-1})} \\ &= \max(a, x^{-1})^{-\lambda} - x^{\lambda} 1_{(0,a^{-1})} \\ &= a^{-\lambda} 1_{[a^{-1},\infty)}(x) \\ &= a^{-\lambda} A(a^{-1})(x)\end{aligned}$$

We check directly that $S^2 A(a) = A(a)$ and

$$(SA(a), SA(a')) = (A(a), A(a')).$$

Since the span of

$$A(a), a > 0,$$

is dense, so S is a unitary operator on $L^2(\mathbb{R}_{\geq 0}, x^{-\lambda-1}dx)$ and $S^2 = 1$. One can check directly that $Sh(a)S = h(a^{-1})$ and then check the relations (2.1)–(2.4). For (2.5), we check it holds on functions $A(a)$, since the span of $A(a)$ is dense, so (2.5) holds. So we obtain the unitary representation of D_λ, it is easy to prove it is irreducible.

2.3.3. *Classification theorem and the Frobenius action.* It is easy to see there are two finite dimensional irreducible unitary representation of $SL_2(\mathbb{R}^{\text{trop}}_{>0})$, both has dimension 1. In the first representation, $h(a)$, $x_+(a)$, $x_-(a)$ all acts as 1. In the second representation, $h(a) = 1$, $x_\pm(a) = 0$.

THEOREM 2.3. *Every infinite dimensional irreducible unitary representation of $SL_2(\mathbb{R}^{\text{trop}}_{>0})$ is isomorphic to H_c or D_λ.*

The semifield $\mathbb{R}^{\text{trop}}_{>0}$ has "Forbenius automorphisms" F_p $(p > 0)$:

$$F_p a = a^p. \tag{2.22}$$

They induce the action of the multiplicative $\mathbb{R}_{>0} = \{p, p > 0\}$ on the monoid $SL_2(\mathbb{R}^{\text{trop}}_{>0})$:

$$[p]\beta(a) = \beta(a^p) \quad \text{for} \quad \beta = x_+, x_-, h.$$

Therefore $[p]$ transforms an irreducible unitary representation H to a new representation $[p]H$: $[p]H = H$ as Hilbert space, $\beta(a)$ acts as $\beta(a^p)$. It is easy to prove

$$[p]H_c \sim H_{c^{p^{-1}}}, \quad [p]D_\lambda \sim D_{p\lambda} \tag{2.23}$$

For example, the isomorphism $T : D_{p\lambda} \to [p]D_\lambda$ is given

$$(Tf)(x) = p^{-\frac{1}{2}} f(x^{\frac{1}{p}}).$$

3. Variation diminishing operators and positivity structures

Let $f(x)$ be a real function defined on $\mathbb{R}_{>0}$. The number $v(f)$, of variations of sign of $f(x)$ in the range $0 < x < \infty$ is defined as follows: if $S : x_1 < x_2 < \cdots < x_n$ is an arbitrary finite sequences on $0 < x < \infty$, then

$$v(f) = \text{Sup}_S\, v(f(x_1), f(x_2), \ldots, f(x_n)) \tag{3.1}$$

where $v(f(x_1), f(x_2), \ldots, f(x_n))$ is the number of sign changes of the sequence $f(x_1), \ldots, f(x_n)$. So $0 \leq v(f) \leq \infty$. We call a linear operator Φ on certain space V of functions on $\mathbb{R}_{>0}$ is variation diminishing if

$$v(\Phi f) \leq v(f)$$

for all $f \in V$. It is clear that the composition of variation diminishing operators is again variation diminishing.

THEOREM 3.1. *On the representation H_w of $SL_2(\mathbb{R}_{>0})$ and representation H_c, D_λ of $SL_2(\mathbb{R}_{>0}^{\text{trop}})$ all the operators from $SL_2(\mathbb{R}_{>0})$ or $SL_2(\mathbb{R}_{>0}^{\text{trop}})$ are variation diminishing operators and they send non-negative functions to non-negative functions.*

Note that a variation diminishing operator may not preserves the non-negative functions, the negative identity operator is such an example.

In the proof of the theorem for the case D_λ, we need the following formula for the action of $x_-(a)$:

LEMMA 3.2. *In the representation D_λ of $SL_2(\mathbb{R}_{>0}^{\text{trop}})$, the action of $x_-(a)$ is given by*

$$(x_-(a)f)(x) = \begin{cases} f(x) & \text{for } x < a^{-1} \\ \lambda a^{-\lambda} \int_{a^{-1}}^{\infty} f(y) y^{-\lambda-1} dy & \text{for } x \geq a^{-1} \end{cases} \tag{3.2}$$

PROOF OF THEOREM 3.1. For the case H_w of $SL_2(\mathbb{R}_{>0})$, it is clear that $x_+(a), h(a)$ are variation diminishing operator. For $x_-(a) = F_w x_+(a) F_w = F_w e^{-\pi a x^2} F_w$, in the notation of [**Hir**],

$$x_-(a) f = e^{-\pi a x^2} \# f$$

which is the Hankel convolution with kernel $e^{-\pi a x^2}$. It is proved in [**Hir**] that the Hankel convolution operator with kernel $G(x)$ is variation diminishing if and only if its Hankel transform $F_w G(x)$ has the form

$$M \left(e^{cx^2} \Pi_{k=1}^{\infty} (1 + \frac{x^2}{a_k^2}) \right)^{-1} \tag{3.3}$$

with M is a non-zero scalar, $c \geq 0$, each a_k are real or ∞ with $\sum_{k=1}^{\infty} a_k^{-2} < \infty$. In our case, $G(x) = e^{-\pi a x^2}$, $F_w G(x) = a^{-\frac{w}{2}} e^{-\pi a^{-1} x^2}$, which is of form (3.3) with all $a_k = \infty$, therefore $x_+(a)$ is a variation diminishing operator. This proves the theorem for H_w. It is clear that $x_+(a)$ and $h(a)$ send non-negative functions to non-negative functions. The similar result for $x_-(a)$ follows from the fact that $e^{-\pi a x^2}$ is a positive function and the Hankel convolution of two non-negative functions is non-negative [**Hir**].

For the case H_c of $SL_2(\mathbb{R}_{>0}^{\text{trop}})$, the operators $x_+(a), x_-(a)$ are the multiplication operators by $1_{[a^{-1},\infty)}(x), 1_{(0,ca^{-1})}(x)$, clearly such operators are variation diminishing. The $h(a)$ is a change variable operator $x \mapsto a^{-2}x$ that preserves the order of $\mathbb{R}_{>0}$, so it is variation diminishing. It is clear that $x_\pm(a)$ and $h(a)$ send non-negative functions to non-negative functions.

For the case D_λ of $SL_2(\mathbb{R}_{>0}^{\text{trop}})$, the operators $x_+(a), h(a)$ are clearly variation diminishing. We use Lemma 3.2 for $x_-(a)$, note that

$$\lambda a^{-\lambda} \int_{a^{-1}}^{\infty} f(y) y^{-\lambda-1} dy = \frac{\int_{a^{-1}}^{\infty} f(y) y^{-\lambda-1} dy}{\int_{a^{-1}}^{\infty} y^{-\lambda-1} dy}$$

is the average of $f(x)$ on $[a^{-1},\infty)$. The operator $x_-(a)$ keeps the value of $f(x)$ for $x < a^{-1}$ and for all $x \geq a^{-1}$, $(x_-(a)f)(x)$ is equal to the average of $f(x)$ over $[a^{-1},\infty)$, from this, we see $x_-(a)$ is variation diminishing. It is clear that $x_+(a)$ and $h(a)$ send non-negative functions to non-negative functions. The similar property for $x_-(a)$ follows from (3.2). □

4. $SL_2(\mathbb{R}_{>0}^{\mathrm{trop}})$ and abstract scattering theory

We recall the abstract scattering theory defined in [**LP**].

DEFINITION. The Lax-Phillips abstract scattering system is defined in a Hilbert space H which contains two distinguished subspaces, $\mathcal{D}_+, \mathcal{D}_-$, called "outgoing" and "incoming". There is a unitary evolution operators $U(t)$, $t > 0$ for which $\mathcal{D}_\pm$ are invariant in the following sense:

$$U(t_1)U(t_2) = U(t_1t_2), \quad U(1) = 1 \tag{4.1}$$

$$U(t)\mathcal{D}_+ \subset \mathcal{D}_+ \quad \text{for } t \geq 1 \tag{4.2}$$

$$U(t)\mathcal{D}_- \subset \mathcal{D}_- \quad \text{for } t \leq 1 \tag{4.3}$$

$$\cap_t U(t)\mathcal{D}_+ = 0 \tag{4.4}$$

$$\cap_t U(t)\mathcal{D}_- = 0 \tag{4.5}$$

$$\overline{\cup_t U(t)\mathcal{D}_+} = H \tag{4.6}$$

$$\overline{\cup_t U(t)\mathcal{D}_-} = H \tag{4.7}$$

Here we changed the additive group $U(t)$ $(t \in \mathbb{R})$ in the original definition to the multiplicative group $U(t)$ $(t \in \mathbb{R}_{>0})$ in order to compare it with the multiplicative subgroup $\{h(a)\}$ of $SL_2(\mathbb{R}_{>0}^{\mathrm{trop}})$.

Note that (4.2) implies that

$$U(t_1)\mathcal{D}_+ \subset U(t_2)\mathcal{D}_+ \qquad \text{for } t_1 > t_2 \geq 1 \tag{4.8}$$

and (4.3) implies that

$$U(t_1)\mathcal{D}_- \subset U(t_2)\mathcal{D}_- \qquad \text{for } t_1 < t_2 \leq 1. \tag{4.9}$$

The abstract scattering system $(H, U(t), \mathcal{D}_+, \mathcal{D}_-)$ is called **casual** if

$$\mathcal{D}_+ \perp \mathcal{D}_-.$$

PROPOSITION 4.1. *If H is an infinite dimensional irreducible unitary representation of $SL_2(\mathbb{R}_{>0}^{\mathrm{trop}})$, set $U(t) = h(t^{\frac{1}{2}})$, $\mathcal{D}_+ = \operatorname{Ker} x_-(1)$, $\mathcal{D}_- = \operatorname{Ker} x_+(1)$, then $(H, U(t), \mathcal{D}_+, \mathcal{D}_-)$ is an abstract scattering system. For H isomorphic to D_λ, the abstract scattering system is causal.*

PROOF. It is clear that (4.1) holds.

The relations (2.3) and (2.4) imply that

$$U(t)x_+(a) = x_+(at)U(t), \quad U(t)x_-(a) = x_+(at^{-1})U(t). \tag{4.10}$$

For $v \in \mathcal{D}_+ = \operatorname{Ker} x_-(1)$, $t \geq 1$,

$$x_-(1)U(t)v = U(t)x_-(t)v = U(t)x_-(t)x_-(1)v = 0,$$

this is proves $U(t)v \in \operatorname{Ker} x_-(1)$, so (4.2) holds. For $v \in \mathcal{D}_- = \operatorname{Ker} x_+(1)$, $t \leq 1$,

$$x_+(1)U(t)v = U(t)x_+(t^{-1})v = U(t)x_+(t^{-1})x_+(1)v = 0,$$

this is proves $U(t)v \in \operatorname{Ker} x_+(1)$, so (4.3) holds.

To prove (4.4), we observe that the closed subspace $\cap_t U(t)\mathcal{D}_+$ is closed under the action of the operator $h(a) = U(a^2)$. We prove $x_-(a)$ acts on $\cap_t U(t)\mathcal{D}_+$ as 0, for $v \in \cap_t U(t)\mathcal{D}_+$, we write $v = U(t)w$ for $t > a^{-1}$ and $w \in \mathcal{D}_+$,

$$x_-(a)U(t)w = U(t)x_-(at)w = U(t)x_-(at)x_-(1)w = 0.$$

By Theorem 2.3, the representation H is H_c or D_λ, for the action formula of $x_-(a)$ in (2.19) or (3.2), we see that element v with $x_-(a)v = 0$ for all $a > 0$ must be 0. This proves (4.4). Similarly we can prove (4.5).

To prove (4.7), we use the classification Theorem 2.3, so H is H_c or D_λ, $\mathcal{D}_-$ is the space of functions supported on $(0, 1]$, $U(t)\mathcal{D}_+$ is the space of functions supported on $(0, t]$, therefore (4.7) holds. (4.6) can be proved similarly.

For $H = D_\lambda$, $\mathcal{D}_-$ is the space of functions supported on $(0, 1]$. Using (3.2), we see that

$$\mathcal{D}_+ = \{f \in D_\lambda \mid f(x) = 0 \quad \text{for } x < 1, \int_1^\infty f(y)y^{-\lambda-1}dy = 0\}$$

It is clear that $\mathcal{D}_- \perp \mathcal{D}_+$, so the system is casual. □

The scattering system for the representation D_λ appear in [**LP**] as the simplest examples of casual systems.

We have the following partial converse of Proposition 4.1.

PROPOSITION 4.2. *If $(H, U(t), \mathcal{D}_+, \mathcal{D}_-)$ is an abstract scattering system, we define operators $h(a)$, $x_+(a)$, $x_-(a)$ $(a > 0)$ on H as follows. The operator $h(a) = U(a^2)$. Let $\mathcal{D}_+^\perp, \mathcal{D}_-^\perp$ be the orthogonal complement of $\mathcal{D}_+, \mathcal{D}_-$. For $a > 0$, set*

$$H_+(a) = U(a)\mathcal{D}_-^\perp, \quad H_-(a) = U(a^{-1})\mathcal{D}_+^\perp \tag{4.11}$$

We define $x_+(a)$ be the orthogonal projection onto the closed subspace $H_+(a)$ and $x_-(a)$ be the orthogonal projection onto the closed subspace $H_-(a)$, then the relations (2.1)–(2.4) hold and if the system is casual, then

$$x_+(a)x_-(b) = x_-(b)x_+(a) \quad \textit{for} \quad ab \leq 1 \tag{4.12}$$

This proposition can be proved directly. The relation (2.5) for the case $SL_2(\mathbb{R}_{>0}^{\text{trop}})$ is (2.20), which implies that

$$x_+(a)x_-(b) = x_-(b)x_+(a) \qquad \text{for } ab \leq 1. \tag{4.13}$$

This means (2.20) hold for the case $ab \leq 1$. However, (2.20) for $ab > 1$ does not hold for general abstract casual scattering systems.

This leads us to define the monoid $S(\mathbb{R}_{>0}^{\text{trop}})$, which is generated by

$$x_+(a), x_-(a), h(a), \quad a > 0,$$

satisfying the following

$$h(1) = 1, \quad h(a)h(b) = h(ab) \tag{4.14}$$

$$x_\pm(b_1)x_\pm(b_2) = x_\pm(\max(b_1, b_2)) \tag{4.15}$$

$$h(a)x_+(b) = x_+(a^2b)h(a) \tag{4.16}$$

$$h(a)x_-(b) = x_-(a^{-2}b)h(a) \tag{4.17}$$

$$x_+(a)x_-(b) = x_-(b)x_+(a) \qquad \text{for } ab \leq 1. \tag{4.18}$$

We introduce $*$-structure as in (2.8) and use it to define the unitary representation of $S(\mathbb{R}_{>0}^{\text{trop}})$ We have

PROPOSITION 4.3. *The abstract casual scattering systems are in one-to-one correspondence with unitary representations of* $S(\mathbb{R}_{>0}^{\text{trop}})$ *with the properties*

$$\lim_{a \to 0+} x_+(a) = \lim_{a \to 0+} x_-(a) = 1, \quad \lim_{a \to \infty} x_+(a) = \lim_{a \to \infty} x_-(a) = 0$$

where the limits are respect to the strong operator topology.

Acknowledgments

We would like to thank Xuhua He, Jun Yu for discussions and Yanze Chen for helping some calculations.

References

[AAR] George E. Andrews, Richard Askey, and Ranjan Roy, *Special functions*, Encyclopedia of Mathematics and its Applications, vol. 71, Cambridge University Press, Cambridge, 1999, DOI 10.1017/CBO9781107325937. MR1688958

[Bateman] A. Erdélyi, W. Magnus, F. Oberhettinger, and F. G. Tricomi, *Tables of integral transforms. Vol. I*, McGraw-Hill Book Co., Inc., New York-Toronto-London, 1954. Based, in part, on notes left by Harry Bateman. MR61695

[FZ] Sergey Fomin and Andrei Zelevinsky, *Cluster algebras. I. Foundations*, J. Amer. Math. Soc. **15** (2002), no. 2, 497–529, DOI 10.1090/S0894-0347-01-00385-X. MR1887642

[GK] F. R. Gantmacher and M. G. Krein, *Sur les matrices oscillatoires*, C. R. Acad. Sci. (Paris) 201, 577–579, 1935.

[HP] Einar Hille and Ralph S. Phillips, *Functional analysis and semi-groups*, American Mathematical Society Colloquium Publications, Vol. 31, American Mathematical Society, Providence, RI, 1957. rev. ed. MR89373

[Hir] I. I. Hirschman Jr., *Hankel transforms and variation diminishing kernels*, Bull. Amer. Math. Soc. **66** (1960), 40–42, DOI 10.1090/S0002-9904-1960-10383-7. MR110926

[Lus1] G. Lusztig, *Total positivity in reductive groups*, Lie theory and geometry, Progr. Math., vol. 123, Birkhäuser Boston, Boston, MA, 1994, pp. 531–568, DOI 10.1007/978-1-4612-0261-5_20. MR1327548

[Lus2] G. Lusztig, *Positive structures in Lie theory*, ICCM Not. **8** (2020), no. 1, 50–54, DOI 10.4310/ICCM.2020.v8.n1.a4. MR4166342

[LP] Peter D. Lax and Ralph S. Phillips, *Scattering theory*, 2nd ed., Pure and Applied Mathematics, vol. 26, Academic Press, Inc., Boston, MA, 1989. With appendices by Cathleen S. Morawetz and Georg Schmidt. MR1037774

[Mo] Th. Motzkin, *Beitrage zur Theorie der linearen Ungleichungen*, Doctoral Dissertation, Basel, 1933. Azriel Press, Jerusalem.

[Pin] Allan Pinkus, *Totally positive matrices*, Cambridge Tracts in Mathematics, vol. 181, Cambridge University Press, Cambridge, 2010. MR2584277

[Pu] L. Pukánszky, *The Plancherel formula for the universal covering group of* SL$(R, 2)$, Math. Ann. **156** (1964), 96–143, DOI 10.1007/BF01359927. MR170981

[Sch1] Isac Schoenberg, *Über variationsvermindernde lineare Transformationen* (German), Math. Z. **32** (1930), no. 1, 321–328, DOI 10.1007/BF01194637. MR1545169

[Sch2] I. J. Schoenberg, *On Pólya frequency functions. II. Variation-diminishing integral operators of the convolution type*, Acta Sci. Math. (Szeged) **12** (1950), 97–106. MR35861

[Su] Masatoshi Suzuki, *A canonical system of differential equations arising from the Riemann zeta-function*, Functions in number theory and their probabilistic aspects, RIMS Kôkyûroku Bessatsu, B34, Res. Inst. Math. Sci. (RIMS), Kyoto, 2012, pp. 397–435. MR3014857

[Titch] E. C. Titchmarsh, *Theory of Fourier Integrals*, Clarendon Press, Oxford, 1948

DEPARTMENT OF MATHEMATICS, THE HONG KONG UNIVERSITY OF SCIENCE AND TECHNOLOGY, CLEAR WATER BAY, KOWLOON, HONG KONG, PEOPLE'S REPUBLIC OF CHINA

SELECTED PUBLISHED TITLES IN THIS SERIES

818 **Howard S. Cohl, Roberto S. Costas-Santos, and Robert S. Maier, Editors,** Classical Hypergeometric Functions and Generalizations, 2025

817 **Mikhail Khovanov, Joshua Sussan, and Anton Zeitlin, Editors,** From Representation Theory to Mathematical Physics and Back, 2025

816 **Luca F. Di Cerbo and Laurenţiu G. Maxim, Editors,** Geometry and Topology of Aspherical Manifolds, 2025

815 **Daniel Orr and Joshua Jeishing Wen, Editors,** Macdonald Theory and Beyond, 2025

814 **Susan Montgomery and Siu-Hung Ng, Editors,** Quantum Groups, Hopf Algebras, and Applications, 2025

813 **Thomas Creutzig, Julia Plavnik, and David Ridout, Editors,** Quantum Symmetries, 2025

812 **Ross Geoghegan, Craig R. Guilbault, and Kim Ruane, Editors,** Topology at Infinity of Discrete Groups, 2025

811 **Krishnendu Gongopadhyay and Shashikant A. Katre, Editors,** Geometry, Groups and Mathematical Philosophy, 2025

810 **Robert J. Berman and Yanir A. Rubinstein, Editors,** Convex and Complex: Perspectives on Positivity in Geometry, 2025

809 **Fabrice Baudoin and Luca Rizzi, Editors,** New Trends in Sub-Riemannian Geometry, 2025

808 **Loring W. Tu, Editor,** Group Actions and Equivariant Cohomology, 2024

807 **Galina Filipuk, Editor,** Recent Progress in Special Functions, 2024

806 **Maximilian Engel, Hildeberto Jardón-Kojakhmetov, and Cinzia Soresina, Editors,** Topics in Multiple Time Scale Dynamics, 2024

805 **Anthony Iarrobino, Pedro Macias Marques, Maria Evelina Rossi, and Jean Vallès, Editors,** Deformation of Artinian Algebras and Jordan Type, 2024

804 **Mahir Bilen Can and Jörg Feldvoss, Editors,** A Glimpse into Geometric Representation Theory, 2024

803 **Peter Gothen, Margarida Melo, and Montserrat Teixidor i Bigas, Editors,** Moduli Spaces and Vector Bundles—New Trends, 2024

802 **Ralph M. Kaufmann, Martin Markl, and Alexander A. Voronov, Editors,** Higher Structures in Topology, Geometry, and Physics, 2024

801 **K. A. Brown, T. J. Hodges, M. Vancliff, and J. J. Zhang, Editors,** Recent Advances in Noncommutative Algebra and Geometry, 2024

800 **Avinoam Mann, Louis H. Rowen, David J. Saltman, Aner Shalev, Lance W. Small, and Uzi Vishne, Editors,** Amitsur Centennial Symposium, 2024

799 **Alberto A. Condori, Elodie Pozzi, William T. Ross, and Alan A. Sola, Editors,** Recent Progress in Function Theory and Operator Theory, 2024

798 **Marat V. Markin, Igor V. Nikolaev, and Carsten Trunk, Editors,** Advances in Functional Analysis and Operator Theory, 2024

797 **Sangita Jha, Mrinal Kanti Roychowdhury, and Saurabh Verma, Editors,** Recent Developments in Fractal Geometry and Dynamical Systems, 2024

796 **John Cremona, John Jones, Jennifer Paulhus, Andrew V. Sutherland, and John Voight, Editors,** LuCaNT: LMFDB, Computation, and Number Theory, 2024

795 **Michael A. Jones, David McCune, and Jennifer M. Wilson, Editors,** Mathematical Analyses of Decisions, Voting and Games, 2024

794 **Jean-Pierre Magnot, Editor,** Recent Advances in Diffeologies and Their Applications, 2024

For a complete list of titles in this series, visit the
AMS Bookstore at **www.ams.org/bookstore/conmseries/**.